ANNUAL REVIEW OF B

ANNUAL REVIEW OF ENTOMOLOGY

RAY F. SMITH, *Editor*
University of California

THOMAS E. MITTLER, *Editor*
University of California

VOLUME 15

1970

PUBLISHED BY
ANNUAL REVIEWS, INC.
IN CO-OPERATION WITH THE
ENTOMOLOGICAL SOCIETY OF AMERICA

ANNUAL REVIEWS, INC.
4139 EL CAMINO WAY, PALO ALTO, CALIFORNIA

ANNUAL REVIEWS, INC.
PALO ALTO, CALIFORNIA, U.S.A.

Standard Book Number 8243-0115-3

Library of Congress Catalog Card Number: A56-5750

FOREIGN AGENCY

Maruzen Company, Limited
6, Tori-Nichome Nihonbashi
Tokyo

PRINTED AND BOUND IN THE UNITED STATES OF AMERICA
BY GEORGE BANTA COMPANY, INC.

PREFACE

Entomology was once the province mainly of specialists devoted to classifying insects and describing their anatomy. Although systematics and morphology will always retain a central role in the field, their unalloyed influence once tended to segregate the study of insects from the rest of biology.

All this has changed during the past half-century. The widening scope of activities now classified as "entomology" had many causes. Increasing concern for the role of insects in human affairs has played an important part. Also, the increased sensitivity of modern experimental techniques has made possible a closer study of insect functions. A central role has been played by many demonstrations that those very specializations identifying insects and responsible for their evolutionary success are also heuristic sources of information about central currents in the stream of biological knowledge. For instance, genetics and our knowledge of the dynamics of animal populations owe much to a small fly. The organized social behavior of Hymenoptera has been the source of much of what we know about animal communication and orientation, and promises a great deal more. Insect sense organs are uniquely accessible to experimentation, and there is now more firm information about chemoreception in this group than in any other class of animals. The specialization of insect muscles for flight is leading to new insights into the mechanism of muscle action and animal movement. Information about chemical signals and chemical defenses suggests new ways of regulating insect pests as well as understanding means of communication lost to man. The parsimonious utilization of a limited neuronal network in insect nervous systems offers opportunities to understand the mechanisms of animal behavior. And, finally, the extreme compartmentalization of insect development and its humoral regulation open interesting possibilities of understanding the control of animal ontogeny as well as suggesting more rational means of insect control.

All of this adds greatly to the widening responsibility of the *Annual Review of Entomology*. One of its functions is to bring authoritative information on these and other topics to those working with insects. Another is to acquaint biologists in general with the latest evidence of diversity, specialization, and unique adaptations in insects, and to point to their suitability for studying basic biological problems. Therefore, the Editorial Committee welcomes suggestions from all biologists regarding topics that seem ripe for future review.

Although our panel of authors make an essential scientific and literary contribution to this volume, their efforts are brought to fruition by the devoted services of our Senior Assistant Editor, Miss Beryl V. Daniel, and by the expertise of our printers, the George Banta Company. We are most grateful to them for their cooperation in the production of this volume of the *Review*.

THE EDITORIAL COMMITTEE

EDWARD A. STEINHAUS

1914–1969

As this volume of the *Annual Review of Entomology* was in the final stages of preparation, we learned with great sadness of the death of Edward A. Steinhaus, first editor of the Annual Review of Entomology. It was Professor Steinhaus who conceived the idea of an Annual Review of Entomology, encouraged its publication and then served as its editor for the first seven volumes (1956–1962).

CONTENTS

RECENT ADVANCES IN INSECT POPULATION DYNAMICS

By G. C. Varley and G. R. Gradwell
Hope Department of Entomology, University of Oxford, Oxford, England

A field study concerning insect abundance begins with the identification of the insects concerned and with a study of their life cycles and natural history. It then progresses to the taking of census counts from which life tables can be derived, as discussed by Harcourt in the last volume of *Annual Review of Entomology*. We wish first to examine the next two stages in the study, the analysis of life tables and the development of population models, and to discuss their properties; second, to stimulate workers to define more precisely their reasons for collecting census figures and help them develop a critical understanding of the analytical methods they apply to them, particularly as there are already in the literature examples of the incorrect use of analytical methods and of their application to unsuitable data.

Census routines must be carefully planned to provide information of sufficient accuracy about all the important factors concerned. Provisional analyses should be used early in the study so that serious gaps are revealed; then, census methods can be improved as soon as possible. Many studies include some routine measurement of weather factors, but if these figures are thought to be important this should be confirmed by physiological studies. If parasites are thought to be important, then a life table must be prepared for them; this will require, as a minimum, separate counts of the parasite's adult and larval population densities in each generation. If predators are thought to be important, their numbers, food requirements and population regulation also must be investigated.

POPULATIONS WITH DISCRETE GENERATIONS

Age-Specific Life Tables

A life table can be derived for a species only from an intensive study of a population in one place. Probably the greatest amount of numerical information on an insect species is that obtained by Canadian workers on the spruce budworm (43). However, this study failed to produce an explanatory model because no single place was sampled continuously. With samples from many different places it was not possible to distinguish between spatial and temporal effects. Morris wisely concluded that "it may be profitable to study intensively, on a few plots, the age specific survivals and mortality factors that determine temporal population changes; and to study extensively, on widely different plots, the factors and processes that determine spatial differences in survival" (45). We consider that if there is a choice between extensive and intensive studies, the intensive study in one place is of prime importance. The part played by spatial differences in the determination of population change and the regulation of population level can be understood only when these differences are included in an explanatory model formulated from an intensive study. In contrast with this extensive study, our own study of the population dynamics of the winter moth on oak in England has been an intensive one (73). Observations have been limited to measuring the population changes taking place on five oak trees within a mixed woodland. In this study it is possible to separate temporal and spatial differences, and life tables can be constructed for the population using either the mean values for the five trees or for each of the trees separately.

The method of setting out a life table which was suggested by Morris & Miller (47) has been widely followed (61, 75). We feel that basic life table information ought to be published so that other workers, if they wish, may check an analysis or make an analysis in a different way. However, these conventional tables occupy a lot of space and, when a large number of tables has been accumulated for a species, often only the analysis of these tables has been published. All the figures normally given in a life table (d_x, 100_{qx} and S_x may easily be derived from the figures for 1_x (the number alive at the beginning of each age interval). However, a figure for S_x or a form of log S_x is extremely useful and if these are given it is not necessary also to give all the figures for 1_x. The life table information for a large number of generations can be concisely summarised on a single page in a set of graphs which show either the values of log 1_x for each age interval (37), or log 1_x at the beginning of each generation and the *k*-values for each age interval (73).

ANALYSIS OF LIFE TABLES

We consider that a logical way to carry out an analysis is first to make a key factor analysis to determine the contributions made by the different mortality factors to population change, and second to attempt to make submodels for these mortalities. The final test of a method of analysis is to combine the different submodels into a single formula and examine how closely the population densities predicted by the formula agree with those observed. We are interested in having a model which is also an explanation of the biological mechanisms involved, which not only predicts population change but also explains the population level about which these changes take place. Our method of analysis is therefore aimed at producing submodels which have a biological meaning.

Conversion of Life Table Data

Harcourt (18) has set out two methods of converting life table data into a form suitable for further analysis. In the first of these (44, 78, 80), the fractional survivals (survival rates $= S_x$) for the several age intervals are related by the formula:

$$\text{Generation survival} = S_G = S_1 \times S_2 \times S_3, \text{ etc.} \cdots \qquad 1.$$

When survival rates have been calculated and correlative methods of analysis are to be used, it is often preferable to use a figure for log S_x rather than S_x; this procedure may reduce variance and improve linearity. In this case the formula becomes:

$$\log S_G = \log S_1 + \log S_2 + \log S_3, \text{ etc.} \cdots \qquad 2.$$

The second conversion method suggested by Harcourt is to make use of k-values (70). A k-value is measured as the difference between the successive values for log l_x and is thus a logarithmic measure of the killing power of a mortality factor. Using k-values the formula is

$$\text{Total Mortality} = K = k_1 + k_2 + k_3, \text{ etc.} \cdots \qquad 3.$$

Formulae 2 and 3 are exactly equivalent and are interconvertible. The relationship between them will be seen from a comparison of Columns 5, 6, and 8 in Table I. Quite clearly, any graphical or statistical analysis which uses either log survivals or k-values should produce equivalent results.

Morris (44) transforms log survivals in the form of Column 5 into a form avoiding negative logarithms by the addition of four to each of the individual log survivals, but he fails to balance the equation by adding the sum of these numbers to the value of log S_G. Individual log S_x values transformed in this way would have no immediately obvious meaning unless all

TABLE I

AN ARTIFICIAL LIFE TABLE TO SHOW THE RELATIONSHIP BETWEEN EXPRESSIONS FOR LOG SURVIVALS AND K-VALUES

COLUMNS							
1	2	3	4	5	6	7	8
x	l_x	100_{qx}	S_x	log S_x	Minus form of log S_x	log l_x	k-value
Eggs	200	50%	0.5	$\bar{1}.7$	−0.3	2.3	0.3
Small larvae	100	90%	0.1	$\bar{1}.0$	−1.0	2.0	1.0
Large larvae	10	50%	0.5	$\bar{1}.7$	−0.3	1.0	0.3
Pupae	5	80%	0.2	$\bar{1}.3$	−0.7	0.7	0.7
Adults	1					0.0	
Generation totals			$S_G=$ 0.005	log $S_G=$ $\bar{3}.7$	log $S_G=$ −2.3		$K=$ 2.3

workers agreed to add the same number to them. Even then, the values for the transformed log S_G would vary depending on the number of terms in the formula. Expressions in the form of either Column 6 or Column 8 avoid these difficulties.

Key Factor Analysis

Life table analysis has usually begun with an attempt to determine the relative importance of the different age intervals by a method (42) in which the criterion has been the degree of simple linear correlation between the separate components (S_x's) and the generation survival (S_G) as the dependent variable. This approach has been criticised on both biological and mathematical grounds. The use of correlation and multiple regression analysis is a valid approach to this problem only when we can guarantee that there is no intercorrelation between the variables. In biological systems such intercorrelations are widespread. For instance, we may expect that increases in population density will be accompanied by increases in the percentage mortality caused by a number of organisms conveniently classified as disease, predators and parasites. Such covariance in factors will largely invalidate the results of any straightforward statistical analysis (49).

One other danger of these kinds of statistical component analyses is that the relative importance of the different age-interval-survivals is assumed to lie in their contribution to population change and, thus, almost from the outset, the analysis is developed along a course leading to a model which predicts one generation ahead.

Even if it were possible to attach to the survival in each age interval some valid mathematical index of its contribution to variations in generation survival, such indices would be largely irrelevant to the further development of models for both determination of population change and regulation of population level. For such models it is essential that each age interval should be described by a submodel in a way which is both mathematically satisfactory and biologically meaningful. For these reasons, we have suggested (70) that this initial stage of an analysis should be a visual comparison of graphs showing the change in the *k*-values (or log survivals) for the different age intervals and the changes in total K (or log S_G) for the generations, each plotted against time. It may be possible from such graphs to identify the "key factor" (42) as that stage in which the variations in *k*-value contribute most to the variations in K. More important, however, inspection shows which *k*-values are so variable that submodels for them are essential, and also which *k*-values are so small or constant that, at least in the initial stages of model-making, they can be considered as constant.

It is important to note that the above definition of a key factor is not the only one in current use. In his original paper, Morris (42) had two quite distinct definitions for this term: first, that it was to be applied to factors which caused "a variable . . . mortality and appeared to be largely responsible for the observed changes in population"; this causal relationship is the one we emphasise. The second definition, that a key factor is "any mortality factor that has a useful predictive value"—implying a statistical rather than a causal relationship—is the sense in which the term has been used by many other workers.

Submodels Based on Multivariate Analysis

Perhaps the best example of this approach to the development of submodels is, again, the work done on the spruce budworm (43). The lack of continuous and detailed life table information from any one place meant that there was no insight into the biological processes involved, and without this understanding there was "no basis upon which to construct explanatory models" (48). Under these circumstances, there was no alternative to a reliance on multivariate analysis and "without exception, each sub-model is purely indicative and empirical" (48), the empirical expressions being selected on the basis of best fit.

For each age interval, measurements were made not only of survivals but also of as many other variables as were conveniently measurable. These variables included not only measurements of temperature and humidity, but also some features of tree size, cumulative defoliation, phenological development, etc. Multivariate analysis was then used to produce a formula for the submodel. That the method would "work" was inevitable, in the sense that the statistical procedure would produce some sort of a formula. Even if changes in an age interval survival had been entirely caused by a variable which was unmeasured (and thus not included in the analysis) the analytical method would, nevertheless, produce a formula expressing these changes in terms of those variables which were included in the analysis. The production of a formula is itself no indication of causal relationship. For the formula to have a biological meaning, the right variables must have been measured. It is not possible to measure every variable which might have some effect on a population, so that the ultimate success of a submodel will depend very much on the planning of the research and on whether or not the planners have the right "hunch" about which factors are going to be important and are able to measure them. The measurements must enter the regression analysis suitably transformed (81); Watt (78) gives some 32 formulae which might be used to describe relationships, and this is by no means an exhaustive list. Often errors in our measurements will not allow us to decide which of two or more formulae best describes the data so that, again, the decision as to how best to enter the measurements into an analysis will depend on the knowledge and experience of the workers concerned.

Submodels Based on Density Relationships

The four terms, density-dependent mortality, inverse density-dependent mortality, delayed density-dependent mortality and density-independent mortality arose from early theoretical ideas about population dynamics. Later, however, the ecological literature contained quite a lot of argument as to how to categorize the effect of climate or of parasites and predators. These arguments were often so vituperative that many later workers have avoided involvement in the arguments by entirely avoiding this terminology. However, we have found that these terms form a useful basis on which to examine the properties of population model components and the sort of interactions which take place between them. The four terms can be defined mathematically as well as verbally, so that field data can be tested statistically for these relationships.

All component mortalities of a life table may be expected to fall into one or other of these categories, but there will be boundaries between them where one kind of relationship is indistinguishable from another. The limited measurements we are able to make of field populations in any particular age interval may mean that the mortality measured confuses the action of factors in two or more of these categories. Such difficulties, however, arise from an inadequacy of our information rather than from an inadequacy in the analytical technique. The possibility that such difficulties may arise serves to emphasise the necessity for making analyses as early in the study as possible.

Density-dependent mortality.—A density-dependent mortality (i.e., direct density-dependent mortality) is defined as one in which an increasing proportion of the population is killed as the population density increases. Equivalent to this are cases in which there is a decrease in fecundity or in fertility with increasing density. Such a relationship between mortality and density occupies a very special place in population theory since, to date, mathematical population models which do not include density-dependent mortalities are unstable. They neither allow the continued existence of a species, nor have they been able to mimic or explain population level. Such models predict that population densities should, in the long term, either increase or decrease indefinitely (36). Most workers now accept that density-dependent mortalities regulate population density, and that it is of prime importance to analyse field data in a way which will allow such mortalities to be identified and quantified. Some, however, would deny this (2, 16). The general form of biological curves suggests that only rarely can we expect to obtain a straight-line relationship between percentage mortality and population numbers per unit area; but a correlation between the k-value for the mortality and the logarithm of the population density may well give a straight line which adequately fits the data (72). This yields a formula for the interval mortality (k) with the general form of $k = a + b\ (\log N)$; where b is the slope of the regression, a the intercept and N the number per unit area. When the independent variable is population density, statistical tests for density dependency may be invalid if they use fractional survivals, or a logarithmic expression of these, as the dependent variable. Such tests are in effect correlating y/x with x or $\log x - \log y$ with $\log x$, and errors in the estimate of x can themselves result in an apparent correlation between the two. This possible source of error is of little importance where empirically derived formulae are concerned, since these do not pretend to mimic biological reality; but where it is hoped to produce a mathematical model which has a biological meaning it is necessary to prove such a relationship. This can be done by showing that a regression of log density of the survivors from the mortality (log y = the independent variable) on log density prior to the action of the mortality (log x) significantly diverges from a slope of $b = 1$. Even this test may not be valid if there are errors in the measurement of x. If reliable estimates are available for the errors to be attached to the measurements of x, one can finally test the relationship by standard statistical techniques (1, 60). If the errors are either very large or cannot be estimated, then a possible test is to consider that all the errors may reside in the estimate of x and test if a regression of log x —now as the independent variable—on log y also has a slope which is significantly different from $b = 1$, and on the same side of this slope as the regression of log y on log x (61, 71, 73).

Unavoidable errors in the measurement of field populations are usually so great that it will be impossible to prove a weak density-dependent relationship. When there is a low value for the slope of k-value on log population density, this is not distinguishable statistically from slopes of $b = 0$. Fortunately, in population models the effect of a density-dependent mortality with a weak slope is not very different from a parameter which treats this mortality either as a constant or as varying within narrow limits. In practice, mortalities in which the slope b is less than about 0.1 have relatively unimportant density-dependent effects. However, this statement needs some qualification. Density-dependent effects can be additive and it is possible for there to be a weak density-dependent mortality acting in each of a number of age intervals, no one of which is statistically significant, which combine to give a very important total effect.

In population models, the greater the slope for the density-dependent mortality between values of 0 and 1.0, the greater the stabilising effect until, with a slope of $b = 1$, we reach a point of absolute stability where all changes in population density are exactly compensated. Most of the published data which demonstrate density-dependent mortality have values for the slope greater than 0.2 but less than 1.0.

If density-dependent mortalities overcompensate, they contribute to population change. Population models which include overcompensating density-dependent mortalities with slopes of $b > 1 < 2$ show densities alternating in successive generations between high and low values; but the relative sizes of these alternations diminish with time so that the population is still tending to attain a stable level. When the slope for the density-dependent mortality is greater than 2 it results in alternations which increase in amplitude with time. Figures and calculations demonstrating these relationships are given by Klomp (38). Ohnesorge (56) has suggested that alternations between large and small population densities in successive generations implies that density-independent factors are mainly responsible for the changes, but this is not necessarily so.

It might be thought that density-dependent mortalities with a slope greater than 1 are so strong that they are unlikely to occur in nature. Duffey's data on the large copper butterfly in England (14) shows that the relationship between the potential number of eggs each year and the number of larvae resulting from them is that expected if a density-dependent mortality—or two successively acting density-dependent mortalities—with an effective slope of $b = 1.8$ is acting on the egg potential. This particular relationship is not shown in Duffey's analysis but comes from our unpublished reanalysis of his published data.

Similarly, the results of Nicholson's experimental studies on blow fly populations (52, 53) can be interpreted as resulting from the imposition of a density-dependent mortality (due to food limitations) with a slope near to the value of 2. Here, the experimental conditions appear to have resulted in a series of distinct generations of the blow fly which alternate between high and low population numbers. For some of the time, these alternations appear to increase in size, suggesting that, during this period, the density-dependent mortality of the feeding larvae has a slope greater than 2. Nicholson considered that these results demonstrated "oscillations" in population density. We feel that there are good reasons for restricting the term, oscillation, to the classical situation in which cyclical changes in population density arise from the action of delayed density-dependent mortalities and not from the series of gaussian distributions of separate generations. The apparent similarity of Nicholson's blow fly "oscillations" and his parasite-host oscillations (54) are the result of different methods of presenting results graphically. The blow fly curves are partial population curves and the parasite-host curves are generation curves (69). When the blow fly data are replotted as generation curves, they appear as alternations. The two methods of graphical expression and the two mechanisms involved are quite distinct; the use of the same term for both can only add to the great semantic difficulties already experienced by all workers on population dynamics.

Morris (46) has introduced a method for assessing the total amount of density dependence operating on a population and also the individual contributions of various components to this. Initially, he plots log density in each generation against that of the previous generation and obtains a regression line to describe these points. This line, he claims, then represents the total amount of density dependence in the system. The method is similar to the reproduction curves used for fish by Ricker (58). Ricker plotted, on a numerical scale, the egg population of generation $N + 1$ against the egg population of generation N and gave examples of graphs which resulted from simple mathematical models in which only one factor—a density-dependent mortality—operated. However, Ricker found it difficult to interpret his own field data using this method. With Morris's method, the regression line through the points has meaning if density-dependent factors act alone, but becomes meaningless whenever the system contains widely variable mortalities due either to density independent or to delayed density-dependent factors. Both Southwood (62) and Hassell & Huffaker (23) have used Morris's method on artificial models which included mortalities other than density-dependent ones; they found that the method failed to give a meaningful analysis.

Even when the factors operating on a population are all density-dependent, Morris's method of assessing this may reveal only the net amount of density dependency rather than its total. If a density-dependent mortality with a slope (*k*-value on log density) greater than 1 is followed, at some other stage in the life cycle, by a density-dependent mortality with a slope of less than 1, the net result is a slope with a value less than that of the first mortality. Thus, it may be possible to find, within a single age interval, a density-dependent mortality with a slope greater than that found by Morris's method for the total generation.

There are now many examples from field studies of density-dependent factors acting on populations. Many of these are effects of intraspecific competition for some limited amount of food or space. They may be revealed by their effects on mortality, e.g., cannibalism (6), on fecundity (13) on "area of discovery" (30) or on emigration (7, 83). Predators too may act as density-dependent factors on their prey (17), as may parasites on their hosts (41)—at least over part of the range of host densities.

Inverse density-dependent mortality.—An inverse density-dependent mortality is defined as one in which the proportion of the population killed decreases with increasing density. Alternatively, there can be an increase of fecundity or of fertility with increasing density. A plot of the *k*-values for this mortality against log population density will show a negative slope. If necessary, a formal test for such a relationship is by a regression of log final on log initial densities which will give a slope significantly greater than 1. Clearly, the effect of any such mortality must be to add to a population's instability, and it will detract from the stabilising effect of any density-dependent mortality. If a density-dependent mortality with a slope (*k*-value on log density) greater than 0 but less than 1 is followed by a weaker inverse mortality, the slope for the combined mortalities is less than that of the density-dependent mortality. However, if the slope for the density-dependent mortality is greater than 1, the net effect with a subsequent inverse density-dependent mortality is to give an even greater slope.

It would seem probable that mortalities which act in an inverse density-dependent way throughout the whole range of a population's density are relatively rare. One possible example is the mortality caused by nonspecific and nonsynchronised parasites of the winter moth at Wytham, England (73). Here, the total effect of the action of a number of parasites shows this relationship. It seems more usual to find that a mortality factor is density-dependent over part of the density range, but is inverse over another part of the range. Such a change in relationships has been suggested for bird predation on caterpillars (66), on phasmids (57) and on psyllids (11). There is also a suggestion (41) that some of the spruce budworm parasites may cause a density-dependent mortality at the lower range of host densities but an inverse density-dependent mortality at the higher part of the range. Such a changing relationship can result when predators or parasites, which themselves have a relatively constant density, react to increases in the lower part of a prey's range of densities by taking an increasing proportion of them, but with further increases in prey density tend to take a constant number and thus a decreasing proportion.

In some aphid populations, fecundity at first increases with increasing density (equivalent to an inverse-density dependent mortality) but further increases in density cause either a density-dependent decrease in fecundity or emigration of adults or both (83). We have noticed a similar change in relationship in the mortality suffered by *Musca* larvae in laboratory experiments; there is a high larval mortality in very low density populations which is at first reduced at intermediate densities but which rises and becomes density-dependent when food is limiting.

We can expect that all insect populations will show an initial inverse density-dependent fertility (an increase in fertility with increasing density) over a range of extremely low population densities where the sexes have difficulty in finding one another.

It is important to realise that the discovery of either a direct or an inverse density-dependent mortality implies that there is a numerical constancy of some sort present in the relationship. Thus, mortality due to intraspecific competition for food will be revealed as density-dependent only when the food supply to the several generations is relatively constant. Similarly, parasites and predators can cause a density-dependent or an inverse density-dependent mortality of their hosts or prey only if their own population density is relatively constant from generation to generation. We can use such observed relationships in a model to mimic the action of predators and parasites, but a biological explanation of these relationships can come only when we have life tables for the predators and parasites which allow us to explain the relative constancy of their populations.

Delayed density-dependent mortality.—The failure to detect a density-dependent or inverse density-dependent mortality does not mean that the mortality is entirely independent of density. It may be a delayed density-dependent mortality. This category was first defined (68) in relation to Nicholson and Bailey's theoretical ideas about parasite-host interactions (51, 54). The theory assumes that it is the parasite's rate of increase and not the host mortality that is proportional to host density. Calculations using this assumption show cyclical changes in host and parasite with the parasite cycle lagging one quarter of a cycle behind that of the host; this means that a maximum host mortality occurs one quarter of a cycle after a peak in host density. Thus, one way of demonstrating this delayed density-dependent mortality graphically is to plot the k-values for this mortality against the log host density and join the points serially; the graph will show an anticlockwise circle or spiral (61, 71). A second and better test is given by a plot of these k-values against parasite density, when a linear relationship appears.

However, it is not so easy to detect a delayed density-dependent mortality from field data. The first graphical method mentioned above reveals a delayed density-dependent mortality only if the mortality is the key factor or is, at least, making a large contribution to population change. Even if the graphical method has suggested the presence of a delayed density-dependent mortality, a proof of this relationship may be difficult, or even impossible, if the densities or the parasite (or predator) have not also been measured. When the cause of the mortality is not known, or if it is known but has not been quantified, the only formal proof of the existence of a delayed density-dependent factor is to show that the particular sequence of mortalities is not one which could have occurred by chance. No author has yet suggested a way of doing this, and we can expect that a long series of observations would be required. This difficulty has arisen in our work on the winter moth. We find heavy mortality between the moth count (egg estimate) and the count of the prepupae, most of which is suffered at the time of egg hatch. A graphic analysis (73) shows that between 1956 and 1966 this mortality appeared to cycle in a way expected of a delayed density-dependent mortality; but we are unable to determine whether this is due to the delayed density-dependent action of some biological agent, or to a fortuitous sequence of density-independent climatic conditions.

If a delayed density-dependent mortality is not making a large contribution to population change it will be impossible to detect this relationship by the first graphic method. In such cases, the delayed relationship is detectable only when the densities of the parasite or predator responsible have also been measured and can be related to this mortality. A delayed density-dependent mortality which cannot be revealed by a graphic method may, nevertheless, be extremely important to a population model, and it may be possible to build a satisfactory mathematical model only if the mortality is included as a delayed density-dependent one. In our model for the winter moth in England (73), the submodel for the mortality caused by the ichneumonid parasite of the winter moth pupae treats this as a delayed density-dependent mortality of the host. Only if we include it in this way will the population model mimic the range of changes in density which we have observed in both the host and parasite populations. A graphic analysis of our field data does not show such a parasite-host relationship because this parasite is not the key factor determining population change. We can suggest that the parasite may be causing a delayed density-dependent mortality of the host only because we have made estimates of the parasite's densities.

There is now quite a lot of evidence from both field observations (3, 42, 73) and from laboratory experiments (10, 31, 67) that the population densities of some specific and synchronised parasites lag behind those of their host, and cause a delayed density-dependent mortality. There is no evidence that nonspecific or nonsynchronised parasites can act in this way. Frank (17) has suggested that the weak delayed density component in the mortality of winter moth pupae in the ground in England is probably caused by the staphylinid beetle *Philonthus*. But *Philonthus* is not a specific predator on winter moth, and pupae are available to it as food for only some six months of the year.

Buckner & Turnock (9) and Holling (24) have shown that the population densities of insectivorous birds and mammals increase in response to a temporary abundance of some insect prey. Nevertheless, these predators depend on other food at other times of the year and from field data it may be difficult to distinguish between apparent changes in predator density due to the adults concentrating in regions of a superabundance of insect food, and real numerical changes which persist until the next generation of that insect in the following year (8, 19).

In some cases we can expect that the quantity or quality of plant food available to the next generation will be affected by the density of the present generation of an insect feeding on it. Thus, food supply, too, can act in a way which causes a delayed density-dependent mortality of an insect. Changes in the growth rates of trees in response to defoliation or heavy damage have been reported (33, 34, 55). Such effects were quantified in the spruce budworm study where a term for the effect of cumulative defoliation is included in the formula for the survival of small larvae (48), and a similar expression is used in the formula for changes in the sex ratio of the adults (48).

Density-independent mortality.—If our methods of analysis of the mortality in a particular age interval have failed to show that it is in any way related to the insect's density then we have to accept that it may be density-independent. We use the word "may" because it is possible for a highly variable density-independent mortality to mask the action of a mortality which is related to density when the two occur close together in time. The category of density-independent factor includes "catastrophic factors" (29), which produce very variable mortalities, as well as those that cause mortalities which vary so little that they can adequately be described by a constant. Often these density-independent mortalities will be caused, directly or indirectly, by the weather through temperature, rainfall, humidity, etc. Food-plants, parasites, predators and disease also can have a density-independent effect on an insect whenever the abundance of these organisms is not related to that of the insect. An explanation of a density-independent mortality is possible only when the causal agents have been measured. In nature, density-independent mortalities are often the key factors causing population change. Their action has to be mimicked in population models which seek to explain and predict population change. However, it is the average effect of a density-independent mortality which contributes to the determination of the average level of a population; and, thus, in models which seek only to explain population level, density-independent mortalities can be represented by their average effect used as a constant.

In the introduction to this section we mentioned the early attempts to generalise about the effects of mortality factors according to whether they were climatic, food supply, parasites or predators. We can see now that these attempts were futile; the biological factors can have effects which come into each of the four categories we have discussed. We know that weather effects must be density-independent because they cannot detect, and thus cannot respond to, change in an insect's density. Nevertheless, their effects could appear to be related to density if they were acting on a population in a diversified habitat, and where increasing density resulted in an increasing part of the population being forced into regions of the habitat which were only marginally suitable for survival.

POPULATION MODELS

The final test of a population model is its ability to predict. But we can seek to predict two quite different things, either the generation-to-generation changes in an insect's population density or the average level about which these changes take place. Ideally, of course, one would like to have an explanatory–predictive model for both change and level. Models published so far do one or the other with differing degrees of success; no published model does both. The method of analysis employed, as well as the sort of model made, depends on which is chosen as the objective.

Prediction of changes in density.—This is the kind of model required by those interested in "chemical control" since the objective is to predict the size of the next generation and thus predict when economically unacceptable damage is likely to occur; that is, when the "economic threshold" (63) is likely to be crossed.

The simplest kind of model which predicts changes in density is based on an identification of the key factor responsible for these changes in density and the development of a submodel only for this mortality. However, since this kind of model does not require a knowledge of the biological mechanism involved, it is not necessary to model the factor causally responsible for changes if other mortalities vary in the same way, and can be used to predict the changes in mortality due to the key factor. An example of this kind of model is that produced by Morris for the black-headed budworm (42) by which a measure of parasitism in year N allows a reasonable prediction of population density in year $N + 1$; although, here, parasitism is causally responsible for rather less than half of the population change. In a somewhat similar predictive model for the spruce budworm, Morris (43) uses, in his formula 18.6, one term for a density-dependent relationship and another for a temperature effect which is assumed to act on the large larval stage. Both terms are empirically derived since the size of the density-dependent relationship is not related to those found when attempting to produce the individual submodels for the different mortalities, nor has it been possible to substantiate the assumed temperature effect on the large larval stage.

In a recent paper, Auer (3) has applied Morris's method of analysis to data on the grey larch moth (*Zeiraphera dineiana* Gm.) collected by Swiss workers. We believe a misinterpretation has arisen here not only through

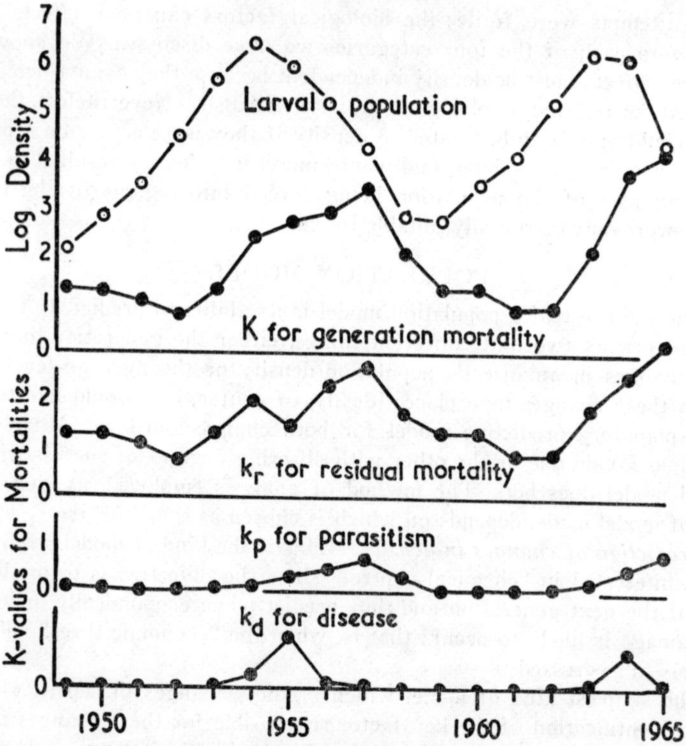

FIG. 1. Key factor analysis for *Zeiraphera diniana.* Data from Auer, C. *Z. Angew. Entomol.*, **62,** 202–35 (1968): Log larval population from Table 3, column 4; or from Figure 4. Generation mortality = K = 2-dy′ (dy′ from Table 2, column 3). Parasitism = k_p, is calculated from Table 2, column 4. Disease = k_d, is calculated from Table 2, column 5. Residual = $k_r = K - k_p - k_d$.

the use of correlative methods, but also because they have been used to analyse incomplete life table data. Auer wanted to find the explanations of the very regular cyclical fluctuations in the larval population densities between the years 1949 and 1966, and examined the effect on these fluctuations of parasitism, disease, damage to needles and weather. He shows that an extremely accurate description of the changes can be derived from measurements of parasitism, disease and temperature. However, by knowing the maximum egg production of females of the grey larch moth, it is possible to complete the life table by calculating a "residual mortality" representing that which must be present though unmeasured. Plots of total *K* and of individual *k*-values for the different mortalities are shown in Figure 1. Quite clearly, the key factor related to population change is in the residual mortality and not in those used in his analysis. Morris's analysis of the black-headed budworm and Auer's analysis of the grey larch moth are remarkably similar. Both insects are species with regular cyclical changes in population density and in both cases changes in density are closely related to changes in parasitism of the larvae. Again, in both cases, the residual mortality is large and varies in a way similar to variations in larval parasitism, and we can surmise that parasites or predators acting on the egg or pupal stages which have not yet been studied are a major cause of both cycles. Morris recognised the important role of the residual mortality, but Auer failed to include the residual mortality in his analysis and appears to have been misled into thinking that the statistical relationship between changes in parasitism and changes in host density meant that parasites of the larvae were causally responsible for the cycle in the host's densities.

Even quite complicated formulations can result in models which predict only population density changes. The spruce budworm study (43) contains two quite distinct approaches to model building; the simple two-term model of Morris which has just been mentioned, and the more ambitious attempt to produce an explanatory model through the development of submodels with biological meaning. However, since it was not possible to construct submodels with a biological meaning, the population model which combined these submodels could be used only to predict changes in density, as did Morris's model.

Prediction of population level.—With this kind of model the objective is to explain why the observed changes in population numbers takes place about a particular average level ["General equilibrium level" (63)]. This is the kind of model which is needed by those interested in "biological control," since now the objective is to predict the long-term consequences of the introduction of additional mortality factors. Biological control methods might either be aimed at reducing the upper ranges of an insect's population fluctuations, and this need not necessarily alter the average level, or the aim might be to reduce a pest's average level and thus alter its status from being a common insect to being a rare one; this need not affect the range of its fluctuations. Our point here can best be made with an example. An insect might have an average abundance of 10 per unit area but fluctuate between 1.0 and 100 per unit area; if we were then to reduce its average level to 1.0 without affecting its range of fluctuations, it would then fluctuate between values of 0.1 and 10 per unit area. In both cases, the range of fluctuations would be one hundred-fold.

In our model for the winter moth in England (73) we have sought to provide an explanation for population level. The model contains five parameters each of which is an oversimplification of the interactions taking place. Nevertheless, as far as possible the parameters represent simple biological relationships, and further development of the model will depend on our better understanding of the best way to model these relationships. The key factor mainly responsible for population change represents the extent to which newly hatched larvae establish themselves in the opening buds. The mortality is density-independent. We have failed to find any satisfactory mathematical description for this and the values used in the models are either those observed (73) or their mean (72).

Our data suggest that nonspecific larval parasites are acting as inverse density-dependent factors. The studies by Hassell (20–22) have provided a mathematical submodel for the effect of the tachinid fly *Cyzenis albicans* whose larva is a specific larval parasite of the winter moth. However, the total effect of all these parasites is so small and varies so little that, in a simple model, they can be considered to have a constant effect. Our model uses a constant to describe them.

The mortality of winter moth in the soil is mainly density-dependent. This relationship has been shown to be significant statistically (73). The degree to which different predatory species contribute to this mortality has been studied by Frank (17), who identifies a weak delayed density-dependent component which he suggests is probably the result of predation by the staphylinid beetle *Philonthus decorus.* The delay, presumably, arises from a numerical response of the beetles to changes in winter moth densities. Our present population model makes use of the density-dependent relationship but does not include the delayed component.

The ichneumonid parasite of winter moth pupae, *Cratichneumon culex,* is modeled in a Nicholsonian way. This, again, is an oversimplification since using it in this way causes the model to generate a parasite-host oscillation in which the parasite cycle lags two generations behind that of the host. From our field data we know that the lag is approximately one generation only. We need to know more about this parasite's behavior before we can put it into the model in a realistic way.

Despite its inadequacies and oversimplification, the model provides a satisfactory explanation for the average level of the winter moth in our area. The model also helps us to understand the way in which different sorts of mortality factors interact with one another. By assuming different values for some of the parameters it is possible to test, in theory, the effects on the average level of the introduction of additional mortalities operating at different times. The use of this simple model has enabled us to make some predictions as to the probable consequences of the introduction of the specific tachinid parasite *Cyzenis* into Nova Scotia, Canada, where winter moth was causing repeated defoliation. Putting the Canadian data (15) into the model resulted in a cyclic interaction between host and parasite which would give defoliating outbreaks of the winter moth at intervals of nine or ten years. The usefulness and adequacy of the model will be tested by whether or not this prediction is realised.

MODELS FOR PARASITES AND PREDATORS

Parasites.—Until recently, the theoretical ideas of Thompson (64, 65) and Nicholson and Bailey (51, 54) about parasite-host interactions have seldom been tested adequately against field data. Thompson suggested that the number of hosts parasitised was related only to the parasite density and the parasite's egg production. Nicholson, on the other hand, suggested that egg production could be considered as unlimited but that the parasite's ability to find hosts would depend on its searching efficiency. This efficiency is the parasite's "area of discovery" which he considered to be constant, and is calculated by the formula

$$\text{area of discovery } 'a' = \frac{2.3}{p}\log\frac{u_1}{u} = \frac{2.3}{p}k \qquad 4.$$

Where p = adult parasite density, u_1 = initial host density, u = host density after parasitism and k = k-value for parasitism.

Watt (77) has combined the ideas of Thompson and Nicholson into a single formulation in which the parasite searches for hosts at random, but where the parasite's efficiency is limited by its egg complement and by the rate at which it can attack hosts during the time these are available. Watt has shown that this interesting concept provides a better description of some host/parasite interactions than do either of the other two simple assumptions. Miller (39, 40) has also shown that the "Watt model" also mimics to a reasonable degree the observed interactions between the host and two of the spruce budworm parasites. However, none of these models can be considered to be entirely satisfactory although Watt's model is more comprehensive since it can also describe either the Nicholsonian or Thompsonian model.

In many cases, the field information will be too variable and inaccurate to show which formulation is the better. But the difficulty in estimating parasite population densities ought not to be used as an excuse for not attempting to make the necessary measurements. A variable efficiency is as useful to a population model as a constant one if we can relate these variations to changes in host or parasite density (30) or their ratio, or to changes in weather conditions (35).

In a model of the interaction between a Nicholsonian parasite and its host, oscillations occur in the population densities of both which increase in amplitude with time. In the field the end result only should now be observed; either the population should undergo big oscillations in density, or the population should have broken down into isolated pockets and have a spotty distribution. Neither of these predictions seems common in the field. We know, from models, that these oscillations can be quenched by suitably powerful density-dependent mortalities acting either on the host or on the parasite, or on both (72). However, even such damped models do not explain all the field observations because, in these models, it is virtually impossible to mimic the coexistence of specific parasites (74) unless we assume that each of the parasites concerned suffers its own density-dependent mortality.

We have therefore been much interested by a recent analysis (30)of a laboratory host-parasite interaction which shows the area of discovery of

Venturia (=*Nemeritis*) decreasing with increasing parasite density. The area of discovery of this parasite is described by the formula

$$\log a = -1.2 - 0.54 (\log p) \qquad 5.$$

This is a density-dependent area of discovery the consequences of which, for the parasite population, are exactly the same as if a density-dependent mortality operated on the population of a Nicholsonian parasite. The cause of this relationship has yet to be determined, but it is likely to be some sort of interference between the adult parasites. Wilson (84) reports a scelionid defending hosts against the attack of parasites of its own species, and Viktorov (76) has described experiments in which the parasite sex ratio alters with changes in parasite density. Either of these mechanisms could result in a calculated area of discovery appearing to be density-dependent.

Substituting the new relationship for a from 5 into Nicholson's relationship 4 we obtain a new formula for parasite/host interaction

$$2.3\ k = Q_p{}^{1-m} \qquad 6$$

where k is the k-value for parasitism; Q is the "quest constant" which equals the value for a when parasite density is 1; p is the parasite adult density and m is a constant representing the "mutual interference" between parasites. Nicholson's formulation that $2.3\ k = ap$ represents the special case when $m = 0$. If field and laboratory studies show that this new relationship is of common occurrence it could resolve many of the contradictions between observations and present theoretical ideas. If each parasite can act as its own density-dependent factor, the coexistence of a number of specific parasites on one host can be explained; also, these parasites could regulate their host's density without increasing oscillations even in the absence of a density-dependent factor acting on the host. However, we presume that this density-dependent area of discovery will not be found in those cases where parasite replacement has been reported.

Predators.—Holling's work (25–28) on developing mathematical models for predator/prey interactions through very detailed experimental studies is most stimulating; but ecologists can apply his work only if the measurements required by his formulae can be obtained from field data. Holling has synthesized an ideal predator from behaviour components of many species. This leaves the precise role of any particular predator in doubt since, depending on the relationships used, the effect of the predator can be density-dependent, inversely density-dependent or delayed density-dependent.

The measurements which Holling has been making so far are possibly not those of prime importance to a predator/prey population model. Studies of field populations of phytophagus insects show that, in many cases, the big variable mortality (the key factor) comes at a time when the eggs are hatching and the young larvae are seeking to establish themselves on their food plants and to obtain their first meal. The survival at this stage is profoundly affected by the availability of suitable food. Similarly, Dixon (12) has shown that a newly hatched coccinellid larva is very inefficient in capturing food, and that the most important feature determining whether or not such a larva establishes itself in the predator population is the ease with which it obtains its first meal. If a predator's abilities to survive, develop and reproduce are mainly determined by the relative abundance of suitably sized prey at the time it seeks its first meal, the interactions between adult predators and their prey may be of secondary importance.

POPULATIONS WITH OVERLAPPING GENERATIONS

The relatively simple methods of analysis which are suitable for insects which have discrete generations, are inapplicable to populations with overlapping generations which are inherently difficult to analyse and model. It is only very recently that attempts have been made to produce life tables for such insects (4, 5). Hughes & Gilbert (32) have recently published an interesting model for the reproductive stage of an aphid population. Logically, it is equally important to make the measurements necessary for modelling the population changes during the season when it is not reproducing so rapidly. Only when this is done will it be possible to produce a population model to explain why this insect is always very abundant.

The development of models for insects of medical importance also has made little progress. Ross (59) in 1910 produced an equation which is probably the first mathematical model published. This described the spread of endemic malaria in terms of five measurable parameters. From a recent book (50), it would seem that, to date, few of these parameters have been seriously investigated. This model has yet to be tested against field data.

RESUME

Up to the present time, the majority of published mathematical population models have relied heavily on multivariate analysis techniques. These models, at their very best, result in crude description but explain nothing. Their use lies in the prediction of population change but, because they are not based on an understanding of the biological mechanisms involved, they cannot be used to suggest ways in which future population changes may be modified. We agree with Watt (79) that "the main advantage of multiple regression analysis is that its procedures are mathematically straightforward rather than that the fitted equations bear a close correspondence to events in biological reality," and that this approach "may have an insidious effect on the development of theoretical work."

The factor limiting the development of mathematical models with a biological meaning is the extent of our knowledge about the behaviour of parasites, predators and disease. We need more knowledge of their behaviour and we need to develop mathematical models to describe this. Workers on field and laboratory studies of behaviour ought not to be content to end the study with merely a verbal description of what seems to be happening; there ought to be also a discussion of the relevance of these studies to our theoretical ideas and of the ways in which the observed behaviour may be modelled mathematically to test its consequences in a population model.

All the present methods of analysis make some simplifying assumptions. These may take the form of assuming that all measurements are made without error, or may assume that concurrently acting mortalities are acting in succession, or that only one kind of mortality is operating. Those who wish to use a particular method must first ensure that they understand the assumptions and the limitations of the method.

We can expect, and hope, that present methods of analysis will be replaced by more effective ones. The only way we know of to test a method of analysis is to try it out on a model. We construct a mathematical population model which contains at least one component from each of the categories we have discussed, and use this model to generate a series of artificial life tables. Multivariate analysis will not deduce from these life tables all the components which went into the model and is of limited use in unravelling the intricacies of biological mechanisms. Submodels based on density relationships are likely to be more successful.

Watt (82) has suggested that, because we now possess large computors which can carry out complex analyses very rapidly, the slowest process in ecological research is the acquisition of data. However, field biologists cannot measure everything. We consider that the slowest process is the development of the theoretical ideas without which we can neither plan what observations to make, nor state what methods will be used to derive biological conclusions from the measurements.

LITERATURE CITED

1. Acton, F. S. *The Analysis of Straight-line data* (Wiley, New York, 267 pp., 1959)
2. Andrewartha, H. G., Birch, L. C. *The Distribution and Abundance of Animals.* (Univ. of Chicago Press, 782 pp., 1954)
3. Auer, C. Erste Ergebnisse einfacher stochastischer Modelluntersuchungen über die Ursachen der Populationsbewegung des grauen Lärchenwicklers *Zeiraphera diniana,* Gm. (= *griseana* Hb.) Oberengadin, 1949/66. *Z. Angew. Entomol.,* **62,** 202–35 (1968)
4. Beaver, R. A. Development and expression of population tables for the bark beetle *Scolytus scolytus* (F.). *J. Animal. Ecol.,* **35,** 27–41 (1966)
5. Berryman, A. A. Development of sampling techniques and life tables for the fir engraver *Scolytus ventralis. Can. Entomologist,* **100,** 1138–47 (1968)
6. Brinkhurst, R. O. Population dynamics of the large pond-skater *Gerris najas* Degeer. *J. Animal. Ecol.,* **35,** 13–25 (1966)
7. Broadhead, E., Wapshere, A. J. *Mesopsocus* populations on larch in England—the distribution and dynamics of two closely-related coexisting species of Psocoptera sharing the same food resource. *Ecol. Monographs,* **36,** 327–88 (1966)
8. Buckner, C. H. The role of vertebrate predators in the biological control of forest insects. *Ann. Rev. Entomol.,* **11,** 449–70 (1966)
9. Buckner, C. H., Turnock, W. J. Avian predation on the larch sawfly, *Pristiphora erichsonii* (Htg.). *Ecology,* **46,** 223–36 (1965)
10. Burnett, T. A. A model of host-parasite interaction. *Proc. Intern. Congr. Entomol., 10th, Montreal,* **2,** 679–86 (1958)
11. Clark, L. R. Predation by birds in relation to population density of *Cardiaspina albitextura. Australian J. Zool.,* **12,** 349–61 (1964)
12. Dixon, A. F. G. An experimental study of the searching behaviour of the predatory coccinellid beetle *Adalia decempuctata* (L.). *J. Animal Ecol.,* **28,** 259–81 (1959)
13. Dixon, A. F. G. The effect of population density and nutritive states of the host on the summer reproductive activity of the sycamore aphid, *Drepanosiphum plantanoides* (Schr.). *J. Animal Ecol.,* **35,** 105–12 (1966)
14. Duffey, E. Ecological studies on the large copper butterfly *Lycaena dispar* Haw. *batavus* Obth. at Woodwalton fen national nature reserve, Huntingdonshire, *J. Appl. Ecol.,* **5,** 69–96 (1968)
15. Embree, D. G. The population dynamics of the winter moth in Nova Scotia, 1954–1962. *Can. Entomol. Mem.,* **46,** 1–57 (1965)
16. Ehrlich, P. R., Birch, L. C. The "balance of nature" and "population control". *Am. Naturalist,* **101,** 97–107 (1967)
17. Frank, J. H. The effect of pupal predators on a population of winter moth *Operophtera brumata* (L.). *J. Animal Ecol.,* **36,** 611–21 **(1967)**
18. Harcourt, D. G. The development and use of life tables in the study of natural insect populations. *Ann. Rev. Entomol.,* **14,** 175–96 (1969)
19. Hassell, M. P. Evaluation of parasite and predator responses. *J. Animal Ecol.,* **35,** 65–75 (1966)
20. Hassell, M. P. The behavioural response of a tachinid fly (*Cyzenis albicans* (Fall.)) to its host, the winter moth (*Operophtera brumata* (L)). *J. Animal Ecol.,* **37,** 627–39 (1968)
21. Hassell, M. P. A study of the mortality factors acting upon *Cyzenis albicans* (Fall.), a tachinid parasite of the winter moth (*Operophtera brumata* (L.). *J. Animal Ecol.,* **38,** 329–39 (1969)
22. Hassell, M. P. A population model for the interaction between *Cyzenis albicans* (Fall.) and *Operophtera brumata* (L). at Wytham. *J. Animal Ecol.,* **38** (In press)
23. Hassell, M. P., Huffaker, C. B. The appraisal of delayed and direct density dependency. *Can. Entomologist,* **101,** 353–61 (1969)
24. Holling, C. S. Components of predation as revealed by a study of small mammal predation of the European pine sawfly. *Can. Entomologist,* **91,** 293–320 (1959)
25. Holling, C. S. The principles of insect

predation. *Ann. Rev. Entomol.*, **6**, 163–82 (1961)
26. Holling, C. S. An experimental component analysis of population processes. *Can. Entomol. Mem.*, **32**, 22–32 (1963)
27. Holling, C. S. The functional response of invertebrate predators to prey density. *Can. Entomol. Mem.*, **48**, 1–86 (1966)
28. Holling, C. S. The tactics of a predator. In *Insect Abundance, Symp. Roy. Entomol. Soc. London, 4th*, 47–58. (Southwood, T. R. E., Ed., 160 pp., 1968)
29. Howard, L. O., Fiske, W. F. The importation into the United States of the parasites of the gypsy moth and the brown-tail moth. *U.S. Dept. Agr. Entomol. Bull.*, **91**, 312 pp. (1911)
30. Huffaker, C. B., Kennett, C. E. Some aspects of assessing efficiency of natural enemies, *Can. Entomologist*, **101**, 425–47 (1969)
31. Huffaker, C. B., Kennett, C. E., Matsumoto, B., White, E. G. Some parameters in the role of enemies in the natural control of insects. In *Insect Abundance, Symp. Roy. Entomol. Soc. London, 4th*, 59–75. (Southwood, T. R. E. Ed., 160 pp., 1968)
32. Hughes, R. D., Gilbert, N. A model of an aphid population—a general statement. *J. Animal Ecol.*, **37**, 553–63 (1968)
33. Ierusalimov, E. N. [Changes in increment of a mixed oak forest after defoliation by insects. *Leon. Zh. Arhangel'sk.*, **8**, 52–55 (1965)], *Forestry Abstr.*, **28**, 316 (1967)
34. Jüttner, O. Ertragsküdliche Untersuchungen in wicklergeschädigten Eichenbeständen. *Forstarchiv.*, **30**, 78–83 (1959)
35. Klomp, H. Infestations of forest insects and the role of parasites. *Proc. Intern. Congr. Zool., 15th*, 797–802 (1959)
36. Klomp, H. The influence of climate and weather on the mean density level, the flunctions and the regulation of animal populations. *Arch. Neerl. Zool.*, **15**, 68–109 (1962)
37. Klomp, H. The dynamics of a field population of the pine looper, *Bupalus piniarius* L. *Advan. Ecol. Res.*, **3**, 207–304 (1966)
38. Klomp, H. The interrelations of some approaches to the concept of density dependency in animal populations. *Meded. Landbhoogesch. Wageningen*, **66**, Pt. 3, 1–10 (1966)
39. Miller, C. A. The interaction of the spruce budworm, *Choristoneura fumiferana* (Clem.), and the parasite *Apanteles fumiferanae* Vier. *Can. Entomologist*, **91**, 457–77 (1959)
40. Miller, C. A. The interraction of the spruce budworm, *Choristoneura fumiferana* (Clem.), and the parasite *Glypta fumiferanae* (Vier.). *Can. Entomologist*, **92**, 839–50 (1960)
41. Miller, C. A. Parasites and the spruce budworm. *Can. Entomol. Mem.*, **31**, 228–44 (1963)
42. Morris, R. F. Single-factor analysis in population dynamics. *Ecology*, **40**, 580–88 (1959)
43. Morris, R. F., Ed. The dynamics of epidemic spruce budworm populations. *Can. Entomol. Mem.*, **31**, 332 pp. (1963)
44. Morris, R. F. The analysis of generation survival in relation to age-interval survivals in the unsprayed area. *Can. Entomol. Mem.*, **31**, 32–37 (1963)
45. Morris, R. F. Résumé. *Can. Entomol. Mem.*, **31**, 311–20 (1963)
46. Morris, R. F. Predictive population equations based on key factors. *Can. Entomol. Mem.*, **32**, 16–21 (1963)
47. Morris, R. F., Miller, C. A. The development of life tables for the spruce budworm. *Can. J. Zool.*, **32**, 283–301 (1954)
48. Mott, D. G. The population model for the unsprayed area. *Can. Entomol. Mem.*, **31**, 99–109 (1963)
49. Mott, D. G. The analysis of determination in population systems. In *Systems Analysis in Ecology*, 179–94. (Watt, K. E. F., Ed., Academic Press, New York, London, 276 pp., 1966)
50. Muirhead-Thompson, R. C. *Ecology of Insect Vector Populations.* (Academic Press, London, New York, 174 pp., 1968)
51. Nicholson, A. J. The balance of animal populations. *J. Animal Ecol.*, **2**, 132–78 (1933)
52. Nicholson, A. J. Compensatory reactions of populations to stresses,

and their evolutionary significance. *Australian J. Zool.*, **2**, 1–74 (1954)

53. Nicholson, A. J. The self-adjustment of population change. *Cold Spring Harbor Symp. Quant. Biol.*, **22**, 153–73 (1957)
54. Nicholson, A. J., Bailey, V. A. The balance of animal populations, Part 1. *Proc. Zool. Soc. London,* 551–98 (1935)
55. Nuoreteva, P. The influence of *Oporinia autumnata* (Bkh.) on the timber-line in subarctic conditions. *Ann. Entomol. Fennici,* **29**, 270–77 (1963)
56. Ohnesorge, B. Beziehungen zwischen Regulationsmechanismus und Massenwechselablauf bei Insekten. *Z. Angew. Zool.*, **40**, 427–83 (1963)
57. Readshaw, J. L. A theory of Phasmatid outbreak release. *Australian J. Zool.,* **13**, 475–90 (1964)
58. Ricker, W. E. Stock and recruitment. *J. Fisheries Res. Board Can.*, **11**, 559–623 (1954)
59. Ross, Sir R. *The Prevention of Malaria.* (Murray, London, 669 pp., 1910)
60. Snedecor, G. W., Cochran, W. G. *Statistical Methods,* 6th ed. (Iowa Univ. Press, 593 pp., 1967)
61. Southwood, T. R. E. *Ecological Methods.* (Methuen, London, 391 pp., 1966)
62. Southwood, T. R. E. The interpretation of population change. *J. Animal Ecol.*, **36**, 519–29 (1967)
63. Stern, V. M., Smith, R. F., van den Bosch, R., Hagen, K. S. The integrated control concept. *Hilgardia,* **29**, 81–101 (1959)
64. Thompson, W. R. Théorie de l'action des parasites entomophages. Les formules mathématiques du parasitisme cyclique. *Compt. Rend. Acad. Sci. Paris,* **174**, 1201–4 (1922)
65. Thompson, W. R. Biological control and the theories of the interaction of populations. *Parasitology,* **31**, 299–388 (1939)
66. Tinbergen, L. The natural control of insects in pine woods, 1. Factors influencing the intensity of predation by songbirds. *Arch. Neerl. Zool.*, **13**, 265–336 (1960)
67. Utida, S. Population fluctuations, an experimental and theoretical approach. *Cold Spring Harbor Symp. Quant. Biol.*, **22**, 139–51 (1957)
68. Varley, G. C. The natural control of population balance in the knapweed gall-fly (*Urophora jaceana*). *J. Animal Ecol.*, **16**, 139–87 (1947)
69. Varley, G. C. General discussion, 2. Oscillations and trends. In *The Exploitation of Natural Populations, Brit. Ecol. Soc.*, 363–68. (Le Cren, E. D., Holdgate, M. W., Eds., Symp. 2, 399 pp., 1962)
70. Varley, G. C., Gradwell, G. R. Key factors in population studies. *J. Animal Ecol.*, **29**, 399–401 (1960)
71. Varley, G. C., Gradwell, G. R. Predatory insects as density dependent mortality factors. *Proc. Intern. Congr. Zool., 16th, Washington,* **1**, 240 (1963)
72. Varley, G. C., Gradwell, G. R. The interpretation of insect population changes. *Proc. Ceylon Assoc. Advan. Sci.*, **18**, 142–56 (1963)
73. Varley, G. C., Gradwell, G. R. Population models for the winter moth, In *Insect Abundance, Symp. Roy. Entomol. Soc., London, 4th,* 132–42. (Southwood, T. R. E., Ed., 160 pp., 1968)
74. Varley, G. C., Gradwell, G. R. Can parasites avoid competitive exclusion? *Proc. Intern. Congr. Entomol. Moscow, 13th.* (In press)
75. Viktorov, G. A. *Problemy dinamiki chislemnosti nasekomykh na primere vrednoi cherepashki* (Nauka, Moscow, 271 pp., 1967)
76. Viktorov, G. A. The influence of the population density upon the sex ratio in *Trissolcus grandis* Thoms. *Zool. Zh.*, **47**, 1035–39 (1968)
77. Watt, K. E. F. A mathematical model for the effect of densities of attacked and attacking species on the number attacked. *Can. Entomologist,* **91**, 129–44 (1959)
78. Watt, K. E. F. Mathematical models for use in insect pest control. *Can. Entomologist, Supp.* 19, 1–62 (1961)
79. Watt, K. E. F. Use of mathematics in population ecology. *Ann. Rev. Entomol.*, **7**, 234–60 (1962)
80. Watt, K. E. F. Mathematical population models for five agricultural crop pests. *Can. Entomol. Mem.*, **32**, 83–91 (1963)
81. Watt, K. E. F. How closely does the model mimic reality? *Can. Entomol. Mem.* **31**, 109–11 (1963)

82. Watt, K. E. F. Ecology in the future. In *Systems Analysis in Ecology,* 253–66. (Watt, K. E. F., Ed., Academic Press, London, New York, 276 pp., 1966)

83. Way, M. J. Intra-specific mechanisms with special reference to aphid populations. In *Insect Abundance, Symp. Roy. Entomol. Soc. London, 4th,* 8–36. (Southwood, T. R. E., Ed., 160 pp., 1968

84. Wilson, F. Adult reproductive behaviour in *Asolcus basalis. Australian J. Zool.,* **9,** 737–51 (1961)

THE ECOLOGY OF STREAM INSECTS

By H. B. N. Hynes
Department of Biology, University of Waterloo, Ontario, Canada

When I was asked to write this review I had just completed a book on the ecology of running waters that surveys the literature to almost the end of 1966 (44). Citations of work earlier than this have therefore been kept to a minimum, and the bibliography given here should be regarded as supplemental to that of the book.

The subject encompassed by the title is enormous; there already exist many hundreds of often lengthy papers. It has thus been necessary to be selective of topics and to confine this review quite literally to streams as opposed to rivers. We are therefore considering only localities where the flow is fairly swift, the water shallow, and the substratum rock, stones or gravel, with only limited areas of sand or silt. This is a very loose definition of the habitat known as the rhithron, although it also includes spring brooks, sometimes called the crenon.

It is, however, a habitat which is usually dominated by insects except in very calcareous waters. Hydrida, Tricladida, Rotifera, Nematoda, Oligochaeta, Copepoda and Parasitengona are nearly always also present, and in hard water they are usually joined by Hirudinea, Isopoda, and Amphipoda and sometimes by Decapoda.

The Insect Fauna of Streams

Perhaps the most striking feature of the rheophile insect fauna is its world-wide uniformity. In this respect it falls short of the biotic uniformity of a few special habitats, such as hot springs or temporary pools but, in contrast to other biocoenoses that are dominated by a large variety of insects, it is remarkably similar on all the continents. Usually one finds many of the same families and life forms and often even the same genera wherever one looks on the planet, with the not unexpected additional fact that isolated land masses, such as New Zealand and Britain and recently glaciated areas, tend to have impoverished faunas. The Pleistocene glaciation of northwestern Europe, because it involved icecaps both to the north and on the Alps to the south, has resulted in some extra limitation of the variety of stream fauna there. There are, for instance, no European Corydalidae or Deuterophlebiidae, and Blepharoceridae are absent from the northern mountains. There is also some dichotomy between the northern and southern hemispheres, which differ almost completely in their families of Plecoptera (45) and in that Leptophlebiidae in Australia and South America occupy some niches which in the north are inhabited by other families of mayflies.

Another particular feature of the stream fauna is that there are many families, or even larger groups, which are virtually confined to running water. These are to be found among the Ephemeroptera, Odonata, Plecoptera and Nematocera, together with some small taxa among the Heteroptera, Coleoptera, Lepidoptera and Brachycera.

These two features lead one to conclude that much of the insect fauna of streams is of ancient lineage, and that it has lived in the same type of environment for a long while. It is not surprising therefore that many genera display obvious adaptational modifications to their peculiar habitat.

Some of these modifications are morphological and involve such specializations as flattening, streamlining, friction discs, close application to the surface of stones and even the presence of hydraulic suckers. The adaptational significance of some of these features has been challenged on the ground that they are not confined to stream animals (72), but that is really a spurious argument. Similar modifications may occur in response to quite different conditions for the same or for different reasons. Even within the stream environment, flattening allows some animals to live on stones, e.g., *Psephenus* larvae, while the flattening of others, e.g., many Heptageniidae, permits them to creep under stones into shelter from the current.

Our understanding of the exact function of many of these morphological adaptations has been greatly clarified by Ambühl's (2) study of flowing water in relation to microhabitats, and his demonstration of a boundary layer of almost static water on stones, and of spaces filled with dead water downstream of and below them. Bournaud (10) has shown that most of the modifications enable animals to avoid many of the effects of flow by keeping their bodies down in the boundary layer away from the force of the current. He also stressed that behaviour is more important than morphology, and that cryptic habits are characteristic of most stream insects. These can also lead to morphological adaptations, such as the long narrow bodies of many Plecoptera and Ephemeroptera (Leuctridae, Capniidae, Chloroperlidae and some Gripopterygidae, Leptophlebiidae and Baetidae). These nymphs burrow into the gravel, and thus avoid any contact with the current. In fact the only insects which regularly expose themselves to fast water and swim against it are the streamlined Baetidae, such as most species of *Baetis* and *Centroptilum*.

Other important changes which are associated with stream life have occurred in respiratory physiology. Normal streams are always well oxygenated because of their turbulent flow; and they are usually cooler, during summer, than bodies of still water. Associated with this many stream insects, all of which must use dissolved oxygen as they cannot go to the surface to obtain gaseous air, are unable to tolerate low oxygen tensions. Moreover, many of them have lost the ability to ventilate, by flapping gills or undulating their abdomens, and have come to rely upon the current to renew the oxygen supply at their body surfaces. Some can still ventilate *in ex-*

tremis, but they are unable to keep it up for long; they are therefore tied to water movement, high oxygen content and low temperature, or at least to any two of these factors, and this confines them to running water (46). Only in the Arctic and at high altitudes does the distinction between stream and lakeshore faunas tend to become blurred because the lakes are very cold at all times (18, 92). But even there, there are some genera, e.g., *Rhyacophila,* which remain restricted to streams, probably because of their respiratory physiology, and others are so confined by their feeding habits, which are considered below.

These modifications, and differences in levels of adaptation between species, go some way toward explaining the close association with particular types of substratum which are so characteristic of many stream insects. For instance, water pennies can attach only to rocks or fairly large stones, the suckers of Blepharoceridae and the friction discs of many other insects can be applied only to fairly smooth surfaces and are rendered useless by a thick covering of silt or diatoms, and Leuctridae need gravel in which to burrow, and are eliminated by any considerable admixture of fine sand.

There are, however, also many unexplained associations with particular types of substratum, and certain species are always found in or on particles of a certain size for no clear reason. A recent study of this type of preference is that of Eriksen (33), who showed that nymphs of *Hexagenia limbata* burrow only into fine substrata, while *Ephemera simulans* chooses gravel, although these two mayflies are superficially fairly similar. The result is that, in the mozaic of types of substratum and areas of shelter on a stream bed, there is a corresponding faunal mozaic (90). This makes quantitative data extremely difficult to obtain, and the now almost classical example of this problem is the study of a single riffle in a Californian stream which demonstrated that an almost impossibly large number of samples is needed to determine the number and weight of animals present per unit area (71).

Differences in temperature tolerance, coupled undoubtedly with competition between allied species, also result in zonation along the lengths of streams, as species replace one another either wholly or in part. This has led to some controversy in recent years about the possibility of defining distinct zonal biocoenoses in terms of their invertebrate faunas. A general conclusion seems to be emerging that such a classification is not justifiable except in the most general way, because specific changes occur for different reasons at various points along a continuum. Nevertheless, the close association of particular species with definable microhabitats is well established, and it often results in a clear zonational distribution along streams within the limits of certain taxa. Several examples have been published during the past three years (18, 64, 65, 93–95).

Generally speaking, it is the most highly adapted insects which are the most abundant and important members of the stream fauna. Such are the Plecoptera, Baetidae, Rhyacophilidae, net-spinning Trichoptera (of which more will be said below), Simuliidae, and Elminthidae, the last of which are remarkable among the water beetles in using plastron respiration. To these we should add many Diamesinae and Orthocladiinae of whose adaptations we know little, although the fact that many of them are confined to streams probably indicates at least respiratory adaptation. Ross (78) has stressed that the primitive families of Trichoptera are confined to streams, and it seems to be true that streams have become refuges for many primitive, but nonetheless highly adapted, types. One has only to consider the Plecoptera, Ephemeroptera, Corydalidae and several highly specialized families of Nematocera to appreciate this point. Most of the insects which live actually out in the current-swept areas are survivors of primitive groups, and they have been joined by a comparatively few more highly evolved insects. The latter include some Rhagionidae, Tabanidae (*Haematopota*) and a very few Schizophora (e.g., *Limnophora*), a few genera of Pyraustidae (e.g., *Cataclysta*) and Hemiptera (some Naucoridae and *Aphelocheirus*), and some beetles, among which only the Elminthidae are important. The widespread Hydraenidae spend only their adult lives out on the stream bed, and the majority of higher Diptera which are associated with streams (e.g., Dolichopodidae, Empidae and most Tabanidae) and Coleoptera (Dytiscidae and Gyrinidae) occur only in sheltered backwaters or at the extreme edge of the water. This perhaps indicates the difficulties presented by the stream environment and the problems faced by more highly evolved forms in replacing well-adapted primitives from niches in a severe habitat. A parallel example is the persistence of the ancient Phyllopoda in temporary pools in the face of competition from the more modern Crustacea.

Particularly striking is the almost complete absence of Hymenoptera which are so common almost everywhere else. Only the genus *Agriotypus,* which parasitizes pupae of Goerinae, has actually succeeded in invading streams, although some stream insects are liable to parasitism by other Aculeata during stages that are not spent under water (11). The absence of parasitoids provides a great contrast with the terrestrial situation.

Among the important groups, the adults of Elminthidae alone are aquatic and long-lived. In nearly all the others adult life is terrestrial, short, and concerned only with breeding and dispersal, although even the latter is in some dispute. Many imagines fly weakly and live for a very short while, and some stoneflies do not fly at all. There is some evidence that black fly adults may move considerable distances (8), but set against this is the fact that at least some species show local chromosomal varieties (22) which indicates little gene exchange between populations. Similarly, many species of stonefly have very limited areas of distribution (76). We clearly need more investigation into the importance of the adult stages in dispersal.

Food

Many stream insects feed on the periphyton growing on stones, which is an unusual diet for insects; but it is available in large amounts only to those species, such as Psephenidae and Glossosomatinae, which spend all or much of their time on exposed, and therefore lighted, surfaces. Even they appear to be not particularly selective as they also eat organic detritus trapped in periphyton, and a great many other insects feed primarily on dead organic matter lodged in the substratum. Egglishaw (25) has demonstrated very elegantly that the micro-distribution of most of the fauna is correlated with the amounts of plant debris present and that the insects seek it out and concentrate upon it; and many studies [e.g. (13, 67, 92)] have emphasized that allochthonous plant material is the most important primary source of energy for the stream fauna, and is made available to higher trophic levels through the activities of detritus-feeding insects. The productive capacity of the valley is therefore more important than the primary production in the stream itself, which may, because of shading, be very low and yet apparently support a dense fauna.

Drifting detritus, plankton and washed-up benthic organisms are also an important source of food for many insects, and this has resulted in a number of specializations which are found only in stream dwellers. They range from the sticky strings of saliva hung between the arms of the hydra-like tubes of *Rheotanytarsus,* through the fringes of hairs on the forelegs, or even the mandibles, of some Ephemeroptera nymphs, to the elaborate nets and associated structures built by many campodeiform Trichoptera larvae. Perhaps the best known of these passive feeders are the larvae of Simuliidae, whose premandibles bear elaborate double fans (37). The fans are held open by the current, up into which the larvae thrust them from their shelter in the boundary layer, and this imposes a lower limit of rate of flow in which the larvae can feed. It is about 20 cm/sec for *Simulium ornatum* (39) and this alone restricts the species to certain types of locality. Feeding is apparently quite unselective, so the quality of the available food affects the population, and where large numbers of larvae are present they remove microseston from the water very rapidly (62). They also catch very tiny particles so, in contrast to most insects, they can complete their life cycles on suspensions of bacteria (34). It seems very probable that, in some situations, bacteria form the bulk of their food, thus giving the stream community almost direct access to organic matter coming from the watershed in solution.

The nets of Trichoptera have been described by many authors; two recent papers are those of Kaiser (48) and Sattler (80); and Gibbs (35) discusses the manner in which some of the more elaborate structures in which they occur may have evolved. In most of them the mesh apertures are quite large, measurable in tens of microns (24), so the food particles caught are large and varied, and the common Hydropsychidae are more or less omnivorous. But in at least one species, *Micrasema ulmeri,* the meshes are only $3 \times 19\ \mu$ (80), so very small particles must be collected. We do not, however, know how common are such very finely meshed nets.

It has been shown that nets, like the fans of black fly larvae, are efficient only in a current, so it is not unexpected that Edington (23, 24) was able to show that when he interfered with the flow in field situations the larvae moved to new locations and constructed new nets. They, like *Simulium,* are restricted to certain current speeds by their feeding mechanisms, and different species are tied to different ranges of speed by the varying structures of their nets (24). At least the genus *Hydropsyche* is also limited by temperature; two species which Kaiser (48) studied became inactive below 8° C. Edington also showed that where *Hydropsyche* is abundant the nets are very evenly spaced downstream of one another, although often closely side by side, in such a way that no net is fishing already filtered water. How this distribution is achieved is not certain, but it may be relevant that the larvae are known to stridulate, an ability which is rather rare among larval insects (47).

Clearly, also, these passive feeders are benefited by conditions which make the water particularly rich in food particles. Several studies have shown that they are often very abundant just below lakes, which are a source of plankton; recent ones are those of Cushing (16) and Ulfstrand (92). It can also readily be observed in Canada that beaver dams and pestilential swarms of black flies go together. As might be expected from our knowledge of seasonal changes in lake plankton, the effect is more noticeable in the summer than at other times, Ulfstrand found that the fauna of lake outflows in arctic Sweden was always much denser by a factor of ten or more, than that of other streams during July and August, but that this did not apply earlier or later in the year. In Sweden, the increase was caused by Simuliidae, and in Canada Cushing recorded a similar, but longer, summertime maximum of campodeiform Trichoptera. It is also interesting to note that some species of black fly seem to occur only near lake outlets. Possibly this restriction is not universal and it may have nothing to do with food. Some mayflies are similarly confined in Swedish Lapland but not elsewhere in their geographical ranges, and others are not found below lakes there although they occur in other streams (92). This indicates that temperature is the operative factor, as lakes warm up more slowly, become warmer and remain warm longer than streams. Possibly the same factors influence some of the species of Simuliidae.

Life Histories

Stream temperatures near springs or ground-water seepages are often remarkably stable, and even far downstream of such sources they are much more uniform than they are on land. This is partly because evaporation and, very often, shade from riparian trees keep temperatures down in hot weather, and partly because, even during severe winters, the water continues to flow under an insulating layer of ice and snow. The habitat of benthic insects therefore never freezes, and Schwoerbel (83) has shown that the winter temperature a few centimetres down in the substratum may be several degrees higher than that of the overlying water. Even during the Canadian winter, after many weeks of freezing weather resulting in 0° C in the water and tens of centimetres of surface ice, we have recorded temperatures as high as 1.3° C 60 cm below the gravel surface.

This lack of frost, coupled with the fact that autumn leaf-fall provides a readily available and abundant supply of food to the biota, has allowed the evolution of many species which are active throughout the winter. These do much of their growing when they are least likely to fall prey to fishes which are rendered sluggish by the cold.

Many investigators, during the last two decades, have demonstrated that there are species of Plecoptera and Ephemeroptera that grow actively during the winter, and Raušer (76) concludes that the threshold temperature for many Plecoptera must be very near 0° C. Ulfstrand (91) found that *Ameletus inopinatus* and *Leuctra hippopus* undergo most of their nymphal development while the streams are iced over in northern Sweden and, although the growth of many species is slowed down or even stopped for a while by intense cold (29, 86), many others continue to grow through the winter. In the mild climate of West Virginia the wing pads of *Epeorus* grow fastest in the coldest weather (66), and in Scotland, which also has mild winters, differences in the severity of different years have apparently little influence on the rates of growth of several stonefly and mayfly species (26). It is clear therefore that in these two orders there are a number of genuine winter species, the adults of which emerge in the early spring. These are extreme examples, but there does seem to be a general tendency towards cold stenothermy among many stream insects. Macan (61) found that *Heptagenia lateralis* is restricted to streams that do not exceed 18° C, because it needs a sufficiently long time between its rather high temperature threshold for growth and its fairly low lethal temperature. So, although the eggs are known to tolerate high temperatures on lake shores, warm streams heat up too fast to allow the species to complete its nymphal development. There are other species whose distribution is probably controlled in a similar manner, and it is becoming clear that in many instances it is the *pattern* of temperature change rather than the absolute temperature attained that is the important factor.

Most of the cold-water insects are univoltine, but in extreme environments some species which are not able to grow adequately at the lowest temperatures have become opportunistic, and make the most of such short slightly warmer periods as occur. Thus, *Diamesa valkanovi* in a Norwegian glacier brook appears to have a two-year life cycle near the glacier but to be univoltine in the lower reaches (79); and *Rhyacophila evoluta* is able to go into cold-induced diapause at any stage of its life history, and so take one, two or three years to develop. This allows it to inhabit a wide variety of streams in the Pyrenees, in contrast to less adaptable animals (18). At the other end of the scale, the cool-water, as opposed to cold-water, bivoltine species of *Baetis* can avoid the effects of very warm water in southern or low-altitude streams by increasing the gap between successive generations and spanning it by remaining in their eggs (75).

Diapausing eggs that avoid the consequences of high temperature seem to be a common device among stream insects, and they are found in many species of Ephemeroptera, Plecoptera and Simuliidae, especially *Prosimulium*. Sometimes indeed the diapause is so long that it allows the nymphs to hatch at the most favourable season for early growth, nearly a year after oviposition. Khoo (52) has shown that this applies to the northern holarctic stonefly *Diura bicaudata,* and that, very interestingly, it seems to be losing this peculiar characteristic in lowland lakes where it is not necessary.

Khoo (50, 51) also showed that *Capnia bifrons,* a winter stonefly, has a nymphal diapause which is induced by temperatures above 9° C, or by increasing day-length at lower temperatures, and that the nymphs avoid the warm weather by remaining deeply buried in the substratum in this condition. We now know that other Capniidae, e.g., *Allocapnia,* behave similarly, and it is a small addition to such aestivation that permits many insects to avoid the consequences of drought. Thus, many cold-water species, such as *Prosimulium gibsoni, Allocapnia pygmaea* and *A. vivipara,* which inhabit temporary streams [(3, 15) and personal observation] survive the dry season as eggs or diapausing nymphs, and some Trichoptera aestivate as prepupae (15). Where, however, drought is very severe and prolonged, as in Rhodesia, it seems that only the last mechanism is possible, and the faunas of temporary and permanent streams differ little because of rapid recolonization (38). This can, however, occur in a warm climate in a way that it could not where the end of a long summer dry period is also the end of the flying season for adults.

In addition to the peculiar cold-water species, there is also a corresponding population of warm-water insects that avoid the winter season as eggs, pupae or very small nongrowing nymphs. The majority of such species are Simuliidae (3, 12), but they include some Trichoptera and Ephemeroptera. Such are *Glossosoma intermedium* in northern Sweden (91) and *Chitonophora* sp. (larvulae) and *Ephemerella ignita* (eggs) in Austria (75). Generally speaking, the number of species of these summer insects is lower than that of the winter species, a fact which often results in the greatest biomass being present in the spring at the end of the growth of the latter (26, 97). There are, however, often two peaks in numbers of individuals, in early winter and early summer, respectively, as the two groups begin their periods of growth. It is also often possible to observe the apparently direct temporal vicarious occurrence of similar species, as did Egglishaw & Mackay (26) between *Leuctra fusca* and *L. inermis,* and *Rhithrogena semicolorata* and *Ecdyonurus* in Scottish Highland streams.

An interesting fact in association with the existence of closely related cold-water, cool-water and warm-water species is that in the Elminthidae there is a decrease in size from the first to the last. Steffan (85) has shown that this applies to several genera in central Europe, and he has suggested that it may be connected with better oxygen relations or longer larval life at the lower temperatures. An analogous situation is known to occur in several species of mayfly in which early-emerging individuals are larger than later-emerging ones of the same or of a subsequent generation. This is probably connected with the temperature at which the nymphs have grown, but there may be complicating factors associated with photoperiod. Khoo (50, 51) found that the adults of *Capnia bifrons,* which emerge from February to May in Wales, become progressively smaller as the season advances, and that exposure of the nymphs to long photoperiods hastens the appearance of adult characters in immature nymphs. The smaller adults also have relatively shorter wings than the larger ones, and this may provide an explanation for the occurrence of short-winged specimens of many insects at high altitudes (79). Presumably, the cold does not permit them to grow as fast as their lowland relatives, but the onset of summer hastens development to the point of emergence before they are large enough to have large wings. It has often been shown that early emerging species emerge later at higher latitudes and altitudes (9, 32, 88); and it is also known that species with long emergence periods at low altitudes have shorter ones at high altitudes, and that altitude makes little difference to the times of emergence of autumnal species (17). These findings would fit with the assumption that emergence times are under the dual control of temperature and day-length, and presumably their relative importance varies from place to place. For example, Minshall (69) concluded that the time of emergence of *Epeorus pleuralis* from near the source of a spring stream in Kentucky must be controlled by photoperiod as the temperature of the water varies very little; and photoperiod is almost certainly the over-riding factor for the winter stoneflies which emerge through small thaw holes in ice-covered streams.

Photoperiod also seems to be important in controlling the rate of maturation of eggs, and in synchronizing the timing of life cycles of species which inhabit varied water bodies. There are two well-worked examples among the Trichoptera. In southern England, *Limnephilus lunatus* inhabits both spring-fed watercress beds and nearby streams. The relatively high winter temperature of the spring water permits earlier emergence from the watercress beds than from the streams. But the females from both types of locality mature their eggs slowly on land in such a way that both groups are ready to oviposit in the autumn (36). In the Pyrenees, *Allogamus auricollis* occurs over a wide range of altitude and, as is usual in such situations, the low-altitude specimens emerge earlier than the high-altitude ones. They then mature their eggs away from the water. The high-altitude specimens are much smaller than their downstream sisters, probably because they spend less time in the fifth larval instar, and they have fewer eggs. But their eggs are mature at the time of emergence, so oviposition is more or less synchronized at all altitudes (18). Control of oogenesis by photoperiod would appear to be the most reasonable hypothesis in both these instances.

Differences in temperature tolerance with particular reference to life histories therefore account in large measure for the altitudinal, longitudinal, and seasonal distribution of species in streams as well as for the often reported relatively great importance of Plecoptera at high altitudes and latitudes (9, 18, 19, 49, 84, 92). They also explain the frequently observed coexistence of closely allied species in the same habitat, as it is possible to avoid direct competitive exclusion by having life cycles out of step with one another. This can occur even when emergence and oviposition times coincide or overlap, because of different eclosion and growth patterns related to different seasons. Such temporal succession may occur even among coexisting multivoltine summer species, as was shown by Obeng (73) among *Simulium* species in Wales. She found, for example, that succeeding generations of *S. reptans* and *S. variegatum* were out of phase, so that small larvae of one accompanied large larvae of the other throughout the summer. This, presumably, reduces competition between the two species and allows them to occur together.

One must, however, be cautious about regarding differences in life history as the only explanation for coexistence of allied species. Often there are differences in preference of microhabitat which are difficult to detect in the fine mozaics of stream beds. A good example is Madsen's recent careful study of *Heptagenia sulphurea* and *H. fuscogrisea* in Denmark (63). These two species look very similar and often occur together, but *H. sulphurea* is more flattened, swims little when detached from the substratum, and takes hold of the bottom as soon as it touches it, settling in the boundary layer. *H. fuscogrisea* swims when released into free water, does not reattach so readily, and seeks out dead water. The two species therefore live only centimetres apart but in quite different microhabitats. Interestingly, Madsen was able to find small morphological differences between them, involving the positions of hair fringes on the legs and the teeth on tarsal claws, which he was able to relate clearly to the habits and habitats of the species. It is pleasant to think that perhaps some day we shall understand the significance of at least some of the trivial morphological characters which we use for distinguishing the young stages of stream insects.

Drift

During the past few years many investigators have concerned themselves with the downstream drift of stream animals as revealed by nets left stationary in the water. It is well established that most stream insects face into the current, and indeed many of them have little option in this as their long tails (Ephemeroptera and Plecoptera) or tapering cases (Trichoptera) sweep them round to face the local direction of flow. Thus, any voluntary move they do make tends to take them upstream. However, any loss of hold sweeps them downstream, and the numbers carried past a given point seem far in excess of any possibility of compensation by merely walking upstream.

When Roos (77) showed that he caught more adult insects flying upstream than were flying downstream in central Sweden the problem seemed to be solved, and the idea of the "colonization cycle" resulted. This postulates that adults fly upstream to oviposit, the young stages drift down and emerge, and the adults again fly upstream. Support was given to this by the finding that even the winter stonefly, *Capnia atra,* which emerges onto the snow when it is too cold to fly in northern Sweden, walks about 100 m directly from the stream and then turns upstream and goes on walking (89). Other workers have also reported upstream flights [e.g. (98)]. Unfortunately for this theory, Elliott (28, 30) was not able to confirm Roos' results in England, where wind direction seems more important than stream direction in controlling movement. And, in Ontario, while we have been able to observe the initial march away from the stream by a winter stonefly, *Allocapnia pygmaea,* the upstream part of the journey is not as clear-cut as reported for *Capnia.* Moreover, about a month elapses in the Capniidae between emergence and oviposition, so unless the females remain close to where they arrive on their initial journey and retain knowledge of the local geography, it is difficult to see how a net upstream movement can result.

Several species of mayfly nymph are known to be negatively phototropic or to be more active in brighter light, and these responses tend to keep them in sheltered places during the day (30, 42, 81, 82). When night falls they wander out of the shelter and are more readily dislodged by the current. As a result, the drifting of mayflies, and also stoneflies, campodeiform Trichoptera and many other insects, is greatest at night, and the same applies in large rivers as in small streams (4, 5, 27, 28, 54, 57, 70, 99). The drift rate of Chironomidae seems, however, not to vary over the 24 hours (4) and some species of Limnephilidae drift by day but not by night (5, 98).

As light is clearly involved in much of this behaviour it is understandable that it has been reported that nocturnal illumination, or even a full moon, reduces the amount of drift, and that artificial darkening before nightfall brings it forward (4, 27). Holt & Waters (41) showed that the critical level of illumination for *Baetis vagans* is about 1 lux, while Elliott (30) reports levels of 5 to 20 lux for the start of activity and 2 to 60 lux for its cessation in a number of other mayflies. These differences may account for some discrepancy in the findings of Anderson (4) and Elliott & Minshall (32) on the effect of moonlight. It seems, indeed, that the effective level of illumination must vary from place to place, or from species to species, as Ulfstrand (92) found that the midnight sun in the Arctic does not totally suppress the diurnal rhythm.

Drift may also be enhanced by freshets, as they increase the flushing effect of the stream. This so-called catastrophic drift has been studied by Anderson & Lehmkuhl (6) and Elliott (28) who found that the diurnal rhythm persisted even where the numbers were greatly increased, and by Weninger (99) who observed little drifting at times of very high water, indicating perhaps that very high discharges cause the animals to seek shelter.

The rhythm, however, is more complex than a simple nocturnal maximum. Usually, the maximum follows soon after sunset, and there is often a later one, or even two, before dawn on long nights (28, 70). A probable explanation for this is provided by a recent study by Chaston (14) who showed that *Simulium, Isoperla* and *Ephemerella* have fluctuating innate rhythms of activity which are suppressed by light. The onset of darkness releases the rhythm and the animals drift most readily at their times of peak activity, of which there may be one or more depending upon the length of the cycle and the duration of the night.

A further complication is that not all size groups of any one species drift in the same proportions as they occur in the population. Anderson (4) found that the drift of larger individuals was more depressed by moonlight than that of smaller ones, and that the early caseless instars of *Hydroptila rono* drift less readily than larger case-bearing ones (5). There are thus indications that drifting is associated with life history (99), and Elliott (29, 31) concludes that in most species the maximum amount of drifting occurs at seasons of maximum growth.

Other studies indicate that temperature may affect the drift rate (74) and that this may be the major controlling factor for daytime drifters (98); that riffles and runs produce more drift than do pools which may indicate a decreasing importance of drift in lower reaches (7); that reductions in flow cause increases in drift (68); and that there is some correlation between the density of the population and the amount of drift (21, 74).

The last point appears to indicate that drift is at least partly an outcome of competition for space, and that it causes the loss of excess production to downstream areas. But in the present state of our knowledge the significance of drift is far from clear. It is even uncertain how far the animals move, and whether there is any significant downstream displacement of the population as a whole. Some earlier studies indicated that individuals travel tens of metres each night, and Carlsson (12) mentions movement of some hundreds of metres by *Simulium* larvae marked with ^{32}P. On the other hand, ^{32}P has been shown to move similar distances upstream, with at least circumstantial evidence that it had been carried by aquatic insects. Recent studies have stressed that only a very small proportion of the population is in the water column at any one time, and that the size composition of the drifting insects varies irregularly (28, 57, 92). This seems to indicate that the distances moved may be quite short and, despite the large numbers caught in drift nets, Elliott (28) has calculated that the loss to the fauna, if it is one, is only 0.37 per cent per day in an English stream. The corresponding figure for an arctic Swedish stream is about 1 per cent (92). These low figures, together with our lack of knowledge of how much upstream movement occurs, possibly indicate that the importance of drift has been overstressed. It certainly remains in need of further study, with particular reference to upstream movement and distances moved, before its ecological importance can be fully assessed.

Avoidance of Disaster

Catastrophic drift caused by high rates of discharge is only one symptom of the difficulties faced by insects in the essentially unstable habitat of the stream bed. Severe floods with their attendant scour, wash-out and abrasion may occur at any time, and only those species which can survive them can successfully inhabit streams. It has been known for a long time that spates reduce the fauna; recent reports of this phenomenon are those of Maitland

& Penney (64) on Simuliidae in Scotland, and Hynes (43) on the insects of a Welsh stream.

One way in which such effects are minimised is the very close attachment of eggs to the substratum, so that even if stones are rolled and abraided some survive. The elaborate attachment mechanisms of the eggs of stoneflies and mayflies have often been described and reports on them continue to appear (20, 53, 55, 56). Moreover, in many insects the eggs continue to hatch over a very long period, often far exceeding the flight period of the adults. Macan (59) has discussed the importance of this phenomenon, and it has been subsequently observed, or inferred from the continued appearance of very small specimens, by many authors. Long hatching periods, coupled with firmly attached eggs, of course ensure that there is a long period during which a spate severe enough to remove nymphs or larvae will not eliminate the species.

Another disaster which may, regularly or infrequently, overtake stream insects is drought, and here again there is evidence that many species survive as eggs, and again the long eclosion period is an insurance. Some invertebrates can burrow deeply into the substratum and escape the dryness, but among insects this ability is confined to Coleoptera and some Diptera. Mayflies and stoneflies, apart from the diapausing nymphs of Capniidae, normally survive only as eggs.

It is not, however, generally appreciated that even when water is present great numbers of the small stages of stream insects, and some large ones also, burrow deeply into the gravel. Schwoerbel (83) has written extensively on this "hyporheic" habitat, and has shown that the insects spread out laterally from the edges of streams and can be collected some distance from the water's edge. Work in my laboratory has confirmed that a dense fauna extends downwards for tens of centimetres below the substratum/water interface in a stream in southern Ontario, and it seems probable that the hyporheal has considerable ecological significance. It seems, for instance, the only reasonable explanation for the steady return of the fauna to a Welsh mountain stream which had been devastated by an unusual flood, because most of the species returned before they had had time to recolonize as flying adults and their size-distribution was the same as it would have been had the flood not occurred (43). The last point indicates that at least most of the recolonization had not resulted from unhatched eggs.

Quantitative Studies

The fashionable trend in modern ecology is towards the study of production and measurement of energy transfer between trophic levels. This is very difficult in streams not only because much of the primary source of energy is allochthonous and irregularly transient and thus very difficult to measure, but because we have no effective methods for quantitative sampling.

We have already seen that very large numbers of samples have to be collected because of the mozaic nature of the stream bed. Those results were based on collections made with a Surber sampler which is, for various reasons, rather inefficient despite its general usefulness. Better samplers are available and have been discussed by Macan (60) and Albrecht (1). None, however, penetrates very deeply into the substratum so, quite apart from any other failings they may have, they do not collect a great deal of the hyporheal. They also all involve the use of some kind of screen, which causes the loss of small specimens if it is too coarse and large ones if it is too fine (60, 87).

Moreover, there are indications that insects move laterally across the beds of streams as they pass through various stages of their life histories (58, 92). In high mountains, the fauna becomes concentrated toward the centre of the bed under the ice (18), and this probably applies in all severe climates, and at least one species of mayfly, *Habroleptoides modesta,* seems to leave the stream altogether for part of its life history, wandering far out into the hyporheal under the banks (83).

It will be clear therefore that all the quantitative data so far obtained are very unsatisfactory, and that even our estimates of standing biomass are almost certainly far too low. This accounts for the paradoxical finding of some fishery workers that fishes seem to eat many times the numbers of insects present at any one time, even though most of those insects are univoltine.

Future Prospects

It will be apparent from this survey of recent work that study of the ecology of stream insects is still largely exploratory and descriptive. We are beginning to understand some of the reasons for particular distributions in relation to substratum and temperature, and also some of the relationships of life cycles to seasonal changes. We suffer, however, very much from our inability to identify immature insects. Wiggins (100) has stressed that only a very small percentage of the species is known and that this type of study is unfashionable. The situation is particularly bad with very early stages, which are often of considerable ecological importance, and the universal and abundant Chironomidae are almost impossible to identify. This lack of taxonomical work is one of the most serious barriers to further progress.

Another serious barrier is the inadequacy of our sampling methods, especially in view of the depth to which many insects apparently penetrate. Until this problem is solved all quantitative work, unless it is strictly comparative, must be accepted with great reservation.

We also need to know much more about the adult stages and what influence their habits and requirements may have on the distribution of species. Van Someren (96) has suggested that the absence of some stream insects at high altitudes on Mount Kenya may be because nightly frosts eliminate adults, and Hartland-Rowe (40) proposes that the distribution of *Rhithrogena* in the Canadian Rockies may be controlled by the oviposition habits of the adults; and the same sort of consideration has been applied to Simuliidae and damselflies (12, 101). The reverse situation may apply to *Hy-*

draena which has aquatic adults but semiterrestrial larvae. It occurs up to only about 1000 m in the Pyrenees, whereas at least some Elminthidae, in which both adults and larvae are aquatic, are found at much higher altitudes (9). Much of this is, however, merely speculation and there is an obvious need for further study.

Finally, there is the almost uninvestigated field of the role of allochthonous organic matter and its associated microflora in the nutrition of the primary consumers in streams. Insects comprise the majority of the species which feed on this organic detritus.

LITERATURE CITED

1. Albrecht, M.-L. Die quantitative Untersuchung der Bodenfauna fliessender Gewässer. *Z. Fisch.*, N.F,. **8**, 481–550 (1959)
2. Ambühl, H. Die Besonderheiten der Wasserströmung in physikalischer, chemischer und biologischer Hinsicht. *Schweiz. Z. Hydrol.*, **24**, 367–82 (1962)
3. Anderson, J. R., Dicke, R. J. Ecology of the immature stages of some Wisconsin black flies. *Ann Entomol. Soc. Am.*, **53**, 386–404 (1960)
4. Anderson, N. H. Depressant effect of moonlight on activity of aquatic insects. *Nature,* **209**, 319–20 (1966)
5. Anderson, N. H. Biology and downstream drift of some Oregon Trichoptera. *Can. Entomologist,* **99**, 507–21 (1967)
6. Anderson, N. H., Lehmkuhl, D. M. Catastrophic drift of insects in a woodland stream. *Ecology,* **49**, 198–206 (1968)
7. Bailey, R. G. Observations on the nature and importance of organic drift in a Devon river. *Hydrobiologia,* **27**, 353–67 (1966)
8. Baldwin, W. J., Allen, J. R., Slater, N. S. A practical field method for the recovery of blackflies labelled with P^{32}. *Nature,* **212**, 959–60 (1966)
9. Berthélemy, C. Sur l'ecologie comparée des plécoptères, des *Hydraena* et des Elminthidae des Pyrénées. *Verhandl. Intern. Ver. Theoret. Angew. Limnol.,* **16**, 1727–30 (1966)
10. Bournaud, M. Le courant, facteur écologique et éthologique de la vie aquatique. *Hydrobiologia,* **21**, 125–65 (1963)
11. Brown, H. P. *Psephenus* parasitized by a new Chalcidoid. II Biology of the parasite. *Ann. Entomol. Soc. Am.*, **61**, 425–55 (1968)
12. Carlsson, G. Environmental factors influencing blackfly populations. *Bull. World Health Organ.*, **37**, 139–50 (1967)
13. Chapman, D. W., Demory, R. Seasonal changes in the food ingested by aquatic insect larvae and nymphs in two Oregon streams. *Ecology,* **44**, 140–46 (1963)
14. Chaston, I. Endogenous activity as a factor in invertebrate drift. *Arch. Hydrobiol.*, **64**, 324–34 (1968)
15. Clifford, H. F. The ecology of invertebrates in an intermittent stream. *Invest. Indiana Lakes Streams,* **7**, 57–98 (1966)
16. Cushing, C. E. Filter-feeding insect distribution and planktonic food in the Montreal River. *Trans. Am. Fisheries Soc.*, **92**, 216–19 (1963)
17. Décamps, H. Introduction à l'étude écologique des trichoptères des Pyrénées. *Ann. Limnol.*, **3**, 101–76 (1967)
18. Décamps, H. Ecologie des trichoptères de la vallée d'Aure (Hautes Pyrénées). *Ann. Limnol.*, **3**, 399-577 (1967)
19. Décamps, H. Vicariances écologiques chez les trichoptères des Pyrénées. *Ann. Limnol.*, **4**, 1–50 (1968)
20. Degrange, C. Recherches sur la reproduction des Ephéméroptères. *Trav. Lab. Hydrobiol. Piscic. Univ. Grenoble,* **50/51**, 7–194 (1960)
21. Dimond, J. B. Evidence that drift of stream benthos is density related. *Ecology,* **48**, 855–57 (1967)
22. Dunbar, R. W. Four sibling species included in *Simulium damnosum* Theobald from Uganda. *Nature,* **209**, 597–99 (1966)
23. Edington, J. M. The effect of water flow on populations of net-spinning

Trichoptera. *Mitt. Intern. Ver. Theoret. Angew. Limnol.,* **13,** 40–48 (1965)

24. Edington, J. M. Habitat preferences in net-spinning caddis larvae with special reference to the influence of running water. *J. Animal Ecol.,* **37,** 675–92 (1968)
25. Egglishaw, H. J. The distributional relationship between the bottom fauna and plant detritus in streams. *J. Animal Ecol.,* **33,** 463–76 (1964)
26. Egglishaw, H. J., Mackay, D. W. A survey of the bottom fauna of streams in the Scottish Highlands. Part III. Seasonal changes in the fauna of three streams. *Hydrobiologia,* **30,** 305–34 (1967)
27. Elliott, J. M. Daily fluctuations of drift invertebrates in a Dartmoor stream. *Nature,* **205,** 1127–29 (1965)
28. Elliott, J. M. Invertebrate drift in a Dartmoor stream. *Arch. Hydrobiol.,* **63,** 202–37 (1967)
29. Elliott, J. M. The life histories and drifting of the Plecoptera and Ephemeroptera in a Dartmoor stream. *J. Animal Ecol.,* **36,** 343–62 (1967)
30. Elliott, J. M. The daily activity patterns of mayfly nymphs *J. Zool. London,* **155,** 201–21 (1968)
31. Elliott, J. M. The life histories and drifting of Trichoptera in a Dartmoor stream. *J. Animal Ecol.,* **37,** 615–25 (1968)
32. Elliott, J. M., Minshall, G. W. The invertebrate drift in the River Duddon, English Lake District. *Oikos,* **19,** 39–52 (1968)
33. Eriksen, C. H. Ecological significance of respiration and substrate for burrowing Ephemeroptera. *Can. J. Zool.,* **46,** 93–103 (1968)
34. Fredeen, F. J. H. Bacteria as food for blackfly larvae in laboratory cultures and in natural streams. *Can. J. Zool.,* **42,** 527–48 (1964)
35. Gibbs, D. G. The larva, dwelling tube and feeding of a species of *Protodipseudopsis. Proc. Roy. Entomol. Soc. London, Ser. A,* **43,** 73–79 (1968)
36. Gower, A. M. A study of *Limnephilus lunatus* Curtis with reference to its life cycle in watercress beds. *Trans. Roy. Entomol. Soc. London,* **119,** 283–302 (1967)
37. Grenier, P. Contribution à l'étude biologique des Simuliides de France. *Physiol. Comp. Oecol.,* **1,** 165–330 (1949)
38. Harrison, A. D. Recolonisation of a Rhodesian stream after drought. *Arch. Hydrobiol.,* **62,** 405–21 (1966)
39. Harrod, J. J. Effect of current speed on the cephalic fans of the larvae of *Simulium ornatum* var. *nitidifrons* Edwards. *Hydrobiologia,* **26,** 8–12 (1965)
40. Hartland-Rowe, R. Factors influencing the life histories of some stream insects in Alberta. *Verhandl. Intern. Ver. Theoret. Angew. Limnol.,* **15,** 917-25(1964)
41. Holt, C. S., Waters, T. F. Effect of light intensity on the drift of stream invertebrates. *Ecology,* **48,** 225–34 (1967)
42. Hughes, D. A. The role of responses to light in the selection and maintenance of microhabitat by the nymphs of two species of mayfly. *Animal Behavior,* **14,** 17–33 (1966)
43. Hynes, H. B. N. Further studies on the invertebrate fauna of a Welsh mountain stream. *Arch. Hydrobiol.,* **65,** 360–79 (1968)
44. Hynes, H. B. N. *The ecology of running waters.* (Liverpool Univ. Press, Liverpool, England, 555 pp., 1969)
45. Illies, J. Phylogeny and zoogeography of the Plecoptera. *Ann. Rev. Entomol.,* **10,** 117–40 (1965)
46. Jaag, O., Ambühl, H. The effect of the current on the composition of biocoenoses in flowing water streams. *Intern. Conf. Water Pollution Res. London,* 33–49 (Pergamon Press, Oxford, 339 pp., 1964)
47. Johnstone, G. W. Stridulation by larval Hydropsychidae. *Proc. Roy. Entomol. Soc. London, Ser. A,* **39,** 146–50 (1964)
48. Kaiser, P. Über Netzbau und Strömungssinn bei der Larven der Gattung *Hydropsyche* Pict. *Intern. Rev. Ges. Hydrobiol. Hydrogr.,* **50,** 169–224 (1965)
49. Kamler, E. Distribution of Plecoptera and Ephemeroptera in relation to altitude above mean sea level and current speed in mountain waters. *Polskie Arch. Hydrobiol.,* **14,** (27), 29–42 (1967)
50. Khoo, S. G. Studies on the biology of *Capnia bifrons* (Newman) and notes on the diapause in the nymphs of this species. *Gewäss. Abwäss,* **34/35,** 23-30 (1964)

51. Khoo, S. G. Experimental studies on diapause in stoneflies. I. Nymphs of *Capnia bifrons* (Newman). *Proc. Roy. Entomol. Soc. London, Ser., A,* **43,** 40–48 (1968)
52. Khoo, S. G. Idem. II. Eggs of *Diura bicaudata* (L.) *Proc. Roy. Entomol. Soc. London, Ser. A,* **43,** 49–56 (1968)
53. Khoo, S. G. Idem. III. Eggs of *Brachyptera risi* (Morton). *Proc. Roy. Entomol. Soc. London, Ser. A,* **43,** 141–46 (1968)
54. Kljutschareva, O. A. On downstream and diurnal vertical migration of benthic invertebrates in the Amur. (Russian, English summary). *Zool. Zh.,* **42,** 1601–12 (1963)
55. Knight, A. W., Nebeker, A. V., Gaufin, A. R. Description of the eggs of common Plecoptera of Western United States. *Entomol. News,* **76,** 105–11 (1965)
56. Knight, A. W., Nebeker, A. V., Gaufin, A. R. Further descriptions of the eggs of Plecoptera of Western United States. *Entomol. News,* **76,** 233–39 (1965)
57. Levanidova, J. M., Levanidov, V. Ya. Diurnal migrations of benthal insect larvae in the river system. 1. Migrations of Ephemeroptera larvae in the Khor River. (Russian, English summary). *Zool. Zh.,* **44,** 373–84 (1965)
58. Lillehammer, A. Bottom fauna investigations in a Norwegian river. The influence of ecological factors. *Nytt May. Zool.,* **13,** 10–29 (1966)
59. Macan, T. T. Causes and effects of short emergence periods in insects. *Verhandl. Intern. Ver. Theoret. Angew. Limnol.,* **13,** 845–49 (1958)
60. Macan, T. T. Methods of sampling the bottom fauna in stony streams. *Mitt. Intern. Ver. Theoret. Angew. Limnol.,* **8,** 21 pp. (1958)
61. Macan, T. T. The occurrence of *Heptagenia lateralis* in streams in the English Lake District. *Wett. Leben,* **12,** 231–34 (1960)
62. Maciolek, J. A., Tunzi, M. G. Microseston dynamics in a simple Sierra Nevada lake-stream system. *Ecology,* **49,** 60–75 (1968)
63. Madsen, B. L. A comparative ecological investigation of two related mayfly nymphs. *Hydrobiologia,* **31,** 37–49 (1968)
64. Maitland, P. S., Penney, M. M. The ecology of the Simuliidae in a Scottish river. *J. Animal Ecol.,* **36,** 179–206 (1967)
65. Marinković-Gopsodnetić, M. The distribution of the caddis-flies populations in a small mountain stream. *Verhandl. Intern. Ver. Theoret. Angew. Limnol.,* **16,** 1693–95 (1966)
66. Maxwell, G. R., Benson, A. Wing pad and tergite growth of mayfly nymphs in winter. *Am. Midland Naturalist,* **69,** 224–30 (1963)
67. Minshall, G. W. Role of allochthonous detritus in the trophic structure of a woodland springbrook community. *Ecology,* **48,** 139–49 (1967)
68. Minshall, G. W., Winger, P. V. The effect of reduction in stream flow on invertebrate drift. *Ecology,* **49,** 580–82 (1968)
69. Minshall, J. N. Life history and ecology of *Epeorus pleuralis* (Banks). *Am. Midland Naturalist,* **78,** 369–88 (1967)
70. Müller, K. Die Tagesperiodik von Fliesswasserorganismen. *Z. Morphol. Oekol. Tiere,* **56,** 93–142 (1966)
71. Needham, P. R., Usinger, R. L. Variability in the macrofauna of a single riffle in Prosser Creek, California, as indicated by the Surber sampler. *Hilgardia,* **24** (14), 383–409 (1956)
72. Nielsen, A. Is dorsoventral flattening of the body an adaptation to torrential life? *Verhandl. Intern. Ver Theoret. Angew. Limnol.,* **11,** 264–67 (1951)
73. Obeng, L. E. Life-history and population studies on the Simuliidae of North Wales. *Ann. Trop. Med. Parasitol.,* **61,** 472–87 (1967)
74. Pearson, W. D., Franklin, D. R. Some further factors affecting drift rates of *Baetis* and Simuliidae in a large river. *Ecology,* **49,** 75–81 (1968)
75. Pleskot, G. Die Periodizität der Ephemeropteren-Fauna einiger österreichischer Fliessgewässer. *Verhandl. Intern. Ver. Theoret. Angew. Limnol.,* **14,** 410–16 (1961)
76. Raušer, J. Zur Verbreitungsgeschichte einer Insektendauergruppe (Plecoptera) in Europa. *Acta Acad. Sci. Cech. Basis Brunen.,* **34,** 281–383 (1962)
77. Roos, T. Studies on upstream migra-

tion in adult stream-dwelling insects. I. *Rept. Inst. Freshwater Res. Drottningholm,* **38,** 167–93 (1957)
78. Ross, H. H. The evolution and past dispersal of the Trichoptera. *Ann. Rev. Entomol.,* **12,** 169–206 (1967)
79. Saether, O. A. Chironomids of the Finse area, Norway, with special reference to their distribution in a glacier brook. *Arch. Hydrobiol.,* **64,** 426–83 (1968)
80. Sattler, W. Weitere Mitteilungen über die Ökethologie einer neotropischen *Macronema*-Larvae. *Amazoniana,* **1,** 211–29 (1968)
81. Scherer, E. Zur Methodik experimenteller Fliesswasser-Ökologie. *Arch. Hydrobiol.,* **61,** 242–48 (1965)
82. Scherer, E. *Analytisch-ökologische Untersuchungen zur Verteilung tierischer Bachbesiedler.* (Doctoral Dissertation, Giessen, Germany, 66 pp., 1965)
83. Schwoerbel, J. Das hyporheische Interstitial als Grenzbiotop zwischen oberirdischem und subterranem Ökosystem und seine Bedeutung für die Primär-Evolution von Kleinsthöhlenbewohnern. *Arch Hydrobiol. Suppl., 33,* 1–62 (1967)
84. Smith, S. D. The *Rhyacophila* of the Salmon River drainage of Idaho with special reference to larvae. *Ann. Entomol. Soc. Am.,* **61,** 655–74 (1968)
85. Steffan, A. W. Bezeihungen zwischen Lebensraum und Körpergrösse bei mitteleuropäischen Elminthidae. *Z. Morphol. Oekol. Tiere,* **53,** 1–21 (1963)
86. Svensson, P.-O. Growth of nymphs of stream living stoneflies in northern Sweden. *Oikos,* **17,** 197–206 (1966)
87. Tanaka, H. On the change of composition of aquatic insects resulting from difference in mesh size of stream bottom-samplers. (Japanese, English summary). *Bull. Freshwater Fisheries Res. Lab., Tokyo,* **17,** 1–16 (1967)
88. Tobias, W. Zur Schlüpfrhythmik von Köcherfliegen. *Oikos,* **18,** 55–75 (1967)
89. Thomas, E. Orientierung der Imagines von *Capnia atra* Morton. *Oikos,* **17,** 278–80 (1966)
90. Ulfstrand, S. Microdistribution of benthic species (Trichoptera, Diptera Simuliidae) in Lapland streams. *Oikos,* **18,** 293–310 (1967)
91. Ulfstrand, S. Life cycles of benthic insects in Lapland streams (Ephemeroptera, Plecoptera, Trichoptera, Diptera Simuliidae). *Oikos,* **19,** 167–90 (1968)
92. Ulfstrand, S. Benthic animal communities in Lapland streams. A field study with particular reference to Ephemeroptera, Plecoptera, Trichoptera and Diptera Simuliidae. *Oikos Suppl., 10,* 120 pp. (1968)
93. Vaillant, F. Utilisation des diptères Psychodidae comme indicateurs de certains caractères des eaux courantes. *Verhandl. Intern. Ver. Theoret. Angew. Limnol.,* **16,** 1721–25 (1966)
94. Vaillant, F. Sur la choix des espèces indicatrices pour une zonation des eaux courantes. *Trav. Lab. Hydrobiol. Piscic. Univ. Grenoble,* **57/58,** 7–15 (1967)
95. Vaillant, F. La répartition des *Wiedemannia* dans les cours d'eau et leur utilisation comme indicateurs des zones écologiques. *Ann. Limnol.,* **3,** 267–93 (1967)
96. Van Someren, V. D. *The biology of trout in Kenya Colony.* (Government Printer, Nairobi, Kenya, 114 pp., 1952)
97. Vincent, E. R. A comparison of riffle insect populations in the Gibbon River above and below the geyser basins, Yellowstone National Park. *Limnol. Oceanogr.,* **12,** 18–26 (1967)
98. Waters, T. F. Diurnal periodicity in the drift of a day-active stream invertebrate. *Ecology,* **49,** 152–53 (1968)
99. Weninger, G. Vergleichende Drift-Untersuchungen an niederösterreichischen Fliessgewässern (Flysch-, Gneis-, Kalkformation). *Schweiz. Z. Hydrol.,* **30,** 138–85 (1968)
100. Wiggins, G. B. The critical problem of systematics in stream ecology. *Spec. Publs. Pymatuning Lab. Fld. Biol.,* **4,** 52–58 (1966)
101. Zahner, R. Über die Bindung der mitteleuropäischen *Calopteryx*-Arten an den Lebensraum des strömenden Wassers II. Der Anteil der Imagines an der Biotopbindung. *Intern. Rev. Ges. Hydrobiol. Hydrogr.,* **45,** 101–23 (1960)

INSECT MIMICRY[1,2]

By Carl W. Rettenmeyer[3]

Department of Entomology, Kansas State University, Manhattan, Kansas

Fisher (82) called the theory of mimicry the "greatest post-Darwinian application of Natural Selection." Since 1862, when Bates first proposed his theory regarding mimicry among insects, there has been a continuous stream of publications describing examples of presumed mimics with evidence, speculation, and personal opinion, supporting or contradicting the validity of mimicry theory. At times, discussions of mimicry have had some of the aspects of a religious cult in which one must accept or reject hypotheses primarily as an act of faith unencumbered by adequate experimental data. Despite the voluminous early literature, mostly summarized by Cott (60), adequately controlled experiments elucidating and substantiating the theory were few until the last 12 years.

Unlike the extremely broad usage of Wickler (200, 201), "mimic" here is restricted to the sense of one animal resembling another. Potential predators confuse the mimic with the model and serve as selective agents for evolution of the mimicry. The resemblance usually is in color, pattern, or form, but it can be solely behavioral; and the mimic occasionally may resemble a group of animals or only some part or aspect of another animal. This usage excludes numerous examples of protective coloration, such as katydids resembling leaves, and other forms of adaptive resemblance. Examples of plants mimicking insects or "pseudocopulation" of flowers by insects are also excluded since recent reviews are available (110, 138). Summaries of the older literature and more detailed discussions of types of mimicry are in Carpenter & Ford (1933), Fisher (1930) (60, 82, 146, 201). Brower's (28) motion picture gives an excellent introduction to mimicry and protective coloration.

Types of Insect Mimicry

Batesian mimicry.—Batesian mimicry has been based on seven premises: (*a*) a species, the model, is unpalatable to predators; (*b*) a second species, the mimic, is palatable to predators but has evolved away from its ancestral

[1] In surveying the literature, references to earlier works in Cott (60) have been cited by author and date to save space.

[2] Preparation of this review was facilitated by National Science Foundation Grant No. GB-6321X with the bibliographic assistance of Lois M. Kadoum.

[3] I thank Stephen B. Fretwell and Lowell Brandner for comments on the manuscript.

appearance until it resembles the model so closely that potential predators are deceived and leave it alone; (*c*) the mimics are less abundant than the models; (*d*) the mimics must be found at the same place and time as the models; (*e*) the model and mimic are conspicuous or readily seen by potential predators; and (*f*) the predators learn or associate unpalatability with color pattern of the model.

There are reasons to doubt the importance of all these premises, but (*a*), (*b*) and (*f*) are probably essential for Batesian mimicry.

Müllerian mimicry.—A somewhat different type of mimicry, first proposed by Müller (1879), is based on two premises: (*a*) two or more species are unpalatable; and (*b*) if two or more species have evolved so that the predators cannot distinguish them, each species will be killed in proportion to its abundance in the habitat. Premises (*d*) to (*f*) listed under Batesian mimicry also apply to Müllerian mimicry.

Müllerian mimicry was originally restricted to distantly related species that have evolved similar color patterns, but Wallace (1889) expanded the concept to include closely related forms or members of the same genus with similar color patterns.

Aggressive mimicry.—The term, aggressive mimicry, is usually attributed to Poulton (1890) who included it in his classification of types of mimicry although he based the term on observations of E. G. Peckham (1889) and previous authors. Poulton used the term for cases of one animal resembling another, in order to approach it without exciting suspicion. Aggressive mimicry differs fundamentally from Batesian and Müllerian mimicry in that the selective agents are simultaneously the models (38). Aggressive mimicry here is restricted to parasitic and predaceous arthropods.

Wasmannian mimicry.— The term, Wasmannian mimicry, is proposed here for resemblances that facilitate a mimic's living with its host. The host species is the selective agent and is usually exploited by the mimic, but the relationship between the two species may be mutualistic or beneficial to both. Such mimicry was most extensively discussed by Eric Wasmann (198). It seems appropriate to name it for him to avoid using "mimicry by commensals" or terms like "ant-mimic" that apply equally to Batesian mimics.

Resemblance Among Batesian Mimics, Models, and Alternate Prey

Resemblance of color and morphology.—Batesian models are usually aposematic or warningly colored and are primarily red or orange, sometimes with bands of black. Poulton (1890) used "pseudaposematic" for Batesian mimics with similar colors. Although bright colors are usually aposematic, they can be cryptic depending upon the habits of the species (30, 60). The predator's association of unpalatability with pattern should be facilitated by a conspicuous pattern. Most Batesian mimics are characterized by a superficial resemblance in color pattern to the model while their basic morphology and wing venation are typical of a group distantly related to the model. Since nothing is known about the relative palatability of most of these insects to any predator, much less to all potential predators, many mimics labeled as Batesian must be Müllerian.

The numerous Batesian mimics of Hymenoptera are striking examples of color patterns that have evolved to give illusions of completely different shapes; Myers & Salt (1926), Nicholson (1927), and Wickler (201) have published good illustrations. Unfortunately, all are paintings or photographs of dead specimens. Natural positions make a great difference in the effectiveness of mimicry (28), so more photographs of insects in their natural habitat are needed. Sexton (167) showed that *Anolis* will refuse mutillid wasps on the basis of shape rather than color pattern by obscuring the pattern with powdered carbon. When elytra and pronota of *Photinus* were cemented to *Tenebrio* adults, seven species of vertebrate predators could distinguish the genera, probably by their shape (166).

Ant-mimicry has arisen many times in spiders: at least four times in Clubionidae, three times in Salticidae, and also in Theridiidae, Araneidae, Thomisidae, Gnaphosidae, Zodariidae, and Eresidae (145). The position and movement of the front legs mimic an ant's antennae, the bodies are typically heavily sclerotized, and they have a "petiole" and sometimes a "postpetiole" (144). No one has made a comprehensive survey of ant-like spiders and insects not found in ant colonies, but personal observations in Central and South American forests indicate that most of these ant-mimics resemble ants with potent stings or defensive chemicals. Both mimics and ant models are seen running conspicuously on vegetation (114, 199). Those observations suggest that the species are typical Batesian mimics (145). However, Peckham (1889) pointed out that ant-like spiders may primarily gain protection from predators such as hummingbirds and solitary wasps that specialize in catching spiders. Thus, these ant-mimics probably are protected from the common predators on spiders as well as from some that avoid preying on ants.

An unusual slug caterpillar (Cochlidae) from Trinidad spins a hard cocoon with four small round holes resembling exit holes of a parasitoid wasp (105). Several other lepidopterous larvae or cocoons show signs of "false parasitism" adequate to fool entomologists (60). Although no experiments have been reported, such resemblances presumably protect the insects from vertebrate predators, not parasitoid insects.

Extent to which predators generalize resemblance.—Schmidt (161, 162) trained chickens to avoid food made distasteful by potassium hydroxide and placed in front of a drawing of a hypothetical model. Various hypothetical mimicking patterns were placed back of a second dish of palatable food. Results showed that chickens generalize, so a small but conspicuous feature of the model, like a patch of red, could protect the incipient mimic. Laboratory experiments provided evidence that jays generalize among two species of *Danaus* and two subspecies of *Limenitis archippus* (18, 20).

Birds may adopt a "searching image" for insects with a specific pattern

(130, 186) or concentrate on only a few species of prey (23). DeRuiter (157) showed that birds will overlook twig-like caterpillars until they find the first one; then the remainder are quickly found. The importance of a searching image is dependent on population density (91); and if several species resemble each other, predation will be spread among them (178). If birds use only one searching image or restrict their feeding to insects with only a few patterns, polymorphism could protect some sympatric morphs, e.g., *Acraea encedon* (135) and ctenuchid moths (15). The number of unpalatable aposematic insects a single bird can learn is unknown, but common red coloration permits generalization. Variation in color pattern may also be selected for so that no two conspecific individuals appear the same (129, 133, 172). Polymorphism also allows a mimic to increase its population without exceeding the population of its various models. Since many insects must be behaviorally or physiologically "polymorphic," the term "polyphenic" would be preferable (48, 127). Different morphs may have different correlated behavior as in protectively colored sphinx caterpillars (65) and aggregations of butterflies (135).

Blest (10) reported that insects have a great range in eyespot patterns, from very realistic copies of vertebrate eyes to those with virtually no resemblance. Some sphingid larvae even have eyespots that appear to wink (28). Realistic eyespots were most effective at repelling birds, but even poor copies had some effect. Eyespots are not considered orthodox Batesian mimics, but similar generalized responses of birds to various mimics would be predicted. Birds without any previous experience with predators or with insects are frightened by eyespots (10), and eyespots have frightened birds in nature (104).

Sexton (166) glued either elytra or pronota, or both, from unpalatable *Photinus* to palatable *Tenebrio* to make mimics of three types. When tested with normal *Photinus,* only mimics with both elytra and pronota of *Photinus* were protected. When tested with *Tenebrio,* mimics with any part of *Photinus* on them were protected. Predators used were seven species of vertebrates (frog, lizards, turtles, fish). Although the experiment seemed to support the hypothesis that partial or incipient mimics are protected, the elytra and pronota may have made *Tenebrio* somewhat unpalatable.

Silverbeak tanagers rejected a blue and yellow butterfly after experience with a distasteful orange species. This was interpreted as generalization based on size and shape (34) but may have been due to previous experience with other brightly colored butterflies, and the birds' reluctance to feed under the experimental conditions.

A predaceous wasp, *Philanthus bicinctus,* preys on several species of *Bombus* and appears to select prey within a certain size range (126). This could result in divergent selection for greater variation in size.

Resemblance of alternate prey.—Holling made a major contribution by demonstrating the importance of alternate prey to the mimic-model relationship (100, 101). His mathematical models show that the more dissimilar both the mimic and model are from alternate prey, the greater is the potential effectiveness of the mimicry.

Sudden versus gradual evolution of mimicry.— It has been stated many times that models are generally conspicuous so it would be disadvantageous for a species to evolve toward a model (become more conspicuous), unless it simultaneously gains some protection from predators. Goldschmidt (92, 93) has maintained that to be effectively protected from predators, mimics must evolve by saltations or macromutations so that initially they closely resemble the model, but Ford (83) refuted that idea; and experiments with artificial, mimic-model pairs demonstrate that small steps toward a mimetic resemblance are protective. The pressure of selection can slowly change an entire population even when selective advantage of a particular genotype is minute (116).

Extensive studies of the genetics of *Papilio dardanus* by Clarke & Sheppard (51–58), and Sheppard (170–173, 175) have been interpreted as supporting the hypothesis that a major mutation may be necessary to protect the new mimic adequately (57). It is clear from their work that *P. dardanus* has a supergene or series of closely linked genes that determine most of the color pattern, but modifying genes improve the resemblance to the model. Fisher (82) stated that mimics arising by a single major mutation resulting in a switch gene would be comparable to females arising from males on the basis of a single mutation because sex is often determined by a single chromosome or a few genes. More likely, mimics evolved gradually and initially resembled a model quite different from the present one. The mimic's resemblance is disadvantageous to the model so its selection is away from the mimic, probably toward a more aposematic form (82). Schmidt (161) states that "Few, if any authentic examples of an incipient mimic to a complex model have been collected." However, many collectors have found what they considered "poor mimics" of many models (60, 68, 178, 201). Most research and descriptions have been on the most realistic mimics. Many species of *Eristalis* have a general bee-like appearance, and some probably evolved a closer resemblance to the honey bee after it was introduced into North America (22).

Disruptive or sexual selection.—The *P. dardanus* complex in Africa now includes nonmimetic males, presumably similar to the ancestral females in color and pattern, and a large series of mimetic and nonmimetic polymorphic females (51, 58, 84). The restriction of polymorphic mimicry to females of some species has been considered a result of heterogametic inheritance, since the males are homogametic in Lepidoptera (82). However, most polymorphism in *P. dardanus* is autosomal (27, 52, 85).

Equal numbers of males and females are probably produced by most mimetic and nonmimetic species (27), but there are a few examples of highly unequal sex ratios. *Danaus chrysippus* and two species of *Acraea* sometimes produce only or mainly females. At other times, *D. chrysippus* produces an excess of males (135). When *A. encedon* produces an excess of females, many remain unmated even though males of that species can mate with at least four females. Both males and females of *A. encedon* are polymorphic with some forms mimicking *D. chrysippus*. Like forms tend to aggregate together (135), suggesting nonrandom mating among forms.

Wynne-Edwards (202) proposed that nonmimetic males permit visual recognition and promote dispersal of males, and females consequently must search for mates. His hypothesis is refuted by abundant aggregations and frequent pursuits of both sexes by males.

Since *Papilio* spp. can be readily hand-paired (50) and reared, offspring from crosses that would never occur in nature can be obtained between various morphs and subspecies. The gene controlling the presence of tails (unlike the tail-less model) is independent of genes for pattern but probably has been retained by disruptive selection (54, 57).

Disruptive selection would be more plausible with better evidence that females select mates by color pattern. Theoretically, female selection of mates is the strongest argument for species that have mimetic females and retain nonmimetic males. However, experiments by Magnus (122–124) and Brower (24, 25, 27, 36, 40, 41) indicate that although males locate females from some meters away solely by vision, females use visual selection very little, if any. Much of that work has been on either nonmimetic or monomorphic species. On the other hand, Stride (179, 180) has shown that males somewhat select nonmimetic aspects of the female's color pattern and that females avoid males with color patterns too different from normal (181). Crane emphasized the importance of visual selection by both males and females of *Heliconius* (62) and stated that old females chase males in addition to the frequent pursuit of females by males (63). There is some contradictory evidence that large amounts of red on *Heliconius* are selected for by mates or predators (193). Visual orientation of female *Arygynnis* toward males also seems important in mating (187).

Prolonged courtship and two interspecific matings occurred between males of *Papilio multicaudatus* and the similarly colored yellow females of *P. glaucus*, but the black mimetic form of *glaucus* elicited only a few weak courtship responses from the *P. multicaudatus* males (24). How much those results depended on responses of the females is unknown. Spermatophore counts indicate that mimetic females of *P. glaucus* mate less frequently than do nonmimetic females (46). However, no virgins were found, and there is no evidence that multiple matings are advantageous.

Swihart's new approach may differentiate responses of males and females to color patterns (182–184). He used electrophysiological techniques to study response of *Heliconius erato* to light and demonstrated that the butterfly has a sharp response peak to narrow red wavelengths that are comparable to the reflectance from the red part of the wings. The response was positively correlated with time of day when the butterflies were most active.

Müllerian mimics may not be able to use visual selection of mates effectively since the species resemble each other; and in Lepidoptera such mimics have developed elaborate scent disseminating organs on the males (27).

Resemblances other than color, pattern, or morphology.—Writers on mimicry, starting with Bates (1862), have reported that behavior of models is often mimicked in addition to color patterns (6, 15, 66, 67, 118, 140, 177). It is difficult to study the role of behavior in mimicry because it cannot be experimentally manipulated as readily as patterns. Palatable North American moths, painted to mimic a model in Trinidad, were released and later trapped (37, 39). The results do not provide good support for the effectiveness of Batesian mimicry but suggest that predation was heavier than expected on the mimic because it flew at different times and higher than the models. Sargent reported that selection of matching backgrounds by cryptic moths is genetically fixed and does not depend on each moth visually matching its circumocular color with the background (159, 160). His results might be questioned on grounds of adult learning or the moth's matching of wing colors with background.

Many Batesian mimics and models may be conspicuous by staying on upper surfaces of leaves rather than underneath as do many cryptic species. Models frequently are slow in flight and do not flee when approached or even touched. Other aposematic species may fly more readily than their cryptic relatives (12). Despite many such observations, no adequately controlled experimental studies show the effect of mimicry of odors, sound, or behavior on predators. Behavioral mimicry, because there is often no conspicuous pattern, may be more important among mimics of Hymenoptera than among butterflies. In a sense, behavior of ants and some wasps is their most conspicuous feature and, therefore, most likely to be mimicked.

Several species of a cerambycid, *Moneilema,* resemble the tenebrionid, *Eleodes,* in color, morphology, and behavior and are found with the model in the same microhabitat. When disturbed, *Eleodes* raises its abdomen high above the substrate, protrudes the red-brown tip, and releases a defensive secretion. The cerambycid mimics the gait and defensive behavior of *Eleodes* but does not have the defensive secretion. Tests of *M. appressum* and *M. armatum* with lizards (*Sceloporus*), a woodrat (*Neotoma*), and two striped skunks (*Mephites*) indicated that mimicry protects the cerambycids from predators (140).

Audio-mimicry of the vespid wasp, *Dolichovespula arenaria,* by the syrphid fly, *Spilomyia hamifera* was described by Gaul (89). The sound produced by the wings of the wasp was 150 Hz; that of the fly, 147 Hz, a difference too slight to be distinguished by a human ear. The sound produced by the drone fly, *Eristalis tenax,* is only a few tones lower than that of a honey bee (47); and the sound of a robber fly, *Mallophora,* greatly enhances its resemblance to a bumble bee (33). Brower & Brower (22) suggested that audio-mimicry might occur between *Eristalis* spp., and *Apis* because toads rejected more drone flies with intact wings than without wings. However, buzzing, presence of wings, or increased movement, rather than sound, could account for the increased rejection.

The death's head moth is said to sound like piping of a queen honey bee, and the sound may enable the moth to enter hives more readily (96). The sound is not completely protective because many moths are stung to death inside hives.

Many ants, mutillid wasps, cerambycid beetles, and other insects produce a generalized buzzing or stridulation when picked up, and these sounds have been assumed to be defensive (47, 96). There are meager data that a hedgehog, *Erinaceus;* polecat, *Poecilogale;* and domestic cats are frightened by stridulation of a scorpion (2). Probably stingless insects like male mutillids gain Batesian advantage by such audio-mimicry.

Distribution and Relative Abundance of Batesian Mimics and Models

Batesian mimicry theory has usually included the provision that mimics and models must be sympatric or at least the distribution of the mimic should not exceed that of the model. Furthermore, mimics should be less abundant than the model so predators are more likely to sample the unpalatable model. Although those two principles have basic validity, field and laboratory data and theoretical considerations have shown that they are less important than was thought earlier.

Although butterflies are generally considered thoroughly collected, it is clear that we do not know nearly enough about the distribution and abundance of even the "best-known" mimicry complex (9, 51, 78, 174, 188, 189). Quantitative collecting is needed to determine relative proportions of all forms of mimics and models at precisely determined localities at known times of the year.

Changes in mimics outside the range of their models.—Geographical distribution data show that mimics are most similar to their models where distributions overlap. When a mimic ranges beyond its model, the color pattern tends to regress toward a nonmimetic form. Excellent examples are found in the United States (17, 32, 139). Frequency of the model *Battus philenor* increases from near 0 per cent in Florida and South Carolina to more than 85 per cent in Tennessee and North Carolina. The mimetic black form of *Papilio glaucus* coincides with areas of abundant *philenor,* and nonmimetic yellow *P. glaucus* are most abundant where the model is rare, regardless of latitude. Thus, the black form is not environmentally induced by increasing or decreasing temperature or humidity (32). Jays experienced with the model ate fewer mimetic forms of *P. glaucus* than did inexperienced control birds based on tests of the undersides of dead butterflies (17).

Papilio troilus also is a close mimic of *B. philenor* where the latter is relatively abundant but is more variable or mimics less accurately where the model is scarce. *Limenitis astyanax* also mimics *B. philenor* but intergrades to *L. arthemis* where the model becomes rare in northern United States.

Platt & Brower (139) have indicated that breeding between the mimetic and nonmimetic forms is at random; there is no evidence of heterogametic inviability, and the two forms are conspecific. One can safely predict that thorough collection data will show more examples of mimicking species extending beyond the range of their models and changing so markedly that they have been considered distinct species.

Geographical variation due to climatic factors.— Bernardi (7–9) analyzed the color patterns of several morphs of *Hypolimnas misippus* and *H. dubia* in Africa and Madagascar. The model *Danaus chrysippus* is found in Africa where mimics are present but is absent on Madagascar. Bernardi has pointed out that parallel clines between model and mimic have darker individuals in more humid parts of the ranges and may be examples of Gloger's rule rather than being due to mimetic pressure. Bernardi also stated that the nonmimetic male-like females of some species have been liberated from mimetic pressure (9) rather than being females not yet subjected to mimetic selection.

Several species of African butterflies have distinctly different wet and dry season color forms that are evidently environmentally determined (134). Both environmental and genetic polymorphisms are known for mimetic and nonmimetic species as well as seasonal fluctuations in relative numbers of the various morphs.

At Legon, Ghana, the population of the model, *Danaus chrysippus,* was found by Edmunds (70) to be greater than that of the mimic, *H. misippus,* during part of the year, but from May 14 to July 7 the models decreased from 31 to 0 per cent of the combined total. At the same time, mimics most similar to the models decreased from 44 to 12 per cent. The latter decrease was interpreted as supporting the hypothesis that nonmimetic forms have an advantage due to sexual selection when models are absent. Environmental effects may have influenced the changes as Bernardi (7–9) suggested, but there is no evidence that any major polymorphic mimicry complex is primarily determined by climatic factors rather than mimetic selection.

Mimicry when model and mimic are separated.—When the model is abundant, birds learn to avoid it. If their learning carried over, mimics most similar to the model would be protected when the latter disappears seasonally. That possibility was suggested by Rothschild (151) for the buff ermine moth, *Spilosoma lutea,* which emerges later than the white ermine, *S. lubricipeda.* Tests with caged experienced predators showed that the buff ermine was always selected when both moths were presented together. Presumably, in nature, predators learn that the white ermine is unpalatable before the buff ermine emerges, and the latter gains some protection. This hypothesis has yet to be tested by a well-planned experiment. There may be many other examples in which prior association of unpalatability with a color pattern may protect some other species at different times or places. Wickler (201) extended that concept to include the possibility of migrating birds not eating mimics in a country or continent different from that inhabited by their models.

Relative abundance of mimic, model, and alternate prey.— For maximum protection of the mimic, the model may have to be much more abundant than the mimic, which could severely limit the population of the mimic. Although there are cases, especially among Lepidoptera (29, 70, 135), in which mimics outnumber their models, the vast majority of realistic, presumed Batesian mimics are much rarer than their models. Some aposematic moths live longer than related cryptic moths (14), and such differences could influence sex ratios and relative abundance of mimics, models, and alternate prey.

If the predator has a moderate rate of forgetting, the model is moderately unpalatable, and alternate prey is available, 30 per cent models will protect 70 per cent mimics, according to Holling's mathematical predictions (99, 101). He also predicted that alternate prey increases the effectiveness of mimicry. In nature, alternate prey usually is much more abundant than mimics and models combined, so mimicry should be more effective than has been demonstrated under laboratory conditions with limited alternate prey. Learning by predators may be slowed by alternate prey, depending on optimal time intervals for maximum reinforcement. A small proportion of mimics to models may also encourage generalization and delay learning the specific pattern of the mimic (161). Using identically painted mealworms as mimics and models but dipping models in quinine dihydrochloride, J. Brower (19) showed that only 10 per cent models with 90 per cent mimics protected 17 per cent of the mimics. Alternate food was given equal to the total of mimics and models presented. The number of mimics rejected or protected stayed about 80 per cent when the birds were given 10, 30, or 60 per cent mimics (balance models). A major difficulty with determining relative abundance in nature is that this must be done with reference to the effective predators, not the entomological collector (82).

Anolis carolinensis apparently did not learn to avoid completely the aposematic models, *Oncopeltus* or *Photinus,* during 14 days. On the 15th day, *Anolis* offered little alternate food thoroughout the 15 days ate more *Photinus* than did those provided many alternate insects (168). No interpretation was offered for the response to *Oncopeltus* not being the same.

Relative Palatability of Model, Mimic, and Alternate Prey

Unpalatable and poisonous chemicals in models.— The model is often characterized as being "distasteful" to the predator, but any property of the model that makes it undesirable could theoretically protect it. Examples are indigestible or poisonous chemicals, stings, urticating hairs, sticky exudates, hard or spiny integuments, impenetrable cases, or effective escape mechanisms. Many insects are eaten by predators with no hesitation or noticeable aftereffects, but later the predators will refuse the same insects (149). Birds may also become accustomed to eating distasteful insects. Vertebrate species killed by poisons may not be effective selective agents for mimics of lethal species such as the black widow spider, *Latrodectus* (115). However, poisonous species are rarely lethal, and selection for predator survival should also occur.

That tastes of birds and men differ is well recognized, yet many individuals have eaten an amazing variety of insects to support statements about palatability (74, 111, 137, 195). Even though some of the assumed models for Batesian mimics are not distasteful to people, no one has ever proposed that man was the selective agent responsible for the evolution of Batesian mimicry. Furthermore, eating some warningly colored or aposematic insects can be dangerous because some are highly toxic to man (112) and, rarely, may be fatal (136).

The monarch, *Danaus plexippus,* contains heart poisons or cardiac glycosides similar to digitalis (136). These have been extensively analyzed in both the monarch and its host plants and found to be primarily calactin and calotropin, but about eight other cardenolides may also be present (141–142).

In addition to cardiac glycosides found in *Danaus,* a few other toxic materials have been partially identified from models or aposematic species: histamine in *Porthesia chrysorrhoea* (154), *Spilosoma lubricipeda* (150), and *Hypocrita jacobaeae* (151); acetylcholine in *Zygaena lonicerae* (151), *Spilosoma lutea* and *S. lubricipeda* (150), and *Arctia caja* (151); calactin, calotropin, and uscharidin in *Poekilocerus bufonius* (141); and hydrocyanic acid in all stages of *Zygaena filipendulae* and *Z. lonicerae* (103). Parasites that attack *Zygaena* possess rhodanese, which can detoxify the poison (103), but it is not known whether it is a specific adaptation for parasitizing *Zygaena* or a more widespread feature of many insects. The latter could help explain why insects unpalatable to birds are accepted by insect predators.

Unpalatable chemicals obtained from plants.— It has long been postulated that distasteful or toxic insects probably obtain their noxious chemicals from food plants on which the adults or larvae feed (33). There are five main groups of butterflies that serve as models (Papilioninae-Troidini, Ithomiinae, Nymphalinae-Heliconiini, Acraeinae, and Danainae), and they are restricted to a narrow group of food plants (72). Representatives of all except Acraeinae were tested by Brower & Brower (33) who found them unpalatable to blue jays. Controls from Satyridae and Nymphalinae were palatable. The unpalatable insects apparently evolved from palatable species and modified their physiology to enable them to survive on toxic materials. Most selection of food plants by Lepidoptera is done by the ovipositing adult, and some of the same toxic and odorous compounds that make the plants and insects distasteful probably are attractants and oviposition stimulants (29). Some ovipositing females may lay eggs on plants related to the normal host but on which their larvae are unable to survive.

The best documented example of an unpalatable insect obtaining toxic materials from plants is that of the monarch, described below. There is slight evidence that the grasshopper, *Poekilocerus bufonius,* may be able to synthesize small amounts of cardenolides when reared on plants that do not contain them (80, 81). Some Lepidoptera, such as female *Arctia caja,* may always be unpalatable regardless of the food plant on which their larvae have fed (152, 154).

Vertebrate herbivores generally avoid poisonous or unpalatable food plants, for example, cattle avoid *Asclepias curassavica* in Costa Rica (29). That way insects able to feed on the plant are protected from being eaten or killed.

Visual and chemotactic signals.— A tasteless poison must be accompanied by some odor, color, or distinctive stimulus for a predator to associate with the poison, or the predator will have no way to learn to avoid the plant or insect. Some strong odors or tastes of plants and insects could serve that function primarily rather than being distasteful per se. Poisonous chemicals in *Asclepias* spp. are not the main attractant or oviposition stimulus for the adult females because they lay eggs on *Asclepias* spp. that have no toxic cardenolides (29, 42).

Brower (29) suggested that some insects may use "chemomimicry" by having no poisonous properties but by having the odor or taste of a poisonous one. Many distasteful insects have odors similar to coccinellids, but it is not clear whether the odors are from the distasteful compounds (95, 149, 153).

Evidence that vertebrates eat models and mimics.— Most experiments concerning predator responses to insects have involved birds, generally assumed to be the main selective agents for mimicry. McAtee (1932) made the most extensive survey of food found in stomachs of birds and reported about 191,000 insects, including many species considered by others to be Batesian or Müllerian mimics or models. He concluded that insects are eaten in approximate proportion to the number of species known in various taxa, and that various protective adaptations such as mimicry and cryptic and aposematic coloration are ineffective. His conclusions have been justly criticized by Nicholson (1932) among others but still are being used (195) against mimicry. That some birds eat unpalatable insects in no way proves that adaptive resemblances and aposematic coloration afford no protective advantage. McAtee's comparisons were on the relative number of species in different taxa, but he had no knowledge of relative numbers of individuals of any group in habitats of the birds.

The mouse, *Onychomys,* has been reported to eat *Danaus plexippus,* and *Peromyscus* has readily eaten *Battus philenor* and *D. plexippus.* Nocturnal habits and lack of color vision by mice support the conclusion that they have not served as selective agents for mimicry of butterflies (21).

Evidence that vertebrates do not eat models and mimics.— Tests with predators indicate that at least a few birds eat the species of insects most consistently refused by birds in general. Within the same species of predator with the degree of hunger controlled there is considerable variability in the acceptance of food, and starving birds may eat almost any insect.

A long series of tests with a tame Shama flycatcher, *Kittacincla malabarica,* showed that, unlike most birds, it would eat coccinellids and monarchs. However, *Pieris brassicae* and *P. rapae* were unpalatable, and *P. napi* was rejected on sight after the first trial (111). *Pieris* is often considered palatable to birds. Hairy larvae were usually refused but "attempts were made to shake off hairs by beak-drumming." Lane (111) suggested that the flycatcher selects males in preference to females of the same species of Lepidoptera, but his data are not convincing. The most noxious insect supplied to the tame flycatcher was a metallic chrysidid wasp. Even its thorax was refused while other metallic insects, *Lucilia* and *Torymus,* were eaten.

Models and mimics have been made by dyeing water various shades of green and shocking chickens when drinking from the darkest shade (69). Highly distasteful models or any mimic somewhat like the models were quickly associated and avoided. When a model was mildly distasteful, selection continued since predators attempted to discriminate. Such data help demonstrate selective advantages of extremely realistic mimicry.

Stinging insects often are considered noxious as a direct result of the sting. Liepelt (117) demonstrated that the venoms of both *Vespa* and *Apis* were distasteful to a flycatcher, *Ficedula,* and a warbler, *Phoenicurus.* The venom irritated the mucous membrane of the birds' beaks, and the birds refused mealworms smeared with venom. Honey bees and bumble bees both have been shown to be noxious to toads, but naive toads would eat bees with stings removed (22, 31, 38). The toads experienced with honey bees refused to eat drone flies. However, Rothschild found certain inexperienced birds rejected both male honey bees and *Eristalis* after only tasting them regardless of which one was sampled first (154).

Ten of 27 stomachs of buff-backed egrets, *Ardea ibis,* contained *Syrphus corollae,* a highly conspicuous, common wasp mimic. Eight birds had eaten only one or two, one ate 27 and one ate 90, suggesting that mimicry was usually effective but two birds learned that the fly was an imposter (105).

Most insects that are highly unpalatable to birds appear to be unpalatable to many other vertebrate predators, including bats (86), cats (111), mice (80), monkeys (14, 15), lizards (22, 120, 166–168), toads (22, 31, 38), frogs (166), and fish (166). Cott (60) has given numerous older records. Lycid beetles appeared to be distasteful to the mouse *Onychomys;* one individual that ate 25 specimens refused them on later days (120).

Viceroy as mimic of the monarch.— The most controversial and well-known "classical example" of Batesian mimicry in the United States is that of the monarch butterfly, *Danaus plexippus,* and its mimic, the viceroy, *Limenitis archippus.* J. Brower (16), using caged jays, showed the monarch to be clearly less edible than the viceroy. She also showed that the same birds apparently generalized from the monarch to the queen, *D. gilippus berenice,* and the Florida viceroy, *L. archippus floridensis,* rejecting these two species on sight (18).

Urquhart (195) has maintained that her experiments and others show no evidence to substantiate the Batesian theory. Petersen (137) has supported Urquhart and demonstrated that the scrub oak jay, pinon jay, and chickadee will eat some monarchs. Half of Petersen's experiment was done in Colorado between December 24 and January 3 when the birds may have been much more likely to eat unpalatable insects for lack of others. Effective rebuttals to many of the claims by Petersen and Urquhart have been made by L. P. Brower (26) and Sheppard (176).

Conflicting evidence regarding the palatability of *D. plexippus* to birds has now been largely resolved (35, 42) by demonstrating that adult palatability depends on the *Asclepias* species the larvae develop on. That genus of milkweeds includes some species with no toxic material and others that contain cardiac glycosides that cause blue jays to vomit. Earlier experiments (35) had established that less than one adult monarch would cause a blue jay to vomit and thus, it is highly unsuitable food because it often causes loss of other food already eaten. Butterflies reared on one species of *Asclepias* were six times as emetic as those reared on the least toxic species. Those results were from force-feeding blue jays with gelatin capsules containing powdered monarchs and then determining the emetic $dose_{50}$ (comparable to an LD_{50}). It is the most objective measure of palatability yet used for insects. One adult of the most poisonous monarchs contained enough toxic material to cause 4.8 blue jays to vomit 50 per cent of the time.

The three species of birds observed by Petersen (137) ate 80 of 103 monarchs, but he had no comparison or control with other insects. Brower (29) reported that 24 per cent of adult monarchs caught in Massachusetts were emetic, essentially the same as the 22 per cent not eaten in Petersen's experiment. In Trinidad, 65 per cent of the monarchs but only 15 per cent of the queens, *Danaus gilippus,* are emetic. Queens in Trinidad look more like monarchs than do the queens in other areas. Thus, queens in that locality are primarily Batesian mimics, and palatable individuals of both queens and monarchs are "automimics" of the unpalatable members of their own species (29, 35, 42). Thus, theoretical objections that mimics cannot be identical in appearance to models (85), no longer apply.

Since monarchs, field collected as adults or laboratory reared, have such a wide range of palatability, there is added interest in more thorough studies of food plant selection by the ovipositing females.

Invertebrate predators that do not eat models and mimics.—There is much less information on responses of invertebrate than of vertebrate predators to mimics and models. Most predaceous arthropods use vision to locate prey, but prey usually has to move before it is attacked (71). Carpenter & Ford (1933) have summarized a few observations showing that mantids and asilids feed on a variety of insects generally unpalatable to vertebrates.

Gelperin (90) found that *Mantis religiosa* tasted and discarded the large milkweed bug, *Oncopeltus fasciatus,* and sometimes regurgitated. Sufficiently starved mantids ate the large milkweed bugs with no evidence of toxic effects. Solpugids and mantids seized lycids but released them immediately, while harvester ants, *Pogonomyrmex,* were repelled by the same beetles. Antlions and lycosids ate the same two species of lycids (120).

Most predaceous arthropods apparently must contact all unpalatable prey before rejecting them, but mantids will associate unpalatability with appearance and not strike at insects such as *Oncopeltus fasciatus* (90). Diptera mimics of inedible Hymenoptera are attacked by spiders, *Xysticus* and *Misumena,* more slowly as the mimicry improves (194).

Learning and Forgetting by Predators

Learning versus innate responses.—Evidence from various experiments with birds shows that avoidance of aposematic insects usually increases with experience, thus supporting learning or association. Furthermore, once a pattern has been learned, reinforcement will occur when the predator sees the pattern without touching the insects (34). Contrary to Emlen's experience (75), it is probably impossible to distinguish between a predator's forgetting prey or misidentifying it.

There is much evidence that aposematic colors are distinctive or attractive to many vertebrates (60, 125). Red is a common intraspecific signal among vertebrates, and most flowers visited by birds are red. It is obvious that attracting the attention of an insectivorous bird is not the way to increase an insect's survival unless the attractiveness is counterbalanced by properties that repel or injure the bird. What appears to be an innate response of predators to red may be primarily increased sensitivity, and the extent that innate or unlearned responses affect predation on mimics and models is unknown.

Since red is not seen by most arthropods, this color, conspicuous to vertebrates, may be cryptic for arthropods and color-blind predators. Probably black spots and stripes combined with red, orange, and yellow are necessary to make patterns distinctive for such predators.

As pointed out above, conspecific birds under controlled conditions of hunger still show considerable variability to insects known to be unpalatable to most predators. In a single trial, a bird will associate a sufficiently noxious insect with its appearance and will reject the same insect or any similar mimic for at least two weeks. A bird can also be induced to eat a somewhat unpalatable insect or a mimic by the response of another bird (109). Crows, *Corvus,* will remember aposematic insects at least nine months, much longer than they will remember cryptic palatable species (152). Two species of tits refused to eat an aposematic moth after 12 months with no reinforcement by aposematic insects during the year. Since field-caught blue jays initially rejected monarchs persistently even when palatable ones were offered, it is possible that the birds had had experience with monarchs.

An alternate hypothesis is that the birds have an innate avoidance response toward orange-red insects. That hypothesis has not been tested to my knowledge, but Blest (15) reported related data indicating that eyespots elicit an innate escape response probably because they are similar to the eyes of vertebrate predators of birds. Sexton (167) showed that lizards avoided aposematic models, *Photinus,* even after 11 months without contacting any warningly colored insects and fed only mealworms most of the time. That suggests that the lizards either remember experience (unknown but possible in nature) with *Photinus* or similar insects or have some innate avoidance of certain aposematic insects. The lizards appear to locate their prey visually. Although there is no evidence that fireflies could be distinguished at a distance by odor, that possibility has not been tested adequately.

Indirect learning from other predators.—The effectiveness of mimicry is enhanced since birds, monkeys, apes, and probably some other predators learn to avoid certain insects without attempting to eat them (5, 88). Thorpe (185), viewing such behavior from the standpoint of the predators, called it "social facilitation." Liepelt reported absence of social facilitation in tests of *Vespa* workers and drones with two species of birds (117).

Although most experimental data have been recorded in terms of "not touched, pecked, killed, or eaten," it is clear that birds show many intermediate responses indicating that the food is distasteful or objectionable. L. P. and J. Brower designed their experiments to eliminate the observing and learning of one bird from another during a trial.

Rothschild & Ford (155) showed that a starling, *Sturnus vulgaris,* imitated the bill-wiping of a mistle thrush, *Turdus viscivorus,* following attack on an aposematic and poisonous grasshopper, *Poekilocerus bufonius.* When the same grasshopper was moved towards the starling, it repeatedly beak-wiped and moved away from the insect. Both birds had been hand-reared and neither had ever been fed any other red orthopteran.

Other birds have been shown to ruffle feathers, raise crests, wipe bills, shake heads, show intention movements (approach but not take food), utter alarm cries, or fly when they seize or even see aposematic or unpalatable prey with which they have had previous direct experience (111, 143, 156).

The tendency of birds to form conspecific and mixed species flocks makes such learning highly probable among many species. Moynihan (131, 132) has discussed such relationships as "social mimicry," defined as association of two or more species of birds with similar habits, color patterns, and alarm signals. To my knowledge, no one has studied how such aggregations of predators respond to insects with various types of adaptive resemblance or aposematic coloration.

Effectiveness of mimicry and aposematic coloration without fatal sampling.—Collectors long ago noticed that aposematic insects, models, and mimics often have harder, more durable bodies than other insects, which was interpreted partly as an adaptation that allowed them to be tasted and still survive (13, 14, 50). Carpenter (1937) found that many butterflies have beak marks on their wings, indicating capture and release. Turner (189) reported comparable data for *Heliconius.* Birds often take ten pecks at the body and ten at wings of a butterfly before it is killed. Older experienced birds kill a butterfly more quickly with less than ten pecks (10), but ample tasting can occur before the prey is killed. Lycid beetles survive considerable injury and field-collected specimens often bear signs of old injuries (120). The above data have been overlooked or discounted by Huheey (102) in discussing the hypothesis that some Müllerian mimics probably evolved from Batesian mimics because slightly distasteful insects would not survive sampling.

Monkeys frequently open their hands to look at captured insects, and then the insects may drop to the forest floor, fly away, or frighten the monkey by eyespots or other defensive behavior (15). Reflex immobilization by such captured insects can be an adaptation for survival since struggling insects are more likely to be killed (11).

Some lizards, *Cnemidophorus* spp., (165, 167) and a frog, *Hyla versicolor,* flicked their tongues at unpalatable insects without moving them on the substrate, but it is unclear whether that behavior is restricted to "doubtful or unfamiliar food."

Toads will puff up, duck, and bat their front legs at models or associated mimics and will "spit out" unpalatable insects which survive (31). Lizards will flee from unpalatable prey they have experienced earlier and will wipe their mouths against the ground or branches after they have released an unpalatable insect (167). It is doubtful that toads and lizards learn to avoid prey by observing other predators.

Crows, *Corvus,* refused to pick up cryptic palatable larvae which were placed near aposematic larvae. "One aposematic larva gave protection to 15 cryptic larvae in this manner" (152). This type of experiment needs repeating under carefully controlled conditions because it suggests that many aposematic insects protect totally dissimilar species from predators. Such protective advantage could be highly significant for Batesian ant mimics found with ants, or for insects in aggregations or on a common food plant.

Birds used as predators in most experiments must adapt to small cages requiring minimal flight. Birds that are too excitable or fly too much often are eliminated (16, 33, 34). Using mimics that were identical to models, Reiskind (143) showed that birds flew more frequently when either model or mimic was presented than when alternate food was offered. Experienced predators pick up models or mimics after a greater time lapse than with alternate food (111, 143, 166, 167, 194). Both flight or delayed responses of the predator allow insects to escape.

Discussing predation only in terms of insects being eaten or not eaten is considering only the extremes of a continuum of predator responses (100). If the rate or intensity of attack is decreased by an insect's appearance, it is adaptive for survival and subject to selection.

Müllerian Mimicry

Differences between Batesian and Müllerian mimicry.—As originally defined, Müllerian mimicry referred to mutual mimicry between two or more distasteful species. Rothschild (150) stated "personally I think all mimicry is Müllerian and that all mimics are slightly distasteful." She later (151) stated that this principle may apply for all predators not only herself. However, as Fisher (82) pointed out, in the strict theoretical sense Müllerian mimicry can apply only when two species are equally unpalatable; and to demonstrate equal palatability with respect to effective predators "would seem to require both natural knowledge and experimental refinement which we do not at present possess."

It is impossible to exclude Müllerian selection from closely related species such as species within the same genus. When Müllerian selection is operating under those conditions, it counteracts the principle of niche diversification, which predicts that species reduce competition by diverging in behavior, selection of food plants, and other aspects (82, 146). Müllerian mimicry also is in direct opposition to the concept of character displacement, which predicts that species in the same habitat will diverge in external appearance as a mechanism to avoid interspecific courtship and hybridization.

From a theoretical standpoint, selection is operating quite differently between Batesian and Müllerian mimicry. In Batesian mimicry, resemblance of the mimic is disadvantageous to the model. Consequently there will be divergent selection away from the mimic. Whereas, in Müllerian mimicry there will be convergent selection of both mimic and model toward a common pattern (158). The rate of selection and convergence toward a common pattern will be unequal and faster for the less abundant of two equally unpalalatable forms (82).

From the standpoint of the predator, Batesian mimicry is detrimental because the predator avoids palatable prey. Whereas, "Müllerian mimicry is advantageous because the predator does not eat or waste time pursuing injurious or unpalatable prey.

The hypothesis that color patterns will converge has been disputed (Marshall, 1908) primarily on the grounds that individuals of an abundant unpalatable species will increase their susceptibility to predation if their color pattern mutates toward that of a less abundant but equally unpalatable species. Fisher (82) stated that since variations are random around the average color pattern, insects with any pattern diverging from that of the less abundant species will be more subject to predation than will those with converging patterns. Both hypotheses are based on assumptions that the predators sample both species in direct proportion to their abundance and learn all patterns equally well.

An important concept that seems to have been overlooked by subsequent authors is that "conspicuously different insects may enjoy the advantages of Müllerian selection provided they display in common any one conspicuous feature" (82). The work of Schmidt (161) supports the view that Müllerian mimics can exist based on one feature such as a patch of red on one pair of wings, while the remainder of the pattern is dissimilar.

Some of the ant-like insects and spiders living outside of ant nests may be Müllerian mimics primarily on the basis of their behavior rather than on morphology or color pattern, but data on predator responses are lacking.

In addition to the mimicry of color pattern, many aposematic and unpalatable insects have similar distinctive odors that probably are warning or recognition signals to predators (149, 153, 154). One of the commonest odors is that of coccinellid beetles; their reflex bleeding protects them against attacks by ants (95). Coccinellids also are not eaten by some vertebrates (95), but it is not known whether other insects with similar odors have analogous chemicals with repellent properties.

Müllerian mimicry among Lepidoptera.—The first examples of this type of mimicry were the danaine and ithomiine butterflies discussed by Müller, but there are few experimental data to establish the extent to which the resemblances are due to Batesian or Müllerian selection or common genetic factors (33). Recent morphological and genetic work by Emsley (77–79) supports the hypothesis that in the genus *Heliconius* the similarities of color pattern are due to close relationship and "explosive evolution" within the genus. On the other hand, comparisons of butterflies collected over 200 years ago with specimens collected recently in the same area, indicate that some morphs have not changed appreciably and others may be hybrids between subspecies (190, 191). There is still no satisfactory hypothesis explaining the tremendous parallel polymorphism of about 30 morphs of *H. melpomene* and *H. erato* (189).

Brower, Brower & Collins (34) have pointed out that similarities in color pattern can be the result of (*a*) convergent evolution due to Müllerian advantage, (*b*) parallel evolution or lack of divergence due to Müllerian advantage, or (*c*) parallel evolution without Müllerian advantage. They tested seven species of heliconiines with 62 caged silverbeak tanagers, *Ramphocelus carbo magnirostris,* in Trinidad. The butterflies were frozen and placed on the floor of the cage with the dorsal surfaces of the wings exposed. The authors pointed out that palatability may have been modified by treatment. Acetylcholine is inactivated by freezing, but it is not known whether it is found in sufficient quantity in heliconiines to affect palatability. Four possible responses of the birds were recorded: not touched, pecked, killed, or eaten.

Differences in palatability in general corresponded with phylogenetic relationship. However, the two species, *Heliconius erato hydara* and *H. melpomene euryades,* considered closely related by Emsley (78) on the basis of morphology, color and genetics, were considered by Brower et al. to be less closely related on the basis of palatability. Although the data are significant, too much weight should not be given to palatability differences based on responses of six to eleven birds of the same species. It is clear that under the experimental conditions the five species of *Heliconious* were all unpalatable but varied in palatability. *Dryas iulia* and *Agraulis vanillae* are somewhat more palatable, but a maximum of 25 per cent was eaten compared with 100 per cent of the satyrid controls. Since the most palatable species were the ones also considered closest to the palatable Nymphalidae on systematic grounds, the data support the hypothesis that unpalatability has evolved more slowly than color pattern resemblances. That is another way of saying that these Müllerian mimics have evolved from Batesian mimics. The unpalatability of these heliconiines is of further interest because they feed on Passifloraceae (passion flowers) which have been considered poisonous (33) and nonpoisonous (34).

Emsley (78) has demonstrated the pronounced geographical variation found in *Heliconius erato* and *H. melpomene* which are grossly sympatric and maintain a close mimetic relationship over their entire range. Polymorphism was attributed to a dynamic equilibrium among (*a*) selective forces of mimicry, (*b*) sexual selection, and (*c*) genetic dominance of red. The similarity of two species in the same area was considered to be due to close systematic relationship. It seems that Müllerian selection was not given enough importance as the main selective force operating to preserve or enhance similarities among sympatric species. Phyletic relationships deduced by Turner (192) from pupal morphology support the hypothesis that *Heliconius* spp. show convergence of pattern due to Müllerian mimicry. Detailed analysis of the pigments in the wings of Lepidoptera shows that colors, appearing identical to a human observer or even to a spectrophotometer, result from a variety of pigments and physical properties of the integument (43, 61, 62). Comprehensive studies of numerous characters are needed to determine the probability and extent of convergence among Müllerian mimics.

Many of the heliconiines form mixed species aggregations as adults, but the larvae of only one of the 11 species found on Trinidad are gregarious (3). The larvae are somewhat synchronous in feeding, with the most nearly synchronous species also being the most gregarious. Nothing is known about the relative palatability of the larvae. Their behavior is considered to be adaptive by limitation of the time during which they are active and more susceptible to predation. Alexander considered the evolution of the larvae and pupae of the heliconiines to be toward cryptic coloration (3, 4).

A weak mimicry or convergence of color pattern has been reported for Hesperiidae based primarily on seasonal synchrony of the mimicking species and their seasonal replacement by other mimics (59). There is no evidence that this mimicry is adaptive or the result of natural selection, and it may be the result of lack of divergence of closely related species. Another doubtful case of seasonal mimicry has been described for two species of Geometridae (169).

Müllerian mimicry among other insects.—Most examples of Müllerian mimicry outside of the Lepidoptera are in the lycid beetle complex (49, 73, 74, 76, 118–120, 165) or among wasps, bees, and ants. Although many examples have been described, few carefully controlled experiments have been designed to test the effectiveness of the Müllerian mimics of Hymenoptera (6). However, the results of experiments with Batesian mimics, especially those with bees or wasps from which stings have been removed, show that Müllerian mimicry must effectively protect many species (22, 31, 38, 117). Van Der Vecht (196) reported parallel variation in color patterns in the vespid wasps, *Eumenes, Pseumenes,* and *Pareumenes,* in various localities from South India to New Guinea. Sympatric species of all three genera have similar distinctive black and yellow patterns that tend to decrease or become monocolored in marginal or isolated populations. The data suggest that the similar contrasting patterns are adaptive, probably by Müllerian selection, but it is more difficult to understand why the patterns are not retained outside the population centers. Perhaps density-dependent predation makes peripheral populations less subject to predator selection. No genetic work has been done with these wasps, but independent changes in color pattern on different parts of the body suggest that different genes determine the entire pattern. The decrease in pattern was possibly a result of decreased genetic variability.

Lane & Rothschild (113) have maintained that the silphid beetle, *Necrophorus investigator,* is a Müllerian mimic of bumble bees. When semitorpid due to cool temperatures or when their colonies are disturbed, some species of bumble bees will roll over on their backs, move the tip of the abdomen in and out, buzz wings, and sometimes squirt material from the anus (87). The silphid has only a crude resemblance to the bee in color pattern but a more similar behavior pattern. Its legs project out to the sides like those of the bee, the tip of the abdomen moves in and out as stinging movements, and the beetle stridulates by rubbing the fifth abdominal tergite against the tips of the elytra. The authors report "definite similarities" between the sounds made by the beetle and the bumble bee, but sound spectrographs show the main pulses or buzzes much farther apart in the bee. Captive tits were very frightened but did not make warning calls when beetles displayed. Most stridulations by other insects must be similar, and there is still no direct evidence that the sound or other behavior pattern has any selective advantage based on mimicry. Alexander (2) has shown that stridulation by scorpions will frighten several kinds of mammals.

Aggregations of mimics.—Conspecific aggregations are common among some aposematic insects such as caterpillars and grasshoppers, and individuals in a group of distasteful insects are presumably better protected from predators than they would be as isolated individuals. One extreme case is an aggregation of short-horned grasshopper nymphs that resembles a large spiny caterpillar about 9 cm long found on the same host plant (94).

Aggregations often contain both Batesian and Müllerian mimics, such as groups of butterflies at mud puddles or in sleeping aggregations. The lycid mimetic complexes in Mexico and southwestern United States commonly include at least two species of Lycidae and one or two species of Cerambycidae (49, 74). At least 200 species in 60 genera and 21 tribes of Cerambycidae are considered to be mimics of lycids (119). All lycid beetles appear to be distasteful to most predators, but the cerambycids probably are palatable. Since lycids are eaten by two species of cerambycids, palatability of the latter may depend upon how recently they have eaten.

Aggregations often are produced when a parent female deposits eggs or offspring in a group that does not disperse (94). The examples cited above are primarily the result of adults coming together to form the aggregation (3, 120). Odor of *Lycus loripes* attracts conspecific individuals and probably other insects to its aggregations (73), whereas vision is assumed to be more important for butterfly flocks. The selective advantage of aggregations must not be outweighed by attacks of specialized predators or parasites that could eliminate the entire group. However, aggregations also reduce contacts with predators (101) and probably facilitate learning by birds with restricted territories. Selection should also favor simultaneous feeding periods for Müllerian mimics (152), but data on all such relationships are lacking.

Aggressive and Wasmannian Mimics

Aggressive mimicry.—No indisputable cases of aggressive mimicry among arthropods are known, but vertebrates furnish some excellent examples (201). Female fireflies, *Photuris,* attract and devour male *Photinus* by mimicking the flash-responses of *Photinus* females (121). It is not known whether male *Photuris* are also eaten by conspecific females. The mantid *Hymenopus* may attract insects to it by resembling a flower, but *Idolum diabolicum* is probably not a case of aggressive resemblance but an example of flash coloration serving to frighten predators (60, 201). Brower, Brower & Westcott reported that the asilid fly, *Mallophora bomboides,* was an effective mimic of the bumble bee, *Bombus americanorum,* in color, pattern, and sound. They concluded that it was probably Batesian rather than aggressive mimicry. The fly ate large numbers of *Bombus,* but it also eats many other insects; and no quantitative data were presented on the relative proportions of prey compared to insect populations available. There were two *Mallophora* for every eight *Bombus* in the population.

It is highly unlikely that the cerambycid mimics of lycid beetles are aggressive mimics because it appears that any predator can approach the lycids without the latter defending itself or fleeing (74, 165).

Wasmannian mimicry.—The type of mimicry in which inquilines or commensals resemble their host is a true deceptive mimicry in which the host species is both model and selective agent. Models may kill mimics but generally do not eat them. The vast majority of cases must be inquilines living within colonies of social insects, but there may be some examples among parasites or parasitoids of solitary species. Solitary and primitively social bees appear to recognize and exclude from their nests foreign or parasitic bees, not on the basis of odor but by visual clues (128). The vision of insects is so poor that some crude resemblances may be effective mimicry if insects are the sole selective agents. Poor vision of the host insects has been used as one of the primary arguments against the hypothesis that hosts are the selective agents, but almost constant physical contact between hosts and inquilines makes chemotactic discrimination probable (198). Inquilines are usually considered to be Wasmannian mimics when they appear to resemble the host, but theoretically Wasmannian selection can affect an insect that looks nothing like its hosts, if some aspect of its morphology or behavior is the important recognition character. In that sense, insects, like the cricket, *Myrmecophila* (98), may be effective ant mimics of the Wasmannian type.

Kloft (108) has proposed that aphids tended by ants are actually ant mimics. The aphids move their hind legs up and down in a manner similar to the movements of an ant's antennae. Aphids palpated by ants excrete honeydew which the ants eat. Moreover, a satiated ant will attempt to feed the posterior end of an aphid or a wax model with a pair of lateral bristles. No evidence has been presented to demonstrate that aphids tended by ants have been selected for improved resemblance.

Wasmannian mimics can be considered somewhat aggressive in the sense that most species undoubtedly compete with adults or immatures of the hosts for the food they provide (1, 106). The extent to which the mimics directly attack their hosts is unknown. Rothschild (111) considered *Volucella bombylans* a Batesian mimic *vis a vis* the bumble bee because the fly presumably must deceive the bee in order to scavenge in its nest. The same fly was called a Müllerian mimic with respect to a flycatcher.

The number of species of insects known to live in ant and termite colonies now is in the thousands, and many have evolved to resemble their hosts in color, form, and behavior. Wasmann was the first to publish extensively on the ant-like inquilines, starting in 1889 and continuing with over 100 papers until 1934. Many of his ideas were incorrect and have been severely criticized (97), but he clearly distinguished the ant-like inquilines living in ant colonies from ant-like Batesian mimics living outside colonies and usually not closely associated with ants (197, 198). Seevers (164) reported at least 33 species in 11 genera of myrmecoid staphylinids found with Old World doryline ants, and 47 species in 27 genera with New World Dorylinae. Many other genera and species could be considered incipient mimics that show similarities in color and punctuation, and the beginning stages of constriction to form a "petiole."

Some termite guests show morphological similarities to termites that may not be obvious after the specimens are preserved (163). The most extreme examples are physogastric staphylinid beetles such as *Coatonachthodes ovambolandicus* (107). The membranous abdomen is bent anteriorly completely covering the head and thorax of the beetle. The tip of the abdomen resembles the head of a termite, and membranous abdominal processes resemble legs. The mimicry must be based on tactic and chemotactic senses since the beetles and their models are in continual darkness.

Akre & Rettenmeyer (1) and Kistner (106) have reported that the ant-like staphylinids are more closely associated behaviorally with their hosts than are the more generalized staphylinids. The ant mimics compete with the ants for food but may have some beneficial effect by cleaning the ants' bodies (1). Such ant-like species have sometimes been considered Batesian mimics, but it is difficult to imagine a vertebrate predator spending hours beside an ant column picking out the occasional guest. Furthermore, some of the adult beetles are about 3 mm long, and a mite, *Perperipes ornithocephala,* 1.0 to 1.3 mm long, is a realistic mimic of an army ant larva (64, 147). Other mites apparently mimic part of an ant's body (148). The most probable selective agents for all the mimics are the hosts themselves. Bequaert (1922, 1930) questioned how an ant-like appearance could protect any insect from predators since there are so many vertebrates that eat ants. His arguments are even more irrelevant than McAtee's (1932) because Wasmannian mimics of ants and termites are seldom exposed to vertebrate predators. The mimics are commonly subterranean and nocturnal, and the presence of their hosts protects them.

Another unstudied problem with implications for the evolution of Wasmannian mimicry is that of parasitic ants that seem to have evolved to resemble their hosts (44, 45).

Wasmannian mimics appear to be rare among the social wasps and bees though it is possible that Wasmannian selection has been important in parasitic bees such as *Psithyrus*. Various authors have expressed conflicting opinions about the degree to which *Psithyrus* spp. resemble their host species of *Bombus* (87). Probably much of the resemblance is due to a common ancestor, but there may be some Müllerian or Wasmannian advantage enhancing the resemblance between the parasite and its host.

The genera *Volucella* (Syrphidae) and *Hyperechia* (Asilidae) include several species that resemble bees or wasps. In at least one case, the asilid preys on the adult carpenter bee, *Xylocopa,* which it resembles, and the fly's larvae feed on the larvae of the bee. Other mimetic species of syrphids and asilids have been considered predators or parasites on their models or scavengers within the hosts' nests, but no studies conclusively demonstrate Wasmannian or aggressive mimicry among these flies [Carpenter & Ford (1933), 87].

Conclusions

When mimicry is carefully viewed, almost every case—even if it can be compressed into one of the well-known categories—presents a fascinating "special situation" that requires clarification [Rothschild (154)]. The validity of that statement does not mean that the nomenclature for types of mimicry is no longer useful. However, responses of all selective agents vary so widely that any label such as "Batesian mimic" applied to an insect must always be considered relative rather than absolute. The same species can range all the way from highly palatable to highly unpalatable depending on its instar, food plant, sex, time of last feeding, etc. The probability that the insect will be eaten depends also on many attributes of the predator, especially its degree of satiation, its adaptation or experience with the potential prey or similar species, and an innate or genetic component. Thus, instead of referring to an insect as a Batesian mimic, it would be more nearly accurate to state that under the specified conditions the insect or some part of its population in a certain locality has a Batesian relationship to a specific predator. The same insect may have a Müllerian or Wasmannian relationship to another predator in the same or different locality.

These hedging expressions in no way are meant to detract from the fact that mimicry has been shown to protect many insects from predators or that insect mimics provide some of the best examples of the effectiveness of natural selection. Although some insects undoubtedly have been incorrectly classified as mimics, future research should show many more species with incipient mimetic relationships, and many of the relationships will depend on behavioral or chemical factors rather than morphology or color.

LITERATURE CITED

1. Akre, R. D., Rettenmeyer, C. W. Behavior of Staphylinidae associated with army ants. *J. Kansas Entomol. Soc.*, **39**, 745–82 (1966)
2. Alexander, A. J. On the stridulation of scorpions. *Behaviour,* **12**, 339–52 (1958)
3. Alexander, A. J. A study of the biology and behavior of the caterpillars, pupae and emerging butterflies of the subfamily Heliconiinae in Trinidad, West Indies. Part I. Some aspects of larval behavior. *Zoologica,* **46**, 1–24 (1961)
4. Alexander, A. J. A study of the biology and behavior of the caterpillars, pupae and emerging butterflies of the subfamily Heliconiinae in Trinidad, West Indies. Part II. Molting, and the behavior of pupae and emerging adults. *Zoologica,* **46**, 105–23 (1961)
5. Armstrong, E. A. The nature and function of animal mimesis. *Bull. Animal Behavior,* **9**, 46–58 (1951)
6. Beebe, W., Kennedy, R. Habits, palatability and mimicry in thirteen ctenuchid moth species from Trinidad, B. W. I. *Zoologica,* **42**, 147–58 (1957)
7. Bernardi, G. Le polymorphisme et le mimétisme de l'*Hypolimnas dubia* Palisot de Beauvois. *Ann. Soc. Entomol. France,* **128**, 141–58 (1960)
8. Bernardi, G. La variation géographique du mimétisme chez les Lépidoptères. *Ann. Soc. Entomol. France,* **130**, 77–94, pl. 2 (1962)
9. Bernardi, G. Quelques aspects zoogéographiques du mimétisme chez les Lépidoptères. *Proc. Intern. Congr. Zool., 16th, Washington,* **4**, 161–66 (1963)
10. Blest, A. D. The function of eyespot patterns in the Lepidoptera. *Behaviour,* **11**, 209–56 (1957)
11. Blest, A. D. The evolution of protective displays in the Saturnioidea and Sphingidae. *Behaviour,* **11**, 257–309 (1957)
12. Blest, A. D. The resting position of *Cerodirphia speciosa* (Cramer): The ritualization of a conflict posture. *Zoologica,* **45**, 81–90 (1960)
13. Blest, A. D. Relations between moths and predators. *Nature,* **197**, 1046–47 (1963)
14. Blest, A. D. Longevity, palatability and natural selection in five species of New World saturniid moth. *Nature,* **197**, 1183–86 (1963)
15. Blest, A. D. Protective display and and sound production in some New World arctiid and ctenuchid moths. *Zoologica,* **49**, 161–81 (1964)
16. Brower, J. V. Z. Experimental studies of mimicry in some North American butterflies. I. *Danaus plexippus* and *Limenitis archippus archippus. Evolution,* **12**, 32–47 (1958)
17. Brower, J. V. Z. Experimental studies of mimicry in some North American butterflies. II. *Battus philenor* and *Papilio troilus, P. polyxenes,* and *P. glaucus. Evolution,* **12**, 123–36 (1958)
18. Brower, J. V. Z. Experimental studies of mimicry in some North American butterflies. Part III. *Danaus gilippus berenice* and *Limenitis archippus floridensis. Evolution,* **12**, 273–85 (1958)
19. Brower, J. V. Z. Experimental studies of mimicry. IV: The reactions of starlings to different proportions of models and mimics. *Am. Naturalist,* **94**, 271–82 (1960)
20. Brower, J. V. Z. Experimental studies and new evidence on the evolution of mimicry in butterflies. *Proc. Intern. Congr. Zool., 16th, Washington,* **4**, 156–61 (1963)
21. Brower, J. V. Z., Brower, L. P. Palatability of North American model and mimic butterflies to caged mice. *J. Lepidopt. Soc.,* **15**, 23–24 (1961)
22. Brower, J. V. Z., Brower, L. P. Experimental studies of mimicry. 8. Further investigations of honeybees (*Apis mellifera*) and their dronefly mimics (*Eristalis* spp.). *Am. Naturalist,* **99**, 173–88 (1965)
23. Brower, L. P. Bird predation and foodplant specificity in closely related procryptic insects. *Am. Naturalist,* **92**, 183–87 (1958)
24. Brower, L. P. Speciation in butterflies of the *Papilio glaucus* group. II. Ecological relationships and interspecific sexual behavior. *Evolution,* **13**, 212–28 (1959)
25. Brower, L. P. Autecological relationships and interspecific sexual behaviour in butterflies of the *Papilio glaucus* group. *Proc. Intern.*

Congr. Zool., 15th, London, 810–12 (1959)
26. Brower, L. P. Biology of the monarch butterfly (review of Urquhart, F. A., 1960, The monarch butterfly). *Ecology,* **43,** 181–82 (1962)
27. Brower, L. P. The evolution of sex-limited mimicry in butterflies. *Proc. Intern. Congr. Zool., 16th, Washington,* **4,** 173–79 (1963)
28. Brower, L. P. *Patterns for survival: A study of mimicry and protective coloration in tropical insects.* (Amherst College, Amherst, Mass., 16 mm color film, 26.5 min, optical sound, 1968)
29. Brower, L. P. Ecological chemistry. *Sci. Am.,* **220**(2), 22–29 (1969)
30. Brower, L. P., Brower, J. V. Z. Cryptic coloration in the anthophilous moth *Rhododipsa masoni. Am. Naturalist,* **90,** 177–82 (1956)
31. Brower, L. P., Brower, J. V. Z. Experimental studies of mimicry. 6. The reaction of toads (*Bufo terrestris*) to honeybees (*Apis mellifera*) and their dronefly mimics (*Eristalis vinetorum*). *Am. Naturalist,* **96,** 297–307 (1962)
32. Brower, L. P., Brower, J. V. Z. The relative abundance of model and mimic butterflies in natural populations of the *Battus philenor* mimicry complex. *Ecology,* **43,** 154–58 (1962)
33. Brower, L. P., Brower, J. V. Z. Birds, butterflies, and plant poisons: A study in ecological chemistry. *Zoologica,* **49,** 137–59 (1964)
34. Brower, L. P., Brower, J. V. Z., Collins, C. T. Experimental studies of mimicry. 7. Relative palatability and Müllerian mimicry among Neotropical butterflies of the subfamily Heliconiinae. *Zoologica,* **48,** 65–84 (1963)
35. Brower, L. P., Brower, J. V. Z., Corvino, J. M. Plant poisons in a terrestrial food chain. *Proc. Natl. Acad. Sci. U.S.,* **57,** 893–98 (1967)
36. Brower, L. P., Brower, J. V. Z., Cranston, F. P. Courtship behavior of the queen butterfly, *Danaus gilippus berenice* (Cramer). *Zoologica,* **50,** 1–39 (1965)
37. Brower, L. P., Brower, J. V. Z., Stiles, F. G., Croze, H. J., Hower, A. S. Mimicry: Differential advantage of color patterns in the natural environment. *Science,* **144,** 183–85 (1964)
38. Brower, L. P., Brower, J. V. Z., Westcott, P. W. Experimental studies of mimicry 5. The reactions of toads (*Bufo terrestris*) to bumblebees (*Bombus americanorum*) and their robberfly mimics (*Mallophora bomboides*), with a discussion of aggressive mimicry. *Am. Naturalist,* **94,** 343–56 (1960)
39. Brower, L. P., Cook, L. M., Croze, H. J. Predator responses to artificial Batesian mimics released in a Neotropical environment. *Evolution,* **21,** 11–23 (1967)
40. Brower, L. P., Cranston, F. P. *Courtship behavior of the queen butterfly.* (Penn. State Univ. Psychological Cinema Register Film 2123K, 18 min. color, 1962)
41. Brower, L. P., Jones, M. A. Precourtship interaction of wing and abdominal sex glands in male *Danaus* butterflies. *Proc. Roy. Entomol. Soc. London, Ser. A,* **40,** 147–51, pl. I-II (1965)
42. Brower, L. P., Ryerson, W. N., Coppinger, L. L., Glazier, S. C. Ecological chemistry and the palatability spectrum. *Science,* **161,** 1349–51 (1968)
43. Brown, K. S. Chemotaxonomy and chemomimicry: The case of 3-hydroxykynurenine. *Systemat. Zool.,* **16,** 213–16 (1967)
44. Brown, W. L., Jr. The first social parasite in the ant tribe Dacetini. *Insectes Sociaux,* **2,** 181–86 (1955)
45. Brown, W. L., Jr., Wilson, E. O. A new parasitic ant of the genus *Monomorium* from Alabama, with consideration of the status of genus *Epixenus* Emery. *Entomol. News,* **68,** 239–46 (1957)
46. Burns, J. M. Preferential mating versus mimicry: Disruptive selection and sex-limited dimorphism in *Papilio glaucus. Science,* **153,** 551–53 (1966)
47. Busnel, R. G. *Acoustic Behavior of Animals.* (Elsevier, New York, xx + 933 pp., 1963)
48. Chance, M. R. A., Russell, W. M. S. Protean displays: A form of allaesthetic behaviour. *Proc. Zool. Soc. London,* **132,** 65–70 (1959)
49. Chemsak, J. A., Linsley, E. G. A revised key to the species of *Elytroleptus* with notes on variation and geographical distribution. *Pan. Pacific Entomol.,* **41,** 193–99 (1965)

50. Clarke, C. A., Sheppard, P. M. Hand-pairing of butterflies. *Lepidopt. News,* **10,** 47–53 (1956)
51. Clarke, C. A., Sheppard, P. M. The genetics of *Papilio darnanus,* Brown. II. Races *dardanus, polytrophus meseres,* and *tibullus. Genetics,* **45,** 439–57 (1960)
52. Clarke, C. A., Sheppard, P. M. The genetics of *Papilio dardanus,* Brown. III. Race *antinorii* from Abyssinia and race *meriones* from Madagascar. *Genetics,* **45,** 683–98 (1960)
53. Clarke, C. A., Sheppard, P. M. The evolution of mimicry in the butterfly *Papilio dardanus,* Brown. *Heredity,* **14,** 163–73, 2 pl. (1960)
54. Clarke, C. A., Sheppard, P. M. The evolution of dominance under disruptive selection. *Heredity,* **14,** 73–87, 3 pl. (1960)
55. Clarke, C. A., Sheppard, P. M. Supergenes and mimicry. *Heredity,* **14,** 175–85 (1960)
56. Clarke, C. A., Sheppard, P. M. Offspring from double matings in swallowtail butterflies. *Entomologist,* **95,** 199–203 (1962)
57. Clarke, C. A., Sheppard, P. M. Disruptive selection and its effect on a metrical character in the butterfly *Papilio dardanus. Evolution,* **16,** 214–26 (1962)
58. Clarke, C. A., Sheppard, P. M. Interactions between major genes and polygenes in the determination of the mimetic patterns of *Papilio dardanus. Evolution,* **17,** 404–13 (1963)
59. Clench, H. K. Temporal dissociation and population regulation in certain hesperiine buttterflies. *Ecology,* **48,** 1000–6 (1967)
60. Cott, H. B. *Adaptive Coloration in Animals.* (Methuen, London, xxxii + 508 pp., 48 pl., 1940)
61. Crane, J. Spectral reflectance characteristics of butterflies from Trinidad, B. W. I. *Zoologica,* **39,** 85–115, pl. I-III (1954)
62. Crane, J. Imaginal behavior of a Trinidad butterfly, *Heliconius erato hydara* Hewitson, with special reference to the social use of color. *Zoologica,* **40,** 167–96 (1955)
63. Crane, J. Imaginal behavior in butterflies of the family Heliconiidae: Changing social patterns and irrelevant actions. *Zoologica,* **42,** 135–45 (1957)
64. Cross, E. A. The generic relationships of the family Pyemotidae. *Univ. Kansas Sci. Bull.,* **45,** 29–275 (1965)
65. Curio, E. Die Schutzanpassungen dreier Raupen eines Schwärmers auf Galapagos. *Zool. Jahrb. Systemat.,* **92,** 487–522, pl. 1 (1965)
66. Curio, E. Ein Falter mit "falschem Kopf." *Natur Museum,* **95,** 43–46 (1965)
67. Curio, E. Die Schlangenmimikry einer südamerikanischen Schwärmerraupe. *Natur Museum,* **95,** 207–11 (1965)
68. Downey, J. C. Mimicry and distribution of *Caenurgina caerulea* Grt. *J. Lepidopt. Soc.,* **19,** 165–70 (1965)
69. Duncan, C. J., Sheppard, P. M. Sensory discrimination and its role in the evolution of Batesian mimicry. *Behaviour,* **24,** 269–82 (1965)
70. Edmunds, M. Natural selection in the mimetic butterfly *Hypolimnas misippus* L. in Ghana. *Nature,* **212,** 1478.
71. Edwards, J. S. Arthropods as predators. *Viewpoints Biol.,* **2,** 85–114 (1963)
72. Ehrlich, P. R., Raven, P. H. Butterflies and plants: A study in coevolution. *Evolution,* **18,** 586–608 (1964)
73. Eisner, T. E., Kafatos, F. C. Defense mechanisms of arthropods. X. A pheromone promoting aggregation in an aposematic distasteful insect. *Psyche,* **69,** 53–61 (1962)
74. Eisner, T., Kafatos, F. C., Linsley, E. G. Lycid predation by mimetic adult Cerambycidae. *Evolution,* **16,** 316–24 (1962)
75. Emlen, J. M. Batesian mimicry: A preliminary theoretical investigation of quantitative aspects. *Am. Naturalist,* **102,** 235–41 (1968)
76. Emmel, T. C. A new mimetic assemblage of lycid and cerambycid beetles in central Chiapas. *Southwestern Naturalist,* **10**(1), 14–16 (1965)
77. Emsley, M. A morphological study of imagine Heliconiinae with consideration of the evolutionary relationships within the group. *Zoologica,* **48,** 85–130, pl. 1 (1963)
78. Emsley, M. G. The geographical distribution of the color-pattern components of *Heliconius erato* and

Heliconius melpomene with genetical evidence for the systematic relationship between the two species. *Zoologica,* **49,** 245–86, 2 pl (1964)
79. Emsley, M. G. Speciation in *Heliconius:* Morphology and geographical distribution. *Zoologica,* **50,** 191–254 (1965)
80. Euw, J. von, Fishelson, L., Parsons, J. A., Reichstein, T., Rothschild, M. Cardenolides (heart poisons) in a grasshopper feeding on milkweeds. *Nature,* **214,** 35–39 (1967)
81. Fishelson, L. The biology and behaviour of *Poekilocerus bufonius* Klug, with special reference to the repellent gland. *Eos (Madrid),* **36,** 41–62 (1960)
82. Fisher, R. A. *The Genetical Theory of Natural Selection,* 2nd revised ed. (Dover, New York, N.Y., 291 pp., 1958) (First edition, 1930)
83. Ford, E. B. The genetics of polymorphism in the Lepidoptera. *Advan. Genet.,* **5,** 43–87 (1953)
84. Ford, E. B. The theory of genetic polymorphism. In *Insect Polymorphism,* 11–19. (Kennedy, J. S., Ed., Roy. Entomol. Soc. London, England, 115 pp., 1961)
85. Ford, E. B. Mimicry. *Proc. Intern. Congr. Zool., 16th, Washington,* **4,** 184–86 (1963)
86. Frazer, J. F. D., Rothschild, M. Defense mechanisms in warningly-coloured moths and other insects. *Verhandl. Intern. Kongr. Entomol., 11th, Wien, 1960,* **3,** 249–56 (1962)
87. Free, J. B., Butler, C. G. *Bumblebees.* (Collins, London, xiv + 208 pp., 1959)
88. Gans, C. Empathic learning and the mimicry of African snakes. *Evolution,* **18,** 705 (1965)
89. Gaul, A. T. Audio mimicry: An adjunct to color mimicry. *Psyche,* **59,** 82–83 (1952)
90. Gelperin, A. Feeding behaviour of the preying mantis: A learned modification. *Nature,* **219,** 399–400 (1968)
91. Gibb, J. A. L. Tinbergen's hypothesis of the role of specific search images. *Ibis,* **104,** 106–11 (1962)
92. Goldschmidt, R. B. Mimetic polymorphism, a controversial chapter of Darwinism. *Quart. Rev. Biol.,* **20,** 147–64, 205–30 (1945)
93. Goldschmidt, R. B. Evolution, as viewed by one geneticist. *Am. Sci.,* **40,** 84–98, 135 (1952)
94. Haas, F. Collective mimicry. *Ecology,* **26,** 412–13 (1945)
95. Happ, G. M., Eisner, T. Hemorrhage in a coccinellid beetle and its repellent effect on ants. *Science,* **134,** 329–31 (1961)
96. Haskell, P. T. *Insect Sounds.* (Witherby, London, England, viii + 189 pp., 1961)
97. Heikertinger, F. *Das Rätsel der Mimikry und seine Lösung: Eine kritische Darstellung des Werdens, des Wesens und der Widerlegung der Tiertrachthypothesen.* (Fischer, Jena., viii + 208 pp., 9 pl., 1954)
98. Hölldobler, K. Gibt es in Deutschland Ameisengäste, die echte Täuscher sind? *Naturwissenschaften,* **40,** 34–35 (1953)
99. Holling, C. S. Principles of insect predation. *Ann. Rev. Entomol.,* **6,** 163–82 (1961)
100. Holling, C. S. Mimicry and predator behavior. *Proc. Intern. Congr. Zool., 16th, Washington,* **4,** 166–72 (1963)
101. Holling, C. S. The functional response of predators to prey density and its role in mimicry and population regulation. *Mem. Entomol. Soc. Can.,* **45,** 1–60 (1965)
102. Huheey, J. E. Studies in warning coloration and mimicry. III. Evolution of Müllerian mimicry. *Evolution,* **15,** 567–68 (1961)
103. Jones, D. A., Parsons, J., Rothschild, M. Release of hydrocyanic acid from crushed tissues of all stages in the life-cycle of species of the Zygaeninae. *Nature,* **193,** 52–53 (1962)
104. Kettlewell, H. B. D. Brazilian insect adaptations. *Endeavour,* **18,** 200–10 (1959)
105. Kirkpatrick, T. W. *Insect Life in the Tropics.* (Longmans, Green, New York, N. Y., xiv + 311 p., 1957)
106. Kistner, D. H. A revision of the African species of the Aleocharine Tribe Dorylomimini II. The genera *Dorylomimus, Dorylonannus, Dorylogaster, Dorylobactrus,* and *Mimanomma,* with notes on their behavior. *Ann. Entomol. Soc. Am.,* **59,** 320–40 (1966)
107. Kistner, D. H. Revision of the African species of the termitophilous tribe Corotocini. I. A new

genus and species from Ovamboland and its zoogeographical significance. *J. N. Y. Entomol. Soc.*, **76**, 213–21 (1968)

108. Kloft, W. Versuch einer Analyse der trophobiotischen Beziehungen von Ameisen zu Aphiden. *Biol. Zentr.*, **78**, 863–70 (1959)

109. Klopfer, P. H. *Behavioral Aspects of Ecology*. (Prentice-Hall, Englewood Cliffs, New Jersey, 173 pp., 1962)

110. Kullenberg, B. Studies in *Ophrys* pollination. *Zool. Bidrag Uppsala*, **34**, 1–340 (1961)

111. Lane, C., Rothschild, M. Preliminary note on insects eaten and rejected by a tame Shama (*Kittacincla malabarica* Gm.), with the suggestion that in certain species of butterflies and moths females are less palatable than males. *Entomol. Monthly Mag.*, **93**, 172–79 (1956)

112. Lane, C., Rothschild, M. A very toxic moth: The Five-spot Burnet (*Zygaena trifolii* Esp.). *Entomol. Monthly Mag.*, **95**, 93–94 (1959)

113. Lane, C., Rothschild, M. A case of Müllerian mimicry of sound. *Proc. Roy. Entomol. Soc. London, Ser. A*, **40**, 156–58 (1965)

114. Lenko, K. On mimicry by the cerambycid *Pertyia sericea* (Perty, 1830) of *Camponotus sericeiventris* (Guerin, 1830). *Papeis Avulsos Dept. Zool. São Paulo*, **16**,(9), 89–95 (1964)

115. Levi, H. W. An unusual case of mimicry. *Evolution*, **19**, 261–62 (1965)

116. Li, C. C. *An Introduction to Population Genetics*. (National Peking Univ. Press, Peiping, vii + 321 p., 1948)

117. Liepelt, W. Zur Schutzwirkung des Stachelgiftes von Bienen und Wespen gegenüber Trauerfliegenschnäpper und Gartenrotschwanz. *Zool. Jahrb., Abt. Allgem. Zool. Physiol. Tiere*, **70**, 167–76 (1963)

118. Linsley, E. G. Mimetic form and coloration in the Cerambicidae. *Ann. Entomol. Soc. Am.*, **52**, 125–31 (1959)

119. Linsley, E. G. Lycidlike Cerambycidae. *Ann. Entomol. Soc. Am.*, **54**, 628–35 (1961)

120. Linsley, E. G., Eisner, T., Klots, A. B. Mimetic assemblages of sibling species of lycid beetles. *Evolution*, **15**, 15–29 (1960)

121. Lloyd, J. E. Aggressive mimicry in *Photuris*: Firefly femmes fatales. *Science*, **149**, 653–54 (1965)

122. Magnus, D. B. E. Experimental analysis of some "overoptimal" sign-stimuli in the mating behaviour of the fritillary butterfly *Argynnis paphia* L. *Proc. Intern. Congr. Entomol., 10th, Montreal, 1956*, **2**, 405–18 (1958)

123. Magnus, D. Experimentelle Untersuchungen zur Bionomie und Ethologie des Kaisermantels *Argynnis paphia* L. I. Über optische Auslöser von Anfliegereaktionen und ihre Bedeutung für das Sichfinden der Geschlechter. *Z. Tierpsychol.*, **15**, 397–426 (1958)

124. Magnus, D. B. E. Sex limited mimicry II—Visual selection in the mate choice of butterflies. *Proc. Intern. Congr. Zool., 16th, Washington*, **4**, 179–83 (1963)

125. Marler, P., Hamilton, W. J. III. *Mechanisms of Animal Behavior*. (Wiley, New York, N.Y., xi + 771 pp., 1966)

126. Mason, L. G. Prey selection by a non-specific predator. *Evolution*, **19**, 259–60 (1965)

127. Michener, C. D. Social polymorphism in Hymenoptera. In *Insect Polymorphism*, 43–56. (Kennedy, J. S., Ed., Roy. Entomol. Soc. London, England, 115 pp., 1961)

128. Michener, C. D. Comparative social behavior of bees. *Ann. Rev. Entomol.*, **14**, 299–342 (1969)

129. Moment, G. B. Reflexive selection: A possible answer to an old puzzle. *Science*, **136**, 262–63 (1962)

130. Mook, J. H., Mook, L. J., Heikens, H. S. Further evidence for the role of "searching images" in the hunting behaviour of titmice. *Arch. Neerl. Zool.*, **13**, 448–65 (1960)

131. Moynihan, M. Some adaptations which help to promote gregariousness. *Proc. Intern. Ornith. Congr., 12th, Helsinki, 1958*, 523–41 (1960)

132. Moynihan, M. Social mimicry; character convergence versus character displacement. *Evolution*, **22**, 315–31 (1968)

133. Owen, D. F. Similar polymorphisms in an insect and a land snail. *Nature*, **198**, 201–3 (1963)

134. Owen, D. F. *Animal Ecology in Tropical Africa*. (Freeman, San

Francisco, Calif., viii + 122 p., 1966)

135. Owen, D. F., Chanter, D. O. Population biology of tropical African butterflies. Sex ratio and genetic variation in *Acraea encedon*. *J. Zool. London,* **157,** 345–74 (1969)

136. Parsons, J. A. A digitalis-like toxin in the monarch butterfly, *Danaus plexippus* L. *J. Physiol. (London),* **178,** 290–304 (1965)

137. Petersen, B. Monarch butterflies are eaten by birds. *J. Lepidopt. Soc.,* **18,** 165–69 (1964)

138. Pijl, L. van der, Dodson, C. H. *Orchid Flowers: Their Pollination and Evolution.* (Univ. Miami Press, Coral Gables, Florida, 214 pp., 1966)

139. Platt, A. P., Brower, L. P. Mimetic versus disruptive coloration in intergrading populations of *Limenitis arthemis* and *astyanax* butterflies. *Evolution,* **22,** 699–718 (1968)

140. Raske, A. G. Morphological and behavioral mimicry among beetles of the genus *Moneilema*. *Pan-Pacific Entomol.,* **43,** 239–44 (1967)

141. Reichstein, T. Cardenolide (herzwirksame Glyoside) als Abwehrstoffe bei Insekten. *Naturw. Rundschau,* **20,** 499–511 (1967)

142. Reichstein, T., Euw, J. von, Parsons, J. A., Rothschild, M. Heart poisons in the monarch butterfly. *Science,* **161,** 861–66 (1968)

143. Reiskind, J. Behaviour of an avian predator in an experiment simulating Batesian mimicry. *Animal Behaviour* **13,** 466–69 (1965)

144. Reiskind, J. The taxonomic problem of sexual dimorphism in spiders and a synonymy in *Myrmecotypus*. *Psyche,* **72,** 279–81 (1966)

145. Reiskind, J., Levi, H. W. *Anatea,* an ant-mimicking theridiid spider from New Caledonia. *Psyche,* **74,** 20–23 (1967)

146. Remington, C. L. Historical backgrounds of mimicry. *Proc. Intern. Congr. Zool., 16th, Washington,* **4,** 145–49 (1963)

147. Rettenmeyer, C. W. Behavior, abundance and host specificity of mites found on Neotropical army ants. *Verhandl. Intern. Kongr. Entomol., 11th, Wien, 1960,* **1,** 610–12, pl. XVII (1962)

148. Rettenmeyer, C. W. Notes on host specificity and behavior of myrmecophilous macrochelid mites. *J. Kansas Entomol. Soc.,* **35,** 358–60 (1962)

149. Rothschild, M. Defensive odours and Müllerian mimicry among insects. *Trans. Roy. Entomol. Soc. London,* **113,** 101–21 (1961)

150. Rothschild, M. (comments following paper by Sheppard, No. 171) (1961)

151. Rothschild, M. Is the Buff Ermine (*Spilosoma lutea* (Huf.)) a mimic of the White Ermine (*Spilosoma lubricipeda* (L.))? *Proc. Roy. Entomol. Soc. London,* **38,** 159–64 (1963)

152. Rothschild, M. An extension of Dr. Lincoln Brower's theory on bird predation and food specificity, together with some observations on bird memory in relation to aposematic colour patterns. *Entomologist,* **97,** 73–78 (1964)

153. Rothschild, M. A note on the evolution of defensive and repellent odors of insects. *Entomologist,* **97,** 276–80 (1964)

154. Rothschild, M. Mimicry: The deceptive way of life. *Natur. Hist.,* **76**(2), 44–51 (1967)

155. Rothschild, M., Ford, E. B. Warning signals from a starling *Sturnus vulgaris* observing a bird rejecting unpalatable prey. *Ibis,* **110,** 104–5 (1968)

156. Rothschild, M., Lane, C. Warning and alarm signals by birds seizing aposematic insects. *Ibis,* **102,** 328–30 (1960)

157. Ruiter, L. de. Some experiments on the camouflage of stick caterpillars. *Behaviour,* **4,** 222–32 (1952)

158. Ruiter, L. de. Some remarks on problems of the ecology and evolution of mimicry. *Arch. Neerl. Zool.,* **13,** 1, Suppl., 351–68 (1958)

159. Sargent, T. D. Background selections of geometrid and noctuid moths. *Science,* **154,** 1674–75 (1966)

160. Sargent, T. D. Cryptic moths: Effects on background selections of painting the circumocular scales. *Science,* **159,** 100–1 (1968)

161. Schmidt, R. S. Behavioral evidence on the evolution of Batesian mimicry. *Animal Behaviour,* **6,** 129–38 (1958)

162. Schmidt, R. S. Predator behaviour

and the perfection of incipient mimetic resemblances. *Behaviour,* **16,** 149–58 (1960)
163. Seevers, C. H. A monograph on the termitophilous Staphylinidae. *Fieldiana, Zool.,* **40,** 1–334 (1957)
164. Seevers, C. H. The systematics, evolution and zoogeography of staphylinid beetles associated with army ants. *Fieldiana, Zool.,* **47,** 137–351 (1965)
165. Selander, R., Miller, J., Mathieu, J. Mimetic associations of lycid and cerambycid beetles in Coahuila, Mexico. *J. Kansas Entomol. Soc.,* **36,** 45–52 (1963)
166. Sexton, O. J. Experimental studies of artificial Batesian mimics. *Behaviour,* **15,** 244–52 (1960)
167. Sexton, O. J. Differential predation by the lizard, *Anolis carolinensis,* upon unicoloured and polycoloured insects after an interval of no contact. *Animal Behaviour,* **12,** 101–10 (1964)
168. Sexton, O. J., Hoger, C., Ortleb, E. *Anolis carolinensis:* Effects of feeding on reaction to aposematic prey. *Science,* **153,** 1140 (1966)
169. Shapiro, A. M. *Antepione thiosaria* and *Xanthotype:* A case of mimicry. *J. Res. Lepidoptera,* **4,** 6–11 (1965)
170. Sheppard, P. M. The evolution of mimicry: A problem in ecology and genetics. *Cold Spring Harbor Symp. Quant. Biol.,* **24,** 131–40 (1959)
171. Sheppard, P. M. Recent genetical work on polymorphic mimetic papilios. In *Insect Polymorphism,* 20–29. (Kennedy, J. S., Ed., *Roy. Entomol. Soc. London, England,* 115 pp., 1961)
172. Sheppard, P. M. Some contributions to population genetics resulting from the study of the Lepidoptera. *Advant. Genet.,* **10,** 165–216 (1961)
173. Sheppard, P. M. Some aspects of the geography, genetics, and taxonomy of a butterfly. *Systemat. Assoc. Publ.,* **4,** 135–52, 1 pl (1962)
174. Sheppard, P. M. Some genetic studies of Müllerian mimics in butterflies of the genus *Heliconius. Zoologica,* **48,** 145–54, 2 pl (1963)
175. Sheppard, P. M. The genetics of mimicry. *Proc. Intern. Congr. Zool., 16th, Washington,* **4,** 150–56 (1963)
176. Sheppard, P. M. The monarch butterfly and mimicry. *J. Lepidopt. Soc.,* **19,** 227–30 (1965)
177. Silberglied, R., Eisner, T. Mimicry of Hymenoptera by beetles with unconventional flight. *Science,* **163,** 486–88 (1968)
178. Someren, V. G. L. van, Jackson, T. H. E. Some comments on protective resemblance amongst African Lepidoptera (Rhopalocera). *J. Lepidopt. Soc.,* **13,** 121–50 (1959)
179. Stride, G. O. On the courtship behaviour of *Hypolimnas misippus* L., with notes on the mimetic association with *Danaus chrysippus* L. *Brit. J. Animal Behaviour,* **4,** 52–68 (1956)
180. Stride, G. O. Investigations into the courtship behaviour of the male of *Hypolimnas misippus* L., with special reference to the role of visual stimuli. *Brit. J. Animal Behaviour,* **5,** 153–67 (1957)
181. Stride, G. O. Further studies on the courtship behaviour of African mimetic butterflies. *Animal Behaviour,* **6,** 224–30 (1958)
182. Swihart, S. L. The electroretinogram of *Heliconius erato* and its possible relation to established behaviour patterns. *Zoologica,* **48,** 155–65, 2 pl (1963)
183. Swihart, S. L. Evoked potentials in the visual pathway of *Heliconius erato. Zoologica,* **50,** 55–62, 3 pl (1965)
184. Swihart, S. L. Neural adaptations in the visual pathway of certain heliconiine butterflies, and related forms, to variations in wing coloration. *Zoologica,* **52,** 1–14 (1967)
185. Thorpe, W. E. *Learning and Instinct in Animals,* 2nd Ed. (Methuen, London, England, x + 558 pp., 1963)
186. Tinbergen, L. The natural control of insects in pinewoods. I. Factors influencing the intensity of predation by songbirds. *Arch. Neerl. Zool.,* **13,** 265–336 (1960)
187. Treusch, H. W. Bisher unbekanntes gezieltes Duftanbieten paarungsbereiter *Argynnis paphia* Weibchen. *Naturwissenschaften,* **54,** 592 (1967)
188. Turner, J. R. G. Geographical varia-

tion and evolution in the males of the butterfly *Papilio dardanus* Brown. ***Trans. Roy. Entomol. Soc. London,* 155,** 239–59 (1963)

189. Turner, J. R. G. Evolution of complex polymorphism and mimicry in distasteful South American butterflies. *Proc. Intern. Congr. Entomol., 12th, London,* **4,** 267 (1965)

190. Turner, J. R. G. Some early works on heliconiine butterflies and their biology. *J. Linnean Soc. London, Zool.,* **46,** 255–66 (1967)

191. Turner, J. R. G. Natural selection for and against a polymorphism which interacts with sex. *Evolution,* **22,** 481–95 (1968)

192. Turner, J. R. G. Some new *Heliconius* pupae: Their taxonomic and evolutionary significance in relation to mimicry. *J. Zool. London,* **155,** 311–25, 4 pl (1968)

193. Turner, J. R. G., Crane, J. The genetics of some polymorphic forms of the butterflies *Heliconius melpomene* Linnaeus and *H. erato* Linnaeus. I. Major genes. *Zoologica,* **47,** 141–52, 1 pl (1962)

194. Tyshchenko, V. P. The relation of some spiders of the Thomisidae family to mimicking insects and their models. (translation) *Vestnik Leningrad. Gosudarst Univ.,* **3,** 133–39 (1961); *Biol. Abstr.,* **41,** No. 4332 (1963)

195. Urquhart, F. A. *The Monarch Butterfly.* (Univ. Toronto Press, Toronto, Canada, xxiv + 361 pp., 1960)

196. Vecht, J. Van Der. Evolution in a group of Indo-Australian *Eumenes. Evolution,* **15,** 468–77 (1961)

197. Wasmann, E. The ants and their guests. *Ann. Rept. Smithsonian Inst., 1912,* 455–74, pl. 1–10 (1913)

198. Wasmann, E. Die Ameisenmimikry. *Abhandl. Theoret. Biol.,* **19,** i-xii + 1–164 p., 3 pl (1925)

199. Wheeler, W. M. The ant *Camponotus* (*Myrmepomis*) *sericeiventris* Guerin, and its mimic. *Psyche,* **38,** 86–98 (1931)

200. Wickler, W. Mimicry and the evolution of animal communication. *Nature,* **208,** 519–21 (1965)

201. Wickler, W. *Mimicry in Plants and Animals.* (Trans. by R. D. Martin. Weidenfeld & Nicolson, World Univ. Library, London, England, 255 pp., 1968)

202. Wynne-Edwards, V. C. *Animal Dispersion in Relation to Social Behaviour.* (Oliver & Boyd, London, England, xi + 653 pp., 1962)

EVOLUTION AND TAXONOMIC SIGNIFICANCE OF REPRODUCTION IN BLATTARIA

By Louis M. Roth[1]

Pioneering Research Laboratory, U. S. Army Natick Laboratories, Natick, Massachusetts

INTRODUCTION

". . . comparison of present-day species can give us a deep insight, with a probability closely approaching certainty, into the evolutionary history of animal species."

Tinbergen (92)

Cockroaches reached their highest development in number of species relative to other forms, during the Upper Carboniferous but have decreased from that period to the present (11). In spite of this decline cockroaches obviously are a highly successful group and have been able to adapt to almost every conceivable type of ecological niche, from desert to aquatic habitats (84). "Reproduction is a phase of cockroach biology that demonstrates the diversity of behavior that has evolved in this relatively ancient group" (74).

Much information has been added to our knowledge of reproduction in cockroaches since the subject was reviewed 15 years ago (74). In particular, advances have been made in analyzing the mechanisms that control various aspects of reproduction (2–8, 14–21, 49–51, 61, 62, 67, 70–72, 88, 90). Most early workers interested in the taxonomy of cockroaches based their classifications almost solely on external morphological characters of dried museum specimens, giving little attention to biological characteristics or internal morphology. Depending on the character or sets of characters used, classifications and species affinities have varied considerably in this order (38).

Ovoviviparity and viviparity are of relatively recent origin and arose from an oviparous method of reproduction (27, 28). Several workers grouped species of Blattaria according to their ovipositional behavior and type of oötheca, and suggested possible pathways for the evolution of internal incubation of the eggs (31, 74, 76, 82, 86, 87). With few exceptions (e.g., Shelford (87)], however, little attention was paid to the possible taxonomic significance of their groupings until McKittrick (38) made reproduction the unifying character of cockroach classification. In this review I shall describe certain characteristics of reproduction in various groups of

[1] I thank Dr. Ashley Gurney and Mr. Marc Roth for reviewing the manuscript.

cockroaches and show how these tend to support McKittrick's taxonomic conclusions regarding the higher categories of the Blattaria.

CLASSIFICATION

"So far we have made but a beginning in solving the involved phylogenetic history of existing Blattidae, . . ."

J. A. G. Rehn (47)

"It would be interesting, and probably decidedly helpful, to correlate the biology of roaches with the classification."

Gurney (23)

If we examine various classifications of cockroaches, we find oviparous and ovoviviparous groups intermingled. In J. W. H. Rehn's (48) system, based principally but not entirely on the evolution of wings (23), the Blattidae includes six oviparous and three ovoviviparous subfamilies (Fig. 1). In addition, ovoviviparous forms are placed in two families, and the one viviparous species, in a third. Princis' (43) system, based on wings, legs, fronto-clypeal region of the head, male subgenital plate, and a few other external characters, defines four suborders; two are strictly oviparous, a third, the Blaberoidea, contains six oviparous, eight ovoviviparous, and one viviparous family, and a fourth suborder, the Epilamproidea, arising from an entirely different line, includes three oviparous and one ovoviviparous fam-

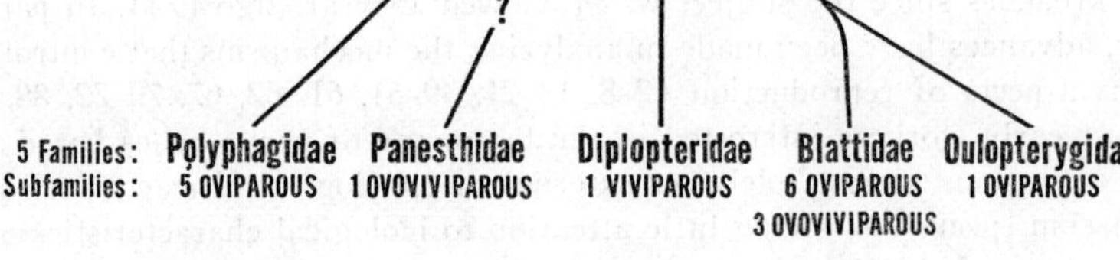

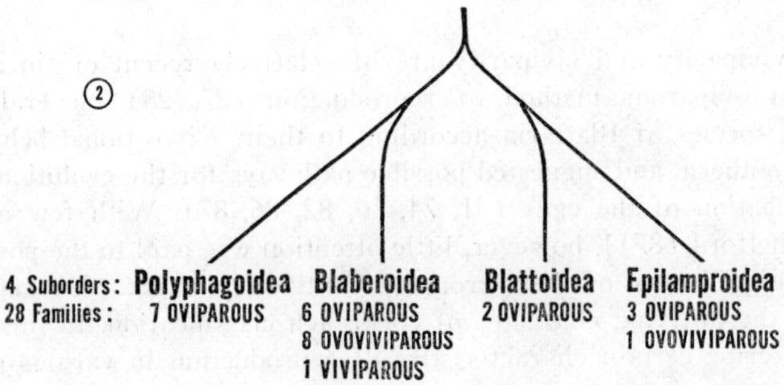

FIGS. 1–2. Classifications of the Blattaria. 1. From Rehn (48). 2. From Princis (43).

ily (Fig. 2). Bey-Bienko (9) used characters similar to those of Princis and his classification consists of three families. The largest (Blattidae) contains 15 contemporary subfamilies of which nine are oviparous, five ovoviviparous, and one is viviparous. Of the other families, the Corydiidae contains three oviparous polyphagid subfamilies, and the Panesthiidae consists of the Panesthiinae (ovoviviparous) and Cryptocercinae (oviparous). The genera in the latter subfamilies inhabit logs, superficially resemble one another, and have been placed close together phyletically by most taxonomists. However, *Panesthia* and *Cryptocercus* are quite different internally and externally (13), and McKittrick (39, 40) separated them widely in her classification.

Princis (43), in criticizing Rehn's classification because it was based chiefly on the phylogeny of wings, pointed out that the phylogeny of the organs of flight does not necessarily reflect the phylogeny of the Blattaria. Based on studies of wing venation and several other external characters, Bey-Bienko (9) concluded that the subfamily Blaberinae represents one of the oldest contemporary groups of cockroaches. However, we now know that the ovoviviparous Blaberinae belong to the most highly evolved family of Blattaria (38) and that oviparous *Cryptocercus* is probably the most primitive living cockroach (39). Princis (43) also criticized all earlier classifications, which he reviewed, in that they contain systematic groupings which cannot be considered monophyletic and consequently cannot be justifiably placed in a phylogenetic system. Singling out Bey-Bienko's and Rehn's Blattidae, he stated, "These groupings plainly cannot be considered as monophyletically related groups." Princis' (43) classification and his monumental catalogues (44) are a vast improvement over previous efforts, but even his system intermingles oviparous and ovoviviparous families, arising from two phyletic lines (Fig. 2).

McKittrick (38) selected four character systems for comparative study: (*a*) female genitalia and their musculature, (*b*) male genitalia, (*c*) proventriculus, and (*d*) oviposition behavior. However, whether or not the group is oviparous or ovoviviparous is the principal unifying character of McKittrick's system. Her classification (order: Dictyoptera; suborder: Blattaria) encompasses two phyletic lineages: in one, the superfamily Blattoidea, species remained oviparous; in the other, the Blaberoidea, species evolved ovoviviparity and viviparity (Fig. 3).

There are about 4000 to 5000 described species of cockroaches (42) and about 440 genera (43). McKittrick's (38) work was based on a study of approximately 100 species in 83 genera (though some were not available for study of all systems) which she placed in 5 families and 21 subfamilies. Future studies may require the erection of additional families and subfamilies. In spite of the relatively small number of species and genera she investigated, the taxonomic importance of McKittrick's classification has been pointed out by Gurney (24) and will provide a base for future comparative taxonomic and behavioral research. McKittrick's conclusions have gained support not only from my studies on various aspects of reproduction, but

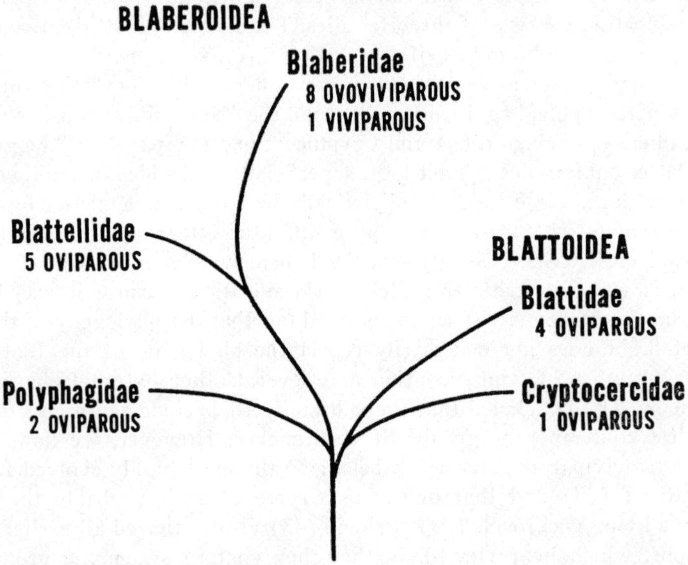

FIG. 3. Classification of the Dictyoptera, suborder Blattaria, according to McKittrick (38) and McKittrick & Mackerras (40). The numerals represent the number of subfamilies.

also from Huber (32) who analyzed her data numerically, and from Leconte et al. (37) who found that differences in the number of groups of Malpighian tubules allowed placement of species into one of three (representatives of the other two families were not available) of McKittrick's families. In addition, Graves (22) discovered marked differences in the spermatophores and their method of retention by the female in the Blattoidea and Blaberoidea, and showed that a bursa copulatrix for retaining the spermatophore in the female is present only in the more highly evolved Blaberidae.

THE EVOLUTION OF REPRODUCTION

> "It is of fundamental importance to all our thinking about reproduction that we realize that reproductive systems have not one but two main functions. The function of reproduction is reproduction, the production of like by like. But most reproductive systems involve *sexual* reproduction, and *the* function of *sexual* reproduction is not reproduction, but the controlled generation of *variety*."
>
> Thoday (91)

MODES OF REPRODUCTION

The following are three basic types of reproduction in cockroaches (82):

Oviparity.—The fertilized eggs develop outside the body of the female. The oötheca (egg case and its enclosed eggs) is carried externally for various periods of time but is often deposited about 24 hours or a few days after its formation. At oviposition, the oötheca may or may not contain sufficient water for development of the eggs. If the amount of water present is insufficient then additional water must be obtained from the substrate. In certain species of a few genera (*Blattella, Chorisia,* and *Lophoblatta*) of Blattellidae, the oötheca is carried externally during the entire developmental period of the eggs and in these species the eggs obtain water from the mother.

Ovoviviparity.—The eggs are first extruded, as in oviparous species, and then retracted into a uterus or brood sac. There they remain until the embryos mature, at which time the oötheca is extruded and hatching nymphs drop free from their embryonic membranes and oötheca. The eggs have sufficient yolk, but must absorb water from the female to complete development.

Viviparity.—Oviposition is similar to ovoviviparous forms except that yolk in the eggs at oviposition is insufficient to allow development. The eggs absorb water and other nutriments since they increase markedly in dry weight during embryogenesis. *Diploptera punctata* (Eschscholtz) is the only known viviparous cockroach (79). Experimental reduction in the number of embryos in the uterus of *Diploptera* results in larger newborn nymphs, whereas this does not occur in ovoviviparous species (68).

Only species of McKittrick's Blaberidae are ovoviviparous and viviparous; all species belonging to the other four of her families are oviparous. It has been suggested (76, 82) that the terms "false ovoviviparity" and "false viviparity" be used to designate types 2 and 3 above because in both groups the oötheca is first extruded to the outside, then retracted into the uterus. To be brief I shall not use the prefix in this review.

Oviposition

More than 50 years have elapsed since Shelford (87) stated that ". . . the egg-laying habits can be of considerable use in any scheme of classification of the Blattidae." At the time, no one had observed complete oviposition by any Blaberidae, although it was known that some blaberids carried oöthecae internally. Before Chopard's (12) description of oviposition in *Gromphadorhina laevigata* Saussure and Zehntner, it was believed that ovoviviparous species simply retained their eggs in a uterus. However, Chopard described the extrusion and retraction of the oötheca, behavior that now has been observed in half a dozen genera of blaberids (41, 74, 79). In addition to the genera listed in these references, I have observed extrusion and retraction of the oötheca in *Epilampra, Lanxoblatta, Rhabdoblatta,* and two undescribed genera; undoubtedly, this behavior is characteristic of all members of the family. It would be safe to say that any species which carries its egg case internally is a blaberid, and oviposits by extruding and retracting the oötheca.

In all cockroaches, mature ovarian eggs are extruded two by two with their micropylar ends facing upwards, surrounded by the initially soft

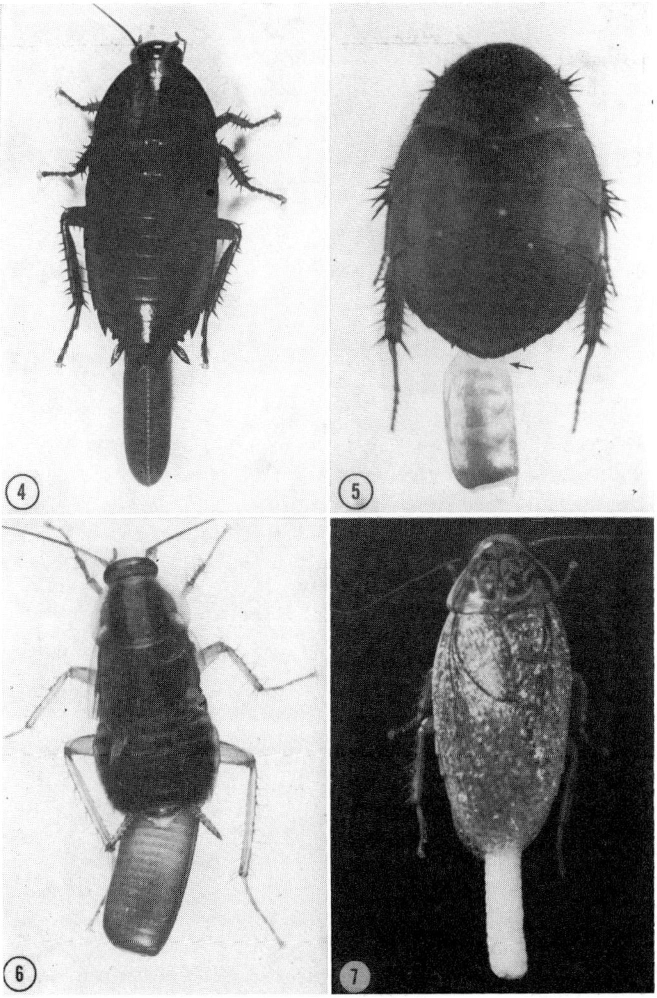

FIGS. 4–7. Method of carrying the oötheca, at the time it is deposited, by different species of cockroaches. 4. *Eurycotis floridana* (Walker) (Blattidae, Polyzosteriinae). No rotation [from (75)]. 5. *Arenivaga* (*Arenivaga*) sp. *A* (Polyphagidae). Primitive rotation; arrow points to flange by which the oötheca is held between the paraprocts [from (52)]. 6. *Blattella germanica* (Blattellidae, Blattellinae). Advanced rotation; the wings have been removed to show the end of the abdomen where the anterior eggs of the oötheca lie within the vestibulum of the female [from (52)]. 7. *Nauphoeta cinerea* (Blaberidae). Advanced rotation prior to retraction of the oötheca into the uterus [from (74)].

oöthecal membrane. When completely formed and still attached to the female, the oötheca is always oriented with the keel pointing dorsally. The position in which the oötheca is carried at the time it is deposited varies and is significant taxonomically (38, 52). Members of some families and subfamilies, hold the oötheca vertically (Fig. 4) at the time they deposit it, whereas others rotate the egg case 90° so that the keel faces laterally (Figs. 5–7). Recently, the taxonomic importance of rotation was pointed out by Gurney & Roth (25) who observed that the blattellid *"Loboptera" thaxteri* Hebard, originally in the Blattellinae, did not rotate its oötheca as do other *Loboptera. Thaxteri* was placed in a new genus and different subfamily (nonrotating Plectopterinae) as a result of their study.

In none of the Blattoidea (Cryptocercidae and Blattidae), the phyletic line which remained oviparous, is the oötheca rotated (Fig. 4). Rotation occurs only in some Blaberoidea (Polyphagidae, Blattellidae, and Blaberidae) (38); it may be considered in the Blattellidae a preadaptation for the evolution of internal incubation in the more highly evolved Blaberidae (52). The adaptive significance of rotation probably differs between the oviparous and ovoviviparous species. In oviparous forms, rotating the oötheca (usually taller than wide) (Fig. 6) probably allows the female to crawl into narrow crevices while carrying it, and reduces the possibility of accidental or premature dislodgement. In oviparous *Lophoblatta brevis* Rehn and *Lophoblatta arlei* Albuquerque, the oötheca, carried vertically by the female until the eggs hatch, is flattened dorsoventrally. Apparently, in these nonrotating plectopterines, reduction in height of the oötheca serves the same function as rotation in other oviparous species (58). In the ovoviviparous Blaberidae rotation (Fig. 7) reorients the eggs so that once retracted their long axes lie in the plane of the cockroach's width. This allows for the eggs' eventual growth, stretching the uterus, principally in a lateral direction, in these relatively flat insects (52).

Two kinds of rotation have evolved (Fig. 8). Primitive rotation occurs in some Polyphagidae, wherein none of the anterior eggs of the oötheca are held inside the female's vestibulum; the oötheca has a flange (Fig. 12) or handle by which it is held between her paraprocts (Fig. 5). Rotation in some Blattellidae (Blattellinae, Ectobiinae, and Nyctiborinae) (Fig. 6) and all Blaberidae (Fig. 7) is of the advanced type because, when rotated, the anterior eggs in the oötheca are closely adpressed to the tissues of the female's vestibulum. Ovoviviparity evolved from a species with advanced rotation because, in intermediate forms like *Blattella* and *Chorisia*, close contact of the oötheca with vestibular tissue was necessary for transfer of water from mother to developing eggs (52).

Oöthecae

The oöthecae of oviparous species usually are hard, rigid structures. The retraction of the eggs into a uterus in internal incubators necessitated dras-

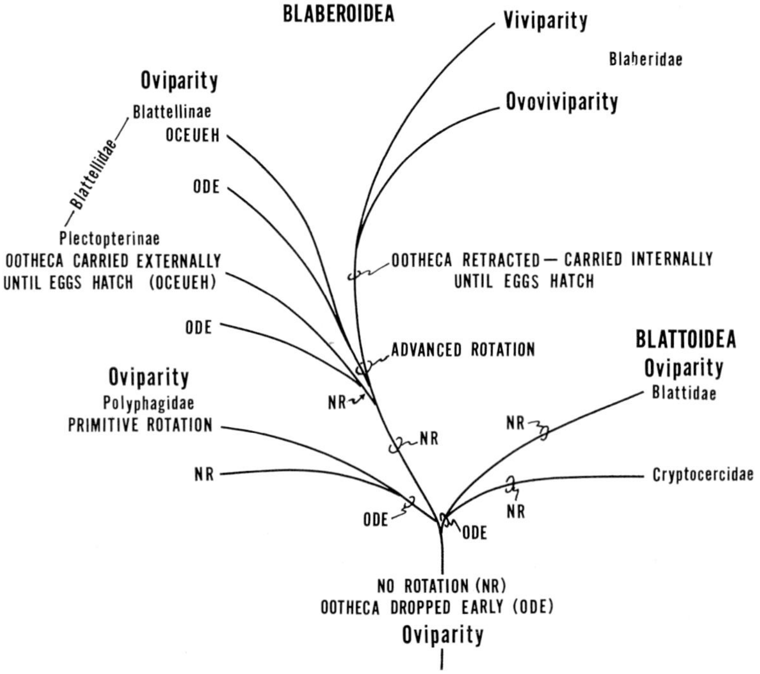

FIG. 8. The evolution of oviposition behavior in the Blattaria [modified from (56)].

tic changes in the oötheca; such changes occur in the Blaberoidea (Polyphagidae, Blattellidae, Blaberidae) but not in the Blattoidea (56).

The oöthecae of all Blattinae and Polyzosteriinae (Blattidae) are remarkably alike. They are hard and rigid with high well-developed keels containing distinct respiratory tubes (Fig. 11). The oötheca of *Lamproblatta* (Fig. 10) resembles that of *Cryptocercus* (Fig. 9) more than other Blattidae. Although the former is in the Blattidae (Lamproblattinae), it has certain characters that indicate an affinity to the Cryptocercidae (39).

In the Blaberoidea, keels and an anterior flange occur in polyphagids. It was in primitive polyphagids that respiratory tubes and denticles arose. The flange (Fig. 12) became markedly reduced in some Polyphagidae and was lost completely in most Blattellidae (Figs. 13–15), allowing close contact of the anterior eggs with the female's vestibular tissues (Fig. 6). In some blattellids the height of the denticles of the seam was reduced, and the seam itself came to lie flat against the micropylar ends of the eggs (Fig. 15). In the Blaberidae the keel usually is nonexistent (Figs. 16, 18), although it may occur in a few species as a reduced relic structure. The oöthecae of all internal incubators are flexible, relatively thin membranes (Fig. 166) compared to those of most oviparous species (56).

The oöthecae of most oviparous species contain dense accumulations of calcium oxalate crystals (10, 26, 56, 89), which apparently contribute to

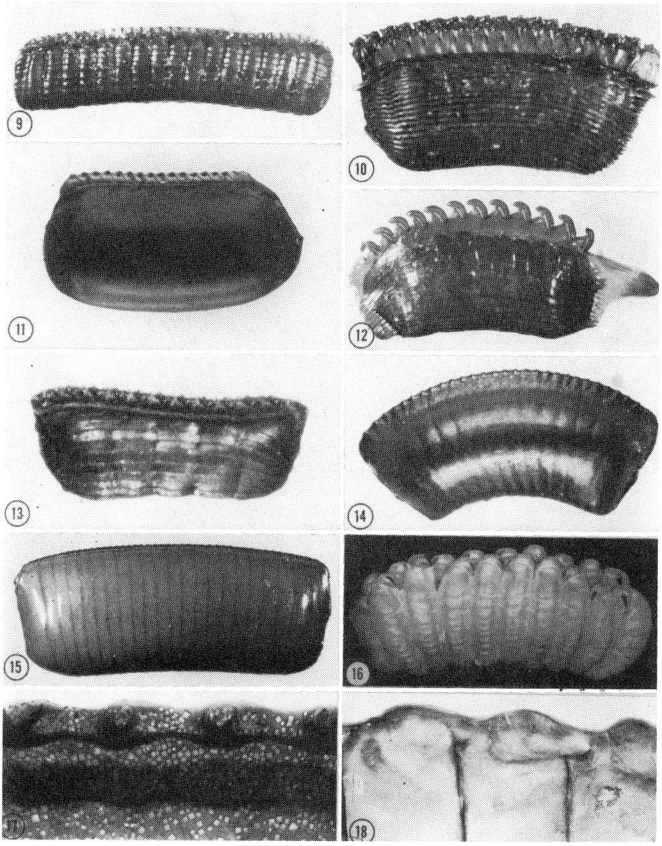

FIG. 9–18. Oöthecae of Blattaria. 9–11. Blattoidea. 9. *Cryptocercus punctulatus* Scudder (Cryptocercidae). 10. *Lamproblatta albipalpus* Hebard (Blattidae, Lamproblattinae). 11. *Blatta orientalis* Linn. (Blattidae, Blattinae). 12–18. Blaberoidea. 12. *Polyphaga aegyptiaca* (Linn.) (Polyphagidae). (Note the elongated flange on the right side.) 13–15. Blattellidae. 13. *Anaplecta hemiscotia* Hebard (Anaplectinae). 14. *Ischnoptera panamae* (Brunner) (Blattellinae). 15. *Blattella germanica* (Linn.) (Blattellinae). 16. Blaberidae. *Capucina patula* (Walker). 17–18. Portions of the keel region of oöthecae. 17. *Ischnoptera deropeltiformis* (Brunner). Note calcium oxalate crystals. 18. *Phortioeca phoraspoides* (Walker) (Blaberidae). Note absence of calcium oxalate crystals. [from (56)].

their hardness (46). In the Blattoidea, the crystals probably arose in the Lamproblattinae and occur in very large numbers in all other Blattidae (Blattinae and Polyzosteriinae). In the Blaberoidea, calcium oxalate crystals (Fig. 17) arose in the Polyphagidae and are numerous in most Blattellidae. Species that carry their oöthecae externally during embryogenesis, however, usually have very few crystals. Species of *Blattella,* for example, can be arranged in a series that run from gradual loss to complete absence of calcium oxalate in the oöthecal walls. The oöthecae of all Blaberidae lack calcium oxalate (Fig. 18). In the oviviviparous line, one can trace in relatively gradual steps changes from a hard rigid oöthecal membrane to a flexible, soft covering that allows the eggs to be retracted into the female's brood sac. Greatest reduction in the oöthecal membrane occurs in the viviparous *Diploptera punctata* (56).

Water changes in the oötheca.—According to Laurentiaux (34–36), the evolution of an oötheca to protect eggs from desiccation was largely responsible for the survival of cockroaches during the disruptive climatic changes of the Permian period. The oöthecae of Blattidae have a waterproofing layer that permits the eggs to develop even at low humidities (76–78). The evolution of ovoviviparity ". . . is meaningful only if the oöthecae of the forerunners of the Blaberidae were, in earlier times, more permeable than that of other forms, . . ." (53). Presumably, an impermeable oötheca would not require for survival a change to ovoviviparity. Some blattellids make hard, thick oöthecae which lose water rapidly unless they are deposited in a high humidity or actually placed in contact with water. Originally, a permeable oötheca may have been deposited in high humidities or on wet substrates where water loss would be negligible. With a change in climatic or microclimatic conditions from wet to dry, survival of the eggs would be dependent on the evolution of a protective device. This could be either the formation of a waterproofing layer over the oötheca, or a change in oviposition behavior whereby the female retained a more permeable oötheca in the vestibule or retracted it into the uterus. It is probable that ovoviviparity arose from a stock having an oötheca incapable of retaining water even in high humidities (53).

The water content of the oötheca at oviposition varies between the two phyletic lines. Almost all Blattidae oöthecae have high water contents of 60 to 68 per cent. In the Blaberoidea, the polyphagids have oöthecae whose initial water content varies from 32 per cent to 42 per cent. In the Blattellidae, the oöthecae of most of Plectopterinae (nonrotators) have less than 50 per cent water, whereas most of the Blattellinae (rotators) have more than 50 per cent. Oöthecae of all ovoviviparous Blaberidae contain 33 to 40 per cent water, whereas the one viviparous species *D. punctata* has 65 per cent water. The eggs of all species that contain less than 50 per cent water at oviposition must obtain additional water from the substrate in order to develop. The shapes of the water uptake curves of blattellid oöthecae kept on a moist substrate vary depending on water content at oviposition. Those having less than 50 per cent water have an S-shaped curve of water uptake

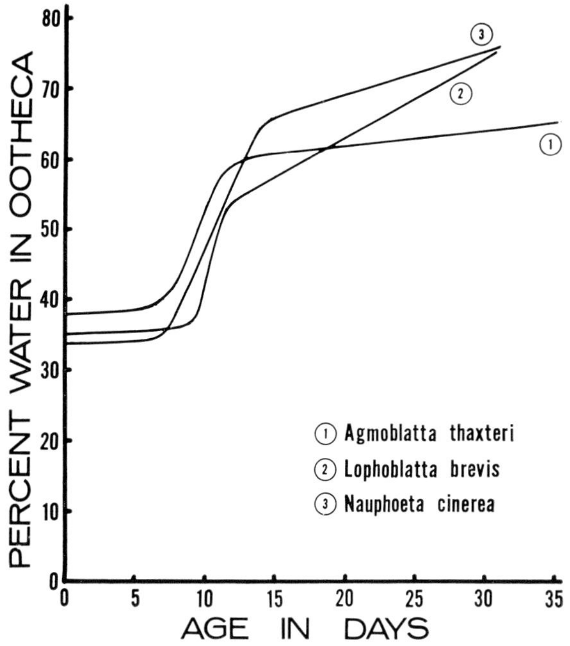

FIG. 19. Changes in water content, with age, in the oöthecae of 3 species of Blattaria with different oviposition behavior. All have less than 40% water initially and the curves are sigmoid-shaped. *A. thaxteri* (Hebard) (Blattellidae, Plectopterinae), deposits its oötheca shortly after forming it [from (53)]. *L. brevis* Rehn (Blattellidae, Plectopterinae), carries its oötheca externally during the entire embryogenetic period [from (58)]. *N. cinerea* (Olivier) (Blaberidae, Oxyhaloinae), is ovoviviparous and carries its oötheca internally during the entire embryogenetic period [from (53)].

(Fig. 19). Those having more than 50 per cent water show a gradual increase in the percentage of water toward the end of embryogenesis (Fig. 20). Similar patterns of water uptake during embryogenesis among species with different oviposition behavior, suggest that ovoviviparous forms arose from a stock whose oöthecae initially had a low percentage of water (Fig. 19). Although viviparity probably evolved from an ovoviviparous form (57), present evidence suggests that it may have arisen from a blattellidlike stock whose oötheca contained more than 50 per cent water, and was carried externally for the entire embryogenetic period (Fig. 20) (53). There are, as yet, no known ovoviviparous blaberids with oöthecae which, when first oviposited, contain nearly as high a percentage of water as *Diploptera*.

OVARIES

A fundamental difference between oviparous and ovoviviparous cockroaches is the frequency with which they oviposit. Oviparous species that

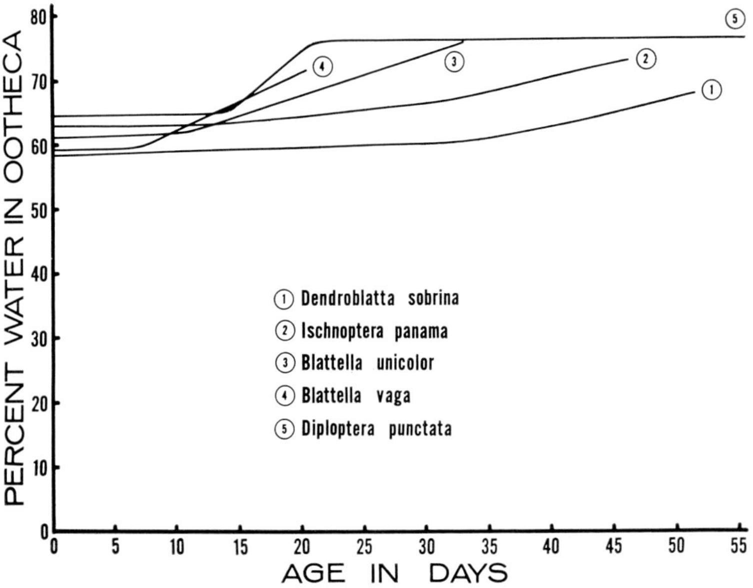

FIG. 20. Changes in water content with age, in the oöthecae of species of Blattaria with 3 different types of oviposition behavior. All these species have more than 50% water initially. *D. sobrina* Rehn and *I. panamae* Hebard (Blattellidae, Blattellinae) deposit their oöthecae shortly after forming them. *B. unicolor* (Brunner), and *B. vaga* Hebard (Blattellidae, Blattellinae), carry their oöthecae externally until the eggs hatch. *D. punctata* (Eschscholtz) (Blaberidae, Diplopterinae) is viviparous [from (53)].

carry their oöthecae for a short time form egg cases rather frequently. Oviparous species that carry their oöthecae externally for the entire embryogenetic period (e.g., *Blattella* spp.), as well as all Blaberidae produce relatively few oöthecae (Fig. 21) because the oöcytes do not mature in the ovaries during the period an oötheca is being carried (70, 72).

There is a fundamental difference in the ovaries of the oviparous and ovoviviparous species which may account, in part, for the differences in frequency of oviposition between species. The ovaries consist of a number of ovarioles each containing oöcytes in various stages of development. Bonhag (9a) divided the ovariole into several zones of which Zone V consists of oöcytes which contain yolk, at the time of oviposition. The number of oöcytes in Zone V varies between some members of the two phyletic lines.

The Blaberoidea exhibit a distinct evolutionary trend toward a decrease in the number of oöcytes in Zone V; the primitive Polyphagidae have 2 or 3, the Blattellidae, 1 or 2. The ovarioles of oviparous species that carry their oöthecae full term [e.g., *Blattella* spp.; *Chorisia trivirgata* (Werner) (= *C. fulvotestacea* Princis, as the latter was used in 52); *Lophoblatta*

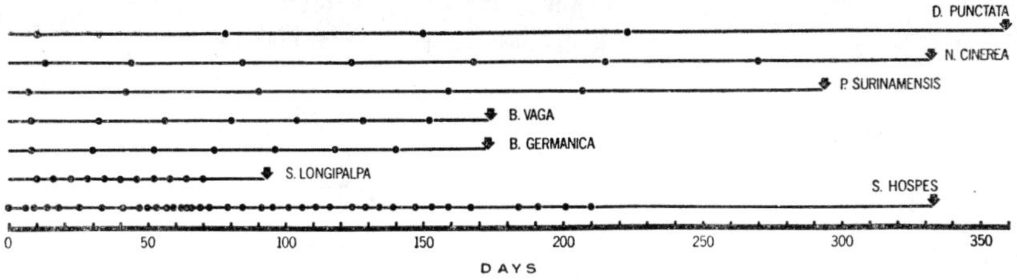

FIG. 21. Frequency of oviposition by individuals of different species of cockroaches. Each circle represents the formation of an oötheca. Arrow indicates time of death. The data for oviparous *Symploce hospes* (Perkins) (Battellidae, Blattellinae) is from Illingworth (33); the female was collected in the field and day 0 represents the day when collected. The data for other species [from (94)] are from females whose time of emergence was known and therefore represents the full lifetime of the individual. *Symploce hospes* and *Supella longipalpa* (Fab.) (Blattellidae, Plectopterinae) are oviparous and drop the oötheca shortly after it is formed. *Blattella germanica* and *B. vaga* carry their oöthecae externally until the eggs hatch. The blaberids *Pycnoscelus surinamensis* (parthenogenetic) and *Nauphoeta cinerea* are ovoviviparous and *Diploptera punctata* is viviparous.

brevis; and *L. arlei*] have few oöcytes per ovariole and only one in Zone V, ovaries which are similar to those of certain blaberids (57, 58). In all Blaberidae there is only one oöcyte in Zone V, and usually a reduced number of oöcytes per ovariole (57).

Oviparous species having more than one oöcyte in Zone V would be expected to oviposit more frequently than species having only one because, at the time the oötheca is completed, the new basal oöcytes already contain yolk and take less time to mature than do oöcytes lacking yolk. This is exemplified by the differences in rate of oviposition between *Symploce hospes* (2 oöcytes in Zone V) and *Supella longipalpa* (1 oöcyte in Zone V); both are oviparous Blattellidae (Fig. 21).

The Blattoidea and Blaberoidea probably arose from a stock in which 3 or more oöcytes were present in Zone V. In the Blattoidea there has been a reduction in Zone V oöcytes from 3 (in Cryptocercidae) to 2 (in all Blattidae), but there are no known Blattidae which have only one oöcyte in Zone V (57).

Two alternate pathways for the evolution of ovarioles in the Blaberoidea are shown in Figure 22. These pathways are based on the hypothesis that ovoviviparity arose from forms whose oöthecae initially contained less than 50 per cent water, and that viviparity evolved from species having more than 50 per cent water in their oöthecae at the time of oviposition. The examples of ovarioles shown in the Blattellidae (Fig. 22) are from species whose oöthecae have $<$ 50 per cent water in the ovoviviparous line and $>$ 50 per cent water in the viviparous line. McKittrick (personal communication) believes that the similarity between the genitalia of viviparous *Diplop-*

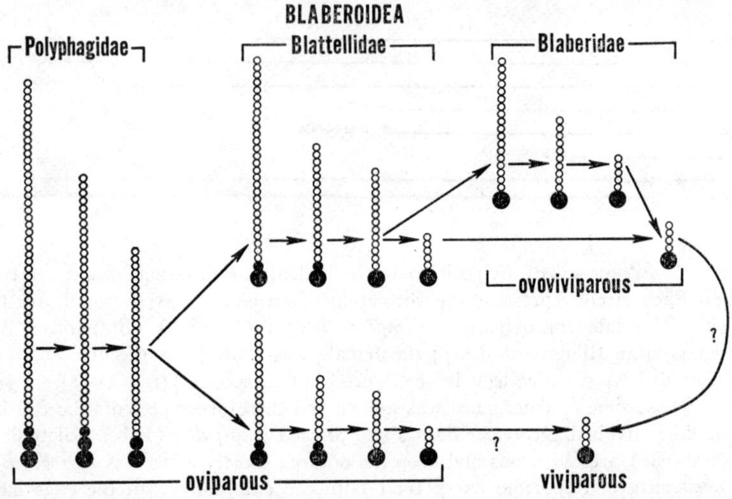

FIG. 22. Possible pathways in the evolution of the ovarioles of the Blaberoidea. Solid circles represent oöcytes in Zone V, and open circles are öocytes in Zone IV. Each symbol represents one oöcyte [modified from (57)].

tera with other ovoviviparous Blaberidae is strong evidence that viviparity evolved from the same stock as the rest of the family.

Testes

The shape of the follicles, their arrangement and point of attachment to the vas deferens differs markedly between members of the Blaberoidea and Blattoidea. The greatest variation in the testes occurs in species of Blattellidae (45).

Uricose Glands

The accessory sex glands in male cockroaches secrete compounds which form the sperm capsule or spermatophore. In some species, however, specialized tubules of the glands store a large amount of uric acid which is poured over the spermatophore after being inserted in the female. In these species, the uricose glands serve as storage excretory organs of uric acid before mating, and as active excretory organs during copulation (65, 66).

Uric acid poured over the spermatophore may have served to protect the sperm capsule from being eaten by the female or other insects. Uricose glands are absent in the Blattoidea; they evolved in the Blaberoidea, probably in plectopterine-blattelline, and blaberid stocks. The glands occur commonly in the Plectopterinae, less frequently in the Blattellinae, and in the Blaberidae, only in some Epilamprinae (54). The female's bursa copulatrix, a deep pouch for housing the spermatophore, evolved only in certain Blaberidae (22). Since there is no need for a protective coating of uric acid for spermatophores deeply inserted in a bursa copulatrix, most Blaberidae lack uricose glands (54).

Male Tergal Glands

On the backs of male Blattaria, a great variety of glandular structures have evolved (60) which serve to maneuver the female into proper precopulatory position (67).

In the Blattoidea (Blattidae) the glands, when present, are almost always on the first abdominal tergite. In the Blaberoidea, on the other hand, there is considerable variation in the number of specialized segments and in the positions on the abdomen they occupy. Tergal glands, rare in the Polyphagidae and Blaberidae, show their greatest development and variation in the Blattellidae. In this family, as many as five segments may be specialized, but the greatest number of blattellid genera have only one tergal gland, usually on segment 7 or segment 1. "The distribution and frequency of occurrence of the tergal glands in the Blattellidae suggests that in the course of evolution there has been a reduction in the number of segments which are specialized and a change in the location of the glands from a posterior to a more anterior position on the abdomen. The "best" position for the tergal glands would be anterior, because when the female places her mouthparts in or palpates a gland located anteriorly on the abdomen, the genitalia of both sexes are relatively close together and would allow the male to grasp the female more readily." (60).

Mating Behavior

With few exceptions, the mating behavior of cockroaches has certain basic elements characteristic of members of both phyletic lines. A sexually stimulated winged male raises his wings, and the responsive female mounts and palpates or "feeds" on his exposed tergites or on tergal gland secretions. When connection is made, in all cockroaches studied, the pair assumes an opposed position with heads facing opposite directions (2, 5, 60, 62, 73, 74, 81, 90).

The few cases of unusual or anomalous mating behavior in cockroaches are found in the highly evolved Blaberidae. There is little preliminary courtship in *Pycnoscelus indicus* (Fab.) wherein the male mounts the female prior to copulation (62, 83). In *Panchlora nivea* (Linn.) (83), *Panchlora irrorata* Hebard (93), and *Gromphadorhina portentosa* (Schaum) (8), the male backs into the female to connect.

Highly developed stridulating structures on the pronotum and tegmina of cockroaches have been found only in blaberid species of the Oxyhaloinae and Panchlorinae (59, 69). Stridulation by the male during courtship, however, has been observed only in *Nauphoeta cinerea* (Olivier); there is no evidence that the male's distinctive sound made when the female is unreceptive, influences her behavior in any way (29, 30).

Parthenogenesis

Although facultative parthenogenesis can occur in several species of cockroaches (80), obligatory parthenogenesis is known in only one, the blaberid *Pycnoscelus surinamensis* (Linn.). At one time, *P. surinamensis* consisted of parthenogenetic and bisexual "strains" (85), but recently I applied the name *Pycnoscelus indicus* (Fab.) to sexually reproducing individuals because the parthenogenetic and sexual species are sexually isolated. The apomictic parthenogenetic female of *surinamensis* cannot reproduce sexually, and the sexual *indicus* cannot reproduce parthenogenetically (55).

Pycnoscelus indicus has chromosomal polymorphism having females with $2n = 36$ and 38, and males with $2n = 35$ and 37. *Pycnoscelus surinamensis* also is cytogenetically polymorphic, may be aneuploid or polyploid, and probably arose polyphyletically from *P. indicus*. The modal numbers of chromosomes of *P. surinamensis* from different geographic areas are 34 (Brazil, Australia, Thailand), 35 (Thailand), 37 (Indonesia), 39 (Brazil), 53 (United States, Brazil), and 54 (Panama, Jamaica, Africa) (63).

P. indicus and *P. surinamensis* are sibling species and often are virtually indistinguishable from one another. Although sometimes the relatively short wings and tegmina can be used to distinguish *indicus* from the long-winged *surinamensis,* this does not always hold true (85). Recently, I have established a culture of *P. indicus* from Thailand which, in external morphology, looks exactly like *surinamensis,* yet reproduces only sexually. Of particular interest is its chromosome number (2n ♂ = 33, ♀ = 34) which differs from our Hawaiian colony of *indicus* (64). Because *Pycnoscelus indicus* is a species with intraspecific variation in chromosome number it should be well suited to studies of how differences in chromosome number have arisen in the course of evolution.

Separation of the sibling species of *Pycnoscelus* would never have been accomplished without an investigation of their reproductive biology. Virgin *P. indicus* drop their newly formed oöthecae without retracting them (55, 85). One reason why *indicus* cannot reproduce parthenogenetically, even if the eggs were capable of doing so, is that sperm in the spermathecae are necessary for normal retraction of the oötheca into the uterus (88). Parthenogenetic *surinamensis* no longer requires sperm for retraction of the oöthcea. Normal oviposition occurs whether or not sperm are present in the spermathecae (55, 85). Females of both species produce a sex attractant that stimulates males of *P. indicus*. However, although sex pheromone production is controlled by the corpora allata in *indicus,* it is not in *surinamensis.* The parthenogenetic females continue to produce sex pheromone even during pregnancy when their corpora allata are inactive (6). The production of sex pheromone by *P. surinamensis,* which has dispensed with the need for males, may be a relic character that persists from the sexual species from which it arose. Loss in hormonal control of attractant production may be the first step in eventual loss of the pheromone (55).

Chromosomes

Cockroaches show considerable cytotaxonomic instability in chromosone numbers. Striking differences exist between species of the same family or those of the same genus. The following range in female chromosome numbers (XX in female; XO in male) occurs in three of the five families studied (64): Blattidae (13 species), $2n = 18$ to 48; Blattellidae (41 species), $2n = 16$ to 50; Blaberidae (39 species), $2n = 24$ to 74.

The diploid number of female chromosomes in eight species of *Blattella* are 24 (4 species), 26, 28, 30, and 50. In the Blaberidae, two species of *Eublaberus* have 2n(♀) = 44 and 66, respectively. In five species of *Blaberus,* the chromosome numbers are 2n(♀) = 38, 40, and 74 (3 species). Unfortunately, chromosome numbers in the primitive Polyphagidae have been determined for only three species, but these indicate a wide range in this family ($2n = 22$, 46, 64?) (64). The greatest number of species having the highest chromosome numbers occur in the highly evolved Blaberidae. Only one blattellid has 50 chromosomes, whereas 14 species of Blaberidae have a diploid number of 50 or more.

CONCLUSIONS

As Anderson (1) points out, " . . . any change from an ancestral to a descendant group of animals involves a gradual series of many changes, each representing a functional intermediate between an old and a new type of morphological organization." Of McKittrick's two phyletic lines, the Blaberoidea shows a series of changes which have led clearly from oviparity to ovoviviparity and viviparity. It is true that not all the changes that have been described are gradual. For example, the change from an oviparous habit of dropping the oötheca, to carrying it during the entire embryogenetic period, to retracting it into a uterus in ovoviviparous species, probably occurred with intermediate steps (76). Additional species showing intermediate steps may yet be found, or they may have become extinct.

The Blattellidae include many more genera (43) than any of the other four families of McKittrick. Not only is it the largest family but also the most variable in several important reproductive features. There can be little doubt that the Blaberidae arose from a blattellidlike ancestor. It is within the Blattellidae that we find variations in (*a*) oviposition behavior (nonrotating and rotating forms, some of which carry the oötheca externally until the eggs hatch; (*b*) types of ovarioles (one or two oöcytes in Zone V); (*c*) water changes in the oötheca (sigmoid-shaped water uptake curves); and (*d*) structure of the oötheca (reduction in thickness, keel height, and loss of calcium oxalate), all features which anticipated characters found in the Blaberidae. Of interest is the fact that Princis (43) included in his Epilamproidea, the Epilampridae, Blattellidae, Nyctiboridae, and Ectobiidae. McKittrick made the latter two families subfamilies of her Blattellidae, and the ovoviviparous Epilampridae a subfamily of the Blaberidae. Some members of the Epilamprinae have uricose glands (54, 66) and keel-like relics of the oötheca (38, 56), emphasizing the relationship between epilamprids and blattellids (60).

Because rotation of the oötheca apparently was a necessary prerequisite to the evolution of ovoviviparity, Blattellinae-like species probably were the forerunners of the Blaberidae (52). *Lophoblatta brevis* carries its oötheca externally during embryogenesis, and its eggs have an S-shaped water uptake curve (Fig. 19). Both characters make it an excellent example linking the oviparous Blattellidae to the ovoviviparous Blaberidae. However, *Lophoblatta* is a plectopterine which does not rotate its oötheca, and the blattelline has yet to be found whose oötheca has a low water content, is carried for the entire embryogenetic period, and whose eggs have an S-shaped water-uptake curve (58).

The *Blattella* type of oviposition probably evolved from a Blattellinae type that retained the rotated oötheca for longer periods of time, culminating in external retention until the eggs hatched. The fact that the plectopterines *Lophoblatta brevis* and *L. arlei* carry their oöthecae vertically until the eggs hatch raises the possibility that *Blattella* oviposition behavior arose from a plectopterine that carried its oötheca for the entire embryogenetic period. However, in *Blattella* and all other Blattellinae, rotation of the oötheca occurs just after it is formed and is then carried in a horizontal position. If rotation in the Blattellinae evolved from a form like *Lophoblatta brevis* one would expect rotation to occur shortly before the eggs hatch and not just after the egg case was formed. It seems likely that in the Blattellidae, increased external retention of the oötheca evolved along parallel lines in both the Plectopterinae and Blattellinae, but rotation evolved only in the latter (and two other subfamilies, the Nyctiborinae and Ectobiinae) and in the Blaberidae. The ovaries of the Plectotopterinae also evolved along lines similar to the Blattellinae, since in both subfamilies, the ovarioles may contain one or two oöcytes with yolk at the time of oviposition.

It is a striking fact that the Blattoidea with relatively little variation in important reproductive characteristics have culminated in a stabilized oviparous type of reproduction. Blattoids show none of the changes which, in the Blaberoidea, can be interpreted as an evolution toward internal incubation of the eggs. Within the Blaberoidea one can see evolution in progress. Differences in behavior and morphology obviously indicate changes which have led from oviparity to ovoviviparity. Of the various classifications proposed for cockroaches, McKittrick's, which stresses reproduction, is logical and supported by considerable biological data. It should remain the basis for all future work on the taxonomy of the Blattaria.

Cockroaches are a fine example of Thoday's (91) statement that ". . . an organism must always be a compromise between present needs and past needs. Present physiology, for example, may well not be fully explicable without consideration of its relation to the past physiology which determined the limiting circumstances under which it evolved."

LITERATURE CITED

1. Anderson, D. T. Morphological integration and animal evolution. *Scientia,* **102,** 1–6 (1967)
2. Barth, R. H., Jr. *Comparative and Experimental Studies on Mating Behavior in Cockroaches.* (Doctoral dissertation, Harvard Univ., Cambridge, Mass., 1961)
3. Barth, R. H., Jr. Hormonal control of sex attractant production in the Cuban cockroach. *Science,* **133,** 1598–99 (1961)
4. Barth, R. H., Jr. The endocrine control of mating behavior in the cockroach *Byrsotria fumigata* (Guérin). *Gen. Comp. Endocrinol.,* **2,** 53–69 (1962)
5. Barth, R. H., Jr. The mating behavior of *Byrsotria fumigata* (Guérin). *Behaviour,* **23,** 1–30 (1964)
6. Barth, R. H., Jr. Insect mating behavior: endocrine control of a chemical communication system. *Science,* **149,** 882–83 (1965)
7. Barth, R. H., Jr. The comparative physiology of reproductive processes in cockroaches. Part I. Mating behaviour and its endocrine control. In *Advances in Reproductive Physiology,* **3,** 167–207 (McLaren, A., Ed., Academic, New York, 1968)
8. Barth, R. H., Jr. The mating behavior of *Gromphadorhina portentosa* (Schaum): an anomalous pattern for a cockroach. *Psyche,* **75,** 124–31 (1968)
9. Bey-Bienko, G. Ia. Fauna of the U.S.S.R. Insects Blattodea. (In Russian) *Zool. Inst. Akad. Nauk, S.S.S.R. Moskva,* n.s. No. 40, 343 pp. (1950) [Translated by Ruth Ericson]

9a. Bonhag, P. F. Histological and histochemical studies on the ovary of the American cockroach *Periplaneta americana* (L.). *Univ. Calif. Publ. Entomol.,* **16,** 81–124 (1959)

10. Brunet, P. C. J. The formation of the oötheca by *Periplaneta americana.* II. The structure and function of the left colleterial gland. *Quart. J. Microscop. Sci.,* **93,** 47–69 (1952)
11. Carpenter, F. M. A review of our present knowledge of the geological history of insects. *Psyche,* **37,** 15–34 (1930)
12. Chopard, L. Sur l'anatomie et le développement d'une blatte vivipare. *Intern. Congr. Entomol., 8th, Stockholm, 1948,* 218–22 (1950)
13. Cleveland, L. R., Hall, S. R., Sanders, E. P., Collier, J. The woodfeeding roach *Cryptocercus,* its Protozoa, and the symbiosis between Protozoa and roach. *Mem. Am. Acad. Arts Sci.,* **17,** 187–342 (1934)
14. Engelmann, F. Die Steuerung der Ovarfunktion bei der ovoviviparen Schabe *Leucophaea maderae* (Fab.). *J. Insect Physiol.,* **1,** 257–78 (1957)
15. Engelmann, F. The control of reproduction in *Diploptera punctata. Biol. Bull.,* **116,** 406–19 (1959)
16. Engelmann, F. Hormonal control of mating behavior in an insect. *Experientia,* **16,** 69–70 (1960)
17. Engelmann, F. Mechanisms controlling reproduction in two viviparous cockroaches. *Ann. N.Y. Acad. Sci.,* **89,** 516–36 (1960)
18. Engelmann, F. Further experiments on the regulation of the sexual cycle in females of *Leucophaea maderae. Gen. Comp. Endocrinol.,* **2,** 183–92 (1962)
19. Engelmann, F. The mode of regulation of the corpus allatum in adult insects. *Arch. Anat. Microscop. Morphol. Exptl.,* **54,** 387–404 (1965)
20. Engelmann, F., Barth, R. H., Jr. Endocrine control of female receptivity in *Leucophaea maderae. Ann. Entomol. Soc. Am.,* **61,** 503–5 (1968)
21. Engelmann, F., Rau, I. Correlation between feeding and the sexual cycle in *Leucophaea maderae. J. Insect Physiol.,* **11,** 53–64 (1965)
22. Graves, P. N. Spermatophores of the Blattaria. *Ann. Entomol. Soc. Am.,* **62,** 595–602 (1969)
23. Gurney, A. Review of "Classification of the Blattaria as indicated by their wings (Orthoptera)" by Rehn, J. W. H. *Entomol. News,* **62,** 202–7 (1951)
24. Gurney, A. Review of "Evolutionary studies of cockroaches" by F. A. McKittrick. *Bull. Entomol. Soc. Am.,* **11,** 101 (1965)
25. Gurney, A. B., Roth, L. M. Two new genera of South American cockroaches superficially resembling *Loboptera,* with notes on bionomics. *Psyche,* **73,** 196–207 (1966)

26. Hackman, R. H., Goldberg, M. Composition of the oöthecae of three Orthoptera. *J. Insect Physiol.*, **5**, 73–78 (1960)
27. Hagan, H. R. *Embryology of the Viviparous Insects.* (Ronald Press Co., New York, 472 pp., 1951)
28. Hagan, H. R. Balanced and unbalanced evolutionary sequences with observations on the hazards to fertilization in insects. *Biol. Rev. City College, New York,* **15**, 19–24 (1953)
29. Hartman, H. B., Roth, L. M. Stridulation by the cockroach *Nauphoeta cinerea* (Olivier) during courtship behaviour. *J. Insect Physiol.*, **13**, 579–86 (1967)
30. Hartman, H. B., Roth, L. M. Stridulation by a cockroach during courtship behaviour. *Nature,* **213**, 1243 (1967)
31 Holmgren, N. Ueber vivipare Inseckten. *Zool. Jahrb. Abt. Syst. Geograph., Biol. Tiere,* **19**, 431–68 (1903)
32. Huber, I. *Numerical Taxonomic Studies of Cockroaches.* (Doctoral dissertation, Univ. of Kansas, Lawrence, Kansas, 1968)
33. Illingworth, J. F. Notes on Hawaiian roaches. *Proc. Hawaiian Entomol. Soc.*, **3**, 136–40 (1915)
34. Laurentiaux, D. Le problème des Blattes Paléozoiques à ovipositeur externe. *Ann. Paléontol.*, **37**, 237–44 (1951)
35. Laurentiaux, D. La reproduction chez les insectes blattaires du Carbonifère : facteurs du panchronisme et classification naturelle de l'ordre. *Bull. Soc. Géol. France,* **1**, 759–66 (1959)
36. Laurentiaux, D. La reproduction chez les Blattes Carbonifères. Essai d'explication du panchronisme des Blattaires et classification sousordinale. *Compt. Rend. Acad. Sci., Paris,* **250**, 1700–2 (1960)
37. Leconte, O., Lefeuvre, J. C., Razet, P. Un nouveau critère taxonomique chez les Blattes : l'insertion des tubes de Malpighi. *Compt. Rend. Acad. Sci., Paris,* **265**, 1397–1400 (1967)
38. McKittrick, F. A. Evolutionary studies of cockroaches. *Cornell Univ. Agr. Expt. Sta. Mem.*, No. 389, 197 pp. (1964)
39. McKittrick, F. A. A contribution to the understanding of cockroach-termite affinities. *Ann. Entomol. Soc. Am.*, **58**, 18–22 (1965)
40. McKittrick, F. A., Mackerras, M. J. Phyletic relationships within the Blattidae. *Ann. Entomol. Soc. Am.*, **58**, 224–30 (1965)
41. Nutting, W. L. Observations on the reproduction of the giant cockroach, *Blaberus craniifera* Burm. *Psyche,* **60**, 6–14 (1953)
42. Princis, K. (Personal communication)
43. Princis, K. Zur Systematik der Blattarien. *Eos,* **36**, 427–49 (1960)
44. Princis, K. *Orthopterorum Catalogus. 's-Gravenhage.* Pars 3, 1–74 (1962) ; Pars 4, 77–172 (1963) ; Pars 6, 174–281 (1964) ; Pars 7, 284–400 (1965) ; Pars 8, 402–614 (1966) ; Pars 11, 616–710 (1967)
45. Quiaoit, E. R. *An investigation of growth, development and dimorphism in cockroaches.* (Doctoral Dissertation, Kansas State Univ., Manhattan, Kansas, 1961)
46. Rajulu, G. S., Renganathan, K. On the stabilization of the oötheca of the cockroach *Periplaneta americana. Naturwissenschaften,* **53**, 136 (1966)
47. Rehn, J. A. G. On apterism and subapterism in the Blattinae. *Entomol. News,* **43**, 201–6 (1932)
48. Rehn, J. W. H. Classification of the Blattaria as indicated by their wings. *Mem. Am. Entomol. Soc.*, **14**, 1–134 (1951)
49. Roth, L. M. Hypersexual activity induced in females of the cockroach *Nauphoeta cinerea. Science,* **138**, 1267–69 (1962)
50. Roth, L. M. Control of reproduction in female cockroaches with special reference to *Nauphoeta cinerea.* I. First preoviposition period. *J. Insect Physiol.*, **10**, 915–45 (1964)
51. Roth, L. M. Control of reproduction in female cockroaches with special reference to *Nauphoeta cinerea.* II. Gestation and postparturition. *Psyche,* **71**, 198–244 (1964)
52. Roth, L. M. The evolutionary significance of rotation of the oötheca in the Blattaria. *Psyche,* **74**, 85–103 (1967)
53. Roth, L. M. Water changes in cockroach oöthecae in relation to the evolution of ovoviviparity and viviparity. *Ann. Entomol. Soc. Am.*, **60**, 928–46 (1967)

54. Roth, L. M. Uricose glands in the accessory sex gland complex of male Blattaria. *Ann. Entomol. Soc. Am.*, **60**, 1203–11 (1967)
55. Roth, L. M. Sexual isolation in parthenogenetic *Pycnoscelus surinamensis* and application of the name *Pycnoscelus indicus* to its bisexual relative. *Ann. Entomol. Soc. Am.*, **60**, 774–79 (1967)
56. Roth, L. M. Oöthecae of the Blattaria. *Ann. Entomol. Soc. Am.*, **61**, 83–111 (1968)
57. Roth, L. M. Ovarioles of the Blattaria. *Ann. Entomol. Soc. Am.*, **61**, 132–40 (1968)
58. Roth, L. M. Oviposition behavior and water changes in the oöthecae of *Lophoblatta brevis*. *Psyche*, **75**, 99–106 (1968)
59. Roth, L. M. Reproduction of some poorly known species of Blattaria. *Ann. Entomol. Soc. Am.*, **61**, 571–79 (1968)
60. Roth, L. M. The evolution of male tergal glands in the Blattaria. *Ann. Entomol. Soc. Am.*, **62**, 176–208 (1969)
61. Roth, L. M., Barth, R. H., Jr. The control of sexual receptivity in female cockroaches. *J. Insect Physiol.*, **10**, 965–75 (1964)
62. Roth, L. M., Barth, R. H., Jr. The sense organs employed by cockroaches in mating behavior. *Behaviour*, **28**, 58–94 (1967)
63. Roth, L. M., Cohen, S. H. Chromosomes of the *Pycnoscelus indicus* and *P. surinamensis* complex. *Psyche*, **75**, 54–76 (1968)
64. Roth, L. M., Cohen, S. H. (Unpublished data)
65. Roth, L. M., Dateo, G. P. Uric acid in the reproductive system of males of the cockroach *Blattella germanica*. *Science*, **146**, 782–84 (1964)
66. Roth, L. M., Dateo, G. P. Uric acid storage and excretion by accessory sex glands of male cockroaches. *J. Insect. Physiol.*, **11**, 1023–29 (1965)
67. Roth, L. M., Dateo, G. P. A sex pheromone produced by males of the cockroach *Nauphoeta cinerea*. *J. Insect Physiol.*, **12**, 255–65 (1966)
68. Roth, L. M., Hahn, W. Size of newborn larvae of cockroaches incubating eggs internally. *J. Insect Physiol.*, **10**, 65–72 (1964)
69. Roth, L. M., Hartman, H. B. Sound production and its evolutionary significance in the Blattaria. *Ann. Entomol. Soc. Am.*, **60**, 740–52 (1967)
70. Roth, L. M., Stay, B. Control of oöcyte development in cockroaches. *Science*, **130**, 271–72 (1959)
71. Roth, L. M., Stay, B. Oöcyte development in *Diploptera punctata* (Eschscholtz). *J. Insect Physiol.*, **7**, 186–202 (1961)
72. Roth, L. M., Stay, B. A comparative study of oöcyte development in false ovoviviparous cockroaches. *Psyche*, **69**, 165–208 (1962)
73. Roth, L. M., Willis, E. R. A study of cockroach behavior. *Am. Midland Naturalist*, **47**, 66–129 (1952)
74. Roth, L. M., Willis, E. R. The reproduction of cockroaches. *Smithsonian Inst. Misc. Collections*, **122**, 1–49 (1954)
75. Roth, L. M., Willis, E. R. *Anastatus floridanus* a new parasite on the eggs of the cockroach *Eurycotis floridana*. *Trans. Am. Entomol. Soc.*, **80**, 29–41 (1954)
76. Roth, L. M., Willis, E. R. The water content of cockroach eggs during embryogenesis in relation to oviposition behavior. *J. Exptl. Zool.*, **128**, 489–509 (1955)
77. Roth, L. M., Willis, E. R. Water relations of cockroach oöthecae. *J. Econ. Entomol.*, **48**, 33–36 (1955)
78. Roth, L. M., Willis, E. R. Relation of water loss to the hatching of eggs from detached oöthecae of *Blattella germanica* (L.). *J. Econ. Entomol.*, **48**, 57–60 (1955)
79. Roth, L. M., Willis, E. R. Intrauterine nutrition of the "beetle-roach" *Diploptera dytiscoides* (Serv.) during embryogenesis, with notes on its biology in the laboratory. *Psyche*, **62**, 55–68 (1955)
80. Roth, L. M., Willis, E. R. Parthenogenesis in cockroaches. *Ann. Entomol. Soc. Am.*, **49**, 195–204 (1956)
81. Roth, L. M., Willis, E. R. Observations on the biology of *Ectobius pallidus* (Olivier). *Trans. Am. Entomol. Soc.*, **83**, 31–37 (1957)
82. Roth, L. M., Willis, E. R. An analysis of oviparity and viviparity in the Blattaria. *Trans. Am. Entomol.*

Soc., **83**, 221–38 (1958)

83. Roth, L. M., Willis, E. R. The biology of *Panchlora nivea* with observations on the eggs of other Blattaria. *Trans. Am. Entomol. Soc.*, **83**, 195–207 (1958)
84. Roth, L. M., Willis, E. R. The biotic associations of cockroaches. *Smithsonian Inst. Misc. Collections*, **141**, 1–470 (1960)
85. Roth, L. M., Willis, E. R. A study of bisexual and parthenogenetic strains of *Pycnoscelus surinamensis. Ann. Entomol. Soc. Am.*, **54**, 12–25 (1961)
86. Shelford, R. Studies of the Blattidae. VI. Viviparity amongst the Blattidae. *Trans. Entomol. Soc. London* (1906), 509–14 (1906)
87. Shelford, R. The oöthecae of Blattidae. *Entomol. Record*, **24**, 283–87 (1912)
88. Stay, B., Gelperin, A. Physiological basis of oviposition behaviour in the false ovoviviparous cockroach *Pycnoscelus surinamensis* (L.). *J. Insect Physiol.*, **12**, 1217–26 (1966)
89. Stay, B., King, A., Roth, L. M. Calcium oxalate in the oöthecae of cockroaches. *Ann. Entomol. Soc. Am.*, **53**, 79–86 (1960)
90. Stay, B., Roth, L. M. The reproductive behavior of *Diploptera punctata. Proc. Intern. Congr. Entomol., 10th*, 1956, **2**, 547–52 (1958)
91. Thoday, J. M. Genetics and the integration of reproductive systems. In *Insect Reproduction. Symp. Roy. Entomol. Soc. (London)*, No. 2, 108–19 (1964)
92. Tinbergen, N. On war and peace in animals and man. An ethologist's approach to the biology of aggression. *Science*, **160**, 1411–18 (1968)
93. Willis, E. R. Biology and behavior of *Panchlora irrorata*, a cockroach adventive on bananas. *Ann. Entomol. Soc. Am.*, **59**, 514–16 (1966)
94. Willis, E. R., Riser, G. R., Roth, L. M. Observations on reproduction and development in cockroaches. *Ann. Entomol. Soc. Am.*, **51**, 53–69 (1958)

INTERPRETATIONS OF QUATERNARY INSECT FOSSILS

By G. R. Coope

Department of Geology, The University of Birmingham, Birmingham, England

Introduction

Within the last fifteen years, there has been an enormous increase in our knowledge of Quaternary insect assemblages. This knowledge has provided a means of raising above the realm of speculation, our interpretations of faunal histories and insect speciation. To the geologist it has provided a sensitive tool with which to unravel the complex environmental changes that are so characteristic of this, the most recent episodes in the earth's history. There can be little doubt that with further work, fossil insect assemblages will facilitate the correlation of the highly complicated strata of this period, though within relatively restricted geographical limits.

Considering the almost overwhelming abundance of fossil insects in many deposits of this period and the wide variety of sediments in which they may be found, it is surprising that they have not attracted more attention in the past. Sporadic investigations have been made, it is true, but there has been no concerted attack on the problem and no continuous development of ideas. The frequent references to insect remains as incidental comments in palaeobotanical studies, testifies to the abundance of these fossils but further than this they were, to a large extent, ignored.

It is not difficult to find the reasons why entomologists were reluctant to involve themselves with Quaternary insect fossils. Two widely held, but erroneous beliefs were, I contend, largely to blame. The first of these was that insect species could not be recognised from isolated fragments because keys to their identification could not be used and the traditional diagnostic features were often missing. This belief is still widely held and the difficulties are indeed very real. Though many fragments must, of course, remain indeterminate for want of adequately distinctive features, by no means all specimens fall into this category, and diagnosis to the species level can be made in a surprisingly large number of cases with a high degree of certainty. The second belief that impeded the progress of Quaternary entomology, was that many species had evolved during this period and that, as a consequence, precise identification was in most cases impossible. This belief was founded largely on studies of the present-day distribution of various species and geographical races. It was reinforced by the work of some of the earlier entomologists who had claimed to recognise new species in even relatively late Quaternary deposits, (Lomnicki, 1894; Scudder, 1900). In the few instances in which the actual specimens upon which new species were based, are available, modern investigations show that in all cases they were ill-founded and that the specimens can be more properly equated with living species (Shotton et al. 1962; Coope 1968b).

I do not propose to give a detailed review here of the earlier work on Quaternary insect faunas mainly because in most cases it has not yet been possible to locate the specimens upon which these studies were based and, in the light of what has been said above, I feel that it is perhaps wisest to defer any discussion of their significance until they have been reinvestigated. A comprehensive bibliography of these earlier papers is given by Henriksen (1933) in his monumental monograph on the late Quaternary fossil insects from Denmark and Scania. Since Henriksen's pioneer work, there has been little systematic investigation of Quaternary insect assemblages until the last fifteen years. References to these studies will be made in the course of this paper.

The Nature and Preservation of Quaternary Insect Fossils

Fragments of insects have been recognised in peat deposits ever since these have been exploited commercially. Usually these were the remains of the larger species of Coleoptera whose shiny and often brilliantly coloured skeletons attracted the eye. However, numerous other deposits also contain insect remains though they are often less conspicuous. Thus, any silt that represents the infilling of a shallow pool or lake margin can usually be made to yield an extraordinary quantity of insect remains when it is treated in the laboratory. Lake sediments deposited at some distance from the shore, however, produce only a limited insect assemblage that is dominated by the larval heads of Chironomidae which, nevertheless, may occur in prodigious numbers. Riverine sediments are, on the whole, less productive, although they also yield fossils in places where a backwater, or the slow rate of flow, has permitted finer deposits to be laid down. In my experience it is organic silts, rather than the more traditional peats, that are the most productive of fossil insects because, on the whole, they are less intractable in the laboratory than are the felted peats. As a general rule, all deposits that contain macroscopic plant remains, in particular seeds, also contain fossil insects.

Almost all the earlier workers and most recent ones as well, relied on macroscopic methods for the collection of fossil insects. Samples of peat were split along the bedding planes and the insect fragments were dissected out. Though this method enabled associated fragments to be kept together, it resulted in the over-representation of the more conspicuous species and the almost total neglect of the small but often more numerous species (Coope, 1961). A cursory glance at most of the earlier faunal lists from Quaternary deposits will immediately betray this error in the sampling. A more representative assemblage can be obtained if the deposit is wet-sieved and the residue held on the sieves, then sorted under a binocular microscope. Of course, this process destroys any association that the fossil fragments may have had in the sediment, but if the disaggregation is done without undue vigor a number of joined heads and thoraces or complete abdomens can often be found. Using flotation methods, it is possible to separate the insect fossils from the bulk of the plant debris with which they are usually associated. This technique has been briefly described by Coope (1968a).

Fossil insects are best kept gummed to cards or loose in tubes of alcohol. Unless treated in this way, they distort badly as they dry out and, though they may be restored by the simple addition of water, they often break up or only partially return to their original shape. If they are left in water for more than a few days they are frequently attacked by fungi which can, in a very short time, enmesh the product of weeks of work inextricably in a mass of hyphae. Delicate structures such as the genital armature do not survive well in a dry condition and are best mounted in balsam or in DMHF (dimethylhydantoin formaldehyde resin). We have recently embedded fossils in various polymerising resins which preserve the colours that are almost invariably lost when the specimens dry. The procedure is, however, time-consuming and rather elaborate, and has been used only in critical cases in which the colours are important and where a permanent mount is needed, say, for transport by post.

Identification of Fossil Insect Fragments

Keys to the identification of modern species cannot be used in the identification of fossil insect remains, since the complete animal is only rarely available for study. In most cases, disarticulated skeletal elements are all that remain and identifications are made by comparing these with modern species. Some fragments are so devoid of diagnostic features that they are impossible of identification, but in many other cases the heads, thoraces and elytra display a wealth of useful characters that enable specific determinations to be made. When dealing with fossil material it is often necessary to examine the details of microsculpture, often adequately visible only at magnifications in excess of one hundred times, which can frequently provide a suite of characters that have often been overlooked by the authors of keys. We have recently used a scanning electron microscope to study details of microsculpture that are near to the limits of our optical apparatus, and such is the perfection of preservation of many fossil insects that magnifications of up to a thousand times can be used with profit. It is not uncommon to find complete abdomens from which it is possible to dissect out well-preserved genitalia and which, if cleared slightly in xylene, frequently show the details of the internal sclerites. Thus, the criteria used in the identification of fossil insects are exactly the same as those used by insect taxonomists on present-day species. The usual argument levelled against the palaeontologist, namely, that his concept of a species is suspect because he is dealing only with hard parts, carries little weight here. The concept of an insect species, whether living or fossil must, for the most part, rely on morphological differences of the exoskeleton. We are thus in a happy, though unfortunately almost unique, position wherein both paleontologists and neontologists use exactly the same bases upon which to establish their species.

When comparing modern with fossil specimens, great care must be taken to recognise postmortem changes that may afflict the latter. These take on a variety of forms. When the fossils dry out on their cards there is a tendency for some specimens to develop deepened puncturation, and on occasion prominences and hollows become somewhat flattened out. Specimens subject to such alterations are best kept in tubes of alcohol. In fossil specimens of some carabid beetles, spurious dimples develop on originally smooth interstices of the elytra and on the pronota. These curious features develop only on dried specimens but are, in some instances, so constantly and regularly displayed that, at first, it was thought that they must represent original structures. Postmortem changes of colour can be quite startling. Structural colours, such as metallic blues, greens and reds are often more vivid in the fossils while they are still wet, than in their modern counterparts, but all colours on fossil material fade to a dark bluish-black when the specimens dry. These colours can be regenerated to some extent if the specimens are wetted again but the colour that returns is often a different shade from that which was seen in the first place. Thus, reds become greens and greens change to blues until, with successive episodes of wetting and drying, the colours gradually fail to reappear and the specimen remains stubbornly black. Fossils of some groups of Coleoptera show spurious colours which may be sufficiently constant to be an aid to their identification. Thus, fragments of *Dytiscus* are usually a brilliant iridescent green which, once seen, is quite unmistakable. All traces of pigment are lost from fossil specimens but areas originally so coloured are represented by patches of thin, sometimes filamentous, cuticle which are occasionally so frail that originally coloured spots are represented merely by holes in the skeleton. It might be expected that fossil insect remains would have lost all traces of hairs and scales but, although this is quite often the case, it is not at all unusual to find specimens in which they have been retained and, particularly for species of weevils, they may be essential for diagnosis. The loss of scales can, however, be turned to good account since it often reveals characteristic microsculpture that is normally hidden in modern specimens but which can be of immense value in specific determinations.

By far the majority of determinable fragments in any assemblage of fossil insects, is of Coleoptera. This is partly because their skeletons are so robust and partly because they are one of the best-documented orders of insect and thus one of the most easy to identify. At times, the Coleoptera are outnumbered by the remains of Diptera, but since these are so delicate they are usually very fragmentary and distorted. However, the larval head capsules of Chironomidae are often very abundant in limnic sediments but, unfortunately, they can be identified only to the generic level which imposes a limit on their evolutionary significance and their value as environmental indicators. Larval heads of Tipulidae are also abundant in many fossil assemblages but to date little work has been done on them. Particularly common are the heads and propodea of Hymenoptera. There is a wealth of detail visible on these pieces but, apart from work being undertaken at the moment by Osborne (personal communication) on late Weichselian (Wisconsin) ants, I am unaware of any other studies in this promising field. Well-preserved fragments of Hemiptera are also common in many Quaternary deposits, some of which can be specifically determined, but again, there has been no concerted attempt to study these remains. Trichoptera are represented by large numbers of larval skins which, though broken up into discrete elements, present such a variety of consistent forms as to suggest that these, too, may be specifically determinable. They often occur in such profusion that statistical studies on faunal structure and dynamics should be possible. Remains of Megaloptera are abundant in limnic sediments laid down in shallow water. They are represented chiefly by larval mandibles and imaginal heads. In contrast to the abundance of the orders mentioned above, others are exceedingly rare in Quaternary deposits. Curiously, in the light of their abundance as fossils in deposits of earlier geological eras, both the Orthoptera and Odonata are rare as Quaternary fossils. Apart from Zeuner's work on the Orthoptera from the late Pleistocene wooly rhinoceros site in Starunia (Zeuner, 1934), records of Orthoptera are almost entirely confined to fragmentary jaws. As fossils, Lepidoptera are known only from the occasional larval heads and jaws.

Until the present time, Coleoptera have dominated the interest of Quaternary entomologists and so most of the discussion which follows will draw upon information from this order. However, I do not believe that the conclusions are only relevant for Coleoptera; rather, they provide a model from which to infer the evolutionary and zoogeographic history of other orders of insects. Verification or refutation of this belief must wait until the other orders of insects that are so well represented in Quaternary deposits, have been more fully investigated. Potentially, at least, the Quaternary history of these orders is within the sphere of objective investigation and need no longer rely on speculations derived by remote extrapolations from the present-day distributions of the animals concerned.

Insect Evolution During the Quaternary Epoch

One of the most outstanding facts to emerge from the study of Quaternary insect fossils is that those skeletal elements which possess distinctive features match precisely those of species that are still living at the present time. Many of these fragments are specifically unique and leave no doubt that they are actually conspecific with existing species. A somewhat unexpected outcome has been that, in my experience and that of my colleagues here at Birmingham, no fossil has yet shown intermediate features between known species, even though many thousands of specimens have been investigated from several dozen Quaternary sites. It should be pointed out, however, that no fauna has been investigated that can be safely said to date from before the Lower Middle Pleistocene, i.e., from older than the Cromerian (Aftonian) interglacial. Nevertheless, it seems inescapable that the first conclusion to be drawn from the fossil evidence is that a great many species have maintained their specific integrity throughout, at least, the upper half of the Quaternary epoch, and that there is no fossil evidence of any morphological evolution among insects during this period. Some selected examples can be cited to illustrate the geological longevity of species of Coleoptera.

A bed of peaty silt, dating from the Cromerian (Aftonian) interglacial period, is occasionally exposed at the foot of the cliffs along the north Norfolk coast in eastern England. Insect remains are not uncommon in this deposit and, although not yet fully investigated, preliminary work shows that it contains species that are exactly comparable with living forms. Only one of these will be dealt with here. Well-preserved elytra were obtained of the weevil *Limnobaris* with the scales on the interstices intact. It is the form of these scales that distinguishes *L. pilistriata* from its morphologically almost identical relative *L. T-album.* The scales on the fossils leave no doubt that it is *pilistriata* that is represented and not *T-album,* and this is further supported by less consistent differences in the puncturation of the second interstice in which the fossils also resemble *pilistriata.* There is no question of contamination of our sample with modern specimens, since the peaty horizon from which it was taken was exposed in a completely fresh section after a land slip had cut back into the cliff by several metres. Also there is no suitable habitat for *L. pilistriata* along that stretch of coast today and, furthermore, the specimens were interbedded within the deposit and had to be dissected out of the felted vegetable matter. Since only the geological longevity of *L. pilistriata* has been established and no fossils have yet been found of *L. T-album,* no time can be set on the separation of these two species.

Many species of aleocharine staphilinid beetles are so similar to one another that they can be separated only on characters of the male genitalia. In this respect, species of the genus *Atheta* are notoriously difficult and as fossils they are almost always indeterminate. Several *Atheta* abdomens were obtained among numerous other insect remains from a layer of organic sand that underlay the interglacial deposits at Hoxne described by West (1956). The fossil insects thus date from the closing phases of the Elster (Kansan) glaciation. Four of the abdomens contained well-preserved male genitalia which, together with the terminal segments, matched exactly those of *Atheta graminicola.* If three generations per annum can be taken as average for this species, it is evident that *Atheta graminicola* has maintained its specific integrity for at least a million generations.

The male genitalia of *Hydraena riparia* is an extraordinarily complex structure. A fossil abdomen with an aedeagus identical to that of the modern species was illustrated by Shotton & Osborne (1965). It came from the Hoxnian (Yarmouthian) interglacial deposits at Nechells, Birmingham.

Again, this species has a close relative, *H. britteni,* but until fossils of this species are found it will not be possible to say how far back in time the separation must have taken place.

Matthews (1968) found, in central Alaska, abundant fossil insects dating from the early part of the Wisconsin glaciation. Among these were many specimens of the carabid beetle *Pterostichus,* subgenus *Cryobius,* whose species are exceedingly difficult to separate. This subgenus has been studied intensively by Ball who believed originally (Ball, 1963) that much of the speciation of the group took place during the Wisconsin glaciation but later he antedated the taxonomic splitting to a date prior to ninety thousand years ago (Ball, 1966). It is fortunate, therefore, that among Matthews' specimens from McGee cut, there were partially articulated specimens of *Cryobius* with genitalia intact. He was thus able to identify with certainty seven species of *Cryobius,* two of which, *P. similis* and *P. parasimilis,* were thought by Ball to have been among the most recent to evolve because of their great morphological similarity. It is evident, then, that these two species which may well be classed as sibling species, were already separate from one another by early Wisconsin times.

I must emphasise that the examples of geological longevity of species of Coleoptera given above are by no means exceptions. They were selected from an ever-growing list of such species which includes not only Arctic species whose evolution might be expected to be relatively slow (Downes, 1964), but also many species which now inhabit temperate latitudes, obtained from deposits laid down in the warm interglacial periods. The fossil record shows that there is no simple correlation between the geological longevity of a species and the climate in which it lives.

The proven constancy of so many species of Coleoptera during the latter half of the Quaternary epoch might seem at first sight to be incompatible with the well-marked tendency shown by many of them to form geographical races at the present time. The formation of such races (subspecies) has frequently been construed as incipient speciation. Unfortunately, there is not yet sufficient evidence to determine to what extent past populations of a species were divided into different races, but there are indications from the fossil record that racial polymorphism was probably just as widespread an occurrence in the past as it is today. The actual recognition of a race is often subtle and requires not only a large number of specimens, but also an expert with an intimate knowledge of the species concerned. Fossil examples that fulfill these requirements and, I may add, experts with adequate time and sympathy toward fossils, are naturally rare. I will discuss one example here.

In an insect assemblage obtained from mid-Weichselian (Wisconsin) deposits at Brandon, Warwickshire (Coope, 1968a), there were a great many fragments of *Helophorus brevipalpis.* I sent specimens representing just over four hundred individuals to R. B. Angus who has made a special study of the species and he reports that there is no doubt that some of them are of the classic style of Siberian specimens. With this in mind, it is interesting to note that in the same assemblage and in several others of the same general age from the English Midlands, there were a number of exclusively eastern Siberian species, suggesting that, if the species had similar physiological characteristics then as they have today, the environment of glacial England might have had similarities to that of modern Siberia. The Siberian style *H. brevipalpis* in glacial England may be a morphological reflection of the physiological adaptations to Siberian style conditions. (It must, of course, be assumed that the Siberian and European races of *H. brevipalpis* are, in fact, conspecific). It is highly unlikely that race formation in *H. brevipalpis* is unique or in any way exceptional—many other species would, I believe, show similar traits if the fossils were given equal attention.

Racial peculiarities are no doubt due chiefly to the fact that differences in the environment in different parts of a species' range exert different selection pressures upon the variability available to species. At any one time, the extent of this variation is limited and this imposes a limit on the subspecific diversity. Rates of change in these limits of variability cannot be inferred from studies based solely on the living representatives of a species but the fossil record can provide, at least in theory, just this type of information. The fact that so many species have remained stable during the last two glacial-interglacial cycles indicates that there has been little change in these limits of variability during this period. The formation of geographical races within these limits of variability does not appear to have resulted in speciation. If this is so, the formation of species and geographical races are not necessarily stages in a continuous process but rather two distinct adaptive responses. The stability of a species during this period may be a consequence of, rather than in spite of, its ability to accommodate itself to environmental differences throughout its range in time as well as in space.

The fact that fossil Coleoptera, after an extensive investigation show no evidence of evolution during the latter half of the Quaternary epoch, does not imply that all speciation must have taken place before that time. New species might have arisen when very small populations became genetically isolated from the stable parent stock. The initial changes in these transitional populations would probably have been rapid, followed soon afterward by a re-establishment of morphological constancy in a stable adaptive relationship within its new environmental niche. The absence of fossil intermediates between closely related species, may merely reflect the smallness of these transitional populations and also the brevity of the time involved from their isolation in the first place to the re-establishment of morphological stability. If the views that have been outlined above are correct I would not expect the fossil record to provide adequate evidence of actual speciation, but rather to provide a minimum age for the origin of a particular species. It also follows that if speciation can occur in the manner just outlined, then the amount of morphological difference between two species is not necessarily a measure of its evolutionary antiquity, an assumption that lies at the bottom of almost all phylogenetic reconstructions.

Extinctions of Species During the Quaternary Epoch

Many fossil insect remains are indeterminate because they are so nondescript that they can be matched equally well with a number of species. Others, though highly distinctive, remain undetermined because no acceptable match can be found for them. Some of the latter may represent extinct species, but more probably they merely betray our own ignorance of the present-day insect fauna. Recent experiences have reinforced the latter.

In deposits of the Saale (Illinoian) glaciation (Osborne & Shotton, 1968) and of the Weichselian (Wisconsin) glaciation (Coope & Sands, 1966; Coope, 1968a; Morgan, 1969) in England, there frequently occurs a very distinctive species of *Helophorus*. This species was first described by D'Orchymont (1927) on the basis of two fossil elytra that dated from the "second glacial period" found near Gostricht, Dresden. Believing that the fossils represented an extinct species he gave them the name *H. wandereri*. The original description was subsequently amplified by us (Coope & Sands 1963) to include the head and thorax. Since this species was abundant in Britain to the end of the glacial period, it seemed remarkable that it should have become completely extinct in so short a time (we have, however, too many examples of the demise of large Pleistocene mammals during the same period, to make such an extinction impossible). I searched in vain in collections of Scandinavian Coleoptera because it was often associated in British deposits with species that now live in Scandinavia and I thought that it might have been masquerading as *H. nubilis* to which it bears a superficial resemblance. Eventually, R. B. Angus recognised that *H. wandereri* was matched perfectly by *H. obscurellus* a species found today in arctic Russia on the Kanin peninsula, on the Lena river in Siberia, in the Altai mountains and in the Tien Shan mountains. The identity of these two species was subsequently confirmed by the discovery of well-preserved abdomens with the male genitalia intact (Coope, 1968a). This example has been dealt with in detail because it was one of the few species in whose extinction we were gradually becoming confident.

Whenever it has been possible to examine specimens of Quaternary fossils that had been considered by earlier workers to represent extinct species, it has almost invariably been found that they correspond to living species. Thus, Angus has pointed out (personal communication) that, not only is D'Orchymont's *H. wandereri* synonymous with *H. obscurellus,* but another allegedly extinct species described by him at the same time as *H. lomnickii* is clearly, from an examination of the type specimen, the north European and Siberian *H. fennicus*. Lindroth (1948) discusses at length the somewhat curious identifications that Mjoberg (1916) gave to the "interglacial" Coleoptera from Harnon in central Sweden. He concludes that Mjoberg must have been confused by postmortem changes in the fossil material and that none of his identifications are acceptable. In the same paper, Lindroth showed that *Notiophilus coriaceus* of Henriksen, the only allegedly extinct species in his extensive fossil list, was, in fact, *N. aquaticus* whose elytral surface had developed a strange post-mortem rugosity. He was able to reproduce the same effect by treating modern elytra with a variety of chemicals and especially with sulphuric acid.

While studying an assemblage of fossil Coleoptera that had been collected at the beginning of this century from the classic Weichselian (Wisconsin) site at Barnwell station, Cambridge, I found on the slides manuscript labels with identifications by P. Lesne, an author of several papers on Quaternary fossil beetles. As far as I am aware these determinations were never published but, nevertheless, they are most instructive because of the insight that they give us into the prevailing climate of entomological opinion of those times. Lesne's identifications included a variety of new taxa ranging from subspecies to new genera. I have interpreted all these specimens as being representatives of living species (Coope, 1968b). It is hard to attribute Lesne's strange diagnoses to postmortem alterations since, for the most part, they show little evidence of such changes. I believe that more likely the readiness to credit fossil specimens with new names, stems from the widely held belief that many Coleoptera had evolved rapidly even in Late Pleistocene times and that extinction was of widespread occurrence during this period. Neither of these opinions is supported by the fossil evidence. There can, in fact, be little doubt that there has been no major extinction of Coleoptera, even of Arctic species, during at least the last two glacial-interglacial cycles. Moreover, I believe that the Coleoptera were not unique in this respect and that mass extinctions are unlikely to have affected the rest of the insect fauna of this period (*contra*—Downes, 1964, p. 304; Howden, 1969). Thus, though some extinction and some speciation cannot be excluded from the latter half of the Quaternary epoch, the fossil record suggests that both events must have been relatively rare. This view is being increasingly reinforced by the ever-growing list of species whose geological longevity can be established.

Physiological Evolution During the Quaternary Epoch

The absence of any evidence of morphological evolution in fossil insect species occurring during the latter half of the Quaternary epoch, does not exclude the possibility that physiological evolution might have taken place without expressing itself morphologically. So far as I am aware, there is no direct way of attacking this problem but it can be approached indirectly. Any physiological evolution would involve alterations in the environmental requirements of a species. If all, or most, species were subject to changes of this sort, we would expect to find that fossil associations of species would become increasingly different from those of the present day, as the ages of the fossil populations were increased. In practice, no such progressive difference has yet been detected. Fossil assemblages from both glacial and interglacial deposits are remarkably consistent ecologically both as to species associations and by comparison with other independent environmental inferences that can be derived from palaeobotanical and geological data. At this point, it should be emphasised that complete ecological homogeneity is not to be expected in fossil assemblages since they are frequently derived from quite an extensive area surrounding their ultimate place of burial and this may have included a number of quite different habitats (Coope, 1967; Matthews, 1968). The occasional association of species that today seem climatically incompatible might be construed as evidence of physiological change. I have pointed out elesewhere, however, that there are more plausable explanations for these anomalous associations (Coope et al. 1961; Coope, 1965a, 1968c), the most important of which is that past climatic conditions may have been at times quite different from those found anywhere in the world today—a brief survey of maps of Quaternary geography should be enough to establish this point without further elaboration. The sporadic occurrence of these curious assemblages of fossils is not inconsistent with the absence of large-scale physiological change.

From the fossil evidence so far available, it seems likely that species of Coleoptera have remained not only morphologically but, in general, physiologically stable, during the latter half of the Quaternary epoch.

Even though the majority of species seems to have remained physiologically stable during this period, individual species could have changed their environmental requirements. In theory, such species should be recognisable by their consistent past association with species which are now incompatible with them. In practice, however, this is not always easy since there are reasons other than physiological evolution to account for anomalous associations. I think that it is too early yet to say unequivocally that any particular species has undergone physiological evolution, but there are a few that so consistently offend against ecological conformity that eventually they will probably be included in this category.

Zoogeographical Implications of Quaternary Insect Fossils

The distribution of a species today cannot be understood purely in terms of the environmental factors governing its existence; historical factors must also be taken into consideration. These historical factors must include past changes in the environment and the associated alterations in the biota as a whole, and also the geographical accidents of space and time that enable a species to take advantage of, or fail to exploit, newly available areas of potential habitat. Although much can be inferred of the past history of a species from studies of its present-day range and recent changes in its distribution, only the fossil record can provide objective data on the past whereabouts of a particular species at any one time. No matter how inadequate this fossil record may be, no historical inferences can be accepted that are at variance with it. It must be pointed out, however, that only the specimens themselves constitute the indisputable facts of the fossil record and that taxonomic determinations based on them must always be a matter of interpretation. When dealing with fossils, the confidence that can be placed on a particular diagnosis varies enormously from species to species and even person to person. For this reason, I would like to make a plea here for the retention of all fossil material upon which interpretations have been based. Unfortunately, this has not always been an invariable practice, even in quite recent studies. Without the actual specimens there is no way of assessing, at a later date, the validity of their diagnosis or the value of the inferences based upon them.

Movements of Insect Species During the Quaternary Epoch

Fossil evidence is available in quantity only from the last two glacial interglacial cycles (Coope, 1965b; Shotton, 1965). During the interglacial episodes, the middle latitudes of the northern hemisphere were largely temperate, at times with somewhat warmer conditions than prevail today. The glacial phases were apparently more complex climatically, with episodes of ice advance (stadials) alternating with periods of ice retreat (interstadials). During some of the latter, at least in Europe and North America, the forest returned to areas previously covered by ice; at other times, even though the climate must have been relatively warm, there was no reforestation perhaps because the trees had been driven far to the south and the warm interval was too short to permit their recolonisation. We will not discuss here the degree to which these changes in the climatic belts extended equatorially, but will consider only the effects of these changes on the insect faunas in the middle latitudes, i.e., in those that are temperate today.

The fossil record presently available leaves no doubt that large-scale changes in the ranges of insect species occurred in the successive glacial and interglacial periods, and that faunal history has been largely a matter of the geographical rearrangement of existing species in response to the extreme fluctuations in the environment. Comparisons of the present distribution of species with the known limits of the Pleistocene ice sheets (usually the last major glaciation) have led some zoogeographers to infer that there was little movement of insect populations in response to these changes in the environment. Thus, Howden (1969) states.

> The conclusion to be drawn from the evidence presently available is that, in relatively sedentary groups, there was no large migration southward in front of the ice, but rather, survival of that portion of the population that was able to survive *in situ* in a habitat already occupied before glaciation.

It is difficult to reconcile this statement with the fossil evidence now available. It is evident that even the most sedentary species of Coleoptera, such as the large flightless weevils and carabids, adjusted their ranges on a grand scale during the Wisconsin and Illinoian glaciations and their associated interglacials. There can be no doubt that these fossils represent flightless individuals since, in the case of the weevils, the elytra are still fused together down the suture. Among the Carabidae such species as *Diacheila polita* have the typically narrow shoulders associated with the rudimentary wings found in modern specimens. These large-scale changes in the ranges of species can be illustrated by some examples drawn mainly from sites in the British Isles. Preliminary work, however, suggests that similar shifts in population also took place in other parts of western Europe and also in North America. The examples will be dealt with in chronological order, starting with the oldest.

Shotton & Osborne (1965) described a temperate climate insect assemblage from a deposit of organic silt dating from the Hoxnian (Yarmouthian) interglacial period, discovered in a foundation excavation here in Birmingham. Most of the species recorded occur in Britain today, but two species of Coleoptera, *Brachytemnus submuricatus* and *Platypus oxyurus* are now absent from these islands and they have a patchy distribution in southern and central Europe. Of particular interest in this fauna is the presence of large numbers of the weevil *Rhynchaenus quercus* which feeds exclusively on oak. In a succession of samples taken vertically through the deposit, this weevil made its first appearance very shortly after the first appearance of *Quercus* pollen in the deposit and it reached a peak of abundance coincidentally with the peak in the count of the *Quercus* pollen. There can be little doubt therefore that the fortunes of this weevil were as much determined by the availability of its host plant then, as they are today.

The insect fauna of the succeeding glacial period, the Saale (Illinoian) is not well known; only one assemblage that dates from this period is known, described by Osborne & Shotton (1968) from another organic silt in the English Midlands. This fauna differed markedly from that of the previous interglacial mentioned above, in spite of the close geographical proximity of the two sites. The more temperate species are absent and in their place there are a number of species whose distribution today is exclusively northern. Two species in this assemblage are of outstanding interest. *Helophorus obscurellus* is an arctic Russian and Siberian species and *Helophorus jacutus* is now restricted to the Trans-Baikal region of Siberia. Both these species reappeared in Britain in considerable numbers during the next glacial period but were absent in any fossil assemblages from the intervening interglacial. Negative evidence of this sort does not prove that these species were absent from Britain during this interglacial period, but had they been present at that time, they would have needed some special physiological dispensation to enable them to live with the warm temperate species known to have lived in Britain at that time.

Two extensive insect faunas are known from the Eemian (Sangamon) interglacial deposits of the southeast of England. They date from the warmest phase of the interglacial which, at that time, was somewhat warmer than is Britain today. Unfortunately, these assemblages have not yet been fully investigated because of the immense number of insect species in the modern fauna of temperate latitudes which must be searched before the fossils can be identified. Some preliminary results may, however, be interesting.

The first of these Eemian interglacial insect faunas came from lenses of organic sandy silt that were associated with the bones of straight-tusked elephant, lion and hippopotamus encountered during the excavations for the foundations of the new building of the Uganda government in Trafalgar Square in central London (Franks et al. 1958). The fauna was dominated by scarabaeid beetles, several of which are found living today exclusively to the south of the British Isles. Among these were the central and southern European species *Caccobius schreberi, Oniticellus fulvus, Onthophagus furcatus,* and *Valgus hemipterus.* In general, the whole assemblage bears a decidedly central European appearance.

The second of these insect faunas came from the interglacial beds at Bobbits Hole near Ipswich, which were described by West (1957). This assemblage was in many ways similar to that taken from Trafalgar Square and came from the same phase of the interglacial. Among the exclusively southern species found at this site was the carabid *Oodes gracilis* whose environmental requirements have been studied in detail by Lindroth (1943). The presence of this species indicates that the summer temperatures must have been at least 3° F warmer than in present-day southern England, a conclusion supported by the associated fossil flora and fauna. Another small assemblage of fossil insects from this interglacial obtained from an outcrop of peat on the foreshore at Selsey Bill on the south coast of England also contained a thermophilous species (Pearson, 1961). The insect fauna of Britain during this interglacial period was thus very different from that of the previous glacial episode and it bore little or no resemblance to that of the next glaciation.

Fossil insect associations from the various phases of the Last (Wisconsin, Weichselian, Wurm) glaciation are better known than are those from the earlier periods of the Quaternary epoch. They have been extensively studied in Britain and, to a lesser extent, on the continent of Europe, and there is now a growing accumulation of information from North America. Only some of the major faunal changes that took place during this period can be reviewed here.

The earliest insect fauna from this glaciation was obtained from a peaty organic mud in a sand pit at Chelford in Cheshire (Coope, 1959). On palynological and radiocarbon grounds, this deposit has been shown to date from the Brörup interstadial in the early part of the glaciation (Simpson & West, 1958). The fauna was similar to that of the boreal forest of northern Europe and contained a few species that are not found in Britain today but which still inhabit northern Europe. Though the fauna was extensively sampled and almost a hundred species recognised, there was no evidence of the thermophilous element so characteristic of the faunas of the preceding interglacial. On the other hand, the Chelford assemblage lacked the more extreme arctic stenothermic species that were soon to become such an important element in the British insect fauna as the more rigorous glacial conditions became established.

A great many insect assemblages have been obtained from various localities in Britain from deposits that were laid down during the long cold period that predated the maximum expansion of the Weichselian ice in western Europe and which took place about twenty thousand years ago. Most of the insect faunas of that period contain a relatively high proportion of species that are today restricted to the far north of Europe, also occurring occasionally in the high mountains further south (Coope, 1962a, 1968a, 1968b; Coope & Sands, 1966; Morgan, 1969). Many of them are typical inhabitants of the arctic tundra. It would be inappropriate to list in this review all the species that are now absent from Britain since the list is so long; in some sites, as much as 30 per cent of the fauna falls into this category. Such species include *Diacheila polita, Pterostichus blandulus, P. kokeili, Helophorus obscurellus, Boreaphilus nordenskioeldi* and *Chrysolina* of the *septentrionalis* group of species, to name just a few of the more abundant Coleoptera that lived in Britain at this time. Among the Hemiptera, *Chiloxanthus stellatus* and *Calacantia tribomi* have withdrawn to the arctic fringes of Europe though they lived in Britain during the Last glaciation. Although the landscape to the south of the ice sheet in western Europe, was almost treeless, these conditions did not always imply a truly tundra climate (Coope, 1969a), and the rather curious faunal assemblage from Upton Warren (Coope et al. 1961) indicated the existence of a more temperate climate than the treeless conditions of the district might have suggested.

It is probable that the so-called interglacial insect faunas described by Lindroth (1948) from central Sweden, also date from this general period (Lundquist, 1967), and a recently described fauna of the same age from Peelo in the Netherlands (Coope, 1969b) contains a number of exclusively Arctic Coleoptera, some of which are common to both the Swedish and British faunas of the period.

In North America, Matthews (1968) has described assemblages of tundra Coleoptera from the Fairbanks district of Alaska that also date from the early stages of the Wisconsin glaciation. In mid-western United States, however, an insect fauna found near Cleveland, Ohio, has been described (Coope, 1968d) which indicates that conditions existing in front of the advancing glacier of the "classical Wisconsin" phase, were more similar to those found now near the northern fringes of the boreal forest, than to those of the tundra. Uncompleted work on a fossil insect fauna from Titusville, Pennsylvania, being carried out at the moment by Totten, here at Birmingham, also suggests that open forest and not tundra conditions preceded an early advance of the Wisconsin ice. To the evidence of these two insect faunas must be added the presence of conifer trunks found in the base of some of the Wisconsin tills in Ohio (Goldthwait, 1958), apparently caught up by the advancing ice while still green. If these scattered observations can be used as bases for generalisations, they would suggest that the tundra zone was greatly contracted in this part of North America even at the height of the Wisconsin glaciation; a situation that contrasts sharply with that in Europe of the same time, when a vast expanse of treeless country lay to the south of the ice sheet. It is clear, however, that this restriction of the tundra zone did not eliminate all the arctic species from this part of North America since such species as *Helophorus splendidus* and *Boreaphilus nordenskioeldi.* (=*nearcticus*) were members of the Titusville fauna.

We know very little about the insect fauna of Britain in the immediate vicinity of the ice front at the height of the Weichselian glaciation. A single impoverished, but decidedly arctic fauna, has been described (Penny et al. 1969) from a bed of laminated silt that underlies more than one hundred feet of till on the Holderness coast of eastern England. The geology of the site and a radiocarbon age of just over eighteen thousand years obtained for the silt, indicate that the fauna only just predates the ice advance over the area. If the maximum expansion of the ice involved still further climatic deterioration, it is likely that the conditions in England, near the ice margin, would have been reduced to the equivalent of polar desert.

With the climatic amelioration that resulted in the final retreat of the ice, a rich and varied insect fauna returned to western Europe (Henriksen, 1933; Pearson, 1962a and b; Coope, 1962b, 1968c; Schweiger, 1967). Initially, these faunas were largely made up of pioneer species and, though they still contain a large number of Arctic species, there is a notable lack of the exotic oriental species so characteristic of earlier Weichselian fossil insect assemblages. The whole fauna of this period bears a well-marked European aspect.

The climatic oscillations occurring during this period have been well documented by palynologists who have established a sequence of zones based on successive changes in the proportions of the different genera in the pollen rain (Godwin, 1965). When, however, the changes in the fossil insect assemblages are compared with contemporary pollen zonation, certain anomalies become apparent. Several unpublished sequences of late Weichselian insect faunas from England and Wales show that the incoming of the thermophilous, nonphytophagous Coleoptera anticipates the changes in the pollen rain that mark the beginning of the Allerød interstadial. In addition, these same species were clearly declining in both variety and number before the pollen shows any evidence of deterioration of the climate toward the close of this interstadial. Although the pollen of the Allerød indicates the existence of an open birch forest over most of the country at this time, the insects point to a summer temperature as warm as, or even slightly warmer, than obtains in Britain today. I believe that these discrepancies stem largely from the different rates of response to climatic changes by the insect fauna when compared to that of plants, particularly trees. There are many reasons why the insect fauna should be able to respond more promptly to changes in the environment. What I wish to emphasise here is the fact that insect faunas can respond with great alacrity to such changes.

The five hundred years that followed the Allerød interstadial saw a return to cold conditions, setting the glaciers advancing once again from the north over areas that they had previously abandoned. During this phase the Arctic species returned in considerable numbers to England, to as far south as London, and there was an almost complete extermination of the thermophilous species that so characterised the early Allerød period.

Though deposits bearing insect remains of post-glacial age are abundant, few systematic studies of these remains have been carried out. Lindroth (1942) has shown that *Calosoma sycophanta* occurred to the north of its present range in Scandinavia during the hypsithermal phase of the post-glacial period. Osborne (Kelly & Osborne, 1964, Osborne, 1965) has found *Rhysodes sulcatus* in post-glacial deposits in Britain, though the species is no longer to be found here. Its restriction in modern times may be largely due to loss of habitat, namely, large decaying trees. A more extensive study of the post-glacial insect faunas would provide a much needed insight into the composition of primitive associations of insects prior to their destruction or severe modification by human activity.

Archaeological sites often produce enormous quantities of insect remains, usually in the infilling of old wells and in refuse pits. Not only do such faunas provide evidence of past living conditions of the human population but they also provide data on the insect pests with which Man had to contend. Osborne (1969) has described an extensive insect assemblage from a Late Bronze Age well near Stonehenge in Wiltshire, which was dominated by dung beetles, including one species *Aphodius quadriguttatus,* which is now absent from Britain. In the filling of a well at the Roman villa near Barnsley Park, Gloucestershire (Coope & Osborne, 1968) and in a Roman refuse pit at Alcester, Worcestershire (Osborne, 1961), remains of Coleoptera were found that are today, and were probably then also, pests of stored grain, such as *Oryzaephilus surinamensis* and *Sitophilus granarius.* From the latter site came a large number of the central and southern European cerambycid beetle *Hesperophanes fasciculatus* which was no doubt imported into Britain at that time in manufactured wooden articles. It is unlikely that this species could have survived in natural populations outside human habitations in the climate of Roman Britain. The effects of human activity, either as an agent of transport or in providing insulated habitats in which exotic species can thrive, is clearly of classical antiquity.

Specific Stability Versus Quaternary Environmental Instability

I have endeavoured to outline above the great changes in the geographical distributions of even the most sedentary of insect species in response to fluctuations of the environment during the latter half of the Quaternary epoch. Such alterations in the ranges of species are to be expected if, as I believe, their physiological requirements have remained more or less stable during this period. The rate of population movement was clearly adequate to enable them to keep pace with the changes in the location of suitable living conditions, so that, paradoxically, the actual environment in which a species lived remained almost constant in spite of the vicissitudes of the Quaternary climate. Thus, the fact that insect species have remained stable during this period requires no reconciliation with these environmental fluctuations.

It is, of course, possible that many species actually owe their constancy to the instability of the Quaternary environment. Many low-temperature, stenothermic species that were widespread in Europe during the Last glacial period, today have very disjunct distributions, being found in high latitudes and in the high mountains of central Europe. This distribution is largely due to the restriction of acceptable thermal conditions in the post-glacial period. At least the northern and southern populations must have been isolated from one another for ten thousand years and probably for even longer, and yet there is no apparent morphological difference between them. (Genetic compatibility has yet to be proved.) Among these species, mere geographical isolation for this order of time has not apparently led to speciation. These mountain top or high latitude evolutionary traps can, however, be easily sprung if climatic deterioration takes place before the marooned populations have become genetically isolated from one another. Since the climatic oscillations giving rise to the sequence of stadials and interstadials, glacials and interglacials, are measured in tens of thousands of years, the duration of isolation imposed by such changes is evidently not adequate in itself to cause speciation. The large-scale movements of insect populations that are so characteristic of the fluctuating Quaternary environment, would have repeatedly broken down the geographical isolation; the gene pool of the species would have been kept well mixed. I must emphasise, however, that these considerations apply only to species inhabiting the larger land masses and their immediately adjacent islands. Those species that inhabit isolated islands or caves, for instance, cannot change location when conditions change, and must therefore endure these changes on the spot. Under such circumstances, rapid but rather atypical speciation is almost the only alternative to extinction. However, even on quite small isolated islands, such as South Georgia, endemic species can be shown to have remained constant for at least six thousand years (Coope, 1963; age determination, R. J. Adie, personal communication).

"Endemic" Species and the Fossil Record

The term "endemic" has been used to describe taxa that are localised within a restricted geographical area and which are thought to have originated there. This is a valuable concept but its application to insects must be tempered with caution in the light of evidence from Quaternary fossils. The specific stability and large-scale changes in the ranges of species that can be demonstrated for many species during the latter half of the Quaternary epoch, mean that the mere fact of present-day geographical restriction of a taxon is not, of itself, adequate evidence for establishing its place of origin. Again, from this discussion must be excluded those taxa restricted to isolated islands, caves or similar situations from which emigration is impossible. The fossil records shows that on the larger land masses the present distribution of an insect taxon must be considered as a stage in a dynamic continuum in which the ranges of species and of groups of species are constantly being adjusted in response to changing conditions. The rate at which these adjustments take place is very rapid in comparison to the rate at which new species can normally be expected to evolve. Some examples will be given here of insect species whose limited geographical distribution today does not necessarily imply that the species originated in that area.

A highly distinctive head of an unknown species of the staphylinid beetle *Oxytelus,* was illustrated (Coope, et al. 1961) from mid-Weichselian (Wisconsin) deposits of Upton Warren, in the hope that, if the species is still living, somebody might recognise it. It has now been identified by W. O. Steele (personal communication) as *O. gibbulus* which is restricted today to a small area in the western Caucasus mountains. It has recently been found in another Weichselian site (Osborne, personal communication) in the southwest of England; so the species must have been widespread in Britain during the Last glaciation. If we can take its present limited distribution as a measure of actual restriction and not of collection failure elsewhere, it is more likely that the western Caucasus represents the last stand of the species rather than its place of birth.

While it is generally accepted that the carabid subgenus *Cryobius* had its centre of dispersal in northeastern Siberia and northwestern America because of the large number of species concentrated there, it is interesting to note that a number of species lived in western Europe during the Weichselian (Wisconsin) glaciation though none is found there at the present time. (Lindroth, 1945; Coope, 1968a and b; 1969a; Morgan, 1969). Fossil remains representing five or six different species of this subgenus were obtained from the Weichselian deposits at Lebenstedt near Brunswick (Coope, unpublished findings) although, because only thoraces and elytra were preserved, no specific determinations were possible. In spite of this, they show that the paucity of species of *Cryobius* in Europe today may not be a true indication of its former status here. The concentration of species in a particular area may, in fact, more reflect their common environmental needs, than their centre of dispersal after relatively recent speciation in that area.

The Fossil Record and the Glacial Refugia Problem

No account of the significance of Quaternary fossil insects would be complete without some discussion of the thorny problem of the possible survival of species in isolated refuges cut off by the rising tide of the ice sheets. It is too early still to discuss the evidence of refugia in North America from fossil data, although Matthews (1968) has shown that there was a rich insect fauna in Alaska during the various phases of the Wisconsin glaciation. In Europe, however, the fossil record can contribute to this problem. Discussions of this aspect have been prolonged and sometimes acrimonious, so it is with some trepidation that I enter the lists. On the one hand, biologists have argued that a considerable part of the fauna and flora of Scandinavia and the North Atlantic islands must have survived at least the Weichselian glaciation *in situ,* but, of course, in very reduced circumstances. On the other hand, geologists have been almost universally sceptical of the possibility that anything other than a very meagre biota could have survived there under full glacial conditions. The arguments seem interminable, primarily because of the almost total lack of any fossil data in the areas in which the refugia are thought to have existed. A recent discussion centering on this general theme has been presented in a compilation of papers edited by Löve & Löve (1963).

Lindroth (1949) has discussed this problem from a largely entomological point of view. He provides convincing evidence that many species of Coleoptera, particularly Carabidae, are most likely to have attained their present distribution in Scandinavia by immigration from the west. His argument is most compelling for those species that today are restricted to northwest Europe, which do not now occur in Britain, and which could not have been expected to have emmigrated from there. For these species at least, the provision of some refugium in western Norway in which to survive the glaciation, seemed an inescapable conclusion.

It has now become apparent that many of these species did, in fact, inhabit Britain during the last glaciation though they have subsequently become extinct there. I have argued (Coope, 1969c) that these species could have contributed to the post-glacial colonisation of Scandinavia by being passively transported in frozen refuse or on the surface of river ice washed to sea at the time of the spring floods. These conditions would have been at their optimum during the last phases of the glaciation and would have ceased to operate as soon as the ice sheets had effectively disappeared. According to this interpretation, the sites of the alleged refugia in Norway should be considered as the landing places; the bridgeheads from which the species subsequently spread eastward.

If I may be permitted one final speculation, this passive transport might also explain the existence in the North Atlantic islands of a puzzlingly consistent biota which has such strong affinities with that of the Palaearctic, even as far west as Greenland. The common occurrence, in these areas, of plants with low dispersal capabilities and flightless, soil-bound insects, has led biogeographers to postulate land bridges connecting these islands at various periods ranging from the Pleistocene to the Tertiary; suggestions that have not on the whole met with geological approval. Such subspecific differences that do exist between the animals on the different islands may merely be morphological reflections of the physiological adjustments that the species have had to make since their arrival there, about ten thousand years ago. This speculation can be put to the test only by a comprehensive investigation of the fossil assemblages from these areas.

ACKNOWLEDGEMENTS

I would like to take this opportunity of thanking the numerous specialists who have been kind enough to aid in the identification of fossil material. In particular, I must thank Prof. C. H. Lindroth of the University of Lund for his inestimable assistance in tracking down arctic species and for proving such a mine of taxonomic and ecological information. Without his sustaining enthusiasm this study would have been impossible. The staff of the Entomology Department of the British Museum of Natural History has been most co-operative in what must seem a highly unorthodox approach to entomology. Prof. P. J. Darlington of the Museum of Comparative Zoology, Harvard University, has kindly read the manuscript. Lastly, I must thank my colleagues in the Quaternary Entomological section of the Department of Geology, The University of Birmingham, Prof. F. W. Shotton, Mr. P. J. Osborne, Mrs. Anne Morgan and Mr. A. C. Ashworth to whom much of the credit is due for what has been written above. I claim proprietary rights only for those views in which I have been mistaken.

LITERATURE CITED

Ball, G. E. The distribution of the species of the subgenus *Cryobius* with special reference to the Bering Land Bridge and Pleistocene refugia. In *Pacific Basin biogeography,* 133–51. (Gressitt, J. L., Ed., Bishop Museum Press, 1963)

Ball, G. E. A revision of the North American species of the subgenus *Cryobius* Chaudoir. *Opuscula Entomol.,* **28,** 1–66 (1966)

Coope, G. R. A late Pleistocene insect fauna from Chelford, Cheshire. *Proc. Roy. Soc. (London), Ser. B,* **151,** 70–86 (1959)

Coope, G. R. On the study of Glacial and Interglacial insect faunas. *Proc. Linnean Soc. London,* **172,** 62–65 (1961)

Coope, G. R. A Pleistocene coleopterous fauna with Arctic affinities from Fladbury, Worcestershire. *Quart. J. Geol. Soc. London,* **118,** 103–23 (1962a)

Coope, G. R. Coleoptera from a peat interbedded between two boulder clays at Burnhead, near Airdire. *Trans. Geol. Soc. Glasgow,* **24,** 279–86 (1962b)

Coope, G. R. The occurrence of the beetle *Hydromedion sparsutum* (Mull.) in a peat profile from Jason Island, South Georgia. *Bull. Brit. Antarctic Survey,* **1,** 25–26 (1963)

Coope, G. R. Fossil insect faunas from Late Quarternary deposits in Britain. *Advan. Sci., London,* **27,** 564–75 (1965a)

Coope, G. R. The response of the British insect fauna to Late Quarternary climatic oscillations. *Proc. Intern. Congr. Entomol., 12th, London,* **173,** 444–45 (1965b)

Coope, G. R. The value of Quarternary insect faunas in the interpretation of ancient ecology and climate. *Congr. I.N.Q.U.A., 7th, Boulder, 1965,* **7,** 359–80 (1967)

Coope, G. R. An insect fauna from Mid-Weichselian deposits at Brandon, Warwickshire. *Phil. Trans. Roy. Soc. London, Ser. B,* **254,** 425–56 (1968a)

Coope, G. R. Coleoptera from the "Arctic Bed" at Barnwell Station, Cambridge. *Geol. Mag.,* **105,** 482–86 (1968b)

Coope, G. R. Fossil beetles collected by James Bennie from Late Glacial silts at Corstorphine, Edinburgh. *Scot. J. Geol.,* **4,** 339–48 (1968c)

Coope, G. R. Insect remains from silts below till at Garfield Heights, Ohio; *Bull. Geol. Soc. Am.,* **79,** 753–56 (1968d)

Coope, G. R. The response of Coleoptera to gross thermal changes during the Mid-Weichselian interstadial *Mitt. Limnol.* (1969a)

Coope, G. R. Insect remains from Weichselian deposits at Peelo, Netherlands. *Mededel. Rijks. Geol. Dienst,* **20** (1969b)

Coope, G. R. The contribution that the Coleoptera of Glacial Britain could have made to the subsequent colonization of Scandinavia. *Opuscula Entomol.,* **34,** 95–108 (1969c)

Coope, G. R. Report on the insect remains from Lebenstedt. (Unpublished data)

Coope, G. R., Osborne, P. J. Report on the coleopterous fauna of the Roman Well at Barnsley Park, Gloucestershire. *Trans. Bristol and Glos. Arch. Soc.,* **86,** 84–87 (1968)

Coope, G. R., Sands, C. H. S. The discovery in British Late Pleistocene deposits, of the extinct species *Helophorus wandereri* d'Orch. *Opuscula Entomol.,* **28,** 94–97 (1963)

Coope, G. R., Sands, C. H. S. Insect faunas of the last glaciation from the Tame Valley, Warwickshire: *Proc. Roy. Soc. (London), Ser. B,* **165,** 389–412 (1966)

Coope, G. R., Shotton, F. W., Strachan, I. A Late Pleistocene fauna and flora from Upton Warren, Worcestershire: *Phil. Trans. Roy. Soc. London, Ser. B,* **244,** 379–421 (1961)

D'Orchymont, A. Über zwei neue diluviale Helophoren—Arten. *Sber. Abh. Naturwiss. Ges. Isis. Dresden,* 1926, 100–4 (1927)

Downes, J. A., Arctic Insect and their environments. *Can. Entomologist,* **96,** 279–307 (1964)

Franks, J. W., Sutcliffe, A. J., Kerney, M. P., Coope, G. R. Haunt of elephant and rhinoceros: The Trafalgar Square of 100,000 years ago—New Discoveries. *Illustrated London News,* 1101 (1958)

Godwin, H. *The History of the British*

Flora. (Cambridge Univ. Press, 1965)

Goldthwait, R. P. Wisconsin age forests in western Ohio. 1, Age and glacial events. *Ohio J. Sci.,* **58,** 209–19 (1958)

Henriksen, K. L. Undersøgelser over Danmark-Skanes kvartaere Insekt-fauna. *Videnskab. Medd. Dansk naturhist. Foren, København,* **96,** 77–325 (1933)

Howden, H. F. Effects of the Pleistocene on North American insects. *Ann. Rev. Entomol.,* **14,** 39–56 (1969)

Kelly, M., Osborne, P. J. Two faunas and floras from the alluvium at Shustoke, Warwickshire. *Proc. Linnean Soc. London,* **176,** 37–65 (1964)

Lindroth, C. H. Ett subfossilfynd av *Calosoma sycaophanta* L. *Popular Biol. Rev.,* **2,** 1–7 (1942)

Lindroth, C. H. *Oodes gracilis* Villa. Eine thermophile Carabide Schwedens. *Notulae Entomol.,* **22,** 109–57 (1943)

Lindroth, C. H. Die fennoskandischen Carabidae. Eine tiergeographische Studie, **1, 2.** *Goteborgs K. Vetensk o. Vittern Samh. Handl., B,* **6** (1945)

Lindroth, C. H. Interglacial insect remains from Sweden. *Arsbok Sveriges Geol. Undersokn., C,* **42,** 1–29 (1948)

Lindroth, C. H. Die fennoskanischen Carabidae. Eine tiergeographische Studie, **1.** *Goteborgs Vetensk. o. Vitterh Samh. Handl., B,* **6** (1949)

Lomnicki, A. M. Pleistocensjie owady z Boryslavia (Fauna pleistocenica insectorum Boryslaviensium) *Mivz. Im. Dzieduszychich Iwow,* **4,** 1–116 (1894)

Löve, A., Löve, D. (Eds.) *North Atlantic Biota and their History.* (Pergamon Press, 1963)

Lundquist, J. Submarana sediment i Jamtlands Lan *Arsbok Sveriges Geol. Undersokn., C,* **618,** 1–267 (1967)

Matthews, J. V. A Palaeoenvironmental analysis of three Late Pleistocene coleopterous assemblanges from Fairbanks, Alaska. *Quaest. Entomol.,* **4,** 202–24 (1968)

Mjoberg, E. Über die Insektenreste der sogen. Härnögyttja im nordlichen Schweden. *Arsbok Sveriges Geol. Undersokn., C,* 268 (1916)

Morgan, M. A. A Pleistocene fauna and flora from Great Billing, Northamptonshire, England. *Opuscula Entomol.,* **34,** 109–29 (1969)

Osborne, P. J. The effect of forest clearance on the distribution of the British insect fauna. *Proc. Intern. Congr. Entomol., 12th, London,* **173,** 37–55 (1965)

Osborne, P. J. An insect fauna of Late Bronze Age date from Wilsford Wiltshire. *J. Animal Ecol.,* 555–66 (1969)

Osborne, P. J. An insect fauna from the Roman site at Alcester, Warwickshire. (Unpublished data)

Osborne, P. J., Shotton, F. W. S. The fauna of the Channel Deposit of Early Saalian Age at Brandon, Warwickshire. *Phil. Trans. Roy. Soc. London, Ser. B,* **254,** 417–24 (1968)

Pearson, R. G. The ecology of the Coleoptera from some Late Quaternary deposits. *Proc. Linnean Soc. London,* **172,** 65–71 (1961)

Pearson, R. G. The Coleoptera from a detritus mud deposit of Full-Glacial age at Colney Heath, Nr. St. Albans. *Proc. Linnean Soc. London,* **173,** 37–55 (1962a)

Pearson, R. G. The Coleoptera from a Late glacial deposit at St. Bees, West Cumberland. *J. Animal Ecol.,* **31,** 129–50 (1962b)

Penny, L. F., Coope, G. R., Catt, J. A. The age and insect fauna of the Dimtington Silts, East Yorkshire. *Nature* (In press, 1969)

Schweiger, H. Über einige subfossile Koleopterenreste aus der Umgebung von Bad Tatzmannsdorf. *Wiss. Arb. Burgerld.,* **38,** 76–91 (1967)

Scudder, S. H. Canadian fossil insects: Canada. *Contrib. Can. Palaeontol.,* **2,** 67–92 (1900)

Shotton, F. W. The geological background to European Pleistocene entomology. *Proc. Intern. Congr. Entomol. 12th, London,* **173,** 452–54 (1965)

Shotton, F. W., Osborne, P. J. The fauna of the Hoxnian Interglacial deposits of Nechells, Birmingham. *Phil. Trans. Roy. Soc. London Ser. B,* **248,** 353–78 (1965)

Shotton, F. W., Sutcliffe, A. J., West, R. G. The flora and fauna from the Brick Pit at Lexden, Essex, *Essex Naturalist,* **31,** pt. 1. (1962)

Simpson, I. M., West, R. G. On the stratigraphy and palaeobotany of a Late Pleistocene organic deposit at Chelford, Cheshire. *New Phytologist,* **57,** 239–50 (1958)

West, R. G. The Quaternary deposits at Hoxne, Suffolk. *Phil. Trans. Roy. Soc. London, Ser. B,* **239,** 265 (1956)

West, R. G. Interglacial deposits at Bobbitshole, Ipswich. *Phil. Trans. Roy. Soc. London, Ser. B,* **241,** 1–31 (1957)

Zeuner, F. Szarańozaki z warstw dyluwjalnych Staruni. Die Orthopteren aus der diluvialen Nashornschicht von Starunia (Polische Karpathen). *Starunia,* **3,** 1–17 (1934)

THE STRUCTURE OF ARTHROPOD CHEMORECEPTORS[1]

By Eleanor H. Slifer

Department of Entomology, Academy of Natural Sciences of Philadelphia, Philadelphia, Pennsylvania

The body surface of most terrestrial arthropods is highly impermeable to water; if it were not, the individuals, nearly all of which are small, would quickly succumb to dehydration. The receptor surfaces of the neural elements of an olfactory or a gustatory sense organ must be located in a position where the stimulating odor or material in solution will reach them promptly. This is accomplished most efficiently when the moist receptor surfaces are exposed directly to the air or other media to be tested. These opposed necessities, then, that the individual 1. minimize water loss and 2. expose the delicate membranous surfaces of the dendrites to the air is a problem with which insects and other terrestrial arthropods are faced.

The problem has been solved in the insects by the exposure of the receptor surfaces at very small openings or pores in the cuticle. Unlike the relatively large apertures of the digestive, respiratory, and reproductive systems and of various glands that are intermittently opened and closed, the pores of the chemoreceptors are continuously open. The openings, however, are extremely small and their combined surface area presents no serious problem in water conservation (86, 87).

The openings, indeed, are so small that they could not be seen by earlier workers. It was not until the electron microscope came into use and adequate methods of preparing materials to be examined with it had been developed that the openings were demonstrated for the first time (86, 87). Before this it had been believed generally that the chemoreceptor nerve endings were covered by a specialized cuticle through which the stimulating substance was able to pass. This view was accepted, at first, by the present writer (61). During the past ten years, special methods for identifying the pores with a solution of crystal violet applied to the external surface of the insect (65), examination of the material in a medium with a suitable refractive index, and the use of silver-stained sections have permitted the identification of the pores in the sense organs of some species with the light microscope. Before the knowledge obtained with the electron microscope became available, the correct interpretation of observations made on such structures with the light microscope was not possible.

[1] Supported in part by a grant from the National Science Foundation GB-7310.

Fortunately, electrophysiological studies on arthropod sense organs that began in 1955 with Hodgson, Lettvin & Roeder (36) and have been continued by many others, have provided abundant evidence confirming the conclusions derived from a study of their structure. It is now possible to state for many species with a high degree of confidence and on structural evidence alone, as could not be done only a few years ago, that a particular sense organ is, or is not, a chemoreceptor. Proof of the presence or absence of pores where the dendrites of sensory neurons are exposed is the basic information needed for a decision. For some receptors this information can be obtained with relative ease with present techniques, but for others only with difficulty or not at all. The specific agent or agents to which a chemoreceptor is sensitive, on the other hand, cannot be decided by inspection and must be determined with physiological or behavioral methods.

Relatively little work has been reported, so far, for aquatic arthropods. In these, the necessity for conserving water is not urgent and some of the species that have been examined with the electron microscope have been found to have chemoreceptors that are covered with a thin cuticle that lacks pores but is highly permeable to materials in solution. The branches of the sensory dendrites lie immediately below this special cuticle (24–26). A chemoreceptor with a single pore has been described for other aquatic species (30, 46).

This paper will be concerned primarily with observations on the structure of arthropod chemoreceptors that have been published during the past fifteen years. References to earlier literature and to physiological and behavioral studies of arthropod chemoreceptors may be found in the reviews of Boeckh, Kaissling & Schneider (7), Bullock & Horridge (11), Butler (12), Dethier (14–18), Dethier & Chadwick (19), Frings & Frings (23), Hocking (31), Hodgson (32–35), Schneider (55, 56), Slifer (66, 69), Snodgrass (89, 90), and Wigglesworth (99). References to single-cell electrophysiological recordings from insect chemoreceptors may be found in an article by Borden (8). Recent papers on the development of insect sense organs are discussed in a review by Gouin (27).

CHEMORECEPTORS OF INSECTS

Information concerning the structure of chemoreceptors obtained with modern techniques is now available for more than 50 species of insects. Because of this, they will be considered first. The chemoreceptors of arthropods belonging to other classes will then be discussed. A typical insect chemoreceptor is composed of (*a*) cuticular parts, (*b*) sensory neurons, and (*c*) sheath cells. These will be taken up in that order.

Cuticular Parts

The cuticular parts of the various kinds of insect chemoreceptors differ conspicuously and may take the form of hairs, pegs, papillae, loops, or plates. This, together with their position on or near the surface of the body wall where they are accessible for study, provides the primary data for a system of classification of these sense organs. The neural and sheath cells associated with them also show differences but these, located below the cuticle, are much more difficult to prepare for examination. In many of the electron micrographs that have been used in recent publications, the cuticle is well-preserved while the cellular parts show poor cytological detail. With our present knowledge, then, the chemoreceptors can best be separated into classes on the basis of their cuticular components. This is not a new approach and has been used by earlier workers for more than a century. However, the large store of new information gained with the electron microscope now necessitates changes in the system used previously.

The selection of the thickness of the wall of the specialized cuticular parts as the first diagnostic property for classification requires some defense, for thickness is, admittedly, a relative character. However, in practice, few occasions have arisen in which it has not been possible to identify a particular structure as thick-walled or thin-walled. Schneider & Kaissling (57) used this character to separate the sense organs of the silkworm, *Bombyx mori,* and Dethier, Larsen & Adams (20) employed it in their study of the olfactory organs of a blow fly, *Phormia regina.* If the chemoreceptor has a thick wall, it usually also has a single opening at its tip where the unbranched dendrites terminate. The number of neurons associated with such a receptor is usually five, or a number not far from it. If the receptor has a thin wall, this is usually perforated by many small openings into which the branched dendrites from 1 to over 60 neurons send clusters of filaments. Some exceptions to these generalizations are known and, as additional information is obtained, changes will become necessary; but for the present this system of classification is a convenient one.

THICK-WALLED CHEMORECEPTORS

These receptors take the form of hairs, pegs, or papillae. They are common on the antennae, mouth-parts, and tarsi of many species but may occur on almost any part of the body surface. In the grasshopper, *Melanoplus differentialis,* they have been found on all external parts except the compound eyes, ocelli, tympanal organs, cervicum, laciniae, mesothoracic tergum, and the pleural and intersegmental membranes of the abdomen (63). The yellow fever mosquite, *Aedes aegypti,* has none on the antennae (79) but they are present on the tarsi (67). There is excellent evidence for some species that they serve as contact chemoreceptors and the proof provided by Dethier and his co-workers is well-known. There is also evidence that the thick-walled hairs or pegs are sensitive to strong odors. They are present on many parts of the body that do not come into contact with materials in solution except accidentally. Hopkins (37) has shown that the thick-walled hairs on the labella of the stable fly, *Stomoxys calcitrans,* can be stimulated by odors. *Romalea microptera,* a large grasshopper, will withdraw a leg when a drop of an essential oil is held near it (62, 64). Another grasshopper, *M. differ-*

entialis, reacts in the same way even after the removal of the head on which most of the olfactory organs are located (74). Examination of the legs of these grasshoppers shows that thick-walled chemoreceptors with an opening at the tip are numerous on them and that no other sense organs likely to have an olfactory function are present (75). It should not be difficult to test the effect of odors with electrophysiological methods on the tarsal and labellar hairs of Diptera but this has not yet been reported.

Surface hairs and pegs.—In some species the thick-walled chemoreceptor is readily separable with high powers of the light microscope from the tactile hair that it resembles; but in others this is difficult or impossible without an electron microscope. Usually the tactile hair has an extremely sharp tip and a lumen that stops well short of it. There is no opening at the distal end where a solution of a dye or other visible material can enter. A cuticular sheath that contains the dendrite of a sensory neuron is attached inside the lumen near the base of the hair (66, 68, 72) . The commonest type of thick-walled chemoreceptor, in contrast, has a tip that is rounded, sometimes inconspicuously, and is provided with a pore where the tips of the dendrites are exposed to the air. The pore in the grasshopper may have a diameter of 2 μ (86). Dye applied externally enters the pore and stains the dendrites (61). Others besides the author and her associates who have succeeded in staining the thick-walled chemoreceptors are Borden & Wood (9), Swartzendruber (94), and Wallis (96). In some species, the tips of the dendrites can be seen just inside the pore or protruding from it (86). Adams, Holbert & Forgash [(2), Fig. 5] have published an electron micrograph of a longitudinal section through such a hair from *Stomoxys calcitrans* in which the dendrites may be seen ending just inside the pore. It is possible that the dendrite tips may be moved inwards or outwards for a short distance in the living animal as hydrostatic pressure inside the body changes; but this has not yet been demonstrated. The dendrites within the lumen of the chemoreceptor are usually enclosed within a cuticular sheath[2] that is invaginated from the periphery of the opening at the tip, extends down the length of the hair or peg and ends below its base [(86), Figs. 25–31]. Sometimes, as in *S. calcitrans* [(2), Figs. 4–12, 17] and *Phormia regina* [(43), Figs. 1, p. 3, 5], the sheath is so intimately attached to the wall that the lumen is divided into two distinct chambers one of which is traversed by the dendrites. The cuticular sheath is continuous with the outer layers of the cuticle, resistant to digestion by the molting fluid and is shed with the exuviae. The inner end of the cuticular sheath is open and the dendrites pass out of it at this point. In the larva of an elaterid beetle, *Ctenicera destructor,* Zachuruk (100) has described a "subcuticular sheath" that extends inward from the proximal end of the cuticular sheath and encloses the cell body and part of its axon. Such a sheath has not, so far, been identified in electron micrographs of the sensilla of other insects. It would be interesting to know more about the fine

[2] This term was introduced in 1957 [(86), 367–68] to replace a long series of other names used by earlier workers before the relationship of the dendrites and the sheath was understood.

structure of the sense organs of *Ctenicera*. In some species, the external surface of the hair or peg is smooth while in others longitudinal grooves spiral around it. Cross sections of the second type have a scalloped periphery (72).

Pegs in pits.—Thick-walled pegs may be set in pits or cavities in the cuticle of the body wall so that the tip of the peg lies below the surface. On the grasshopper antenna, the cavities containing the pegs are provided with a thin, transparent roof that has a small circular opening in it above the tip of the peg (87). In some species, the peg is situated at the end of a long and slender canal that connects it to the surface. Such sense organs are common in Hymenoptera. A cuticular sheath invaginated from the peg tip encloses the dendrites of the sensory cells as it does in the surface hairs and pegs. It was suggested by earlier workers that chemoreceptors set below the surface react to an odorous material after the receptors on the surface have become fatigued, since passage into the cavity would require a longer time. Steiner (92) reported that pegs set in pits are the only sense organs present on the antennae of several species of Odonata. Representatives of this order should be examined with the electron microscope. It is not yet known whether the pegs of the Odonata are thick-walled or thin-walled.

Papillae.—Chemoreceptors in the form of papillae about 0.5 μ high are present in the dorsal wall of the food canal of aphids (97). Each has a pore at its center where the tips of dendrites from three, four, or five neurons end. The dendrites are enclosed within a cuticular sheath that is invaginated from the pore. Papillae 10 μ high and innervated by four neurons are present between the pseudotracheae of the labellum of the blow fly (44). Moulins (48, 49) found somewhat similar papillae on the hypopharynx of a roach, *Blabera craniifer*. Each of the 50 papillae is innervated by 5 neurons that end just inside a pore that is from 0.1 to 0.5 μ in diameter. The dendrites are enclosed within a cuticular sheath.

THIN-WALLED CHEMORECEPTORS

Thin-walled chemoreceptors are abundant on the antennae of many species of insects from the Collembola (51) to the Diptera (20, 21, 79, 81). A few have been reported on the palps of the mouth-parts of larval Lepidoptera (60) but their detailed structure has not been studied. Evidence obtained with modern techniques for their presence elsewhere on the body is lacking. As yet no species of insect that has been examined thoroughly has been found in which these are absent on the antennae. According to Steiner (92), pegs in pits are the only sense organs present on the antennae of Odonata but it is not known whether these are thick- or thin-walled. Structurally, the thin-walled receptors are the most elaborate of the antennal sense organs of most insects. They may take the form of hairs, pegs, circumfila, or plates. They may arise from the surface or be set in pits.

Surface hairs and pegs.—A thin-walled hair or peg may be nearly straight, slightly curved, or abruptly curved. The tip is usually rounded although this may be apparent only after examination with the highest powers of the light microscope. The wall is usually colorless, transparent, and may be less than 0.1 μ thick. Except at the base, it is perforated by a large number of excessively small pores. In *Lygaeus kalmii,* the small milkweed bug, an estimate of 1500 pores for a single peg has been reported (80). In a grasshopper, *Romalea microptera,* in which the pores were first seen, they are simple tubular openings with a diameter at the surface of about 0.1 μ (87). In the flesh fly, *Sarcophaga argyrostroma,* the pores may be nearer 0.03 μ in diameter at the surface. Each pore expands below the surface to form a small chamber with a delicate floor at which the filaments extending from the dendrites end and where their tips are exposed to the air within the chamber (81). It is this increase in diameter below the surface that permits the detection of these pores, in some species, with the light microscope. In still other species, there is a constriction a short distance below the surface and the filaments from the dendrites end just below it (80). A list of sixteen species of insects in which thin-walled pegs with pores in their wall have been reported is given in a recent review (69). To this list we may now add *Danaus gilippus berenice,* the queen butterfly (50); an ambrosia beetle, *Trypodendrum lineatum* (47); three species of face flies belonging to the genus *Hippelates* (21); and the sorghum midge, *Contarinia sorghicola* (85). All of these have been examined with the electron microscope. In addition, the presence of the pores is indicated by studies with the light microscope of the thin-walled pegs of a walkingstick, *Carausius morosus* (68); several species of earwigs, including *Forficula auricularia* (70); three species of mantids belonging to the genus *Tenodera* (71); *Gromphadorhina portentosa* and five other species of roaches (72); and a parasitic wasp, *Nasonia vitripennis* (73).

When it was first found that the thick-walled chemoreceptors of insects were penetrated by stain when a solution of acid fuchsin was applied to the external surface, the thin-walled chemoreceptors remained uncolored (61). It was not until crystal violet was substituted for acid fuchsin that color could be detected in the latter (65). In living thin-walled pegs the stain diffuses rapidly through the fluids within the peg lumen and is lost by dilution; but in preserved material the dye moves more slowly, tends to accumulate and is easier to detect. The thin-walled chemoreceptors of many species have been found to take up the stain (1, 9, 53, 68, 70–73, 76, 78, 80–82, 85, 88).

The cuticle of the thin-walled hair or peg is continuous with the epicuticle of the antennal surface and approaches it in thickness. It is not affected by the molting fluid and is shed with the exuviae with little or no change in its dimensions. When Campbell's test for chitin is applied to the antenna of the grasshopper, the thin-walled pegs disappear completely (87). This indicates that they contain no chitin. It has been objected that the membrane at the base of the peg may contain so little chitin that it may disintegrate in the hot KOH solution used and the peg be detached and lost (56)). To test this, grasshopper antennae were placed in the solution used by Campbell but, instead of heating it to 160° C, with which results are obtained in 20 minutes, the treatment was carried out in an incubator at 37° C. Here, the antennae remained for four months. Examination then showed the pegs to be no longer present either on the antennae or at the bottom of the container.

In the grasshopper, a cuticular sheath is invaginated from the base of the peg. Near its distal end the dendrites leave the sheath through minute openings in its wall and pass upward into the peg lumen (83, 87). In the small milkweed bug, a number of cuticular sheaths are attached around the periphery of the base of the peg. Eight or ten are common and as many as thirteen have been seen for a single sense organ. Each contains a few dendrites from the large mass of neurons lying below the peg. The dendrites leave the sheaths through openings at their outer ends (80). In some species, no cuticular sheath is present and the dendrites pass directly into the lumen of the peg.

Pegs in pits.—The thin-walled pegs that have been studied most thoroughly with the electron microscope are situated on the surface of the antenna. Others occur that are set in pits opening onto the antennal surface. In the higher Diptera, these may occur in groups within single pits or in multiple pits that have a common outlet to the exterior. In *Sarcophaga argyrostoma,* the thin-walled pegs in pits may be of several kinds in different parts of the antenna or different kinds may be together in the same pit (81). One type is a simple, short, round-tipped structure, another has a sharp tip, and a third narrows suddenly about halfway between its base and tip. Pit pegs of several kinds were described by DuBose & Axtell (21) for the antennae of *Hippelates pusio, H. pallipes,* and *H. bishoppi.* One of these increases abruptly in diameter about halfway between its base and tip.

Grooved pegs.—In addition to the types of thin-walled chemoreceptors described above, others with an elaborately sculptured cuticle are known. Some occur on the surface while others are located in pits. They have been reported for *Aedes aegypti* [(79), Figs. 19–24]; *Phormia regina* (20), Pl. I, Fig. 1; Pl. III, Fig. 4; Pl. IV, Fig. 5]; *Apis mellifera* [(78), Figs. 46, 47, 49]; *Sarcophaga argyrostoma* [(81), Figs. 37–44]; three species of *Hippelates* [(21), Figs. 15–23, 26], and *Danaus gilippus berenice* [(50), Figs. 19, 23, 24]. They have been referred to as stellate pegs and coronal pegs because of their appearance in cross sections (20). These sense organs are infrequently seen in electron micrographs and seem to occur in small numbers in the species examined. A complete description of one of these sensilla is still lacking. The wall of the peg is elaborately grooved or otherwise ornamented, and may appear to be double wtih a space between the two parts that is interrupted by bridges of cuticle. Narrow canals pass through the wall, sometimes in a sinuous course. These end at small pores in the surface. Occasionally, a thin filament may be seen extending from a dendrite in the peg lumen and entering one of the canals [(81), Fig. 38]. Presumably, these filaments traverse the canals and their tips are exposed in the open pores. Satisfactory evidence to establish this is difficult to obtain because of the scarcity of the receptors and because the canals within the cuticle are convoluted. An entire canal is seldom present in a single section. As yet, pegs of this type have been seen, with few exceptions, only in electron micrographs. They are difficult to recognize in light microscope preparations. If a species were found that possessed grooved pegs in large numbers it would be especially favorable for a thorough study of these curious structures.

Circumfila.—The transparent, thin-walled circumfila on the antennae of males of the dipteran family Cecidomyidae encircle the antenna in a series of loops. In the sorghum midge, *Contarinia sorghicola,* two such series of loops encircle each subsegment of the antennal flagellum. The loops arise from a short cylindrical structure about 1 μ long. This branches to form two and each branch is joined to its neighbor on its own side. In the female, these are also present but they lie close to the antennal surface and are attached to it in a complex pattern. It is difficult to imagine how such structures develop in the pupa. In electron micrographs, the circumfila are seen to have the structure of thin-walled pegs (85). The lumen is filled with dendrite branches and the wall is perforated by a very large number of openings about 100 Å in diameter. The dendrites arise from a few neurons, possibly only two. The circumfila stain with crystal violet when a solution of this dye is applied to the external surface of the antenna.

Plate organs.—Plate organs are rounded, oval, or elongate areas in the surface of the antenna. They may be nearly flat, elevated only slightly above the surface, or rise above it to form a low dome. They occur in large numbers on the antennae of many species of Hymenoptera, some species of Coleoptera, and some species of Homoptera. Their function has long been a matter of controversy but recent experiments by Lacher & Schneider (42) and Lacher (41) on the honey bee, *A. mellifera,* give strong support to those who believe that they have an olfactory function. In the aphid, *Megoura vicae,* the structure of the plate wall is very similar to that of the wall of the thin-walled hairs and pegs of other insects (88). Krause (40) reported that small perforations were present in the outer surface of KOH preparations of the plate organ of a scarabaeid beetle, *Phyllopertha horticollis,* and Ivanov [(39), Figs. 4 I, 4 II] shows electron micrographs of sections through the plate organ of a dytiscid beetle, *Acilius sulcatus.* Distinct pores are present with dendrite branches near their inner openings.

The entire surface of the elongated plate in *Nasonia vitripennis,* one of the parasitic wasps, is thin-walled and perforated uniformly by small openings. They stain rapidly when a solution of crystal violet is applied to the external surface (73). The oval plate organ of the honey bee has a different structure. Here the entire central portion of the plate is transparent, about 0.5 μ thick and has no openings in it. The plate is attached to the antennal wall by a thin, narrow membrane, about 0.1 μ in depth, that has in it radially arranged rows of pores. Each pore has a diameter of less than 100 Å. The dendrites from the sensory neurons lie just below the thin membrane and extend around the entire periphery of the plate but do not lie below the plate itself (78). It is possible that fine pore filaments extend from the dendrites to the small pores but this has not yet been proven. The diameter of the pores is less than the thickness of the usual thin section prepared for the electron microscope. This makes them difficult to follow in the dense cuticle. The work on the honey bee plate organs was done before the introduction of improved methods of fixation and special stains for material to be studied with the electron microscope. The honey bee plate organs should be reinvestigated with these more modern techniques.

Sensory Neurons

For many years, it was thought that the mechanoreceptors of insects were provided with a single bipolar sensory neuron and chemoreceptors with a group of such neurons (14, 90). For most of the sense organs examined so far, this generalization still holds but recently a few exceptions have been reported. A single neuron is described as present for the thin-walled pegs of *Blabera craniifer* by Urvoy (95), for several species of *Necrophorus* by Boeckh (5), for *Phormia regina* by Dethier, Larsen & Adams (20), for *Telea polyphemus* by Boeckh, Kaissling & Schneider (6), and for the larvae of *Protoparce sexta* by Schoonhoven & Dethier (60). In the last one listed, the receptors are located on the maxillary palps. At the other extreme, 50 neurons may be associated with a single thin-walled sense organ on the antenna of the grasshopper (87), and 60 or more with those of the small milkweed bug (80). When only a few neurons are present they may lie entirely within the antennal epidermis but when large numbers occur, the mass may bulge below the epidermal layer. In the honey bee, so many neurons are present that those of one sense organ are in close contact with those of their neighbors on all sides and together they form a thick, dense layer beneath the cuticle.

An axon extends below each neuron and joins a branch of the antennal nerve that lies in the antennal lumen. Whether the axons from several neurons fuse has long been a question of interest to physiologists (98). Attempts to find the answer by comparing, with the light microscope, the number of neuron cell bodies in the antenna and the number of axons in cross sections of the antennal nerve at the proximal end of the flagellum, are hindered by the size of the smaller axons. Some are less than 0.1 μ in diameter. Even with the electron microscope, such a count is not easy because of the difficulty of deciding whether a small circular structure is or is not an axon. Moeck (47) found for a beetle, *Trypodendron lineatum,* that the number of neurons and the number of axons in a single antenna are both not far from 2000, and that there is no evidence for fusion. In fact, his axon count was slightly higher than his neuron count. Axons contain mitochondria, endoplasmic reticulum, microtubules, and other cytoplasmic elements such as are commonly found in the axons of other animals. The cell body of the neuron, likewise, contains the expected cytoplasmic organelles. The nucleus is usually rounded and the chromatin may be more diffuse than that in nearby epidermal and sheath cells.

The dendrite of the sensory neuron, in contrast to the axon and cell body, presents many unusual features. As it leaves the cell body it contains the usual cytoplasmic constituents but it narrows a short distance above the nucleus and there assumes the structure of a cilium. The presence of the ciliary region in the dendrites of the plate organs of the honey bee, *A. mellifera,* was reported independently in 1960 by Slifer & Sekhon in the United States (77) and Krause in Germany (40). Such a ciliary region was first described for an insect by Gray & Pumphrey in the dendrites of the auditory organ of the locust (29). Since then, ciliary regions have been reported for the sensory dendrites of many species of insects. The ciliary region may be only a few microns long and is easily missed in electron micrographs of thin sections. In a few species, it has been seen as a thin filament with the light microscope (70) but its nature cannot be correctly understood without knowledge acquired with the electron microscope. Below the ciliary region lies a basal body[3] from which rootlets with a periodic structure extend deep into the proximal portion of the dendrite. Frequently, a second basal body lies below the first and the rootlets pass around it. When two are present, they have been referred to as the distal and proximal basal bodies. It is not known whether two are always present or whether, sometimes, there may be only one. The basal bodies have the expected structure and in cross section show nine sets of triplet tubules at the periphery. Two of each of the nine triplets continue into the cilium to become the nine pairs of doublets. There is no central pair in the cilium although cross sections of cilia are sometimes seen with one or more additional tubules. Arms on one member of the doublet, such as are seen in motile cilia, have not been found in the ciliary region of insect dendrites and it is possible that both tubules are alike. It should be noted that of the two types of thick-walled chemoreceptors on the antenna of an earwig, *Forficula auricularia,* one has a dendrite that expands to form a sphere, about 1 μ across, immediately below the ciliary region (70, 84). Such a structure has not been seen on the sensory dendrite of any other insect so far examined.

Above the short ciliary region, the dendrite widens again and here the pairs of ciliary tubules separate and become indistinguishable from the microtubules present elsewhere in the neuron and in neighboring sheath cells. Distally, the number of microtubules increases beyond the eighteen that would result from separation of the ciliary tubules and there is some evidence that the ciliary tubules and the microtubules are continuous structures (84). The increase in number appears to be the result of branching. Occasionally, cross sections of dendrites in this region are obtained that are sufficiently clear for the microtubules to be counted. Numbers close to some multiple of 18—such as 36, 72, or 144—are then obtained. Exact fits

[3] Basal bodies were seen by earlier workers with the light microscope and were called Riechsstäbchen, Sinnesstäbchen, sensory rods, etc. For a discussion of this see Slifer & Sekhon (78).

would not be expected since the tubules seem to branch at different levels and it is possible that all may not branch the same number of times. This has also been noted for *Sarcophaga argyrostoma* [(81), Figs. 31, 32].

Distal to the base of the ciliary region, mitochondria, Golgi complexes, endoplasmic reticulum, etc., are completely absent. The dendrite here consists only of cell membrane, microtubules, and the structureless material, probably a fluid, that surrounds the microtubules. Sometimes, vesicles are present within the dendrite [(83), Figs. 11, 14, 15, 17] in or beyond the ciliary region and these may consist of nutritive material that is passing outward. This has been suggested for somewhat similar vesicles in the developing cilia of tissues and organ cultures of the lungs of foetal rats (91). Moulins [(48), Figs. 3, 6, 10] noted multivesicular bodies in the ciliary segment of the chemoreceptors of the hypopharynx of *Blabera craniifer*. The absence of mitochondria, etc., within the outer ends of the dendrites poses some interesting problems for the physiologist.

In those sense organs in which a cuticular sheath is present, the dendrites enter its inner, open end at or near the point where the short ciliary region ends and the wider part containing only microtubules begins. If the hair or peg is of the thick-walled type, the dendrites traverse the cuticular sheath without branching and end at the single opening at the tip. If the sense organ is thin-walled and has one or more cuticular sheaths attached inside its base, the dendrites leave through openings in the sheath wall and enter the peg lumen (83, 87). Branching may occur at the base of the peg, as in the grasshopper (83, 87), or at different levels, as in the flesh fly (81). In the former, the dendrites in a cross section are all of nearly the same diameter while in the latter both large and small dendrites may be seen. The smallest dendrite branches approach 400 Å in diameter and contain a single microtubule (85). Moulins (49)) has noted that one of the five dendrites in the hypopharyngeal sense organs of the roach differs structurally from the other four.

Since the thick-walled tarsal and labellar hairs of Diptera respond to mechanical as well as to chemical stimuli, it was proposed by Grobowski & Dethier (28) that a dendrite from one of the neurons of the several innervating the hair was attached to the base while the others continued to the tip. Adams, Holbert & Forgash (2) suggest that a filament that extends from the group of dendrites shown in Figure 11 of their paper may be the dendrite sensitive to movement. However, the structure indicated is not sufficiently well-preserved to be sure that it is a dendrite. Larsen (43) also reports such a dendrite but, again, the material is in too poor condition to permit a decision. The problem awaits further study.

The first thin-walled pegs to be examined with the electron microscope were those of the grasshopper (87). Sections through the peg wall showed it to be perforated by large numbers of small openings. A single opening contained about two dozen tubular structures, each measuring from 100 to

200 Å in diameter. These lie parallel to one another and their outer ends are in contact with the air. They were thought to be the subdivided ends of the smallest branches of the dendrites. Later, after improved methods had been introduced for fixing and staining materials to be studied with the electron microscope, the thin-walled pegs of the grasshopper were re-examined (83). It was then found that the structures within the pores were not the tips of the dendrites but were, instead, clusters of fine filaments that arose from the sides of the dendrites. Meanwhile, the thin-walled receptors of other species of insects had been studied and their pores also found to contain filaments that could be traced back to the dendrites. They were given the name "pore filaments" (81) and have been reported in species of Hemiptera (80), Homoptera (88), Lepidoptera (50, 58), Coleoptera (59), and Diptera (21, 54, 81). Because of their extreme delicacy and because they are suspended between the cuticle and the dendrites where damage may easily occur during fixation and later handling, the filaments are difficult to demonstrate, and good electron micrographs showing them not easy to obtain. The pore filaments are less than 200 Å in diameter. This is about half that of the smallest dendrite branches. They have a surface membrane that is thinner than that of the dendrite (81). No internal detail has yet been seen in them and the material of which they are composed is unknown. Several suggestions as to their nature have been proposed (69).

In those chemoreceptors in which no cuticular sheath is present, the dendrites pass upward to the cuticular parts of the sense organ without it. Certain of the olfactory organs of Homoptera and Hymenoptera lack such a sheath. The plate organs of *Megoura, Nasonia,* and *Apis* are of this type. In all three, the outer plate, together with a second but incomplete cuticular layer below it, encloses a chamber into which the dendrites pass through one or more openings in the inner layer. The dendrites of *Megoura,* after branching within the outer chamber, send pore filaments into the small openings in the plate at the outer surface of the sense organ (89). *Nasonia* has not been studied with the electron microscope so whether it has pore filaments is not known (73). The work on *Apis* was done before adequate methods had been discovered for staining sections to be examined with the electron microscope. It is probable that pore filaments will be found in both *Nasonia* and *Apis.*

In the first description of the dendrites of the thin-walled peg of the grasshopper it was pointed out that molting presented a special problem since the dendrites, after passing through the cuticular sheath, leave the sheath through small openings in its wall (87). Either the dendrites must be retracted into the new sheath or else their outer ends torn off as the old sheath is pulled out with the rest of the exuviae. No evidence was obtained at that time to permit a choice between the two possibilities. Recently, Wensler & Filshie (97) have obtained electron micrographs of molted cuticular sheaths of thick-walled chemoreceptors from the food canal of aphids and in these the distal ends of the dendrites are clearly present. Thus, the second of the two possibilities is shown to be the correct one and we now know that the outer ends of the dendrites may be lost with the exuviae. It was suggested earlier that the basal bodies may be necessary for the regeneration of the outer ends of the dendrites (83).

Sheath Cells

The specialized epidermal cell that produces the hair or peg of a sense organ is known as the trichogen cell, and the one that produces the membrane at its base is the tormogen cell. These terms have been in use for many years. Difficulties arise, however, when it is found that more than two such cells are present in a sense organ. The terms, moreover, are not appropriate when structures such as plate organs are involved. Earlier workers designated these as cap cells, envelop cells, etc. Collectively, all such cells will be referred to here as sheath cells but the more limited terms will be used when the identity of a particular cell can be determined. As many as four cells, one encircling the other, have been seen surrounding the dendrites [(81), Fig. 26]. Moulins (49) describes a distal and a proximal tormogen cell, as well as a trichogen cell, for a chemoreceptor in *Blabera craniifer*. Sheath cells are usually large and conspicuous in insects preparing to molt or in those that have recently done so. Later they decrease in size.

After the trichogen cell has secreted the wall of the hair or peg, including the cuticular sheath if one is present, it withdraws leaving the lumen filled with fluid. Whether the dendrites and trichogen cell are all inside the hair or peg as it is formed or whether the neural elements enter after the trichogen cell pulls back remains to be discovered. The tormogen cell, likewise, withdraws from the membrane that it has secreted. In the fully developed chemoreceptor, a cross section just above the ciliary region shows the dendrites occupying the central area. These may or may not be enclosed within a cuticular sheath. The trichogen cell forms a closed ring that encircles the dendrites and its membranes from opposite sides meet in a well-defined junction. Outside this the tormogen cell is wrapped around the trichogen cell and its opposite edges make contact at a point some distance from that where the edges of the trichogen cell meet [(70), Fig. 17]. Additional sheath cells may be arranged concentrically around these two. The boundaries between sheath cells may range from relatively simple to extremely intricate [(70), Figs. 15–18]. Septate desmosomes are often well-developed where the various cell membranes come into contact.

A cross section through the chemoreceptor above the region just described but still below the inner surface of the antennal cuticle, usually shows a space between the trichogen and tormogen cells. This is filled with a fluid that leaves a fine coagulum in fixed material and is continuous with the fluid inside the peg. The surfaces of the trichogen and tormogen cells that are exposed to the fluid are covered with microvilli and other extensions of a less regular shape. The fluid between the two cells is almost certainly secreted by them; at least, there is no other nearby source for it.

These are the cells that earlier secreted the cuticular parts of the sensillum, and secretory granules are present in them in later stages. A sheath cell may be wrapped around a fluid-filled space in such a way that the space, when seen in a cross section, appears to be inside the cell. In a longitudinal section, however, the space is clearly seen to be outside it [(80), Fig. 24]. Larsen [(43), Fig. 8] states that the tormogen cell in *Phormia* contains a large vacuole but this appears to be a fluid-filled space surrounded, in part, by the cell but not actually within it. The "intracellular fibrillae" that he describes are actually microvilli on the external surface of the cell.

The fluid mentioned in the preceding paragraph may be continuous, either directly or indirectly, with that surrounding the ciliary region of the dendrites, with that within the peg lumen and with that within the cuticular sheath, when one is present. Since mitochondria and most other cell organelles are absent in the distal portions of the dendrites, it is probable that the fluid plays an important role in the nutritive and excretory functions of the dendrites .It must also serve to keep the neural elements moist where they are exposed to the air. It may also play a part in the transmission of electric changes along the dendrites following stimulation. It would be interesting to know the ionic composition of the fluid. Evidence was obtained earlier that it contains proteins that are readily denatured [(86), p. 372]. The fluid can be forced from the living thick-walled chemoreceptor by gentle pressure and if extruded into mineral oil, for example, might be picked up with a micropipette for analysis. It should also be noted that the cuticular sheath is permeable to dyes in aqueous solution and so, presumably, to other small molecules [(86), p. 367]. Stürckow (93) has recently reported that a viscid fluid may be extruded from the tip of the thick-walled chemoreceptors of the blow fly. Moulins (49) states that the fluids inside the cuticular sheath of the hypopharyngeal papillae of *Blabera* and those in the vacuole between the trichogen and tormogen cell are in separate compartments. It would be interesting to know whether the cuticular sheath containing the dendrites in *Blabera* is permeable to the materials in the two spaces.

An unusual feature that has been noted in some sheath cells in species studied with the electron microscope is a labyrinth of conspicuous interconnected vesicles and canals within the cytoplasm (49). These are especially well-developed in the antennae of the sorghum midge, *Contarinia sorghicola* (85). Some of them appear to open into the lumen of the antenna and others into the vacuole between the trichogen and tormogen cells. These may form a continuous pathway between the antennal lumen and the vacuole; however, this seems unlikely. Also, as was suggested by Moulins (49), it is not known whether these are a permanent feature of the cell or appear and disappear according to its functional state. Like the fluid within the lumen of the peg, that inside the cuticular sheath and that between the trichogen and tormogen cells are both filled with a fine granular residue in fixed material. The hemolymph in the antennal lumen leaves a similar residue.

CHEMORECEPTORS OF OTHER ARTHROPODS

The chemoreceptors of arthropods other than insects have received little attention but a few important papers have been published and will be discussed here.

MEROSTOMATA

Hayes (30) has examined the chemoreceptors present on the coxal gnathobases of the walking legs, chilarial appendages, and chelae of all the limbs of *Limulus polyphemus*. The only part visible externally for each of these sense organs is a pore. From 6 to 15 neurons send dendrites through a cuticular sheath to the surface where their tips are exposed to the surrounding sea water. No ciliary region was found but this part of a dendrite may be easily missed in thin sections since it is often only a few microns long. In cross sections of the dendrites (Fig. 26) from 18 to 28 microtubules can be counted and this hints that a ciliary region will be found. The chemoreceptor of *Limulus,* in its general features, resembles the gustatory papillae described for insects by Larsen (44), Moulins (48, 49), and Wensler & Filshie (97).

ARACHNIDA

In 1966, Farish & Axtell (22) concluded, from experimental studies, that olfactory sense organs are probably present on the tarsi of the first pair of legs of *Machrocheles muscaedomesticae,* a mite that is associated with house fly eggs and adults. Coon & Axtell (13) have now examined the tarsi of these mites with the phase microscope and transmission and scanning electron microscopes. Eight of the sense organs on each of the fore tarsi have a blunt tip, many pores of an unusual shape in their thin wall, and a lumen that contains dendrites. Their resemblance to the thin-walled olfactory sensilla of insects is striking.

CRUSTACEA

Laverack (45) and Laverack & Ardill (46) described the aesthetascs of the crustacean antennule. These are the sense organs for which a chemoreceptor function has been reported by physiologists (25). The species studied by Laverack & Ardill was a marine form, *Panulirus argus,* and their work included examination with the electron microscope. The aesthetascs of *Panulirus* are numerous elongate hairs that are concentrated in an area along one side of the antennule. Laverack & Ardill state that a pore is present at the distal end of the aesthetasc and that the dendrites of many sensory cells come into contact with the sea water which enters and leaves through the pore. According to them, the dendrites do not extend into the distal third of the aesthetasc and this condition is not a fixation artifact. The aesthetascs are remarkable for the large number of neurons associated with each. More than a hundred may be present. A ciliary region is present in their dendrites.

The work of Ghiradella, Case & Cronshaw (24–26) is especially interesting. They have compared the aesthetascs of six species of marine and terrestrial Crustacea. In *Pagurus hirsutiusculus,* a marine hermit crab, the aesthetasc is long and slender while in *Coenobita compressus* and *C. brevimanus,* both terrestrial hermit crabs, it is short. A comparison of the cuticle covering the distal end of the structure shows that the aesthetascs of the aquatic species are covered on all sides by a very thin cuticle in which no pores can be detected either at the tip or in the cuticle itself when examined with the light and electron microscopes. The cuticle of the terrestrial species, in contrast, has thin cuticle along a part of one side only while the rest is covered with a heavier cuticle. When the antennule is treated with a solution of crystal violet applied to the outer surface, the dye penetrates the thin cuticle within seconds and colors the contents of the aesthetasc deeply. Elsewhere, the heavier cuticle is impermeable to it. The rapid passage of the dye into the sense organ contrasts strongly with the much slower entrance of stain into the insect chemoreceptor. The three other species of Crustacea examined—*Panulirus interruptus, Cancer antennarius,* and *C. productus*—are all marine and their aesthetascs are similar to those of *Pagurus.*

Several other unusual features have been noted by Ghiradella and her co-workers. As many as 500 neurons may be associated with a single aesthetasc. This is far more than has been reported for any insect chemoreceptor. A single neuron has a dendrite from which two ciliary regions arise, instead of one as in the insect. In *Pagurus* the ciliary region lies within the aesthetasc while in *Coenobita* it lies below the base of the sense organ as it does in insects. Each ciliary region has its own basal bodies, rootlets, etc. Another feature, so far not noted in insects, is the beading of the outer portions of the dendrites. At intervals the dendrites expand and then decrease. Microtubules may be seen running through the expanded portion. Whether such expansions are present in the living dendrite is not known. Ochs (52), who was interested in the flow of axoplasm in the sciatic nerves of mammals, states that swellings occur on axons that have been stretched slightly at the time of fixation. His illustrations resemble those shown for the dendrites of the crustacean aesthetasc.

Bouligand [(10), Plate VIII], in his study of the development of the integument of copepods, includes an electron micrograph of a portion of the antennule of *Lamippe rubicunda* which shows cross sections of about 15 dendrites. Ciliary fibrils can be recognized in some of them.

Diplopoda

The fine structure of a pair of temporal organs on the head of a millipede, *Glomeris romana,* has been studied by Bedini & Mirolli (3, 4). The organs are located near the base of each antenna and are thought to be chemoreceptors. Physiological or behavioral evidence for this, however, is lacking. In several respects the structures resemble the plate organs of aphids. The cuticle at the surface is thin and the branches of dendrites from the sensory neurons lie below it. Water placed on the surface is quickly absorbed. An unusual feature is the presence of two cilia on each dendrite. This recalls the similar situation in Crustacea for which two cilia have been reported for each dendrite of the aesthetasc (25).

In summary, the chemoreceptors of one or more species from each of twelve orders of insects have now been studied with (*a*) the electron microscope, (*b*) with dyes applied to the external surface, or (*c*) with both techniques. Species from all of the larger orders of insects are included but seventeen smaller orders, as listed by Imms (38), have received no attention. Information concerning arthropods other than insects is even more scanty. There are many opportunities here for interesting work.

Note added in proof: An important paper by K.-D. Ernst (*Z. Zellforsch. Mikroskop. Anat.,* **94,** 72–102, 1969) appeared shortly after the present article was completed. Ernst has obtained evidence that the pore filaments of the olfactory organs of a beetle, *Necrophorus,* are cuticular canals and not extensions of the dendrite branches. He reports that they are present in the pupa before the dendrites enter the peg. The relation of the dendrites to the inner ends of the filaments or tubules remains obscure.

LITERATURE CITED

1. Abushama, F. T. The olfactory receptors on the antenna of the damp-wood termite *Zootermopsis angusticollis* (Hagen). *Entom. Monthly Mag.*, **100,** 145–47 (1965)
2. Adams, J. R., Holbert, P. E., Forgash, A. J. Electron microscopy of the contact chemoreceptors of the stable fly, *Stomoxys calcitrans.* (Diptera; Muscidae). *Ann. Entomol. Soc. Am.*, **58,** 909–17 (1965)
3. Bedini, C., Mirolli, M. Sensory cilia in the temporal organs of *Glomeris* (Myriapoda, Diplopoda). *Naturwissenschaften,* 373–74 (1967)
4. Bedini, C., Mirolli, M. The fine structure of the temporal organs of a pill millipede, *Glomeris romana* Verhoeff. *Monitore Zool. Ital.*, *1,* 41–63 (1967)
5. Boeckh, J. Elektrophysiologische Untersuchungen an einzelnen Geruchsrezeptoren auf den Antennen des Totengräbers (*Necrophorus,* Coleoptera). *Z. Vergleich. Physiol.*, **46,** 212–48 (1962)
6. Boeckh, J., Kaissling, K.-E., Schneider, D. Sensillen und Bau der Antennengeissel von *Telea polyphemus* (Vergleiche mit weiteren Saturniden: *Antheraea, Platysamia* und *Philosamia*). *Zool. Jahrb. Abt. Anat. Ontog. Tiere,* **78,** 559–84 (1960)
7. Boeckh, J., Kaissling, K. E., Schneider, D. Insect olfactory receptors. In *Cold Spring Harbor Symp. Quant. Biol., Sensory Receptors,* **30,** 263–80 (1965)
8. Borden, J. H. Antennal morphology of *Ips confusus* (Coleoptera: Scolytidae). *Ann. Entomol. Soc. Am.*, **61,** 10–13 (1968)
9. Borden, J. H., Wood, D. L. The antennal receptors and olfactory response of *Ips confusus* (Coleoptera: Scolytidae) to male sex attractant in the laboratory. *Ann. Entomol. Soc. Am.*, **59,** 253–61 (1966)
10. Bouligand, Y. Le tégument quelques copépodes et ses dépendances musculaires et sensorielles. *Mem. Museum Natl. Hist. Nat. (Paris),* **40,** 189–206 (1966)
11. Bullock, T. H., Horridge, G. A. Arthropoda: receptors other than eyes, in *Structure and Function in the Nervous System of Invertebrates,* **2,** Chap. 18, 1005–62. (Freeman & Co., San Francisco and London, 1965)
12. Butler, C. G. Insect pheromones. *Biol. Rev.*, **42,** 42–87 (1967)
13. Coons, L. B., Axtell, R. C. (Personal communication, 1969)
14. Dethier, V. G. Chemoreception, in *Insect Physiology,* 544–76. (Roeder, K. D., Ed., John Wiley & Sons, Inc., New York, 1953)
15. Dethier, V. G. The physiology of olfaction in insects. *Ann. New York Acad. Sci.*, **58,** 139–57 (1954)
16. Dethier, V. G. The physiology and histology of the contact chemoreceptors of the blowfly. *Quart. Rev. Biol.*, **30,** 348–71 (1955)
17. Dethier, V. G. *The Physiology of Insect Senses.* (John Wiley & Sons, Inc., New York, 266 pp., 1963)
18. Dethier, V. G. Feeding behaviour. *Symp. Roy. Entomol. Soc. London,* **3,** 46–58 (1966)
19. Dethier, V. G., Chadwick, L. E. Chemoreception in insects. *Physiol. Rev.*, **28,** 220–54 (1947)
20. Dethier, V. G., Larsen, J. R., Adams, J. R. The fine structure of the olfactory receptors of the blowfly. *Olfaction and Taste. Proc. Intern. Symp. Wenner-Gren Center, 1st, Stockholm,* **1,** 105–10. (Zotterman, Y., Ed., MacMillan Co., New York, 1963)
21. DuBose, W. P., Axtell, R. C. Sensilla on the antennal flagella of *Hippelates* eye gnats. *Ann. Entomol. Soc. Am.*, **61,** 1547–61 (1968)
22. Farish, D. J., Axtell, R. C. Sensory functions of the palps and first tarsi of *Macrocheles muscaedomesticae* (Acarina: Macrochelidae), a predator of the house fly. *Ann. Entomol. Soc. Am.*, **59,** 165–70 (1966)
23. Frings, H., Frings, M. The loci of contact chemoreceptors in insects. *Am. Midland Naturalist,* **41,** 602–58 (1949)
24. Ghiradella, H., Case, J., Cronshaw, J. Fine structure of the aesthetasc hairs of *Coenobita compressus* Edwards. *J. Morphol.*, **124,** 361–86 (1968)
25. Ghiradella, H., Case, J. F., Cronshaw,

J. Structure of aesthetascs in selected marine and terrestrial decapods: Chemoreceptor morphology and environment. *Am. Zool.*, **8**, 603–21 (1968)

26. Ghiradella, H., Cronshaw, J., Case, J. Fine structure of the aesthetasc hairs of *Pagurus hirsutiusculus* Dana. *Protoplasma*, **66**, 1–20 (1968)
27. Gouin, F. J. Morphologie, Histologie und Entwicklungsgeschichte der Myriapoden und Insekten. III. Das Nervensystem und die neurocrinen Systeme. *Fortschr. Zool.*, **17**, 189–237 (1965)
28. Grabowski, C. T., Dethier, V. G. The structure of the tarsal chemoreceptors of the blowfly, *Phormia regina* Meigen. *J. Morphol.*, **94**, 1–19 (1954)
29. Gray, E. G., Pumphrey, R. J. Ultrastructure of the insect ear. *Nature*, **181**, 618 (1958)
30. Hayes, W. F. Chemoreceptor sensillum structure in *Limulus*. *J. Morphol.*, **119**, 121–42 (1966)
31. Hocking, B. Smell in insects: A bibliography with abstracts. *Defense Research Board, Canada. EP Tech. Rept. 8*, 266 pp. (1960)
32. Hodgson, E. S. Problems in invertebrate chemoreceptors. *Quart. Rev. Biol.*, **30**, 331–47 (1955)
33. Hodgson, E. S. Chemoreception in arthropods. *Ann. Rev. Entomol.*, **3**, 19–36 (1958)
34. Hodgson, E. S. Chemoreception. In *The Physiology of Insecta.* (Rockstein, M., Ed., Academic Press, New York, 1964)
35. Hodgson, E. S. Chemical senses in the invertebrates. In *The Chemical Senses and Nutrition*, 7–18. (Kare, M. R., Maller, O., Eds., Johns Hopkins Univ. Press, 1967)
36. Hodgson, E. S., Lettvin, J. Y., Roeder, K. D. Physiology of a primary chemoreceptor unit. *Science*, **122**, 417–18 (1955)
37. Hopkins, B. A. The probing response of *Stomoxys calcitrans* (L.) (the stable fly) to vapours. *Animal Behaviour*, **12**, 513–24 (1964)
38. Imms, A. D. *A General Textbook of Entomology*, 9th ed. (Revised by Richards, O. W., Davies, R. G., Methuen & Co., Ltd.. London, 886 pp., 1957)
39. Ivanov, V. B. Ultrastructural organization of chemoreceptive antennal sensilles of the beetle *Acilius sulcatus*. *Zh. Evoliutsionnoi Biokhim. Fiziol.*, **2**, 462–72 (1966) (In Russian with English summary)
40. Krause, B. Elektronenmikroskopische Untersuchungen an den Plattensensillen des Insektenfühlers. *Zool. Beitr.*, N. F., **6**, 161–205 (1960)
41. Lacher, V. Elektrophysiologische Untersuchungen an einzelnen Rezeptoren für Geruch, Kohlendioxyd, Luftfeuchtigkeit und Temperatur auf den Antennen der Arbeitsbiene und der Drohne (*Apis mellifica* L.). *Z. Vergleich. Physiol.*, **48**, 587–623 (1964)
42. Lacher, V., Schneider, D. Elektrophysiologischer Nachweis der Riechfunktion von Porenplatten (Sensilla placodea) auf den Antennen der Drohne und Arbeitsbiene (*Apis mellifica* L.). *Z. Vergleich. Physiol.*, **47**, 274–78 (1963)
43. Larsen, J. R. The fine structure of the labellar chemosensory hairs of the blowfly, *Phormia regina* Meig. *J. Insect Physiol.*, **8**, 683–91 (1962)
44. Larsen, J. R. Fine structure of the interpseudotracheal papillae of the blowfly. *Science*, **139**, 347 (1963)
45. Laverack, M. S. The antennular sense organs of *Panulirus argus*. *Comp. Biochem. Physiol.*, **13**, 301–21 (1964)
46. Laverack, M. S., Ardill, D. J. The innervation of the aesthetasc hairs of *Panulirus argus*. *Quart. J. Microscop. Sci.*, **106**, 45–60 (1965)
47. Moeck, H. A. Electron microscopic studies of antennal sensilla in the ambrosia beetle *Trypodendron lineatum* (Olivier) (Scolytidae). *Can. J. Zool.*, **46**, 521–56 (1968)
48. Moulins, M. Les cellules sensorielles de l'organe hypopharyngien de *Blabera craniifer Burm.* (Insecta, Dictyoptera). Étude du segment ciliare et des structures associées. *Compt. Rend. Acad. Sci., Paris*, **265**, 44–47 (1967)
49. Moulins, M. Les sensilles de l'organe hypopharyngien de *Blabera craniifer* Burm. (Insecta, Dictyoptera). *J. Ultrastruct. Res.*, **21**, 474–513 (1968)
50. Myers, J. The structure of the antennae of the Florida queen butterfly, *Danaus gilippus berenice*

(Cramer). *J. Morphol.*, **125,** 315–28 (1968)
51. Noble-Nesbitt, J. The fully formed intermoult cuticle and associated structures of *Podura aquatica* (Collembola). *Quart. J. Microscop. Sci.*, **104,** 253–70 (1963)
52. Ochs, S. Beading phenomena of mammalian myelinated nerve fibers. *Science,* **139,** 599–600 (1963)
53. Prestage, J. J., Slifer, E. H., Stephens, L. B. Thin-walled sensory pegs on the antenna of the termite worker, *Reticulotermes flavipes. Ann. Entomol. Soc. Am.*, **56,** 874–78 (1963)
54. Richter, S. Unmittelbarer Kontakt der Sinneszellen cuticularer Sinnesorgane mit der Aussenwelt. Eine Licht- und Elektronenmikroskopische Untersuchung der Chemorezeptorischen Antennensinnesorgane der *Calliphora*-larven. *Z. Morphol. Oekol. Tiere,* **52,** 171–96 (1962)
55. Schneider, D. Electrophysiological investigation of insect olfaction. Olfaction and Taste. *Proc. Intern. Symp. Wenner-Gren Center, 1st, Stockholm,* **1,** 85–103. (Zottermann, Y., Ed., MacMillan Co., New York, 1963)
56. Schneider, D. Insect antennae. *Ann. Rev. Entomol.*, **9,** 103–22 (1964)
57. Schneider, D., Kaissling, K.-E. Der Bau der Antenne des Seidenspinners *Bombyx mori* L. II. Sensillen, cuticulare Bildungen und innerer Bau. *Zool. Jahrb. Abt. (Anat. Ontog. Tiere*), **76,** 223–50 (1957)
58. Schneider, D., Lacher, V., Kaissling, K.-E. Die Reaktionsweise und der Reaktionsspektrum von Riechzellen bei *Antheraea pernyi* (Lepidoptera. Saturniidae). *Z. Vergleich. Physiol.*, **48,** 632–62 (1964)
59. Schneider, D., Steinbrecht, R. A., Ernst, K. D. Cover picture on *Naturw. Rundschau,* **19** (3) (March 1966)
60. Schoonhoven, L. M., Dethier, V. G. Sensory aspects of host-plant discrimination by lepidopterous larvae. *Arch. Neerl. Zool.*, **16,** 497–530 (1966)
61. Slifer, E. H. The permeability of the sensory pegs on the antennae of the grasshopper (Orthoptera, Acrididae). *Biol. Bull.*, **106,** 122–28 (1954)
62. Slifer, E. H. The reaction of a grasshopper to an odorous material held near one of its feet (Orthoptera: Acrididae). *Proc. Roy. Entomol. Soc. London, Ser. A,* **29,** 177–79 (1954)
63. Slifer, E. H. The distribution of permeable sensory pegs on the body of the grasshopper (Orthoptera: Acrididae). *Entomol. News,* **66,** 1–5 (1955)
64. Slifer, E. H. The response of a grasshopper, *Romalea microptera* (Beauvois), to strong odours following amputation of the metathoracic leg at different levels. *Proc. Roy Entomol. Soc. London, Ser. A,* **31,** 95–98 (1956)
65. Slifer, E. H. A rapid and sensitive method for identifying permeable areas in the body wall of insects. *Entomol. News,* **71,** 179–82 (1960)
66. Slifer, E. H. The fine structure of insect sense organs. *Intern. Rev. Cytol.*, **11,** 125–59 (1961)
67. Slifer, E. H. Sensory hairs with permeable tips on the tarsi of the yellow-fever mosquito, *Aedes aegypti. Ann. Entomol. Soc. Am.*, **55,** 531–35 (1962)
68. Slifer, E. H. Sense organs on the antennal flagellum of a walkingstick *Carausius morosus* Brünner (Phasmida). *J. Morphol.*, **120,** 189–202 (1966)
69. Slifer, E. H. The thin-walled olfactory sense organs on insect antennae. In *Insects and Physiology,* 233–45. (Beament, J. W. L., Treherne, J. E., Eds., Oliver & Boyd, Edinburgh, 1967)
70. Slifer, E. H. Sense organs on the antennal flagella of earwigs (Dermaptera) with special reference to those of *Forficula auricularia. J. Morphol.*, **122,** 63–79 (1967)
71. Slifer, E. H. Sense organs on the antennal flagellum of a praying mantis, *Tenodera angustipennis,* and of two related species (Mantodea). *J. Morphol.*, **124,** 105–16 (1968)
72. Slifer, E. H. Sense organs on the antennal flagellum of a giant cockroach, *Gromphadorhina portentosa,* and a comparison with those of several other species (Dictyoptera, Blattaria). *J. Morphol.*, **126,** 19–30 (1968)
73. Slifer, E. H. Sense organs on the

antenna of a parasitic wasp, *Nasonia vitripennis* (Hymenoptera, Pteromalidae). *Biol. Bull.,* **136,** 253–63 (1969)
74. Slifer, E. H. (Unpublished observations)
75. Slifer, E. H. (Unpublished observations)
76. Slifer, E. H., Brescia, V. T. Permeable sense organs on the antenna of the yellow fever mosquito, *Aedes aegypti* (Linnaeus). *Entomol. News,* **71,** 221–25 (1960)
77. Slifer, E. H., Sekhon, S. S. The fine structure of the plate organs on the antenna of the honey bee, *Apis mellifera* Linnaeus. *Exptl. Cell Res.,* **19,** 410–14 (1960)
78. Slifer, E. H., Sekhon, S. S. Fine structure of the sense organs on the antennal flagellum of the honey bee, *Apis mellifera* Linnaeus. *J. Morphol.,* **109,** 351–81 (1961)
79. Slifer, E. H., Sekhon, S. S. The fine structure of the sense organs on the antennal flagellum of the yellow fever mosquito, *Aedes aegypti* (Linneaeus). *J. Morphol.,* **111,** 49–67 (1962)
80. Slifer, E. H., Sekhon, S. S. Sense organs on the antenna of the small milkweed bug, *Lygaeus kalmii* Stal (Hemiptera, Lygaeidae). *J. Morphol.,* **112,** 165–93 (1963)
81. Slifer, E. H., Sekhon, S. S. Fine structure of the sense organs on the antennal flagellum of a flesh fly, *Sarcophaga argyrostoma* R.-D. (Diptera, Sarcophagidae). *J. Morphol.,* **114,** 185–207 (1964)
82. Slifer, E. H., Sekhon, S. S. Fine structure of the thin-walled sensory pegs on the antenna of a beetle, *Popilius disjunctus* (Coleoptera; Passalidae). *Ann. Entomol. Soc. Am.,* **57,** 541–48 (1964)
83. Slifer, E. H., Sekhon, S. S. The dendrites of the thin-walled olfactory pegs of the grasshopper (Orthoptera, Acrididae). *J. Morphol.,* **114,** 393–410 (1964)
84. Slifer, E. H., Sekhon, S. S. Some evidence for the continuity of ciliary fibrils and microtubules in the insect sensory dendrite. *J. Cell Sci.,* **4,** 527–40 (1969)
85. Slifer, E. H., Sekhon, S. S. Circumfila and other sense organs on the antenna of a male gall midge, *Contarinia sorghicola* (Diptera, Cecidomyidae). (In preparation)
86. Slifer, E. H., Prestage, J. J., Beams, H. W. The fine structure of the long basiconic sensory pegs of the grasshopper (Orthoptera, Acrididae) with special reference to those on the antenna. *J. Morphol.,* **101,** 359–97 (1957)
87. Slifer, E. H., Prestage, J. J., Beams, H. W. The chemoreceptors and other sense organs on the antennal flagellum of the grasshopper (Orthoptera: Acrididae). *J. Morphol.,* **105,** 145–91 (1959)
88. Slifer, E. H., Sekhon, S. S., Lees, A. D. The sense organs on the antennal flagellum of aphids (Homoptera), with special reference to the plate organs. *Quart. J. Microscop. Sci.,* **105,** 21–29 (1964)
89. Snodgrass, R. E. The morphology of insect sense organs and the sensory nervous system. *Smithsonian Inst. Misc. Collections,* **77,** 1–80 (1926)
90. Snodgrass, R. E. *Principles of Insect Morphology.* (McGraw-Hill Book Co., Inc., New York, 637 pp., 1935)
91. Sorokin, S. P. Reconstructions of centriole formation and ciliogenesis in mammalian lungs. *J. Cell Sci.,* **3,** 207–30 (1968)
92. Steiner, H. Die Bindung der Hochmoorlibelle *Leucorrhinia dubia* Vand. an ihren Biotop. *Zool. Jahrb. (Syst. Ökol. Geograph.),* **78,** 65–96 (1948)
93. Stürckow, B. Occurrence of a viscous substance at the tip of the labellar taste hair of the blowfly. In Olfaction and Taste. *Proc. Intern. Symp. Wenner-Gren Center, 2nd, Stockholm,* **8,** 707–20 (1967)
94. Swartzendruber, D. C. The permeability of certain of the sensory hairs on the antennae of lacewings (Neuroptera, Chrysopidae). *Entomol. News,* **69,** 253–58 (1958)
95. Urvoy, J. Étude anatomo-fonctionelle de la patte et l'antenne de la blatte, *Blabera craniifer* Burmeister. *Ann. Sci. Nat. Zool. Biol. Animale,* **5,** 287–514 (1963)
96. Wallis, D. I. The sense organs on the ovipositor of the blowfly, *Phormia regina* Meigen. *J. Insect Physiol.,* **8,** 453–67 (1962)
97. Wensler, R. J., Filshie, B. K. Sense organs in the food canal of some

aphids. (Personal communication, 1969)

98. Wigglesworth, V. B. The histology of the nervous system of an insect, *Rhodnius prolixus* (Hemiptera). I. The peripheral nervous system. *Quart. J. Microscop. Sci.*, **100,** 285–98 (1959)

99. Wigglesworth, V. B. *The Principles of Insect Physiology,* 6th ed. (Methuen & Co., Ltd., London, 1965)

100. Zacharuk, R. Y. Exuvial sheaths of sensory neurones in the larva of *Ctenicera destructor* (Brown) (Coleoptera, Elateridae). *J. Morphol.,* **111,** 35–47 (1962)

HONEY BEE NUTRITION[1]

By Mykola H. Haydak[2]

Department of Entomology, Fisheries & Wildlife, University of Minnesota, St. Paul, Minnesota

In discussing honey bee nutrition, one has to differentiate, as with many other insects, between the nutrition of adults and that of the immature instars.

Nutrition of Adult Bees

The food of adult worker bees consists of pollen and nectar or honey. The nutritive value of pollen from different plants varies considerably (61, 72, 108, 117). Mixed pollens brought into the hives have a high nutritive value (115) and supply all the necessary materials for proper development of young animals (4, 6, 68, 109). When dried, pollen quickly loses its nutritive value on storage at room temperature (42, 73, 112), therefore, in studying the nutritive value of pollens for bees, pollens of the same age, preferably freshly collected, should be used to preclude erroneous interpretations of results.

When bees don't have access to pollen, they may be offered a pollen supplement—foods mixed with pollen (18) or pollen substitutes—foods intended to replace pollen completely. The most widely used substitute consists of a mixture of soybean flour, dried brewers' yeast, and dry skim milk. When commercial casein and dried egg yolk are added, the nutritive value approaches that of mixed pollens brought fresh to the hive by bees (36, 111, 117, 118). Such pollen substitutes do not adversely influence the quality of larval food produced (41), nor do they have any deleterious influence on the activity of the enzymes of the midgut (27).

Nectar and honey contribute mostly mono and oligo saccharides (75, 101, 102, 123) to the food of bees.

Criteria in studying honey bee nutrition.—1. The longevity of emerging bees kept in cages and offered the food being tested is compared with controls (71). 2. The development of various internal organs of emerged bees fed the diet tested is observed or measured (79, 88). 3. Growth (changes in weight and N content), the building activity of bees, the quality and quantity of reared bees (weight and N content), and the mortality of the bees are ascertained (28).

Nutrition of worker bees.—After emerging, some worker bees begin con-

[1] Paper No. 6907, Scientific Journal Series, Minnesota Agricultural Experiment Station, St. Paul, Minnesota.

[2] The author is greatly indebted to Dr. B. Furgala for his careful reading of the manuscript and for his editorial suggestions and corrections in the text.

suming pollen during the first 1 to 2 hours of their presence in the colony. At 12 hours after emergence, 50 per cent or more of the workers have started eating pollen in small amounts (13, 26). However, mass consumption begins when bees are 42 to 52 hours old (26) and reaches a maximum when they are five days old (81, 88). The mandibles are used in crushing clumps of pollen or bee bread (52). For reliable results in experimental feeding, pollen or pollen substitutes should be fed separately as such or in candy form, and not added to a sugar solution (88). Growth begins when emerging bees commence eating pollen. Simultaneously, their hypopharyngeal glands, fat body, and other internal organs develop (72, 83). The degree of these changes, however, depends on general conditions, such as the state, the requirements, and strength of the colony, brood rearing, presence of queen, incoming nectar and pollen, weather (19, 60, 93). Proper nutrition is one of the most important factors influencing the longevity of emerged bees (10, 70, 73, 114). However, the presence of the queen (98) and group rather than isolated living (104) also increases the life span, other conditions being equal.

When bees are 8 to 10 days old, their pollen consumption diminishes (66). This can also be determined by the diminishing weight and N content of their digestive tracts (29). Under abnormal conditions, such as when they are forced to rear brood continuously, proteinaceous food is consumed for a longer period (43, 44).

In addition to feeding themselves, bees feed each other. This is called "food transmission" (21, 87, 90) and is observed only over the brood area (17). Workers may also obtain food from the queen (90); drone food is acquired by ingesting material regurgitated by drones (22, 53), but only in some cases is direct feeding observed (12). Such food transmission is believed to play an important role in maintaining cohesion of the colony.

Nutrition of drones.—The study of nutrition of drones has been neglected and only recently have some observations and analyses been made. Young drones (1 to 8 days old) are fed mainly by younger workers with food which resembles modified worker jelly—a mixture of glandular secretions, pollen, and honey. In some cases, food for drones is derived from the honey stomach of feeding workers. This type of feeding is followed by a period in which the drones feed themselves on honey from the combs, with occasional feeding by workers. The food of flying drones (12 to 26 days old) consists mostly of honey which they take from cells. Only rarely do they receive it from the workers (22, 80).

There is a distinct growth of drones after emergence, an increase amounting to 28 per cent in dry weight, and 38 to 62 per cent in N content for drones four days old (11). Thoraces of nine-day-old animals increase as much as 53 per cent in dry weight and 57 per cent in N content (40).

Nutrition of queens.—There is no clear-cut knowledge of the nutrition of the queen bee. It was previously assumed that during the active season her food consists of royal jelly given by nurse bees, and honey during the winter period. However, chromatographic analyses of the contents of honey stomachs and of ventricular lumina of wintering queens have shown the presence of 17 amino acids but only traces of glucose and fructose. Apparently, wintering queens are fed larval food by workers, and neither feed themselves on honey nor are fed honey (20). Isolated queens kept in cages without workers can feed themselves on sugar-honey candy, provided water is available to them. Under these conditions, most of the queens live more than two weeks but some continue to feed for more than 48 days (122).

Queens also grow after emergence. The greatest increase in thoracic dry weight (47 per cent) and in N content (52 per cent) occurs in queens two years old (40).

Food Requirements of Adult Bees

Water requirement.—Water plays an important role in the life of bees. The form of food has little influence on the longevity provided water is available (71). Isolated queens kept in cages and offered sugar candy and water are known to live on an average of two weeks or more; kept without water they survive only three or four days (122). Water economy in bees is influenced by secretions from the corpora allata and corpora cardiaca, the first increasing water consumption, the latter decreasing it (2). It is a well-known fact that a colony of bees utilizes large amounts of water during the active season to dilute honey and to regulate temperature in the brood nest (63).

Protein requirement.—Under natural conditions, pollen supplies the necessary proteins for bees. As mentioned earlier, growth begins when emerging bees start eating pollen. Within five days the N content increases by 93 per cent in the head, by 76 per cent in the abdomen, and by 37 per cent in the thorax (29). Simultaneously, the hypopharyngeal glands, fat body, and other organs also develop (72). The dry weight and N content of heads and thoraces of queens and drones increase, reaching a peak at certain periods of the life of these castes (queen—2 years, drones—14 days) and then decline (40). Under adverse conditions, when supplies of pollen are lacking for long periods, bees use honey (mostly carbohydrate) as their only food. Under experimental conditions, a colony of bees kept on a pure carbohydrate diet will start rearing brood. However, for this the bees utilize materials of their own bodies, with a consequent loss of weight and diminished N content which is greatest in the abdomen. The resulting young bees have less body nitrogen than do normally reared bees, with the greatest decrease also occurring in the abdomen (19 per cent) (30). Such bees have 62 per cent less thiamine than do normal emerging bees (35). These results are significant in that they suggest that food components can be stored in body tissues during immature development, and demonstrate the importance of an abundant, balanced diet of nurse bees to the well-being of the colony.

When freshly emerged bees are kept on a pure carbohydrate diet, the N content of their bodies diminishes and mortality greatly increases. However, when, even after 30 days on a pure sugar diet, protein-starved bees are offered pollen normal development is re-established and the young bees reared by them are normal (31). When the diet consumed by emerged bees is inadequate, weight and N content increases very slightly. If these bees, even after 60 days on such a ration, receive a proper diet, their growth becomes normal (28). This phenomenon is of great importance. It indicates the tremendous ability of a colony of bees to adjust itself to adverse circumstances and to recover when normal conditions return.

This raises the question of nutrition and aging. Older bees have considerably smaller amounts of vitamins of the B group in muscular tissues than do emerging bees—a condition similar to that found in other animals (50). A colony consisting of bees 47 days old kept on pure sugar diet for 189 days, maintained normal flying activity; but the bees lost 33 per cent of body weight and 22 per cent of N, the greatest N loss (44.7 per cent) occurring in the abdomen (32). Apparently, older bees need only a supply of carbohydrate for energy, deriving all the necessary materials for repair of vital organs by catabolizing the body stores deposited during earlier periods of growth.

When adult bees are forced to rear brood they continue to consume pollen well past the normal nurse age. Under these conditions, as high as 70 per cent of the hypopharyngeal glands of bees 75 to 83 days old may remain fully active (44, 82). However, larval food produced by such bees is of a watery, rather than the usual milky consistency, and contains considerably smaller amount of vitamins of the B group (43). The intestines of emerging bees produced by nurses 50 days old and older are very fragile and the longevity of such bees diminishes as the age of the nurse bees increases (44). The weight and N content of queens produced under similar conditions is considerably lower than that of queens reared by young bees (49).

The amount of protein necessary for a certain growth rate is dependent on the quantitative amino acid composition of the protein concerned and on the requirement of the organism for each of the amino acids. The following amino acids and their proportional ratio are essential for growth of adult honey bees: arginine (3.0), histidine (1.5), lysine (3.0), tryptophan (1.0), phenylalanine (2.5), methionine (1.5), threonine (3.0), leucine (4.5), isoleucine (4.0), and valine (4.0)—the same 10 which are essential for normal growth of rats. Serine, glycine, and proline, though not essential for growth, exert a stimulating effect at suboptimal growth levels. In the study of quantitative requirements for essential amino acids, it was demonstrated that there is a nutritional surplus of each of the essential amino acids in the natural food of bees (11).

Carbohydrate requirement.—Carbohydrate requirements of honey bees have been determined by feeding various sugar solutions to bees and comparing their longevity with that of those receiving pure water. In case of "unsweet" sugars, the latter were mixed with a certain quantity of sucrose solution which is barely enough to sustain bees longer than on pure water alone, but which imparts taste to the "unsweet" sugars and makes the bees consume it. By this method it was found (116) that bees can utilize the following "sweet" sugars: glucose, fructose, saccharose (sucrose), maltose, trehalose, melezitose; those unsweet: arabinose, xylose, galactose, cellobiose, raffinose, mannitol, sorbitol. They cannot utilize rhamnose, fucose, mannose, sorbose, lactose, melibiose, dulcitol, erythritol, or inositol. Mannose is decidedly poisonous to honey bees (110). However, it is found in royal jelly bound in glycopeptide form (100). These findings are supported by results of study of the enzymes of the hypopharyngeal glands and the midgut (73, 74, 76). Bees can utilize dextrins well but only those starches which are biologically important to them (pollen starches). Intact starch grains are not affected by the bee's diastase, because they are protected by a shell of amylopectin (65).

The requirements of adult bees for lipids, vitamins, and minerals have not been studied to any great extent. Although lipase is secreted by the ventriculus of honey bees (125), apparently for growth and development of the hypopharyngeal glands, adult bees do not need extra lipids (47). Nor do they need extra vitamins (5, 11, 47, 72) or minerals (47). It is possible that the minute quantities of these materials in test diets are sufficient for normal adult growth, or that they are stored in sufficient quantity during the immature stadia and can be utilized for growth of adults, provided the diet of emerged bees contains enough good quality proteins for the task (117). The significance of Fe deposited in various tissues of pollen-fed bees (66) is unknown.

Nutrition of Larvae

Larvae of honey bees are fed a special food. Several intensive reviews of analyses of this food are available (3, 9, 46, 55, 56, 94, 95, 113) and interested readers are referred to these sources. It may be stated here that this food supplies all the necessary materials for complete development of all three castes of honey bee larvae.

General.—Upon hatching the bee larva begins to feed on food supplied by nurse bees. Actual observations (62) showed that each feeding is preceded by an inspection during which the nurse bee makes sure where the head of the larva is located. She then turns herself in such a way that the points of her mandibles lie very near the head of the larva. The mandibles open and start to vibrate with minute motions. After one or two seconds a drop begins to appear between the mandibles and is left near the larva. The drop is generally spread out slightly with the mandibles forming a little pool around the larva. The time taken for one feeding, including inspection, is variable, usually from one-half to two minutes.

Nutrition of queen larvae.—Queen larvae, reared in special cells, are supplied throughout their larval life with an abundance of royal jelly. Even after they are sealed in the cells they have food to consume. Nurse bees feeding the larvae deposit two types of secretions: watery-clear and milky-opaque, the former, secreted by nurses averaging 17 ± 2 days of age, while the latter by those of 12 ± 2 days of age. The ratio of these components is approximately 1:1. However, this ratio is dependent on the age of the nurses, the older nurses providing less of the white component. The number of feedings per hour is increased as larvae grow older (1 day old—13 feedings; 3rd day—16; and on the 4th day—25 feedings). The average length of individual feeding also increases with the age of the larva. Total number for the whole larval period is 1600 feedings, lasting 17 hours. In a normal colony, larvae one-fourth to one-half day old receive about 1.13 mg of clear and 0.81 mg of milky white components per feeding. The total amount of food provided one larva is about 1.5 g (57). There is relatively little variation in the composition of royal jelly fed to younger and older larvae (51, 99). However, one analysis showed a considerable difference in vitamin content (64). Only traces of pollen are found in royal jelly (33, 105).

Queen larvae, up to the age of three days, are fed more of the white secretion, while those four days and older, receive more of the clear component. Judging from the pH of these secretions it may be concluded that the clear substance is a mixture of hypopharyngeal gland secretion and honey, while the milky-white mixture is the secretion of the mandibular glands. The milky-white secretion gives a light blue fluorescence on chromatographic analysis as do the secretions of the mandibular glands. Paper chromatographic separations showed that both secretions contained ninhydrin-positive substances (57). It was established (89) that the proteins of royal jelly are derived from the secretion of the hypopharyngeal glands. The protein content of the clear component is 110.5 mg/g and that of the white 140.5 mg/g. Consequently, the milky-white secretion is a mixture of the secretion of the mandibular and the hypopharyngeal glands (57). These conclusions are supported by the fact that when the secretion of the mandibular glands comes in contact with the lobules of the hypopharyngeal glands, the latter become milky-opaque, condensation appearing inside the cells (38). During the first three or four days of their lives, queen larvae grow more slowly than do the worker larvae of corresponding age (16, 119). They then increase their rate of growth and overtake the worker larvae, and reach a weight of 300 to 322 mg. However, there is considerable variation in the weight of larvae of the same age. The differences in rate of growth may be found between colonies, as well as between individual larvae within a colony (105).

Nutrition of worker larvae.—Young (newly hatched to 2.5 days old) larvae are always surrounded by, or even float on an excessive amount of food material which is uniformly grayish-white and of pastelike consistency (worker jelly). Although a certain amount of penetration of dissolved substances may take place through the body wall (105), the amount absorbed is very small and is not enough to account for increase in the weight of larvae. The food consumed by larvae is the most important factor. Young worker larvae receive, as do queen larvae, two different food components—water-clear and milky-white, the proportion being about 3:1 or 4:1. Only worker larvae from which emergency queens are reared receive the food components in a 1:1 proportion (57). The number of feedings is considerably fewer than in the case of queen larvae, only 143, lasting 1 hour, 50 minutes for the whole larval period (62). Older (over 3 days old) larvae receive, in addition to the clear secretion, a yellowish, pollen-containing food (modified worker jelly). Feeding with the white secretion is seldom observed. The ages of the nurses feeding young and older worker larvae do not show any significant differences, being, on an average, about 11 to 13 days (57).

The significance of the addition of pollen to the modified worker jelly is not known. Pollen does not supply more than about one tenth of the N requirement of larvae (103). Furthermore, pollen is not an essential constituent of the food of worker larvae because normal colonies, deprived of pollen for a short period of time, can rear worker brood if given only sugar solution (30), or when fed a number of pollen substitutes (36). Pollen grains are probably incorporated into modified worker jelly when the nurses add to it the sugary material from the honey stomach. This material could be contaminated with pollen grains (103). However, royal jelly, although containing an admixture of honey from the honey stomach, shows only traces of pollen (105) and older drone larvae usually have considerably more pollen in their food than worker larvae of the same age. Nurse bees apparently recognize the sex (39) and the caste (120) of larvae and may exercise some choice in feeding pollen to larvae of different castes.

The bees do recognize the quantity of food present in the cells, which is evident by the fact that all larvae of approximately the same age and position on the comb have about the same amount of food at all times. In spite of such care and feeding, there are great individual variations in the daily rate of growth as manifested in the differences between weight of individuals of the same age in the same lot. Differences between the smallest and the largest individual amount to nearly 100 per cent (86).

In addition to more sugar, modified worker jelly has considerably more dry matter and considerably less proteins, lipids, minerals, and vitamins (51, 99). An addition of pollen appears to be relatively unimportant, but the addition of honey has an important dilution effect (99).

Nutrition of drone larvae.—Drone larvae grow larger (384 mg versus 159 mg for worker larvae) (107) and receive considerably more food during their development than do worker larvae (avg. 9.6 mg versus 1.7 mg per cell) (92). The food of young drone larvae (drone jelly = DJ) is milky-white. Microscopic examination shows no pollen, or the presence of occasional single grains (37). As in the case of queen and worker larvae, DJ consists of the mixture of water-clear and milky-white components (57). Its composition is similar to that of other jellies. The food of older drone larvae (modified drone jelly = MDJ) is a dirty-yellow-brown color. Numerous pollen grains are present. The changes in the composition of MDJ follow closely those occurring in modified worker jelly. (37). Actually, normal drones can be produced by feeding drone larvae with the food of worker larvae of corresponding age. This would indicate that both foods are physiologically equivalent (96).

Food Requirements of Larvae

Little work has been done to ascertain the fundamental requirements for growth of honey bee larvae. Up to the present no chemically defined diet which can be used for rearing larvae from hatching to the adult stage has been developed. One can, to a certain extent, judge the requirements of larvae from the results of various laboratory brood rearing experiments.

Water and mineral requirements.—Water is indispensable for growth and development of living organisms. That the requirement of water for growth of honey bee larvae is specific is obvious from the results of those investigators who reared honey bee larvae under laboratory conditions. Larvae grow and are more likely to pupate successfully and become adults on jellies that are diluted with water (91, 105). There is a gradual decrease in the percentage of water in the food of older worker and drone larvae (33, 37, 105). In royal jelly this change is opposite: the moisture content increases in the food of older larvae. These gradual opposite changes in the moisture content of food of worker and queen larvae could be related to the different character of growth of those larvae and may be instrumental in initiating the caste differentiation in honey bees (15). Thus, the influence of moisture changes in the food or larvae may be very significant.

Practically nothing is known about the mineral requirements of larvae. There are some indications (8) that cobalt added to sugar solution fed to colonies of bees increases the growth rate of the larvae reared.

Protein requirements.—Larval food of the early stages of all three castes of the honey bee is abundant in proteins. The food of the young worker larva is especially rich. Undoubtedly, proteins play an important part in the nutrition of larvae. Calculations from feeding experiments indicate that between 4 and 6 mg of N is used to rear one larva, depending on the diet fed to the nurses (1, 34). When bees are forced to utilize the stores of their own bodies only 3.1 mg of N is used (30). However, even starved larvae can develop into smaller, but still normal-looking adult individuals (54). This would indicate a considerable adaptability of naturally growing larvae to adverse conditions. The proteins in larval food contain all the essential amino acids necessary for the development of emerging bees (10, 121). That larvae also need complete proteins for their development is suggested by the fact that colonies of newly emerged bees fed two-year-old pollen do not start normal brood rearing until 30 days after the beginning of an experiment. They begin to rear brood several times but the larvae are eaten by adult bees when they reach two or three days of age. Those offered fresh pollen start normal brood rearing in six days (45). The nutritive value of old stored pollen can be restored to the quality of fresh pollen by the addition of lysine and arginine (14). This suggests that these two amino acids are required by larvae.

The results of studying utilization of sugars uniformly labeled with ^{14}C showed that cystine, aspartic acid, asparagine, isoleucine, leucine, phenylalanine, methionine, tryptophan, valine, threonine, taurine, thyrosine, lysine, histidine, and arginine are essential for the growth of honey bee larvae. In this case, amino acids in the haemolymph of larvae which had become labeled with ^{14}C from the sugars ingested were considered nonessential and those not so labeled as essential (67). However, until a method is devised to rear honey bee larvae on synthetic diets in which the presence or absence of certain amino acids can be regulated, the exact protein requirement of larvae cannot be established.

Carbohydrate requirements.—For most insects carbohydrates serve as a convenient source of energy. The addition of sugars (glucose and fructose mixture) to worker jelly also increases larval weight, probably due to an increased synthesis of fats (16). The suitability of various carbohydrates in the nutrition of insects is assessed by comparing the average length of life of the insect fed the carbohydrate in question with that of an insect fed water alone. By this technique it was established (7) that larvae are able to utilize the following carbohydrates, named in order of their apparent value: sucrose, fructose, maltose, melizitose, glucose, trehalose, dextrins, galactose, and lactose.

Lipid requirement.—Little is known about the lipid requirement of honey bee larvae. Larvae can synthesize fats from carbohydrates (16). Ether-extracted royal jelly supported larval growth remarkably well. However, mortality was high on pupation, and there was a reduction in the amount and quality of silk spun. Many larvae failed to spin silk at all. There was very little difference in the amount of fat laid down by the larvae on the ether-extracted diet as compared to normal jelly and the fats were found to be identical in structure. The ether-extractable lipids in the diet do not appear to be essential for fat synthesis in the tissues, or to affect the composition of the fats laid down (106). Honey bee larvae do not need acetone-extractable lipids for their growth and development into adults. However, when acetone- and ether-extracted royal jelly is fed, the larvae grow normally and some change to pupae, but the latter die one or two days before emergence at the time when they are already well pigmented (91). More controlled experiments are needed to solve these problems.

Vitamin requirement.—Again, we can judge larval needs for vitamins only from inference. On vitamin-free casein, minerals, and invert sugar diet, emerging bees develop their bodies and their hypopharyngeal glands normally. However, although the food in their cells is abundant and of normal color and consistency, the larvae do not grow beyond two to three days of age (according to their size) and then disappear. When the B vitamins and cholesterol are added to this diet, four 10-day cycles of normal brood are produced (47). These findings emphasize that in the study of bee nutrition it is necessary to follow more than one factor. The development of the hypopharyngeal glands alone cannot serve as an indication of suitability of any food for brood rearing, since such well-developed glands may secrete a product that is deficient in a factor or factors essential for normal development of growing individuals. Of those vitamins studied pyridoxine (48) and inositol (85) appear to be definitely required for normal brood rearing.

Not only the proper nutritional balance of food elements is essential for growth of honey bee larvae (16) but the proper balance of certain unknown substances is important. This can be demonstrated from the fact that it is impossible to rear larvae to pupation using only the food of young queen or worker larvae (97, 105). There are also indications of improved brood rearing when the food of nurse bees contains heteroauxins (23), oil growth factor, antibiotics, cobalt (24), or giberellic acid (84).

In conclusion, another aspect of larval nutrition should be mentioned. Experiments in the Soviet Union indicate that larval food secreted by the nurse bees has an influence on the hereditary characteristics of bees produced. The food is supposed to have a direct influence on the larvae reared by the nurse bees, as well as through the laying queen fed the same food (59). In this manner the nurses, through their brood food, influence the type of cappings placed over honey (25), the orientating ability of bees produced (58), or morphological characters (78). It is of interest to note that some of these changes are claimed to be hereditary (77). And even the sex of larvae is changed by the nurse bees (79). The latter investigators explained these results by qualitative differences in the composition of the food secreted by the nurses for feeding queens and drone larvae. However, in this latter case no evidence is given that the authors took into consideration the possibility of haploid females (69) or diploid males (124) appearing in the colony.

The study of larval nutrition is still in the beginning stages and much more work is needed to solve the problem. Further valuable information on the subject of honey bee nutrition can be found in chapters written by a number of research workers in *Traité de Biologie de l'Abeille,* published under the editorship of R. Chauvin, Masson & Cie., Paris, 1968.

LITERATURE CITED

1. Alfonsus, E. C. Zum Pollenverbrauch des Bienenvolkes. *Arch. Bienenknd.,* **14,** 220–23 (1933)
2. Altmann, G. Hormonale Regelung des Wasserhaushalt der Honigbiene. *Z. Bienenforsch.,* **2,** 11–16 (1953)
3. Armbruster, L. Gelee Royale. *Arch. Bienenknd.,* **37,** (1), 1–39 (1960)
4. Auclair, J. L., Jamieson, C. C. A quantitative analysis of amino acids in pollen collected by bees. *Science,* **108,** 357–58 (1948)
5. Back, E. Einfluss der in Pollen enthaltenen Vitamine auf Lebensdauer, Ausbildung der Pharynxdrüsen and Brutfähigkeit der Honigbiene. *Insectes Sociaux,* **3,** 285–91 (1956)
6. Barbier, M., Hugel, M. F., Lederer, E. Isolement du 24-méthylène-cholesterol a partir du pollen de différentes plantes. *Bull. Soc. Chim. Biol. Paris,* **42,** 91–97 (1960)
7. Bertholf, L. M. The utilization of carbohydrates as food by honeybee larvae. *J. Agr. Res.,* **35,** 429–52 (1927)
8. Burtov, V. Ya. Effect of cobalt on reproduction in honeybees. *Sel. Khoz. Sev. Kaukaza,* **2,** 77–79 (1958) (In Russian)
9. Chauvin, R. La gelee royal II. Composition biochimique de la gelee royal. *Apiculteur* (Sect. Sci.), **100,** 45–56 (1956)
10. DeGroot, A. P. Effect of a protein containing diet on the longevity of caged bees. *Proc. Koninkl. Ned. Akad. Wetenschap., Ser. C,* **54**(3), 272–74 (1951)
11. DeGroot, A. P. Protein and amino acid requirements of the honeybee (*Apis mellifera* L.). *Physiol. Comparata et Oecolog.,* **3,** fasc. 2 & 3, 90 pp. (1953)
12. Delvert-Salleron, F. Étude au moyen de radioisotopes, des échanges de nourritures entre reines, mâle et ouvrières d'*Apis mellifica* L. *Ann. Abeille,* **6,** 201–27 (1963)
13. Dietz, A. Initiation of pollen consumption and pollen movement through the alimentary canal of newly emerged honey bees. *J. Econ. Entomol.,* **62,** 43–46 (1969)
14. Dietz, A., Haydak, M. H. Causes of nutrient deficiency in stored pollen for development of newly emerged honey bees. *Proc. Intern. Jubilee Beekeeping Congr., 20th, Bucharest,* 222–25 (1965)
15. Dietz, A., Haydak, M. H. Caste determination in honey bees: The significance of moisture in larval food. *Proc. Intern. Beekeeping Congr., 21st, Maryland, 1967,* 470 (Abstr.)
16. Dixon, S. E., Shuel, R. W. Studies on the mode of action of royal jelly in honey bee development. III. The effect of experimental variation in diet on growth and metabolism of honey bee larvae. *Can. J. Zool.,* **41,** 733–39 (1963)
17. Douault, P. Influence du nid a couvain sur les échanges de nourriture entre ouvrières d'abeilles. (*Apis mellifica*). *Compt. Rend. Acad. Sci. Paris,* **264,** 1092–95 (1967)

18. Farrar, C. L. The overwintering of productive colonies. *The Hive and The Honey Bee,* Chapt. XIII, 341–68. (Grout, R. A., Ed., Dadant and Sons, Hamilton, Ill., 556 pp., **1963)**
19. Filipovic-Moskovlevic, U. Influence of normal and changed social structure on the development of pharyngeal glands and on the work of bees (*Apis mellifera* L.). *Serbian Acad. Sci.,* **262,** 101 (1956) (Div. of Natural-Math, Sci., Book 14. In Serbian, extended English summary)
20. Foti, N., Dobre, V., Crisau, I. Research on the composition of food for queens during the winter season. *Proc. Intern. Beekeeping Congr., 21st, Maryland, 1967,* 262–67 (1967)
21. Free, J. B. The transmission of food between worker honeybees. *Brit. J. Animal Behaviour,* **5,** 41–47 (1957)
22. Free, J. B. The food of adult drone honey bees. *Brit. J. Animal Behaviour,* **5,** 7–11 (1957)
23. Glushkov, N. M. The influence of nutrition on the growth and on the development of honey bee. *Mezhdunar. Kongr. po Pchelovodstvu., 19th, Prague, Moskva 1963,* 65–74 (In Russian with English summary)
24. Glushkov, N. M., Yakovlev, A. S. New data on how to use growth stimulators in apiculture. *Proc. Intern. Beekeeping Jubilee Congr., 20th, Bucharest,* 114–17 (1965)
25. Gubin, A. F., Khalifman, I. A. The influence of food on racial characters of the honey bee. *Pchelovodstvo,* **30**(3), 22–32 (1953) (In Russian)
26. Hagedorn, H. H., Moeller, F. E. The rate of pollen consumption by newly emerged honey bees. *J. Apicult. Res.,* 6, 159–62 (1967)
27. Hartwig, A. Histochemical investigations of the midgut of worker honey bees which were fed pollen substitutes. *Pszczel. Zeszyty Nauk.,* **9**(1–3), 69–77 (1967) (In Polish)
28. Haydak, M. H. Der Nährwert von Pollenersatzstoffen bei Bienen. *Arch. Bienenknd.,* **14,** 185–219 (1933)
29. Haydak, M. H. Changes in total nitrogen content during the life of the imago of the worker honey bee. *J. Agr. Res.,* **49,** 21–28 (1934)
30. Haydak, M. H. Brood rearing by honey bees confined to a pure carbohydrate diet. *J. Econ. Entomol.,* **28,** 657–60 (1935)
31. Haydak, M. H. The influence of a pure carbohydrate diet on newly emerged honey bees. *Ann. Entomol. Soc. Am.,* **30,** 258–62 (1937)
32. Haydak, M. H. Changes in weight and nitrogen content of adult worker bees on a protein-free diet. *J. Agr. Res.,* **54,** 791–96 (1937)
33. Haydak, M. H. Larval food and development of castes in the honeybee. *J. Econ. Entomol.,* **36,** 778–92 (1943)
34. Haydak, M. H. Causes of deficiency of soybean flour as a pollen substitute for honeybees. *J. Econ. Entomol.,* **42,** 573–79 (1949)
35. Haydak, M. H. Changes in thiamine content of adult worker honey bees on protein-free diet. *Ann. Entomol. Soc. Am.,* **47,** 548–52 (1954)
36. Haydak, M. H. Pollen substitutes. *Proc. Intern. Congr. Entomol., 10th, Montreal,* **4,** 1053–56 (1956)
37. Haydak, M. H. The food of drone larvae. *Ann. Entomol. Soc. Am.,* **50,** 73–75 (1957)
38. Haydak, M. H. Changes with age in the appearance of some internal organs of the honey bee. *Bee World,* **38,** 197–207 (1957)
39. Haydak, M. H. Do bees recognize the sex of the larvae? *Science,* **127,** 1113 (1958)
40. Haydak, M. H. Changes with age in weight and nitrogen content of honey bees. *Bee World,* **40,** 225–29 (1959)
41. Haydak, M. H. Vitamin content of royal jelly from honey bee colonies fed normal diet and from those fed pollen substitutes. *Ann. Entomol. Soc. Am.,* **53,** 695 (1960)
42. Haydak, M. H. Influence of storage on nutritive value of pollens for newly emerged honey bees. *Am. Bee J.,* **10,** 354–55 (1961)
43. Haydak, M. H. The changes in the vitamin content of royal jelly produced by nurse bees of various ages in confinement. *Bee World,* **42,** 57–59 (1961)
44. Haydak, M. H. Age of nurse bees and brood rearing. *J. Apicult. Res.,* **2,** 101–3 (1963)
45. Haydak, M. H. Influence of storage on nutritive value of pollen for brood rearing by honey bees. *J. Apicult. Res.,* **2,** 105–7 (1963)
46. Haydak, M. H. Nutrition des larves d'abeilles. *Traité de Biologie de l'abeille,* **1,** 302–33. (Sous la direction de Chauvin, R., Masson & Cie., Paris, 538 pp., 1968)
47. Haydak, M. H., Dietz, A. Influence of the diet on the development and

brood rearing of honey bees. *Proc. Intern. Beekeeping Jubilee Congr., 20th, Bucharest,* 158–62 (1965)
48. Haydak, M. H., Dietz, A. Cholesterol, pantothenic acid, pyridoxine and thiamine requirements of honey bees for brood rearing. *Proc. Intern. Beekeeping Congr., 21st, Maryland, 1967,* 469 (Abstr.)
49. Haydak, M. H., Patel, N. G., Dietz, A. Queen rearing and the age of nurse bees. *Ann. Entomol. Soc. Am.,* **57,** 262–63 (1964)
50. Haydak, M. H., Vivino, A. E. Changes in vitamin content during the life of the worker honey bee. *Arch. Biochem.,* **2,** 201–7 (1943)
51. Haydak, M. H., Vivino, A. E. The changes in the thiamine, riboflavine, niacin and pantothenic acid content in the food of female honey bees during growth with a note on the vitamin K activity of royal jelly and bee bread. *Ann. Entomol. Soc. Am.,* **43,** 361–67 (1950)
52. Hejtmanek, J. Digestion of pollen in honey bee. *Statní Výzk. Ustav Včel Liptovský Hrádek.,* 53 pp. (1943) (Slovakian with German summary)
53. Hoffman, I. Gibt es bei Drohnen von *Apis mellifica* L. ein echtes Futtern or nur eine Futterabgabe? *Z. Bienenforsch.,* **8,** 249–55 (1966)
54. Jay, S. C. Starvation studies of larval honeybees. *Can. J. Zool.,* **42,** 455–62 (1964)
55. Johansson, T. S. K. Royal jelly. *Bee World,* **36,** 3–13, 21–32 (1955)
56. Johansson, T. S. K., Johansson, M. P. Royal jelly. II. *Bee World,* **39,** 254–64, 277–86 (1958)
57. Jung-Hoffmann, I. Die Determination von Königin und Arbeiterin der Honingbiene. *Z. Bienenforsch.,* **8,** 296–322 (1966)
58. Khalifman, I. A. The influence of feeding on the orienting instinct of honey bees. *Izvest. Akad. Nauk, USSR. Ser. Biol.,* No. 3, 52–62 (1951) (In Russian)
59. Khalifman, I. A., Gubin, A. F. New fact and the old arguments. *Agrobiologiya,* No. 6, 149–54 (1954) (In Russian)
60. Levin, M. D., Haydak, M. H. Seasonal variations in weight and ovarian development in worker honey bees. *J. Econ. Entomol.,* **44,** 54–57 (1951)
61. Levin, M. D., Haydak, M. H. Comparative value of different pollens in the nutrition of *Osmia lignaria. Bee World,* **38,** 221–26 (1957)
62. Lindauer, M. Ein Beitrag zur Frage der Arbeitsteilung in Bienenstaat. *Z. Vergleich. Physiol.,* **34,** 299–345 (1952)
63. Lindauer, M. The water economy and temperature regulation of the honey bee colony. *Bee World,* **36,** 62–72, 81–92, 105–11 (1955)
64. Lingens, K., Rembold, H. Ueber den Weiselzellenfuttersaft der Honigbiene. II. Vitamingehalt von Königinen und Arberterinnen Futtersaft. *Z. Physiol. Chem.,* **314,** 141–46 (1959)
65. Lotmar, R. Abbau und Verwertung von Stärke und Dextrin durch die Honigbiene. *Arch. Bienenknd.,* **16,** 195–204 (1935)
66. Lotmar, R. Untersuchungen ueber den Eisenstoffwechsel der Insekten besonders der Honigbiene. *Rev. Suisse Zool.,* **45,** 237–71 (1938)
67. Lue, P. T., Dixon, S. E. Studies on mode of action of royal jelly in honey bee development. VIII. The utilization of sugar uniformly labelled with ^{14}C and aspartic L-^{14}C acid. *Can. J. Zool.,* **45,** 595–99 (1967)
68. Lunden, R. A short introduction to the literature on pollen chemistry. *Svensk Kem. Tidskr.,* **66,** 201–13 (1954)
69. Mackensen, O. The occurrence of parthenogenetic females in some strains of honeybee. *J. Econ. Entomol.,* **36,** 465–67 (1943)
70. Malashenko, P. V. Influence of pollens of various origin on the longevity of bees. *Nauk. Praci. Ukrainian Expt. Sta. Beekeeping,* **3,** 44–48 (1961) (In Ukrainian).
71. Maurizio, A. Beobachtungen ueber die Lebensdauer und den Futterverbrauch gefangen gehaltenen Bienen. *Beih. Schweiz. Bienenztg.,* **2,** 1–48 (1946)
72. Maurizio, A. Pollenernährung und Lebensvorgänge bei der Honigbiene (*Apis mellifica* L). *Landwirtsch. Jahrb. Schweiz.,* **68,** 115–82 (1954)
73. Maurizio, A. Breakdown of sugars by inverting enzymes in the pharyngeal glands and midgut of the honeybee. 2. Winter bees. *Bee World,* **40,** 275–83 (1959)
74. Maurizio, A. Zuckerabbau unter der Einwirkung der invertierenden Fermente in Pharynxdrüsen und Mitteldarm der Honigbiene (*Apis mellifica* L). 3. Fermentwirkung

während der Ueberwinterung bei Bienen der Ligustica-Rasse. *Insectes Sociaux,* **8,** 125–75 (1961)

75. Maurizio, A. From the raw material to the finished product: Honey. *Bee World,* **43,** 66–81 (1962)
76. Maurizio, A. Zuckerabbau unter der Einwirkung der invertierenden Fermente in Pharynxdrüsen und Mitteldarm der Honigbiene (*Apis mellifica* L). 4. Sommerbienen der Italienischen, Kaukazischen und Grichischen Rasse. *Insectes Sociaux,* **9,** 39–72 (1962)
77. Melnichenko, A. N. Inheritance of changes arising in queen bees and drones under the influence of the nurse bees of another race. *Uchen. Zap. Gor'kovsk. Gos. Univ. Ser. Biol.,* **55,** 128–43 (1962) (In Russian)
78. Melnichenko, A., Burmistrova, N. Directed change of heredity in a bee colony and its biochemical basis. *Mezhdunar. Kongr. po Pchelovodstvu, 19th, Selkhozizdat, Moscow.,* 28–42 (1963) (In Russian, English summary)
79. Melnichenko, A., Burmistrova, N. D., Trishina, A. C. Directed change of heredity and of the sex of queens and drones in bee colony. *Jubileiny Mezhdunar. Kongr. po Pchelovodstu, 20th,* Ministerstvo Selskogo Khoz. Izdat "Kolos". Moscow. 346–55 (1965) (In Russian, English summary)
80. Mindt, B. Untersuchungen ueber das Leben der Drohnen, insbesondere Ernährung und Geschlechtsreife. *Z. Bienenforsch.,* **6,** 9–33 (1962)
81. Morton, K. The food of worker bees of different ages. *Yaikoot Hamichveret,* **4** (1950) (Bee World Abstracta 158/51)
82. Moskovlevic, V. Zh. About the ability of regeneration of pharyngeal glands in nurse bees. *Novy Pchelar,* **2** (8–9), 152–54 (1938) (In Serbian, English summary)
83. Moskovlevic-Filipovic, V. The development of the pharyngeal glands of the honey bee in a normal bee colony. *Bull. Acad. Sci. IV, Sci. Natur.,* No. 2, 257–62 (1952) (In Serbian, English summary)
84. Nation, J. L., Robinson, F. A. Gibberellic acid: effect of feeding in an artificial diet for honey bees. *Science,* **152,** 1765–66 (1966)
85. Nation, J. L. Robinson, F. A. Brood rearing by caged honey bees in response to inositol and certain pollen fractions of the diet. *Ann. Entomol. Soc. Am.,* **61,** 514–17 (1968)
86. Nelson, J. A., Sturtevant, A. P., Lineburg, B. Growth and feeding of honeybee larvae. *U.S. Dept. Agr. Bull. 1222,* 37 pp. (1924)
87. Nixon, H. L., Ribbands, C. R. Food transmission within the honey bee community. *Proc. Roy Soc. Ser. B,* **140,** 43–50 (1952)
88. Pain, J. Sur quelques facteurs alimentaires, accélérateurs du développement des oeufs dans les ovaries des ouvrières d'abeilles (*Apis mellifica* L). *Insectes Sociaux,* **8,** 31–93 (1961)
89. Patel, N. G., Haydak, M. H., Gochnauer, T. A. Electrophoretic components of the proteins in honey bee larval food. *Nature,* **186,** 633–34 (1960)
90. Pershad, S. Analyse de diflérents facteurs conditionant les échanges alimentaires dans une colonie d'abeilles *Apis mellifica* L. au moyen du radioisotope P^{32}. *Ann. Abeille,* **10**(3), 139–97 (1967)
91. Petit, J. Étude 'in vitro' de la croissance des larves d'abeilles (*Apis mellifica* L). *Ann. Abeille,* **6**(1), 35–52 (1963)
92. Planta, A. Ueber den Futtersaft der Bienen. *Z. Physiol. Chem.,* **12,** 327–54 (1888)
93. Poteikina, E. How winter the bees that fed and did not feed brood in the fall. *Pchelovodstvo,* **38**(11), 15–16 (1951) (In Russian)
94. Rembold, H. Ueber den Weiselzellenfuttersaft der Honigbiene. *Proc. Intern. Congr. Entomol., 11th, Vienna, 1960, B,* **III,** 77–81 (1960)
95. Rembold, H. Biologically active substances in royal jelly. *Vitamines Hormones,* **23,** 359–82 (1965)
96. Rhein, W. Von. Ueber die Ernährung der Drohnenmaden. *Z. Bienenforsch.,* **1**(4), 63–66 (1951)
97. Rhein, W. Von. Ueber die Ernährung der Arbeitermade von *Apis mellifica* L. insbesondere in der Altersperiode. *Insectes Sociaux,* **3**(1), 203–12 (1956)
98. Roger, B., Pain, J. L'influence de la reine d'abeille (*Apis mellifera* L) sur le taux de mortalité des ouvrières accompagnatrices. *Ann. Abeille,* **9,** 5–36 (1966)
99. Shuel, R. W., Dixon, S. E. Studies in the mode of action of royal jelly in honeybee development. II.

Respiration of newly emerged larvae on various substrates. *Can. J. Zool.*, **37**, 803–13 (1959)
100. Siddiqui, I. R., Furgala, B. The structure of the carbohydrate moiety of a glycopeptide from royal jelly. *J. Apicult. Res.*, **5**, 113–20 (1966)
101. Siddiqui, I. R., Furgala, B. Isolation and characterization of oligosaccharides from honey. Part I. Disaccharides. *J. Apicult. Res.*, **6**, 139–45 (1967)
102. Siddiqui, I. R., Furgala, B. Isolation and characterization of oligosaccharides from honey. Part II. Trisaccharides. *J. Apicult. Res.*, **7**, 51–59 (1968)
103. Simpson, J. The significance of the presence of pollen in the food of worker larvae of honeybees. *Quart. J. Microscop. Sci.*, **96**, 117–20 (1955)
104. Sitbon, G. L'effet de groupe chez l'abeille. 1. L'abeille d'hiver, survie et consommation de candi des abeilles isolées ou groupées. *Ann. Abeille*, **10**, 67–82 (1967)
105. Smith, M. V. Queen differentiation and biological testing of royal jelly. *Cornell Univ. Agr. Expt. Sta. Mem. 356*, 1–56 (1959)
106. Smith, M. V. Effect of experimental feeding on growth and dimorphism in the female honey bee. Symposium on Female Dimorphism and Colony Organization in Social Hymenoptera. Papers presented at Meeting of *Entomol. Soc. Am., Atlantic City*, **1960**, 29–31 (1960)
107. Stabe, H. A. The rate of growth of worker, drone and queen larvae of honey bee. *J. Econ. Entomol.*, **23**, 447–53 (1930)
108. Standifer, L. N. A comparison of the protein quality of pollens for growth stimulation of the hypopharyngeal glands and longevity of honey bees, *Apis mellifera* L. *Insectes Sociaux*, **14**, 415–25 (1967)
109. Standifer, L. N., Devys, M., Barbier, M. Pollen sterols—a mass spectrographic survey. *Phytochem. Phytobiol.*, **7**, 1361–65 (1968)
110. Staudenmayer, T. Die Giftigkeit der Mannose für Bienen und andere Insekten. *Z. Vergleich. Physiol.*, **26**, 644–68 (1939)
111. Stroikov, S. A. Digestibility of pollen substitutes by bees. *Pchelovodstvo*, **84**(3), 32–33 (1964) (In Russian)
112. Stroikov, S. A. About the digestibility of natural proteinaceous foods by bees. *Tr. Nauch.-Isscedovatelskogo Inst. Pchelovodstva. Moskowsky Rabochi*, 44–77 (1966) (In Russian)
113. Townsend, G. F., Shuel, R. W. Some recent advances in apicultural research. *Ann. Rev. Entomol.*, **7**, 481–500 (1962)
114. Veselý, V. The importance of hereditarily established longevity in bees. *Vědecké Práce Výzk. Úst. Včel. v Dole*, **4**, 185–92 (1965) (In Czechoslovakian, English summary)
115. Vivino, A. E., Palmer, L. S. The chemical composition and nutritive value of pollens collected by bees. *Arch. Biochem.*, **4**, 129–136 (1944)
116. Vogel, B. Ueber die Beziehung zwischen Süssgeschmack und Nährwert von Zuckeralkoholen bei der Honigbiene. *Z. Vergleich. Physiol.*, **14**, 273–348 (1931)
117. Wahl, D. Vergleichende Untersuchungen ueber den Nährwert von Pollen, Hefe, Soyamehl und Trockenmilch für die Honigbiene (*Apis mellifica*). *Z. Bienenforsch.*, **6**, 209–80 (1963)
118. Wahl, D. Le Nourrissement. *Traite de Biologie de l'Abeille*, **III**, 161–80. (Sous la direction de Chauvin, R., Masson & Cie, Paris, 389 pp., 1968)
119. Wang, Der-I. Growth rates of young queen and worker honeybee larvae. *J. Apicult. Res.*, **4**, 3–5 (1965)
120. Weaver, N. Physiology of caste determination. *Ann. Rev. Entomol.*, **11**, 79–102 (1966)
121. Weaver, N., Kuiken, K. A. Quantitative analysis of the essential amino acids of royal jelly and some pollens. *J. Econ. Entomol.*, **44**, 635–38 (1951)
122. Weiss, K. Zur vergleichenden Gewichtsbestimmung von Bienenköniginnen. *Z. Bienenforsch.*, **9**, 1–21 (1967)
123. White, J. W. The composition of honey. *Bee World*, **38**, 57–66 (1957)
124. Woyke, J. Drone larvae from fertilized eggs of the honeybee. *J. Apicult. Res.*, **2**, 19–24 (1963)
125. Zherebkin, M. V. Secretion of digestive enzymes in the ventriculus of the honeybee. *Proc. Intern. Beekeeping Congr., 21st., Maryland, 1967* (Apicultural Abstracts 689/67)

BIOSYSTEMATICS OF THE EMBIOPTERA

By Edward S. Ross[1]

California Academy of Sciences, San Francisco, California

Embioptera comprise a well-defined, lesser-known insect order. Their survival potential and anatomical peculiarites are intimately linked with life in labyrinths of silk galleries. Although silk production and gallery life occur in many other insects, such silk is almost invariably a product of labial or rectal glands of larvae. Embiids are unique in that they produce silk in all developmental stages and throughout the adult life of both sexes. Even more remarkable, the silk strands issue from numerous glands located in the fore tarsi. This function is apparently equally developed in all species of the order, and the galleries shelter all life activities except short dispersal movements of adults in the open environment.

Some entomologists have regarded the order as an ancient, relict group past its zenith. There can be little doubt that it had an ancient origin but the insects are probably as abundant now as they ever were. Many taxonomic sections "behave" in a most unrelict manner—breaking up into clusters of races, or weak species, and young genera. The writer's extensive field work indicates that the order is moderate in size, perhaps as large as the Plecoptera, or the Isoptera, and in its own way, as diversified. It may never be rated as an important order, however, because of the infrequency with which most entomologists encounter these insects. Also, Embioptera seldom develop large populations and play only minor ecological and food chain roles; furthermore, no species has significant economic importance.

Basic studies of the group, however, promise to add much to evolutionary and zoogeographic theory. Male Embioptera may also prove to be ideal subjects for demonstrating the generalized anatomy of Pterygota. Easily maintained laboratory cultures can yield a dependable supply of specimens and such ease of propagation should encourage the use of Embioptera in general studies of insect physiology and behavior.

When the writer commenced his studies of the order it was soon evident that a significant collection of Embioptera had not yet been assembled. Much time which he might have devoted to laboratory studies had to be spent organizing, funding, and conducting extensive surveys in the major evolutionary centers of the order. Five safaris to Africa, including Madagascar and Ethiopia, totalling about 120,000 road miles were made. Tropical Asia and Australia were covered during an eighteen-month trip by road.

[1] Part of the author's research reported in this review was supported by grants from the National Science Foundation and the National Geographic Society.

South America was visited three times and shorter excursions have been made to many other regions. As a result of these efforts, a collection totalling at least 130,000 specimens has been accumulated at the California Academy of Sciences. In this virtual monopoly of study material, about 800 species are represented, of which approximately 650 are new, as well as all levels of higher categories. In addition, most of the collections of other museums have been studied, including the specimens used by H. A. Hagen, H. A. Krauss, G. Enderlein, L. Navás, and Consett Davis as the basis of past revisional studies.

Between excursions afield, much progress has been made in the production of manuscript and illustrations for a world-scope monograph. However, the rapid accrual of new specimens and new taxons has made the project difficult to culminate. As a compromise, the writer has recently completed manuscript for an initial volume covering general information and higher classification down to genus. This should appear during 1970. Species and other details will be treated in a series of regional monographs, the smallest of which, dealing with the Mediterranean region, has already appeared [Ross (15)].

The present paper is intended to preview some of the ideas which will be more fully supported in the first volume. Unfortunately, the names of the many new higher categories have not yet been validated. An attempt will be made at this time, however, to present a general picture of evolution and classification of the order without prematurely introducing new names.

ORIGIN OF THE EMBIOPTERA

Hypothetical ancestor.—A slender, alate, completely terrestrial, perloid type of insect with the following characteristics: (*a*) Cranium and its appendages as in males of Recent Embioptera; mandibles adapted for chewing. (*b*) Thoracic segments well separated, dorso-ventrally compressed; legs moderately long, adapted for crawling, probably similar to those of Plecoptera; tarsi three-segmented. (*c*) Alate in both sexes, fore and hind wings subequal in size and shape, vannal area reduced except for slight lobing in hind wings; attachments widely spaced, as in termites; flight a poorly directed flutter. (*d*) Terrestrial in all stages, occupying bark crevices and leaf litter; feeding on vegetative debris, outer bark surfaces, lichens, moss, etc. (*e*) Defense chiefly by concealment in crevices and curled leaves; running from danger; flight ineffective as an escape due to slow takeoff; lacking ability to bite defensively, or to exude objectionable secretions. (*f*) Distributed in warm portions of Gonwanaland.

During Carboniferous and Permian times, there must have existed many insects fitting the above description. Unfortunately, no fossils of true Embioptera older than Baltic amber have been found. However, as in most other insects, the amber Embioptera are highly specialized, Recent types. This indicates that the order's ancestry is deeply rooted in the Cretaceous, or much earlier. As will be indicated later, the order was probably highly diversified and widely distributed in Gonwanaland long before this land mass split into several continents.

Relationships.—Embioptera were first placed near the termites and these orders may indeed have had a remote common ancestry but they are worlds apart today. The latest placement is that of Rohdendorf (10) who, with the concurrence of Illies (3), assigned the Embioptera to the superorder Plecopteroidea with the Plecoptera as the only other Recent order. The fossil orders, Paraplecoptera and Miomoptera, are the only other components of the superorder. It is possible that some of the Miomoptera fossils actually represent early Embioptera.

Evolution of the order.—The generalized ancestors of the Embioptera must have inhabited cracks and crevices of tree bark and cautiously extended their feeding to immediate surroundings. Finding food probably never was a problem because almost any decaying bark surface underfoot was, as it is today, acceptable as food. Also, moss, lichens, and decaying leaves must have been highly palatable. However, predators certainly lurked in this environment and exposed embiid activities must have been hazardous. The most effective escape probably was a quick scurrying, often backward, into a crevice too narrow to be penetrated by a predator.

Related insects, such as Plecoptera, had remained in an amphibiotic way of life, their nymphs utilizing the cover of submerged objects in fresh-water environments. The probably high mortality rate of defenseless Plecoptera adults was perhaps offset by the great individual fecundity of the surviving females.

Ancestral termites developed another means of survival—burrowing in wood and using the excavated substance as food, thanks to concurrent evolution of intestinal biota capable of digesting cellulose. Foraging termites also extended their feeding forays under the cover of galleries made of a mortar of saliva-moistened soil. Such masonary skill was also applied to the construction of mounds which offer the advantage of thermal and moisture control, as well as physically enclosing the society. With the added advantages of workers and soldier castes, exposure of most termites to the open environment was limited to massive nuptual flights of adults whose sheer numbers exceed the immediate predatory capacity of the environment.

If ancestral Embioptera were to survive through the ages in a world made increasingly hazardous by highly effective predators, such as ants, they would have to have something better than an eat-and-run way of life. Several evolutionary accommodations could have been made, such as fleeter movement or flight, a defensive bite, repugnant secretions, or evasion of predators by an improved protective cover. Embioptera were destined to adopt a protective coverway defense, but not coverways made of ordinary environmental materials. They were to use silk, a by-product of their own metabolism, and to produce it in a most unusual manner thanks to a singular, aberrant stroke of evolution.

An ability to produce silk galleries by action of their forelegs was the "key innovation" which started a novel evolutionary trend. Anatomical and behavioral adaptations associated with life in silk galleries have become the order-defining characters. The tarsal spinning organ must have had a modest beginning. Probably a single mutation having simultaneous expression on many epidermal gland cells of the basal segment of the fore tarsi started it all. Perhaps, at first, only a crude sheet cover was produced but this was enough to encourage a trend worthy of steady improvement by natural selection. Within the fore tarsi of modern Embioptera, each ball-like, syncytial gland is derived from an invagination of the hypodermis. The seta-like projections of the plantar surface of the tarsus which conduct the gland ducts to the exterior, appear to be evaginations of the exocuticle and thus unique structures, not setae [Barth (1)]. They are best termed silk-ejectors.

Such silk-shielded, primordial Embioptera must have a definite survival edge over their relatives without a silk-production capacity. Also, successive generations were increasingly likely to remain within a common cover and inbreed mutations favorable to gallery life. There was probably preadaptation for a gregarious way of life due to a habit of laying eggs in a cluster and of the clumping of at least early instar nymphs in the vicinity of the parent female.

The biggest problem, one inherent to any innovative evolutionary trend is to justify the selective value of the earliest stage of silk gland evolution. Once functional, however, it is easy to appreciate how the tarsal glands could have gradually become perfected as the elaborate, highly efficient glands possessed by all Recent Embioptera.

At first, the silk must have been used to produce a simple sheet cover over nutritious bark surfaces surrounding a crevice retreat which also served as a brood chamber. Later, as a means of extending the feeding range, and effecting a more economical use of materials, the silk was shaped into branching tubular galleries. The form of these became the standardized or universal factor conditioning a series of special adaptions which characterize the order, as follows:

1. Slender, supple body form. Undoubtedly a primordial condition of primitive Pterygota which constituted a preadaptation for movement in narrow galleries.
2. Prognathism of the head and sclerotization of its gular surface.
3. Shortening of the legs and specialization of their function.
4. Improvement of an ability to run backward. Reverse movement is the most rapid way of escaping predators which are most likely to be encountered at gallery extremities farthest from the solid-walled protective crevice. This was accomplished by increased development of the tibial depressor muscle and consequent enlargement of the hind femora.
5. Development of a highly tactile function in the short, two-segmented cerci to guide the backward movement.
6. Special modification of wings or their elimination, or both, through neoteny to more perfectly adapt the insects for gallery life.

7. Development of a gregarious way of life in interconnected galleries but lacking true social organization, castes, or division of labor.

WING MODIFICATIONS AND LOSS

Special adaptations in the wings are of sufficient interest to warrant more detailed attention. The earliest Embioptera undoubtedly were alate in both sexes and must have remained so for a considerable period in the order's evolution. During this more vagile period, the order must have gained its primary distribution throughout Gonwanaland. Later, the females became universally apterous and their movement was limited to the very short distance they could safely walk without the protection of galleries. It would therefore be difficult to explain the present wide distribution of the order if the females had been apterous before the fragmentation of the southern land mass. This factor is additional evidence of the great antiquity of the order.

During the long period when both sexes were alate and yet depending on silk galleries as the principal aid to survival, there still remained a disadvantage in having these stiff alar projections rubbing and snagging against opposing gallery walls during critical reverse movement. Early termites and ants must have experienced the same disadvantages in their earthern and wooden galleries. Both groups reduced such disadvantages by developing means of dropping off their wings following nuptual flights and by having most of the society's work done by members of a more agile, apterous, worker caste or nymphs (in the case of primitive termite species).

Embioptera reduced the disadvantage of stiff wings by evolving greater wing flexibility which enabled these appendages to fold or crumple forward, often over the head, during backward movement and thereby reverse the axis of friction or barb effect. Flexibility is increased in most species by reduction in the strength of most of the longitudinal veins. In many species such veins are reduced to mere lines of setae. Thus, it might appear that the wings have sacrificed much of their effectiveness as organs of flight in satisfying the need for quick movement in the galleries. However, a compensating evolutionary trend occurred which made possible temporary wing stiffening for flight.

This adjustment took place in the form of alar blood sinuses, with dark sclerotized walls. The most important of these sinuses follows the full course of the first radius vein with a secondary sinus angling caudad in the cubital region. Both taper terminally and thus constitute almost completely closed sacs. With increased blood pressure as the wings are expanded in flight, the turgid sinuses stiffen the wings. When the wings return to the repose position over the back, the sinuses flatten and the wings are again flexible. Such sinus veins are universally present in all alate male Embioptera and serve as a major recognition feature of the order. This should be remembered by insect paleontologists who attempt to identify the order in early formations. Of course, there must have been a long period in the early evolution of the order when the insects did not have the blood sinus adaptation. In such cases, the fore tarsal specialization would be the best ordinal recognition character. It is quite possible that many Permian and Carboniferous insects, known only from wing imprints, and assigned to other orders, were actually Embioptera.

The wing specializations are so complex that they would be difficult to justify simply as adaptations to increase the survival potential of the short-lived adult males. They must have evolved to serve the needs of adult females which need to live long enough to insure sufficient time for maturation of eggs within their body, for oviposition, and for protection of the egg mass and early instar nymphs. In other words, the wing specializations evolved in a species in which both sexes were alate and before the order became fully distributed. It is unlikely that such complexity could have appeared independently on distinct evolutionary lines.

During this universally alate period it is probable that nuptual swarms took place as in termites and ants. Alate male Embioptera still tend to have distinct times of flight from scattered parent colonies. These usually are triggered by concurrent maturation during a limited period each year and the stimulus of special meterological conditions, such as the first rains of a season. The writer has noticed that virgin females also exhibit a kind of epigamic swarming excitement in culture jars. In unison they may crawl out of the galleries and run about the laboratory container for a period and then return to the galleries. Inasmuch as populations of Embioptera are never very large, it might be considered disadvantageous for females of a species to have wings and fly away from their protective galleries, exposing themselves to the many hazards of the external world. As environmental conditions become more marginal (e.g., through aridity), flight would become more costly to a species.

Thus, not only were the wings useless appendages in the gallery-bound lives of females, they also undoubtedly continued to slow evasive movements in spite of the evolution of increased flexibility. Furthermore, an ability to fly probably resulted in considerable loss of females to the population of a species. In short, the females would better survive hazards both inside of or out of their galleries if they were apterous.

Thus, when mutations occurred effecting apterism in females, they were readily accepted, and today females universally are apterous. It is probable that such apterism occurred independently in several distinct evolutionary lines as is currently happening in males of many species living in scattered marginal environments. Universal apterism in females was accomplished through mutations effecting neonteny, for all Recent females simply exhibit the anatomy of an early instar nymph before the appearance of even a trace of wing pads. During their nymphal development females merely increase in size while maturing their internal reproductive organs. For this reason the females are not as useful as males in the recognition of species and higher categories.

Males undoubtedly mate very soon after becoming adults and quite often must do so with sisters or close relatives within the parent colony. Thus, it matters little how imperfectly they are adapted for prolonged survival once they have transmitted their spermatophores. Males therefore retain more ancestral anatomy and behavior and consequently are much more useful than females in determining inter- and intra-ordinal relationships. Flight of males is probably of value, however, in the dispersal of genetic materials in a species, but the spread of a species is limited to the hazardous short distances the females can walk. There is, however, a strong tendency in most evolutionary lines for males to become apterous as well. This trend seems to correlate with climate—males of species inhabiting the benign climate of wet forests are universally alate, those of arid regions, or localities with a long dry season, frequently are flightless through various degrees of brachyptery and, finally, are completely apterous, as in females. The head and external genitalic structures of apterous males almost always remain distinctly adult and serve as the most important characters in systematics. In some species, both alate and apteroid males occur in the same species, but in many genera, e.g., *Metoligotoma* Davis, males invariably are completely apterous. The Baltic Amber species, *Electroembia antiqua* (Pictet), had completely apterous males and this strongly indicates that its environment was subject to a prolonged dry season [Ross (12)].

The most extreme case of male apterism appears in a recently discovered new genus and species from Afghanistan. In this case, the male is completely nymphaform *including the abdominal terminalia!* Only a slight, inner-basal lobe on the left cercus and a more gross cranium differentiates its facies from that of a nymph or a female. Such a species is difficult to place and it seems better to create a new family for it than to assign it without evidence to one of the established families.

In eastern and southern Africa where there are many completely apterous species forming several apparently young genera, apterism obviously has increased the rate of speciation by encouraging inbreeding of small, isolated populations. There are many other examples of this in different regions and on diverse evolutionary lines, e.g., *Metoligotoma* Davis (2) of Australia and *Chelicerca* Ross in its arid zone range of the Neotropical region.

Because wings are of such slight value in the survival of embiids and in their routine activities, they are subject to much intraspecific and asymmetrical, individual variation. Kinds of venational variation and anomalies may occur which, in many another group of insects such as Diptera and Apoidea in which flight is of critical functional value, would be considered to be family-level characters. This suggests a law which may be expressed as follows: The more vital an organ or structure is to the daily needs and survival of a creature, the less likely it is to vary intraspecifically.

The most remarkable case of anomalous variation in an embiid is found in the wings of a new species of *Oligembia* from the lower Amazon. In various populations of the one species, the hind wings may be reduced to haltere-like rudiments, or be completely absent. In the latter, the metathorax is reduced to a tiny segment no longer than the first abdominal. With the fore wings remaining perfectly normal in size, this is perhaps the most dipterous insect.

CLASSIFICATION

No attempts will be made at this time to discuss the history of Embioptera classification since the first literature reference in 1825. The great numbers of new species and higher categories known to the writer make all old classifications obsolete. It seems best, therefore, to summarize at this time the new higher classification as it appears on the eve of publication of a general treatment of the order.

Embioptera do not display a wide diversity of facies reflecting adaptations to varied ecological niches. In spite of the great antiquity of the group, the general form of all species has been uniformly shaped in a common "mold" or die-cast, in the physical monotony of silk galleries. Thus, Embioptera groups cannot be recognized on sight, even to family, although a specialist can make good educated guesses based on specimen size and geographic origin. Sight recognition of higher groups is also made difficult by repetition of color patterns, apterism, wing venation, and certain genital structures in almost every distinct evolutionary line.

The abdominal terminalia of adult males provide the most useful characters for defining and recognizing Embioptera taxa. This is due to their complexity and functional independence from daily environmental influences. Such characters must be used with due caution, however, because of convergence. If a set of phenetic characters is accepted without tracing back the serial ontogeny of each individual structure, great systematic errors could result. It is therefore important to supplement terminalia characters with those of other body parts but with caution based on awareness of convergences in these structures as well. In the cranial and mandibular characters, there is great diversity and yet much convergence because the head of male Embioptera is, in effect, a secondary sexual organ. Very rarely do males ingest food; their mouthparts apparently are almost exclusively used for grasping the female's head prior to copulation and the mandibles are often especially modified to serve this function.

Although the less elaborate genitalic structures of females may one day prove useful aids in classification, it is almost certain that the more conspicuous and intricate male terminalia will always remain the primary basis of classification. In a sense, the evolutionary diversity of Embioptera is a reflection of a diversity of copulatory mechanics.

There are certain consistencies, however. For some unexplained reason copulating males invariably slide the tip of their abdomen down the right side of the female's abdomen and then curve it leftward and upward to reach the vulva. Although males of the most archaic species of the order,

Clothoda nobilis (Gerstaecker), have almost perfectly symmetrical terminalia, very early there developed a "need" to improve the male's genital mechanics through asymmetry directed toward the left. This was based on a need to extend the "reach" of the terminalia as well as to secure a grip by means of tergal and sternal processes and hooks. The clasp of the male's left cercus against the right side of the female's genital segments is of primary importance in most species.

Even within the genus *Clothoda* and related genera of the primitive family Clothodidae, it is possible to see early steps in evolution of terminalia complexity to improve copulatory efficiency. All clothodid species maintain an archaic symmetry of the left and right cerci but in the most primitive genera of the next family, the Embiidae, we see the beginning of a major trend to improve the grip of the left cercus by means of a lobe development on the inner side of the basal segment. The effectiveness of the lobe is further enhanced by shortening of setae on its surface to become short pegs (echinulations). On several distinct embioid lines, the terminal of the two left cercus segments is "absorbed" in varying degrees into the basal segment and this improves the clasper function by making the cercus longer and unjointed. In the most extreme case, *Dactylocerca rubra* (Ross), the cercus is a long, C-shaped, tubular arm which generously embraces the female's left side. Embioptera, which usually utilize the full length of the basal segment of the left cercus as a clasper, form several distinct families.

Within the primitive family Clothodidae, there are also species representing early stages of another trend in copulatory mechanics. In this the clasping area of the left cercus has "migrated" to the cercus' extreme inner base. This extremity often is developed as a sclerotic ring with one or more inner lobes and processes which, until this writing, have been identified as the left cercus-basipodite. At the beginning of this trend, *Aposthonia* Krauss, the entire basal segment still functions as a weak, nonspiculate, clasper but at the most evolved extremes, *Oligotoma* Westwood, and all of the Teratembiidae, the ring-like cercus base is often highly complex. This trend also appears convergently in several unrelated lines of evolution and therefore other terminalia structures must be given due consideration.

The family composition of the Embioptera may now be outlined and discussed. The many categories which have no validated name will be referred to by symbol.

SUBORDER EMBIOPTERA

Family Clothodidae Enderlein

This small family represents more of a trend than a clearly defined category. It includes the most generalized of all Embioptera, *Clothoda nobilis* (Gerstaecker), which has almost perfectly symmetrical male terminalia, very strong and complete wing venation, as well as archaic head and thoracic structure. In spite of this, its silk spinning organs and gallery architecture are as advanced as in the most specialized species of the order. The second known species of *Clothoda* shows the beginning of terminalia cleavage and asymmetry and, through a series of species in the other two genera of the family, there is a steady increase in specialization, as well as reduction in body size and strength and furcation of wing veins. In such species there appears to be the beginning of two basic types of sclerotal cleavage of the tenth abdominal tergite, reflecting distinct kinds of copulatory mechanics.

All clothodids are confined to tropical forest zones of the Amazon Basin, northern South America, and Trinidad. They have the greatest chromosome number and largest body size (body length 25 mm) as is characteristic of primitive representatives of insect orders. However, in the most specialized evolutionary extreme of each large family of Embioptera, the species are smallest in size with narrowest wings and greatest reduction in venation. In the order's smallest species (body length about 3 mm), the minute wings are almost unveined and have long marginal setae. This is but one more example of the thysanoptery which appears in the smallest species of many orders, e.g., Trichopterygidae (Coleoptera), Trichogrammatidae (Hymenoptera), certain Microlepidoptera, and the entire order Thysanoptera.

Family Embiidae Burmeister

This very large assemblage is perhaps a catch-all for several discrete generic groups which might be considered families were it not for intermediate species. It seems best to regard such generic groups as subfamilies.

SUBFAMILY A. This is strictly a South American group, only slightly more specialized than the Clothodidae but with males distinguished by possession of a left cercus lobe, process formation on the tenth tergum, reduced wing vein strength, and other characters. Its oldest known species is *Embia batesi* (McLachlan), 1877, of the Amazon Basin which has a fair number of relatives in the forest zones of eastern Brazil. Most of these await description and a new generic name to replace the use of *Embolyntha* Davis which must be restricted to its type species *brasiliensis* (Gray), a member of the next subfamily. The extensive anatomical studies of Lacombe (4–9) were based on "*Embolyntha batesi*" from the Rio de Janeiro area which actually is a new species of the proposed new genus.

Batesi has the broadest wing form of any embiid and a definite vannal lobe which is a primitive condition not found in any clothodid wing. Otherwise, in its venation, it displays fairly advanced wing characters. The second genus to be assigned to the subfamily is *Calamoclostes* Enderlein which is confined to the northern Andes, often at high altitude, and northwestern regions of South America. One of the many new species occurs in the deserts of Ecuador's Santa Elena peninsula.

SUBFAMILY B. This is a very large complex of genera with a wide distribution and diversity in the Neotropical and Ethiopian regions. Typical genera are *Embolyntha* Davis, *Rhagodochir* Enderlein, *Pararhagadochir* Davis, *Scelembia* Ross, and *Apterembia* Ross. The Baltic Amber species, *Elec-*

troembia antiqua (Pictet), also belongs to this group and is thus not closely related to modern Mediterranean species of *Embia*.

Genera of this subfamily range southward from the lowlands of Mexico to the subtropical zones of Argentina. Africa appears to have an even richer fauna extending from subsaharan savanna zones and the highlands and lowlands of Ethiopia to the Cape. Members of this family show definite transatlantic relationships but, for the present, no genus is considered to have representation on both sides of the Atlantic.

Strangely, this large group apparently is not represented east of Africa. However, the subfamily appears to have had the same ancestry as the family Notoligotomidae which is centered in southeastern Asia and also has representation in Australia. There is also one newly discovered notoligotomid in southeastern India. In fact, the relationship of subfamily B is so close to that of the Notoligotomidae that its components were tentatively assigned to that family [Ross (14)].

SUBFAMILY C. This new category will be created for a very small, aberrant species from the upper Amazon Basin.

SUBFAMILY EMBIINAE. This is an old category which is now to be restricted to a group of genera related to the large genus *Embia* Latreille (e.g., *Dihybocercus* Enderlein, *Parembia* Davis, *Donaconethis* Enderlein, *Metembia* Davis, as well as several other old and new genera). Males have in common a robust form; elongate mandibles with abruptly, apically curled, overlapping dentation; a narrowly cleft tenth tergum; a weak right cercus; and many other features. It might have been treated as a distinct family were it not for genera, such as *Machadoembia* Ross, which link it with subfamily B.

Embiinae are strictly Old World with their greatest concentrations in Africa and the Indian region. No components have been found in southeastern Asia or Australia. However, occurrence as far east as Burma may be expected inasmuch as the writer has found new species in the adjacent tropical forests of Assam and Pakistan (East). The genus *Embia* and its Asian counterpart, *Parembia,* range into the order's ecologically marginal zones of southern Europe and the Middle East. Most Embiinae prefer savanna and savanna-woodland habitats.

SUBFAMILY D. This interesting category comprises several genera characterized by large pale species whose males have short, coarsely dentate mandibles, and peculiar terminalia structure. Some of its typical genera are: *Berlandembia* Davis, *Dinembia* Davis, and *Pseudembia* Davis. Several new genera will be defined, most of which occur in savanna zones south of the Congo region in Africa, as well as in comparable zones of peninsular India. No species occurs in tropical rain forests and the group is absent outside of Africa and India. Again, as in Embiinae, if it were not for intermediate genera (e.g., *Leptembia* Krauss), this subfamily might be regarded as a distinct family.

SUBFAMILY E. Males of this group have mandibles similar to those characterizing the preceding subfamily, but they have distinct terminalia characters and dark pigmentation.

All five component genera are new except *Parthenembia* Ross (13), created for a common parthenogenetic species of southern Congo and Angola. All of these genera are confined to tropical zones of southern Africa where the typical habitat is bark surfaces in *Brachystegia* woodland of south-central Africa. Several species range, however, into the arid zones of the inner Namib Desert.

SUBFAMILY F. This category will be created for a single African species, *Embia ramosa* Navás, or its species complex, which ranges eastward from the Kalahari through the savanna zones of Rhodesia, Mozambique, and Transvaal. The new generic and subfamily concepts will be based on many peculiarities, including an invaginated sac located on the right side of the medial cleft of the tenth abdominal tergite of the male.

FAMILY NOTOLIGOTOMIDAE DAVIS

This family has been recently redefined [Ross (14)] but, as indicated in the discussion of subfamily B, it seems best to limit the group to certain tropical Asian genera: *Ptilocerembia* Roepke, a related new genus from Thailand, a distinct new genus from southeastern India, and *Notoligotoma* Davis of Australia which represents the family's most extreme evolutionary development as well as that of the subfamily B evolutionary trend.

FAMILY EMBONYCHIDAE NAVÁS

This family is based on a single known species, *Embonycha interrupta* Navás, from northern North Vietnam. It appears to be a divergent product of the same stock which gave rise to the Notoligotomidae. The presence of such an unusual species hints that the warmer mountainous zones of northern Burma, Laos, Vietnam, and southern China promise to have a most peculiar Embioptera fauna. When these regions are surveyed by a specialist, more components of the Embonychidae should be discovered.

NEW FAMILY A

To be created for a minute new species discovered in northern Thailand which has very distinctive embioid male terminalia, including a one-segmented left cercus, and oligotomoid head and wing characters.

NEW FAMILY B

To be created for one or more new species of a new genus known only from rain forest zones of southern Cameroun. These are very large embiids with broad, extensively veined wings, and terminalia convergent to the type found in *Enveja* Navás.

NEW FAMILY C

Will comprise two new genera and several new species recently discovered in the cloud forest zones of the Andes of Peru, Ecuador, and Colombia.

Family Anisembiidae Davis

This is strictly a New World family, all males of which have non-dentate mandibles and oligotomoid wing venation in combination with a special complex of terminalia characters. There is a tremendous amount of evolutionary diversity within this rather large family which ranges from southwestern United States southward into Argentina. Within the group are males with almost symmetrical terminalia (e.g., *Saussurembia* Davis), and all stages of increasing complexity to *Dactylocerca* Ross occurring in the northern extreme of the order's New World range. The genus *Chelicerca* Ross has many apterous components and Mexico is particularly rich in species due to its varied ecological zones. The only known Galapagos embiid belongs to this family [Ross (16)].

NEW SUBORDER A

Family Australembiidae Ross

This recently named family [Ross (14)], includes two peculiar genera, *Australembia* Ross and *Metoligotoma* Davis (2), found only in eastern Australia. At least in *Metoligotoma,* these remarkable insects are undergoing intensive speciation which appears to be accelerated by universal apterism and the lack of dispersing "incentive"; i.e., there is unlimited food supply in the form of the leaf litter within which the insects live. Thick layers of fallen *Eucalyptus* leaves constitute the typical niche. It is possible that the size of gene pools may also be locally reduced by frequent bush fires.

Australembiids have many archaic features but very complex, distinctive, male terminalia. At times, females display the largest valvifer rudiments of any embiid. It is regrettable that no species has alate males for wing, and pterothoracic characters might throw much light on the affinities of the australembiids to other families of the order.

New Family D

This will be created to include a peculiar Burmese amber fossil species of presumed Eocene age. It is a very highly specialized, small embiid with a distinct type of reduced wing venation. It is placed tentatively in this suborder because its terminalia, especially the left cercus, suggest a kinship with australembiids.

NEW SUBORDER B

This will be created for a new family to include *Enveja bequaerti* Navás and a complex of striking, closely related species known only from the dense *Brachystegia* woodlands of Katanga and northern Zambia. These are very large, often colorful embiids with distinctive features in almost every structure. Furthermore, they have peculiar habits, living underground during the dry season and in silk galleries which web foliage of low shrubs and annuals during the wet season. In the dry season there are no evidences of the presence of the species in the environment. The group has no close relatives.

NEW SUBORDER C

Family Oligotomidae Enderlein

This old family includes many of the most common Embioptera, several of which have become widely distributed through commerce. Curiously, except for a western extension to the Mediterranean and Black Sea regions of one genus, *Haploembia* Verhoeff [Stefani (17); Ross (16)], the family is endemic to Asia and Australia. The many species occupy niches also commonly found in the Ethiopian and Neotropical regions and this makes the restricted distribution all the more surprising.

The oligotomids have male head and terminalia characters very similar to those of the Teratembiidae. They are consistently larger, however, and invariably the wings have MA unforked, and do not always have specialized clasping structures confined to the left cercus base. In the oligotomids the inner processes and lobes of this cercus base are less complex and never bear claw-like spines. Some of the oligotomids also have two hind basitarsal papillae, whereas the teratembiids have only one.

The largest genera of the family are *Oligotoma* Westwood, which probably differentiated on peninsular India; and *Aposthonia* Krauss, which is most richly represented in southeastern Asia, Indonesia, and Australia. A number of interesting genera awaiting description reveal the great evolutionary diversification which has taken place in the family.

New Family E

This family will be named to include undescribed, large-sized embiids from Afghanistan which are neotinic in both sexes. Superficially, the apex of the adult male's abdomen is identical to that of nymphs. However, close examination reveals a tiny flap-like lobe at the inner left cercus base. The cranium is also more massive and sclerotic than that of nymphs. The pale color (essentially the basic color of chitin) indicates that the species are subterranean in habit and must avoid exposure to sunlight. Its colonies were found under stones near a stream in a desert area.

The new family category is to be created because of the lack of evidence for placement of the species in any other family and because it represents a seemingly ultimate extreme in specialization. It is placed next to the Oligotomidae only for circumstantial reasons. The writer suspects that it is a specialized end-product of *Haploembia* stock.

Family Teratembiidae Krauss

Until the identity of Krauss' *Teratembia geniculata* was determined [Ross (11)], the components of this family were assigned to Davis' now synonomous family Oligembiidae. At first it was thought that this was an exclusively New World family but the writer's fieldwork reveals that Africa has a very rich fauna with perhaps a larger number of genera. More recently, a few species were collected in India and Thailand but it is apparent that the family is poorly represented east of Africa. The large numbers of new species recently discovered in the Neotropical and Ethiopian regions indicate that this should prove to be the largest family of the order. Some of the species appear to be very old and stable, and in *Diradius* Friederichs we have the only transatlantic representation of an Embioptera genus, now that African new species have been discovered in the Nigerian region.

Teratembiids are consistently small and include the diminutive representatives of the order (body length about 3 mm). All of the New World species have a two-branched MA wing vein but many African species have this vein simple. The hind basitarsus of all species has a single ventral papilla. The male terminalia are oligotomoid with a large median sclerite fused to the inner sides of the tenth tergite's hermitergites and thus forming a single plate which generally triangulately projects beneath the ninth tergum. The right tergal process is usually completely, transversely cleft from the hemitergite and the epiproct sclerite is closely appressed to its inner basal side. The clasping area of the basal segment of the left cercus is formed as a sclerotic ring which has one or more inner lobes or processes which at times have one or two claw-like developments. In the previous literature this basal area has been treated as the left cercus-basipodite.

LITERATURE CITED

1. Barth, R. Untersuchungen an den Tarsaldrüsen von *Embolyntha batesi* MacLachlan 1877 (Embioidea). *Zool. Jahrb. (Anatomie) Jena,* **74,** 172–88 (1954)
2. Davis, C. Studies in Australian Embioptera. Part III: Revision of the genus *Metoligotoma,* with descriptions of new species, and other notes on the family Oligotomidae. *Proc. Linnean Soc. N. S. Wales,* **63,** 226–72 (1938)
3. Illies, J. Phylogeny and zoogeography of the Plecoptera. *Ann. Rev. Entomol.,* **10,** 117–40 (1965)
4. Lacombe, D. Contribuição ao Estudo dos Embiidae. III. Aparelho Respiratorio de *Embolyntha batesi* MacLachlan, 1877 (Embiidina). *Studia Entomol.,* **1,** 177–95 (1958)
5. Lacombe, D. Contribuição ao Estudo dos Embiídeos. IV. Polimorfismo sexual da região cefálica de *Embolyntha batesi* MacLachlan, 1877 (Embiidina, Embiidae). *Mem. Inst. Oswaldo Cruz,* **56,** 655–81 (1958)
6. Lacombe, D. Contribuição ao Estudo dos Embiídeos. VI Parte—Diferenças Anatômicas e Histológicas. No Aparelho Digestivo de *Embolyntha batesi* MacLachlan, 1877 (Embioptera). *Bol. Museu Nacl. Zool.,* **219,** 1–16 (1960)
7. Lacombe, D. Contribuição ao Estudo dos Embiidina. VIII Parte: Sistema Nervosa de *Embolyntha batesi* MacLachlan, 1877. *Anais Acad. Brasil. Ciênc.,* **35,** 393–411 (1963)
8. Lacombe, D. Contribuição ao Estudo dos Embiidae. VII Parte: Musculatura da região cefálica de *Embolyntha batesi* MacLachlan, 1877 (Embiidina). *Bol. Museu Nacl. Zool.,* **245,** 1–20 (1964)
9. Lacombe, D. Contribuição ao Estudo dos Embiidina. VIII Parte: Anatomia, Histologia e Excreção de Corantes pelos Tubos de Malpighi de *Embolyntha Batesi* MacLachlan, 1877 (Embioptera). *Anais Acad. Brasil. Ciênc.,* **37,** 503–17 (1965)
10. Rohdendorf, B. B. Die Palaoentomologie in der USSR. *Verhandl. Intern. Kongr. Entomol., 11th, Vienna,* **1,** 313–18 (1961)
11. Ross, E. S. The identity of *Teratembia geniculata* Krauss, and a new status for the family Teraembiidae (Embioptera). *Wasmann J. Biol.,* **10,** 225–34 (1952)
12. Ross, E. S. A new genus of Embioptera from Baltic Amber. *Mitt. Geol. Staatsinst. Hamburg,* **25,** 76–81 (1956)
13. Ross, E. S. Parthenogenetic African Embioptera. *Wasmann J. Biol.,* **18,** 297–304 (1961)
14. Ross, E. S. The families of Australian Embioptera, with descriptions of a new family, genus, and species. *Wasmann J. Biol.,* **21,** 121–36 (1963)
15. Ross, E. S. The Embioptera of Europe and the Mediterranean region. *Bull. Brit. Museum,* **17,** 273–326 (1966)
16. Ross, E. S. A new species of Embioptera from the Galapagos Islands. *Proc. Calif. Acad. Sci.,* **34,** 499–504 (1966)
17. Stefani, R. Revisione del genera *Haploembia* Verh. e descrizione di una nuova species (*Haploembia palaui* n. sp.) (Embioptera, Oligotomidae). *Boll. Soc. Entomol. Ital.,* **85,** 110–20 (1955)

PHYSIOLOGY OF INSECT HEARTS[1]

By Frances V. McCann
*Department of Physiology, Dartmouth Medical School,
Hanover, New Hampshire, U.S.A.*

Introduction

The physiology of insect hearts was the subject of two comprehensive compendia, by Beard in 1953 (9) and by Jones in 1964 (64). From a study of these reviews and others since (18, 22, 37), it is clear that while insect hearts have been studied for some three hundred years, much of the work has been descriptive and qualitative; thus little real understanding of basic mechanisms and fundamental processes has ensued. I have, therefore, chosen to discuss certain areas of insect heart function from the perspective of comparative physiology, at the level of basic cellular processes.

Physiology of insect hearts at the cellular level is a relatively unexplored and somewhat novel area. It may seem premature to review such a fledgling topic, but the existing data has in large part not been examined from this point of view. This article will evaluate primarily those reports that explore insect myocardia at the cellular level and will also examine fundamental properties as they relate to excitable tissues.

Current hypotheses of mechanistic principles of cardiac cell function emanate almost exclusively from studies on vertebrate hearts. Such interpretations have been made possible by the rigorous application of precise quantitative techniques such as electron microscopy, tracer methods, ultracentrifugation, and microelectrode analyses. Intense attention has focused on such basic considerations as the mechanisms operative in impulse origination, impulse propagation ("cable properties" of the cell), impulse transmission from cell to cell, coupling of the impulse to the contractile machinery and the correlation of subcellular structures with specialized function. This experimental approach seeks to elaborate cellular properties from a critical analysis of quantitative data. The application of this technology to insect hearts is now essential if quantitative data are to be obtained.

This review will, therefore, be directed not so much to what is known about insect hearts but rather to what is not known, not so much to what has been done but rather to what needs to be done, and not so much to individual reports but rather to an interpretation of the *issues* these reports

[1] This review was prepared during the tenure of an Established Investigatorship of the American Heart Association and supported in part by the Vermont and New Hampshire Heart Associations and the National Heart Institute of the National Institutes of Health.

have provoked. The acute need for a more fundamental approach to insect heart physiology will be stressed and the application of physiological methods in the search for mechanisms of action will be encouraged. Since attention will be focused on broad fundamental problems, many of the references will not be obviously pertinent to insect hearts alone. It is hoped that this discussion will stimulate the application of new techniques that will promote a more precise definition of cardiac function in insects and will extend the perspectives of those working in this area.

ULTRASTRUCTURE—FUNCTION

The advent of electron microscopy has provided great impetus in the search for intracellular structural inclusions that can be correlated with specialized functions. Since gross anatomic and histologic features of insect hearts have been extensively documented (36, 111), attention will be focused on the more current topic of ultrastructure of single cells as it relates to function.

The cellular geometry of insect hearts apparently varies from a single layer composition to more intricate fibrillar arrangements (137), such as the interlacing network described in *Aeshna* (142) and the mixture of spiral and longitudinal fibrils observed in *Nepa cinerea* L. (39). These distinctions, however, may not be as definitive as has been supposed, for insect material is notoriously difficult to fix both for light and electron microscopy. As a result, detail lost or altered in the fixing process, the interposition of alary muscles and extensive basement membranes complicate interpretations. Recent improvements in fixation methods should enhance studies with the electron microscope and help clarify discrepant observations.

The most detailed ultrastructure–function studies on an insect heart available at this date are those reported for the myocardium of the moth *Hyalophora cecropia* (88, 115). The moth myocardium is composed of a single layer of striated muscle cells oriented in a helical pattern with respect to the lumen similar to that described for mosquito hearts (63, 140) and fly heart (35).

Sarcomere units in *cecropia* myocardia are bounded by clearly defined Z-bands composed of discontinuous dense bodies. Z-bands have been reported in the heart of the cockroach *Blattella germanica* (32) and *Periplaneta americana* (95). However, the fixation of these hearts does not allow clear resolution of the fine structure of the Z-bands, i.e., whether they are composed of densely packed continuous material or discontinuous dense bodies. The beaded Z-band was also observed in the alary muscles as well as in the muscles that allow the moth nerve cord to thrash (114). Beaded Z-bands have been reported in cardiac muscle of the snail *Helix* (105) and dense bodies similar in appearance were found in molluscan smooth muscle (40). Both continuous and beaded Z-bands have been reported in the barnacle *Balanus* (48) where the former serves the resting condition and the latter the contracted state, i.e., when the muscle is fully contracted, the thick filaments can pass between the dense bodies. Fixation of the moth heart both at rest and fully contracted lengths revealed that the discontinuous structure is the permanent state of the Z-band and that the thick filaments pass between the component dense bodies. This structure, the beaded Z-band, may allow the heart to continue beating while the whole body length is remarkably shortened (114).

Other components of sarcomere structure of the insect heart have been reported for only a few species. A-bands have been identified and measured in myocardia of only two insects, *H. cecropia* [1.8 μ (115)] and *P. americana* [2 μ (95)]. Although only Z-bands were reported to be present in the heart of the roach *B. germanica* (32), it should be emphasized that glutaraldehyde fixation was not then known, and tissues could not be well preserved for examination in the electron microscope.

Cross sections of A-band regions where thin and thick filaments of contractile proteins interdigitate according to the classical picture of sarcomeric structure (57) reveal the presence of both thick and thin filaments. A ring of 10 to 12 thin (60 Å) filaments orbits each thick (200 Å) filament in *cecropia* heart (115). While comparable measurements from other insect hearts are not available, this distribution has been reported in other insect muscles, viz., ventral intersegmental abdominal muscle of *Rhodnius prolixus* (131), intersegmental muscles of *P. americana* (121) and flight muscle (5). All of these muscles are characterized as *slow,* and an inverse relationship between the number of thin-to-thick filaments and the frequency of contraction has been supported by studies on insect flight muscle, where slowly contracting fibers have 9-12 thin filaments around each thick filament and fast, only six. Z-bands also differed in the slower fibers (5).

Details of the fine structure of the sarcoplasmic reticulum (SR) and transverse tubular system (TTS) in insect hearts are again limited to only a few reports. Transverse tubules in moth (88, 115) and roach (95) hearts are irregularly arranged, unlike the orderly orientation in vertebrates. The poorly developed SR and TTS seem characteristic of slow or tonic muscles in other insect muscles (19, 32).

Characterization of muscle function on the basis of distinctive structural features has been attempted with vertebrate and invertebrate material. Thus, the presence of M-bands, straight Z-bands with a filamentous substructure, an extensive well-developed system of transverse tubules and sarcoplasmic reticulum characterize fast contracting muscles (42, 107, 108), while opposite features characterize slow muscle fibers. In accordance with this classification, the insect heart would be aligned with slow or tonic muscle fibers on a structural basis.

As in most tissues that are rhythmically active, mitochondria are abundant. Numerous mitochondria are localized and concentrated in outpocketings of the sarcolemma such that the protrusions extend into the lumen of the heart on one side of the cell and on the other side into the pericardial space (95, 115).

Because of branching, measurements of cell size in the insect myocardium, as in the vertebrate myocardium, are difficult to establish. Measuring between two intercalated discs, an average value of 30 μ in width and 100 μ in length was given for moth myocardial cells (115) with the electron microscope, and only 1 to 1.2 μ for *Dysdercus fasciatus* (43) using a light microscope. In this instance (43), Z-bands apparently have been mistakenly identified as intercalated discs. An accurate evaluation of cell size is of critical importance if one is to attempt physiological measurements. That there should be such a great difference in the size of insect myocardial cells is striking and should be further verified.

Information regarding the structure and function of intercellular boundaries has assumed prime importance in attempts to explain mechanistic principles involved in impulse transmission from cell to cell. Intercalated discs, appearing as dense bands transecting mammalian myocardial fibers, were identified as intercellular boundaries by classical histologists working with the light microscope. The clearer resolution now afforded by the electron microscope has enabled a detailed examination of this structure (see 27). Three distinct components have been identified in the mammalian heart: interfibrillar junction, desmosome, and nexus (8, 62). The nexus, or tight junction, has been proposed as the site of electrical transmission between cells (7, 8).

In insect hearts the demonstration of intercalated discs with the light microscope and the identification of their component structure with the electron microscope have only recently been successful. Intercalated discs were erroneously reported in the hearts as well as alary muscles of several species of Hemiptera; detailed photographs were not presented (43). Intercalated discs were reported in roach heart *B. germanica* (32) and were described as identical to those in vertebrate cardiac muscle. Since the nexus and its physiological significance were not recognized at this time, the fine structure of the disc will require re-examination.

Intercalated discs have been demonstrated with both the light and electron microscopes in the myocardium of *H. cecropia* (88, 115). Two components have been identified: interfibrillar junction and septate desmosome. Only the interfibrillar junction is identical to that in the mammalian heart, while the septate desmosome is distinctively different from the desmosome and nexus of vertebrate hearts.

Of particular interest is the observation that materials appear to pass between cells through septate desmosomes (79, 80, 136). However, the structural integrity of septate desmosomes in salivary cells of *Chironomus* remained unaltered after permeability of the junctional membrane had been reduced by raising intracellular calcium (15). These authors have emphasized that the assignment of a transmission function to junctional membranes on the basis of structural evidence is at this time unwarranted. If the septate desmosome is an area where low electrical resistance (high conductance to ions) prevails, one may speculate that this structure may be a functional analogue of the vertebrate nexus.

It is obviously premature to characterize the ultrastructure of the typical insect myocardium in view of the severe limitation of available data. A summary of the observations at this time would picture the insect heart as a tube of striated muscle, a single layer in thickness with a helical arrangement of fibers. Sarcomere structure is composed of thick and thin filaments in differing numbers. A-bands are conventional, I-bands very small, Z-bands may be discontinuous dense bodies, M-bands are absent. Sarcoplasmic reticulum is relatively poorly developed, as it is in vertebrate hearts. The transverse tubular system is not aligned with Z-bands. A thick basement membrane surrounds the cells, mitochondria are abundant but located in clusters, and multiple invaginations of the sarcolemma increase the cellular surface area.

Of particular significance is the conclusion that there is no structural evidence for the existence of different types of cells throughout the myocardium (115). The cells appear to be anatomically homogeneous and no one has reported any data that would identify morphologically specialized tissues in the myocardium. That the cells may not be alike physiologically (35, 58, 84) serves to emphasize that functional specializations cannot be elaborated solely by studying the architecture of subcellular particles but rather circumspect attention must focus on the complex phenomena secreted in the cell membrane.

Origin of the Heart Beat

While great attention has been paid to the type of tissue in which impulse origination occurs, elucidation of the mechanism by which this occurs seems to have been neglected. General statements have pervaded the literature that attempt to classify insect hearts on the basis of whether they are neurogenic (impulse originates in nervous tissue) or myogenic (impulse originates in muscle or modified "specialized" muscle tissue) (64, 87, 104, 110). The difference between neurogenic and myogenic beats seems something of an artificial distinction, for reliable quantitative criteria have been difficult to formulate from which to conclude such an origin. The chronotropic and inotropic responses to acetylcholine and epinephrine have long been used as evidence for the neurogenicity or myogenicity of the heart, but this seems a totally inadequate criterion, and provides little insight into the physiological mechanisms involved.

Histological demonstrations of neural elements in insect hearts have led to confusing conclusions regarding their function. Neurogenicity does not necessarily derive from the fact that nerves or ganglia are present, nor does myogenicity prevail if the heart continues to beat in isolation after separation from the animal. That ganglia can be a part of the myocardium or that neural twigs persist after its excision makes true and complete denervation difficult to establish. Thus, conclusions regarding the myogenic origin of the heart beat in certain insects from denervation experiments and from histological studies in which nerves cannot be demonstrated require judicious evaluation.

The presence or absence of a neural innervation to the insect dorsal vessel is difficult to establish. Not only does neural innervation apparently vary from one species to another, but neural tissue can be easily overlooked. Other tissues absorbing the stains may be incorrectly identified as nerves.

Patterns of innervation for insect hearts range from the complete absence of neural elements as in mosquitoes (17, 63) to an elaborate network as in cockroaches where electron microscopy has been employed to demonstrate the detail of axons and neuromuscular junctions (32, 61, 98). No nerves at all could be demonstrated anywhere along the moth dorsal vessel even when it was serially sectioned and examined in the electron microscope (115). Conventional staining techniques were used for preliminary observations but only alary muscles took the stain. This is indeed puzzling since the nerve supply to the lepidopteran dorsal vessel has been described in detail (73, 90). Unless alary muscles absorbed the stain in these studies and were mistaken for nerves, we are unable at this time to resolve this important discrepancy.

While nerves may provide a mechanism of regulation and control as well as initiation, it is interesting that the presence of inhibitory nerves to insect myocardia has not been confirmed. Krijgsman (71) has confirmed that inhibitory influence can be produced by stimulation of the cerebral ganglion (74). Tachycardia, induced by electrical stimulation, is followed by a period of cardiac arrest in the *cecropia* heart (87). The membrane potential appears very slightly (2 to 3 mv) hyperpolarized and is strikingly reminiscent of vagal inhibition in vertebrate hearts (20, 55). Whether inhibitory influences proceed by way of humoral or neural pathways remains to be determined.

Direct studies on single cells of insect myocardia have as yet provided little information regarding the mechanism whereby activity is actually induced. Microelectrode studies on single cells of moth (59, 81, 84), fly (35), cockroach (94), and cicada (58) myocardia demonstrate the presence of an initial, slowly rising component of some intracellular action potentials. The general conclusion is that this represents pacemaker activity and is universally regarded as a pacemaker potential, designated as phase 4 in vertebrate myocardia (45). At first glance this would seem to furnish the needed criterion to distinguish between myogenic and neurogenic activity. However, that the mere presence of a prepotential should be regarded as evidence of a dominant or primary pacemaker is regarded as inadequate (25, 123, 124). True pacemakers are identified as those that increase their rate of firing when subjected to transmembrane pulses of depolarizing current and slow it when hyperpolarized. Dormant and latent pacemakers are those that maintain an unchanged rate of firing in response to depolarizing and hyperpolarizing pulses although the contour of the action potential may change (77). The application of depolarizing and hyperpolarizing pulses of constant current to moth heart cells produced distinctive changes in phase 4 and phase 2, but the heart rate did not change (85). This suggests that, although the cells being studied did generate prepotentials, their role is that of latent or dormant pacemakers and not dominant. Slight increases in heart rate of the moth have been induced by depolarizing pulses but the direct effect on phase 4 is indistinct (87). By indirect methods, such as when heart rate activity is increased by driving the whole myocardium or by the application of drugs (epinephrine), prepotentials do change, primarily by an increase in the rate of rise of the initial phase [phase 4 (87)].

Assuming that the factors that initiate and control the rate of firing in pacemakers of insect hearts are similar or identical to the mechanisms proposed for vertebrate hearts, i.e., change in the slope of the prepotential, the value of threshold potential or the initial level of resting potential (45), evidence that these factors control pacemaker activity in insect hearts must be regarded as preliminary.

Since many of the intracellular action potentials recorded from the moth myocardium have exhibited a distinctive slow phase 4, McCann (84, 87) has proposed that probably all of the myocardial cells of adult *H. cecropia* retain pacemaker properties, either latent or dormant, and that the heart is normally driven by a dominant pacemaker. From electrograms recorded at multiple sites on the moth myocardium, Tenney (128) concluded that dominant activity is localized in the caudal tip of the heart with latent pacemakers at the cephalic end of the heart. In transection experiments, Irisawa et al. (58) demonstrated that segment 7 of the cicada heart was the dominant pacemaker and segment 2 the latent, and a similar conclusion was reached by Jones (63) on *Anopheles quadrimaculatus*. No such localization was concluded from similar experiments on the fly heart (6). Primary or dominant pacemaker activity may be characteristic of individual cells, and whether there is a topographical localization remains to be confirmed.

The literature that has characterized the cockroach heart as a classical example of a neurogenic heart is well known and stems primarily from evidence of an elaborate nervous innervation (3, 90) and an increased heart rate in response to acetylcholine (71, 93).

Recently, however, this issue has been opened for reconsideration. Correlating neural burst activity with heart rate and finding no intracellular pacemaker potentials in the myocardium, Senff (119) concludes a neurogenic origin. He also states that removal of the nerve causes cessation of the heart beat. These observations are the antithesis of those reported by Miller (94, 95, 97), who reports the continued beating of the heart after denervation and the presence of pacemaker-type potentials in myocardial cells. Smith (122) also concludes a myogenic origin of the heart beat from experiments in which tetrodotoxin selectively abolished electrical activity in the ganglia without inhibiting the heart beat.

The resolution of such conflicting observations lies perhaps in establishing a clearer perspective of underlying mechanisms. First, it is agreed that there is an elaborate nerve supply, and the elemental issue is whether under normal circumstances the impulse originates in the nerve or in the muscle. Denervation probably cannot be effectively complete, for neural twigs remain as well as neuromuscular junctions. This same situation prevails when the question regarding mammalian heart is the topic. The question is whether every cell is capable of initiating activity or is a remnant, however small, of specialized tissue necessary for initiating spontaneous depolarization. The latter choice is the more widely accepted (46, 106).

If the primary pacemaker is to be that cell which fires first, and, since pacemaker-type potentials have been recorded in the cockroach heart (95), it may be that while they retain pacemaker potentiality, this feature is not revealed until the dominant driving pacemaker in the nerve is removed. Thus, discrepancy in observations here may be a reflection of what is happening under normal conditions and what the heart is capable of doing when stripped of its normal source of impulses.

On the other hand, if the impulse does originate in the myocardium, the neural innervation then becomes a complex regulatory mechanism and perhaps it could even compete with the myocardium for dominance, achieving it under some propitious circumstances and yielding it under others. Thus, the distinction between the two origins could be even more complicated. Obviously, these are complex issues and will not be easily settled, but the mechanism by which the spontaneous depolarization occurs would seem to be the issue of primary importance.

If the frequency of firing of the pacemaker-type potentials in the cockroach myocardium can be altered by manipulation of the membrane potential, then good evidence for primary pacemaker activity would be on hand, but more direct experiments are needed to resolve these issues.

It becomes particularly significant to refer here to the localized pacemaker activity demonstrated in vertebrate myocardia for, in this instance, it is generally accepted that the presence of specialized tissue is prerequisite to pacemaker activity. The nature of this specialization remains obscure, but the ultrastructural detail of the mammalian sinoatrial node may be pictured as a delimited area in which muscle cells are diminished in number and myofibril outlines are indistinct and somewhat obscure (113). No such cellular differences have been identified in cells comprising insect myocardia.

A suggestion has been proffered that pacemaker activity of a cell may be directly related to the relative amount of surface area of the plasma membrane to cell volume (113). The large amount of surface area associated with individual heart cells (32, 88, 115), coupled with the physiological evidence of ubiquitous pacemaker activity, is consistent with this suggestion. Until more data are available from a variety of hearts, such conclusions must be relegated to the realm of speculation.

ELECTROPHYSIOLOGY

Despite the introduction of microelectrodes some twenty years ago (78), few have applied this technique to insect hearts. This method allows the direct study of signal generation in individual cardiac cells and eliminates activity patterns of other cells, e.g., alary muscles. The direct intracellular method of recording electrophysiological events now makes surface electrograms, as they have been used, obsolete. Suction electrodes are useful for certain studies (87), but microelectrodes should be used where possible.

The diversity of transmembrane ionic concentration ratios in certain insects makes them particularly attractive as test media for conventional concepts of cellular excitability proposed in the classic studies of Hodgkin & Huxley (44). Of particular interest are those herbivorous insects whose hemolymph is bizarre in composition when compared to other animals (14, 34, 126). The hemolymph, practically lacking in sodium content, is characterized by extraordinary concentrations of potassium and magnesium, and thus resembles plant sap more closely than animal plasma. The problem of electrogenesis in the myocardium of insects in general or in those groups mentioned above has received only scant attention and only few reports are available. Measured transmembrane resting potentials of −40 mv, action potentials of 55 mv, with overshoot values of 15 mv are reported for the silk moth *Bombyx mori* (59). Resting potentials measured in *H. cecropia* heart cells average −65 mv, action potentials 85 mv, with overshoot of +20 mv (81, 84, 87). These values stand in striking contrast to those ($E_{K^+} = -11$ mv: $E_{Na^+} = +12$ mv) predicted by transmembrane values of potassium and sodium inserted into the Nernst equation:

$$E = \frac{RT}{nF} \ln \frac{[\text{ion}]_o}{[\text{ion}]_i}$$

E, equilibrium potential; *R*, universal gas constant; *T*, absolute temperature; *n*, valency of ion; *F*, Faraday; ion concentration outside and inside cell, respectively.

There is some question as to the sign of the sodium equilibrium potential, for a transmembrane ratio of less than unity has been reported for Lepidoptera (16, 50, 51), but it must be emphasized that the values of transmembrane sodium concentrations are very small and the detection of significant differences at this level of resolution is uncertain (89). The important observation is that there is a great discrepancy between measured and calculated values of electrical potential.

Resting and action potentials have been recorded in the heart of the fly *Musca domestica* for which a maximum of 50 mv and a mean of only 20.3 mv were reported, but this was not correlated with values predicted by Nernst equation calculations (35). These action potentials were directly affected by sodium manipulation in a conventional manner. Transmembrane action potentials in the heart of *P. americana* are reported to be insensitive to alterations in transmembrane sodium, even though sodium is required for neural activity (95). Apparently sodium functions as a current-carrying ion in some insect hearts and it may be that only in those orders where unusual ionic patterns exist (particularly Lepidoptera) do ions other than sodium and potassium function as major carriers of transmembrane currents. However, many more studies will be required to justify such a conclusion.

How electrical and mechanical activity is generated and maintained in the hearts of those insects that are exposed to an environment of such remarkably unorthodox composition is a topic of very broad interest and is a complex subject in itself. It should be mentioned that this extremely important problem of the genesis of electrical potentials in different ionic media has received notable attention from those working on the same problem but with insect muscles other than the heart. A spectrum of interesting and unusual membrane electrode properties as related to transmembrane ionic concentrations is now emerging from the studies of various insect muscle fibers (10, 49, 52, 53, 139) and an as yet amorphous multi-ionic current-carrying system awaits definition.

Despite the lack of correlation between the predicted value of the membrane potential (−11 mv) and the measured value (−65 mv) in *Hyalaphora cecropia,* the membrane potential indicates that it shares certain fundamental characteristics with other membranes that behave as conventional potassium electrode systems (81, 84, 87, 88). Metabolic inhibitors such as dinitrophenol reveal two components of the resting potential, one sensitive to the metabolic poison, the other insensitive, so that approximately 30 per cent of the original level of polarization is maintained in the presence of this poison (86). This has also been reported for lepidopteran skeletal muscles (54) and it now seems certain that there are two processes that maintain the resting potential, as in mammalian tissues (26, 78, 91), one portion dependent on oxidative phosphorylation to energize an active transport mechanism, and the other on passive diffusion.

The observation that anions of large diameter could apparently penetrate the membrane of cardiac muscle cells with ease in a herbivorous insect (moth) suggests that the permeability qualities of the membrane may be unique (83). Hyperpolarization of the membrane resulted from soaking in a solution in which all anions were replaced by acetate ion. This was further supported by confirmation of the very low transmembrane ionic concentration ratios that indicate that the membrane is a poorly selective barrier (89). Ferritin, a protein substance consisting of very large diameter particles (100 Å) and used frequently on vertebrate muscles to mark the sarcotubular system (56, 107), apparently penetrated the cell membrane with ease. That this was not a simple breakdown of the membrane from a toxic effect was demonstrated with coincident intracellular electrical recordings. The electrical and mechanical performance of the cell remained unaltered. There was no evidence of pinocytosis (88). The permeability properties of the moth heart cell as indicated by anion (83) and ferritin (88) studies and the apparent absence of a tissue barrier when examined in the electron microscope (115), lead one to conclude from the present evidence that the tissues are directly exposed to these bizarre ionic concentrations. Similar investigations have not been carried out on any other insect hearts, but such studies may reveal further unique membrane phenomena.

Examination of the *H. cecropia* heart cell membrane with pulses of constant current gave a polarization resistance of $3.8 \pm 1\ M\Omega$ and a specific membrane resistance of $1360\ \Omega cm^2$. The action potential appears to be a postsynaptic potential by these experiments. The magnitude is a linear function of membrane voltage and the membrane itself behaves as an ohmic resistance (85). Conductance increases rapidly during the upstroke (phase 0) and resistance increases during the plateau (phase 2), and decreases again during the final states of repolarization (phase 3). These results are similar to those observed in mammalian cardiac cells (45).

H. cecropia heart cells exhibit a refractory period of about 200 msec at a heart rate of about 60 beats per minute under normal conditions (87), unlike *P. americana* which can be tetanized (96, 141). Like vertebrate hearts, the plateau phase of the moth heart is the most labile, being altered by changes in pH, temperature, drugs, and rate (87). Within a range, total duration of the action potential is a linear function of heart interval (87).

The relationship between conduction velocity and heart rate is not yet resolved. Although it appears that conduction velocity increases with heart rate when latent periods are measured (82), this method does not take into account that activation rates may also increase. Conduction processes through the myocardium remain to be more clearly evaluated. Some insect cells appear to function in a number of respects like vertebrate heart cells, but more data must be obtained from a variety of insects. The diversity of physiological adaptation found in insects makes them singularly attractive for such studies.

Reversal

The physiological mechanisms that precipitate and implement periodic reversal of the direction of heart beat continue to elude definition. Reversal has been reported in a number of insects at all stages of development [see reviews (9, 64)], and has been studied extensively in the heart of tunicates (4, 68, 69, 102), blood vessels of vertebrates as well as invertebrates, and embryonic heart tubes (101). A tubular structure is common to these organs and the mechanisms that underlie the process of reversal probably share a common denominator. It seems appropriate, therefore, to consider the fundamental problem of reversal as it pertains, not only to insect hearts, but tubular hearts in general.

In those hearts in which reversal has been studied recently, a conclusion with which most authors seem to agree is that under normal circumstances pacemaker activity is localized at either pole of the heart, one being dominant over the other, and that each cell in the myocardium has the potential of spontaneous depolarization, the latter being a quality that is normally suppressed (4, 84, 100, 102, 128).

During reversed beating, the heart rate is often slower than in the normal forward direction (24, 69, 128) as would be predicted when pacemaker activity is shifted from a dominant to a latent site which, by definition, is that one which would operate at an inherently slower rate. The velocity at which the contractile wave proceeds, however, is quite another matter and several reports point out that conduction velocity in the reversed direction is slowed. Hecht (41) reported conduction velocity measurements of 21.2 mm/sec in the forward direction as compared with 17.6 mm/sec in the reversed direction in the ascidian heart. Tenney (128) reported 40 mm/sec forward and 30 mm/sec retrograde. Conduction velocities also vary on a segmental basis, the slowest occurring in the central segments and terminal areas (128). Conduction velocities are slowest in the apical regions of *Ciona* heart (67), but Anderson (4) reported that the differences were inconsistent and variable. McCann (82) recorded alternately from two myocardial cells situated at equal distances (100 μ) from a centrally located stimulating suction electrode. This allowed an impulse to be driven in either orthograde or retrograde directions. This method of measurement did not reveal any statistically significant directional differences in conduction velocity. Measurements of conduction velocity over a distance of several cell lengths have assumed a linear and uniform pathway. The helical arrangement of the fibers and their profuse branching make this assumption untenable (88, 115), and may help to explain these discrepant observations.

That conduction velocity is facilitated in one direction over the other implies that a preferential direction exists in the pathway for conduction. Such specialized pathways are well known in vertebrates. Bundles of His and Purkinje fibers are distinguishable on the basis of histological (132) as well as physiological evidence (2, 106, 118). The cells of the moth myocardium appear to be homogeneous; there is no evidence by examination in the electron microscope that cells exhibit varying diameters or that one end is tapered so as to provide the geometric complexity that differing conduction velocities would require. Neither can any specialized tissues be identified (115). Therefore, it must be concluded that at this time there is no demonstrable evidence which would account for a preferred direction of conduction.

The very diversity of the factors suggested as effectors of reversal clearly demonstrates that there is little agreement on a common cause, and emphasizes that many changes in the external environment may precipitate the phenomenon. Reversal has been induced by many alterations of the cellular environment of the heart, viz., changes in temperature and pressure (68, 69), production of extrasystoles (101, 102), and changes in partial pressure of CO_2 (13). Anderson (4) succinctly proposes that all modifications of the cellular environment ultimately affect the intrinsic pacemaker properties of the cells.

Undoubtedly, one of the explanations for the multitude of seemingly unrelated experimental manipulations of the cellular environment that have induced reversal, e.g., changes in pressure, acidity, temperature, anoxia, drugs, hormones, age, to cite a few, is that ultimately these factors all exert their effect at the level of the individual cell membrane and this, in turn, includes not only enhancement or suppression of spontaneous depolarization but also alteration of those factors that define their excitability (cable characteristics). These features of the cell would then determine the pattern of wave propagation.

In an attempt to demonstrate this concept we have formulated a computer model of an insect heart in the same manner as has been done to describe normal and arrhythmic patterns of activity in mammalian hearts (103). The model of an insect heart embodied in this program for the computer is designed to relate specific observations with simple logical concepts. As in all science, the objective is to make a particular aspect of the observations comprehensible. We concentrate on a few features of the phenomena and attempt to formulate minimum postulates from which these features can be deduced. Other conceivable postulates or properties are intentionally excluded.

In the insect heart model reported here, we concentrate on wave propagation, with special attention to reversals, and find that the observed propagation can be deduced from postulates that ignore details of action-potential forms.

The models embodied in programs MOTH-HT3 and MOTH-HT4 differ in the inclusion and exclusion, respectively, of a simple postulate which effects reversal. In each program, the heart is represented as a set of 15 cells. Each cell has an internal potential which is defined periodically with period T. For cell #J, the potential at the start of a time period is A(J), and that at the end of the period is B(J). Each cell exists in a state symbolized by A$(J) at the start of any time period, and B$(J) at the end. The states may be FIRED, REFRACTORY, or EXCITABLE.

The time interval is chosen so that a wave travels from one cell to the next in one time interval, T. When a cell is FIRED during one interval, either neighboring cell or both, if EXCITABLE, is FIRED in the next interval. During a REFRACTORY period, a cell potential changes at the rate R, a negative number, and becomes more negative by $T \cdot R$ millivolts per time interval. During an EXCITABLE period, the potential rises at a rate $S(J)$, until (*a*) the potential reaches a threshold, $T(J)$, or (*b*) a neighboring cell has fired in the preceding time interval.

In either case (*a*) or (*b*), the state of the cell changes to FIRED and the action potential reaches a maximum. In the time interval following that in which a cell is FIRED, it becomes REFRACTORY and so remains until the

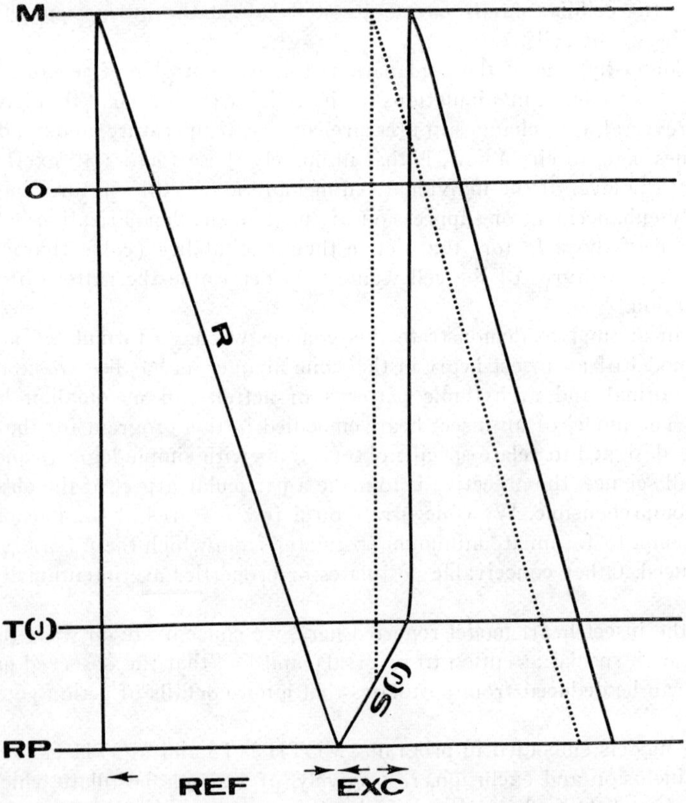

FIG. 1A. A schematic of the parameters used here to simulate the excitability characteristics of heart cells. T(J), threshold; S(J), slope of pacemaker; RP, resting potential; 0, zero potential; M, maximum potential; R, repolarization; REF, refractory period; EXC, excitable. Dashed line indicates course of action potential when cell is fired prematurely. See text for discussion.

FIG. 1B. Patterns of wave initiation and propagation predicted by a computer program. Far right, MOTH-HT3; left MOTH-HT4. Symbols used: • , excitable cell; #, fired; H , refractory.

HT3 tail cell (T) has lower threshold than head (H). Note wave pattern as tail assumes dominance over head. No reversal.

HT4 is a pattern of reversal when head and tail cells are identical. Note extrasystoles that occur as a result of excitability characteristics of cell rather than as a cause of reversal. See text for detail.

MOTH-HT4 04/17/69 10:56
H
T
MOTH-HT3 04/17/69 14:03
H
T

potential falls to the resting value D(J) for that cell. When this value is reached, the cell becomes EXCITABLE.

For both MOTH-HT3 and MOTH-HT4, the rates of depolarization S(J), the thresholds T(J) and the resting potentials D(J) are taken to be identical for all except perhaps the head and tail cells. In MOTH-HT3, the action potential reaches a value M that is not determined by its value prior to firing, whereas in MOTH-HT4, the maximum for any firing is a constant plus a fraction of the potential at the instant firing was initiated.

The run of MOTH-HT3 (right column, Fig. 1B) starts with randomized values of the cell potentials, but with somewhat lower thresholds T(J) at the head end (H), and still lower threshold at the tail (T) cell. It is seen that waves eventually tend to travel from the tail to the head in the figure. In general, any cell which happens to have the lowest threshold, whether the cell is at an end or not, automatically becomes the pacemaker; nothing other than a higher natural frequency of firing distinguishes the pacemaker from the other cells.

It is impossible for this model to exhibit reversal, except when external conditions alter the threshold (or rate of depolarization, etc., with equivalent changes in frequency). The model of MOTH-HT4, however, does exhibit reversals. Note that the distinction between the two models lies in the added postulate for MOTH-HT4, i.e., that the potential on firing is affected by the potential when firing was initiated. The cell thus retains a "memory" that extends beyond the present time interval.

The run of MOTH-HT4 (left column, Fig. 1B) has been started with exactly the same conditions as was that of MOTH-HT3, except that the head and tail cells are identical. Note, however, that there is now a tendency for reversals to occur, even though the two end cells are identical in all respects.

It will be noted, also, that extrasystoles appear in MOTH-HT4, but not in MOTH-HT3. This behavior is a consequence of this added postulate, and would be difficult to produce without it unless the end cell were assumed to differ radically from the others in the parameters diagramed in Figure 1A.

With this computer program we have attempted to demonstrate that it is unnecessary to invoke a multitude of factors to explain the phenomenon of reversal. Patterns of wave propagation can arise simply from inequalities in any of the fundamental properties that determine the natural firing period. Reversals can occur when the firing period has been altered by previous activity.

There has been little consideration of the possible physiological requirement fulfilled by reversing the direction of blood flow. Perhaps reversal is a biologically useful event and if so, the basic functional requirement being fulfilled must be defined. It may be a compensatory mechanism, a built-in safety feature that operates only when some stress has been imposed that threatens the homeostasis of the *intérieur milieu*. Another consideration may be that it is an abnormal event and occurs only when the normal sequence is interrupted by some pathological alteration of the tissue or its media, or it may be a sporadic event of no consequence, an accident. These uncertainties await resolution and until more quantitative data are available the phenomenon of reversal remains at this time a perplexing enigma.

Pericardial Cells

Pericardial cells, together with connective tissue and alary muscles, comprise in large part the dorsal diaphragm of the insect heart (22). These cells cluster along the alary muscles and are in some instances directly linked to the myocardium by strands of connective tissue (135). Their ingestion of foreign particles has suggested a phagocytic role (22, 64, 66, 125) in a manner likened to the reticuloendothelial system of vertebrates (99, 137).

Subcellular inclusions identified as coated vesicles and tubular elements led Bowers (12) to conclude a protein uptake function, based on proposals for similar structures in other cells (33, 92, 112). Hoffman (47) did not observe iron particles in the coated vesicles of locust pericardial cells when they were exposed to a solution of iron saccharate, but Sanger & McCann (117) demonstrated ferritin particles in both coated vesicles and tubular elements and concluded that pericardial cells fulfill a role in protein uptake and transport.

The inosculation of pericardial cells with the myocardium and alary muscles has prompted study and speculation relative to their involvement in cardiac activity. Davis (23) considered the possibility that they could be the pacemakers of the heart, but when the heart beat persisted after he selectively destroyed individual cells, he discounted this proposal.

That they have secretory characteristics has been clearly demonstrated with the electron microscope (114). A cardio-regulatory control function has been assigned by Davey who has concluded from a number of studies that the pericardial cells produce a cardio-accelerator substance from an inactive precursor when stimulated by a peptide-like substance released from the corpus cardiacum (22). Kater (65) attempted to show that release of the cardio-accelerator substance is mediated via a neural pathway from the brain to the corpora cardiaca. However, he considered the action of the cardio-accelerator substance to be directly on the heart and did not consider pericardial cells as possible intermediaries.

Pericardial cells are generally diminished in number and in many cases are absent around the anterior region of the dorsal vessel designated as aorta. The accumulation of large amounts of neurosecretory material in the aortal wall of *Ranatra elongata* has led to the conclusion that it serves as a storage-and-release organ for such material and essentially resembles the corpora cardiaca (29). Dogra (30) considers the aorta a neurohemal organ. Intracellular action potentials recorded in the aorta of the *H. cecropia* moth are identical to those recorded in the other regions of the heart (87). The aorta is also pulsatile and appears to be an extension of the tubular heart. How much variation is due to species differences will have to be resolved by more studies. Further examination of this area in other insects with the electron microscope may provide evidence of neurosecretory granules if such activity is a common factor.

In those insect hearts that have been described as aneural (17, 63, 115), the release of accelerator and decelerator chemicals from closely apposed sites, viz., the pericardial cells, would provide a ready source of cardio-regulatory chemicals. The presence of pericardial cells and alary muscles clearly presents added complications in assessing pharmacological responses, for a direct effect on the heart muscle itself will require critical and definitive experimental techniques.

Alary Muscles

A summary of the literature regarding alary muscles is included in this review because they appear to be coupled to the myocardium both structurally and functionally. A critical analysis of a few reports will be presented in order to evaluate current concepts regarding their structure and their physiological contribution to the cardiac cycle.

Alary fibers, the long filamentous muscles that suspend the heart in the dorsal region of the abdominal cavity, are present in most insects at some stage of development. Even in those insects in which the heart is reduced to only one segment, alary fibers are present in their characteristic fan-like overlay (43). Alary fibers apparently approach the heart by way of three routes (43, 116). Proceeding from the lateral body wall, they may run along the surface of the heart parallel to the long axis for a distance before emerging to proceed to the other side; they may pass over or under the heart and continue to the other side; or they may make direct contact with a cardiac cell (43, 116).

Although alary fibers have long been described as striated muscle (11, 72), this fact has at various times been obscured. Thus, their assignment to a definite histological category has been somewhat complicated by conclusions drawn from descriptions relative to their apparent behavior rather than from histological studies. They have been variously described as muscle-like threads (127), elastic fibers (23, 75, 133, 134), permanently contracted muscle and noncontractile muscle (31), and even nerves (1). That they are, in fact, striated muscle has been confirmed with the electron microscope (116).

Only one detailed study of the ultrastructure of alary muscle is currently available (116). Despite their striated appearance in the light microscope, their greatly exaggerated A-bands (5.5 μ) make them appear strikingly dissimilar to heart cells. Clearly defined sarcomere units bounded by Z-bands confirm the striated nature of the fibers. Certain ultrastructural features suggest that, like the heart, they may be functionally similar to vertebrate slow muscle, viz., a reduced sarcoplasmic reticulum and transverse tubular system, ten to twelve thin filaments around each thick filament, beaded Z-bands, and the absence of M-bands.

Most striking of the component structures of these fibers is the presence of an abundant number of microtubules. These structures have been implicated in the intracellular transport of water and ions (120), cytoplasmic streaming (76), and particularly in the development and maintenance of the skeleton or shape of the cell (129, 130). Their presence as well as their abundance is at present unexplained, but it may be that they played a role in the development of the fiber and remained after the muscle was formed (114, 116).

Alary fibers, if they are to be advised of cardiac activity, must in some way contact the heart. An anatomical link between the alary fibers and the myocardium has been alluded to, but generally by a diagrammatic representation that interposes an undefined type of tissue between the two closely apposed muscle structures (11, 43, 72). A point of contact between an alary cell and a cardiac cell, termed a myo-muscular junction, has been demonstrated with both light and electron microscopes (116). The two structurally distinctive muscles are joined by an intercalated disc. This junction, examined in the electron microscope, presents the same ultrastructural detail as that of the intercalated discs that link the cells of the myocardium. The function of this structure as a possible transmission relay site may be considered.

The intercalated discs in the moth heart may be presumed to play some role in cell-to-cell communication since electrical signals must be transmitted across these boundaries. The presence of the same disc structure with septate desmosomes at the alary-cardiac junction has been considered as presumptive evidence for a transmission function at this site (88), since the latter have been suggested as analogous to the nexus.

Nexuses, or tight junctions, are currently regarded as pathways of low electrical resistance that allow current flow between adjoining cells (79). They are part of the intercalated disc in vertebrate hearts (7, 27) and exist without either the disc or the desmosome in *Ciona* heart (70).

If the alary-cardiac intercalated disc serves as a physiological junction, then the mechanism of transmission must be explained. If a message is to be transmitted, its nature and direction must be defined. If, on the other hand, the junction point is simply an architectural reinforcement between two muscles so that stretch may be effectively transmitted, this feature may have broader application as a mechanism of impulse transmission throughout the myocardium (88).

There is no general agreement as to how alary muscles function or what effective contribution they make to the cardiac cycle. That they actively and rhythmically contract to assist in re-expanding the heart following systole has been the most widely supported hypothesis (11, 28, 88). They are even termed "cardiac dilator fibers" in the cockroach (37). Others, however, have proposed that they actually initiate contraction of the heart (31), or that they have no active function, i.e., they do not contract, and serve only a passive role as an elastic band in holding the heart in the mid-dorsal region (23, 75, 133, 141). Hinks has proposed (43) that there is a "division of labor" among the alary fibers. They may not all do the same thing at the same time.

Technical difficulties have contributed greatly to the erroneous conclusions and conflicting interpretations regarding alary muscle function. Their small size, 5 to 15 μ in diameter (116), translucent composition and intimate association with the myocardium have all but precluded the application of quantitative techniques. It has not so far been feasible to study them in isolation or completely separated from the heart. Thus, the sparse information available derives from visual observations of their activity (141), electrical records superimposed on surface electrograms of the heart (60, 109), or mechanical records that register one wave form from two sources (28, 96). Visual observation is inadequate and no quantitative data can be critically evaluated. Surface electrograms of the heart are in themselves indirect and complex to interpret without the added complication of a second event (21). Lever systems introduce inaccuracies in themselves, especially time lags. This is particularly critical where phase relationships are being measured. Perhaps the laser beam system recently introduced can be extended to these muscles (138). Recording a single variable cannot give valid information about the separate events; rather it indicates the resultant. It must be recognized that two systems will be required to register independently the activity of alary and cardiac fibers if a correlation between their functions is to be valid.

Whether alary muscles can be stimulated by the indirect application of electrical pulses is still unclear. Alary muscles of *Cossus cossus* appeared to contract in response to a faradic stimulus (28). *Periplaneta* alary fibers were unresponsive in the experiments of Yeager (141) and DeWilde (28), but Miller & Metcalf (96) reported their stimulation.

Selective stimulation and inhibition of alary muscles with drugs have been reported. Hamilton (38) reports stimulation of alary fibers with acetylcholine, Miller & Metcalf (96) report inhibition of alary activity with 2-pyridine aldoxime methodide, and Pouzat et al. (109) recorded inhibition of alary muscle activity with curare.

In the absence of accurate measuring devices with which to record their responses, such observations are of questionable validity. Direct stimulation with a microelectrode and electrophoretic application of drugs would resolve this point.

Direct investigations of single alary muscle fibers in the moth *H. cecropia* have provided presumptive evidence for a stretch-sensitive membrane and a mechanistic explanation of their activity (88). Muscle fibers, immobilized by a microsuction device, produced very low (−15 mv) resting potentials that were interrupted by rhythmic slow waves of small (10 to 15 mv) amplitude. Manipulation of the membrane voltage by passing hyperpolarizing pulses of constant current through the microelectrode revealed the presence of rhythmic, cardiac-type action potentials about 60 mv in amplitude. Interestingly, these potentials could be fractionated into voltage-dependent slow and fast components. A slow initial rise suggests that the fibers may have autorhythmic properties, but this evidence is admittedly tenuous and only suggestive.

Why the cell should appear to be depolarized before hyperpolarizing pulses are applied may result from one of two possibilities. One, it may be that the cells are so small that injury produced by the microelectrode penetration inactivates the membrane (all carriers of polarizing and depolarizing current are inactive), or two, the application of the suction effectively depolarizes a stretch-sensitive membrane. This latter possibility allows the assumption that the mechanism for activation of alary muscles is the mechanical deformation of the membrane induced by stretch imposed by cardiac systole. Such a mechanism was generally implied by the studies of DeWilde (28).

If an alary muscle is to function as a conventional muscle, it must be activated along its length so that the whole fiber contracts synchronously. The pronounced length of these fibers suggests that impulses must be propagated at high velocity or that an impulse be transmitted to multiple sites along the fiber simultaneously. This latter suggestion would require the presence of a profuse multiterminal neural innervation. Serial sections of moth alary fibers examined in the electron microscope deny the existence of any other nerve terminals along the fiber. Multiterminal neuronal innervation is not present (88, 116).

Histological evidence for the presence of a neural innervation to alary muscles has been described in the cockroach (3, 90) and mosquito (17, 63, 140). Fibers from the ventral nerve cord innervate the base of the fibers. They are localized at the point of attachment of the alary muscles to the lateral body wall. The mechanism for conduction probably does not involve a propagated action potential, for the membrane behaves like an ohmic resistance and the response exhibits the characteristics of a post-synaptic potential (88).

The diagram in Figure 2 has been constructed to summarize current knowledge and to propose a scheme whereby alary muscles may be activated. There seems little appreciation of the possibility that the pathway by which alary fibers are activated as well as the role they perform may be quite complex. The presence of a nervous innervation intuitively precludes a passive suspensory function. Assuming that at systole, the cardiac fibers (H) shorten, stretch may be transmitted across the myo-muscular junction to the alary fibers (AM) that are then depolarized. The shortening of the alary muscle has two effects, first, pulling against the heart fibers it effectively assists diastole (dashed line) and second, it stimulates the nerve ending (N). Impulses set up in this nerve would be coded, i.e., the alary muscle-nerve portion of the circuit could be a stretch receptor and this informa-

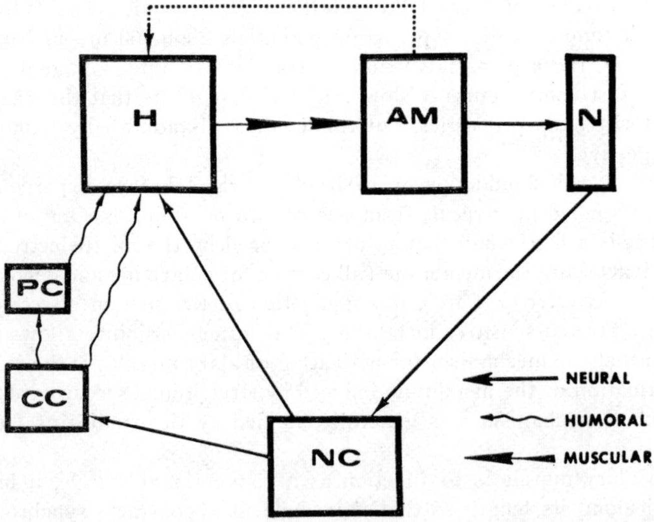

FIG. 2. A schematic proposal for heart-alary muscle functional relationship. Direction of arrows is only speculative; activity could proceed in reversed direction. See text for detail.

tion then fed into the nervous system for integration (NC = neural center). The effector message that emerges could then proceed by either or both of two possible routes, directly via a neural connection to the heart where present, or through a humoral path that involves the liberation of a cardio-active substance from the corpus cardiacum (CC) directly to the heart or through the intermediary auspices of the pericardial cells (PC). The latter route would be necessary for those hearts that are aneural. If such a system prevails, then alary muscles could be part of a frequency-amplitude modulation circuit.

It should be pointed out that this scheme shows only elementary components that have so far been described. This diagram is purely speculative and one may well question a basic assumption, viz., the direction of the arrows that describe the sequence of activation. The direction could be reversed. The "message" could originate in the neural center (NC), no definition of which is intended here, only that it be an integrating center. There remains the possibility that the alary muscles could be activated by the nerve, but their function would then seem more obscure. These proposals require rigorous testing and must remain, for the present, attractive hypotheses.

ACKNOWLEDGMENTS

I wish to thank Dr. George Stibitz for many helpful discussions and the computer program. I thank Mrs. Abby McLaughlin for her very able assistance.

LITERATURE CITED

1. Akehurst, S. C. Larva of *Chaoborus crystallinus* (De Geer). *J. Roy. Microscop. Soc.*, **42**, 341–72 (1922)
2. Alanis, J. Propagation of impulses through the specialized tissues of the mammalian heart. In *The Specialized Tissues of the Heart,* 175–201. (Paes de Carvalho, A., De Mello, W., Hoffman, B., Eds., Elsevier, Amsterdam, 218 pp., 1961)
3. Alexandrowicz, J. S. The innervation of the heart of the cockroach (*Periplaneta orientalis*). *J. Comp. Neurol.*, **41**, 291–309 (1926)
4. Anderson, M. Electrophysiological studies on initiation and reversal of the heart beat in *Ciona intestinalis. J. Exptl. Biol.*, **49**, 363–85 (1968)
5. Auber, J. Distribution of the two kinds of myofilaments in insect muscles. *Am. Zool.*, **7**, 451–56 (1967)
6. Ballard, R. C., Holcomb, B. An investigation of the house fly heartbeat through transection experiments. *Ann. Entomol. Soc. Am.*, **58**, 608–11 (1965)
7. Barr, L. Electrical transmission between the cells of vertebrate cardiac muscle. In *Comparative Physiology of the Heart: Current Trends. Experientia Suppl. 15.* (McCann, F. V., Ed., Birkhauser, Basel, in press)
8. Barr, L., Dewey, M. M., Berger, W. Propagation of action potentials and the structure of the nexus in cardiac muscle. *J. Gen. Physiol.*, **48**, 797–823 (1965)
9. Beard, R. L. Circulation. In *Insect Physiology,* Chap. 9, 232–72. (Roeder, K. D., Ed., John Wiley and Sons, New York, 1100 pp., 1953)
10. Belton, P., Grundfest, H. Potassium activation and K spikes in muscle fibers of the mealworm larva (*Tenebrio molitor*). *Am. J. Physiol.*, **203**, 588–94 (1962)
11. Bethe, A. Vergleichende physiologie der blutbewegung. In *Handbuch der Normalen und Pathologischen Physiologie,* **VII**/1, Chap. 1, 3–62. (Bethe, A., Bergmann, G. v., Embden, G., Ellinger, A., Eds., Verlag von Julius Springer, Berlin, 862 pp., 1926)
12. Bowers, B. Coated vesicles in the pericardial cells of the aphid (*Myzus persicae* Sulz). *Protoplasma,* **59**, 351–67 (1964)
13. Brocas, J., Thiebold, J., Saudray, Y., Perrin, J. Recherche d'un stimulus de l'inversion cardiaque chez un tunicien: *Ciona intestinalis* L. *Compt. Rend. (Ser. D),* **262**, 790–92 (1966)
14. Buck, J. B. Physical properties and chemical composition of insect blood. In *Insect Physiology,* Chap. 6, 147–90 (Roeder, K. D., Ed., John Wiley & Sons, New York, 1100 pp., 1953)
15. Bullivant, S., Loewenstein, W. R. Structure of coupled and uncoupled cell junctions. *J. Cell Biol.*, **37**, 621–32 (1968)
16. Carrington, C. B., Tenney, S. M. Chemical constituents of haemolymph and tissue in *Telea polyphemus* Cram. with particular reference to the question of ion binding. *J. Insect Physiol.*, **3**, 402–13 (1959)
17. Christophers, S. R. *Aedes aegypti* (L.) *The Yellow Fever Mosquito. Its Life History, Bionomics and Structure.* (Cambridge Univ. Press, London and New York, 738 pp., 1960)
18. Clements, A. N. *The Physiology of Mosquitoes,* **17.** (Intern. Ser. Monographs on Pure and Applied Biology, Division: Zoology. Pergamon Press, Macmillan Co., New York, 393 pp., 1963)
19. Cochrane, D. G., Elder, H. Y. Usherwood, P. N. R. Electrical, mechanical and ultrastructural properties of tonic and phasic muscle fibers in the locust (*Schistocerca gregaria*). *J. Physiol.*, **200**, 68P–69P (1969)
20. Coraboeuf, E., Suekane, K., Breton, D. Some effects of strong stimulations on the electrical and mechanical properties of isolated heart. In *Electrophysiology of the Heart,* 133–45. (Taccardi, B., Marchetti, G., Eds., Pergamon Press, New York, 344 pp., 1965)
21. Crescitelli, F., Jahn, T. L. Electrical and mechanical aspects of the grasshopper cardiac cycle. *J. Cellular Comp. Physiol.*, **11**, 359–76 (1938)
22. Davey, K. G. The control of visceral muscles in insects. *Advan. Insect Physiol.*, **2**, 219–45 (1964)
23. Davis, C. C. Experiments upon the

heart of *Chaoborus albipes. J. Cellular Comp. Physiol.,* **47,** 449–68 (1956)

24. Davis, C. C. Periodic reversal of heart beat in the prolarva of a gyrinid. *J. Insect Physiol.,* **7,** 1–4 (1961)
25. DeHaan, R. L., Gottlieb, S. H. The electrical activity of embryonic chick heart cells isolated in tissue culture singly or in interconnected cell sheets. *J. Gen. Physiol.,* **52,** 643–65 (1968)
26. De Mello, W. C. Metabolism and electrical activity of the heart: action of 2-4-dinitrophenol and ATP. *Am. J. Physiol.,* **196,** 377–80 (1959)
27. Dewey, M. M. The structure and function of the intercalated disc in vertebrate cardiac muscle. In *Comparative Physiology of the Heart: Current Trends. Experientia Suppl. 15.* (McCann, F. V., Ed., Birkhauser, Basel, in press)
28. de Wilde, J. Contributions to the physiology of the heart of insects, with special reference to the alary muscles. *Arch. Neerl. Physiol.,* **28,** 531–42 (1948)
29. Dogra, G. S. Studies on the neurosecretory system and the functional significance of NSM in the aortal wall of the bug, *Dysdercus koenigii. J. Insect Physiol.,* **13,** 1895–1906 (1967)
30. Dogra, G. S. Studies on the neurosecretory system of *Ranatra elongata* Fabricus with reference to the distal fate of NCC I and II. *J. Morphol.,* **121,** 223–40 (1967)
31. Dubuisson, M. Contribution à l'étude de la physiologie du muscle cardiaque des invertébrés. Des causes qui declanchent et entretiennent les pulsations cardiaques chez les insects. *Arch. Biol. (Liege),* **39,** 247–70 (1929)
32. Edwards, G. A., Challice, C. E. The ultrastructure of the heart of the cockroach, *Blattella germanica. Ann. Entomol. Soc. Am.,* **53,** 369–83 (1960)
33. Fawcett, D. W. Surface specialization of absorbing cells. *J. Histochem. Cytochem.,* **13,** 75–91 (1965)
34. Florkin, M., Jeuniaux, C. Hemolymph: composition. In *The Physiology of Insecta,* **3,** Chap. 2, 109–52. (Rockstein, M., Ed., Academic Press, New York and London, 692 pp., 1964)
35. Fourcroy, S. J. *Microelectrode Studies of the Housefly Heart, Musca domestica* L. (Master's thesis, San Jose State College, 1967)
36. Grassé, P.-P. (Ed.) Insectes. *Traité de Zoologie. Anatomie, Systématique, Biologie,* **IX** and **X,** Part. I and II. (Masson et Cie., Paris, 1117 pp., 975 pp., 1948 pp., 1949–1951)
37. Guthrie, D. M., Tindall, A. R. *The Biology of the Cockroach.* (St. Martin's Press, New York, 408 pp., 1968)
38. Hamilton, H. L. The action of acetylcholine, atropine and nicotine on the heart of the grasshopper, *Melanoplus differentialis. J. Cellular Comp. Physiol.,* **13,** 91–103 (1939)
39. Hamilton, M. A. Morphology of the water scorpion, *Nepa cinerea* Linn. *Proc. Zool. Soc. London,* **104,** 1067–136 (1931)
40. Hanson, J., Lowy, J. The structure of the muscle fibers in the translucent part of the adductor of the oyster, *Crassostrea angulata. Proc. Roy. Soc. (London), Ser. B,* **154,** 173–96 (1961)
41. Hecht, S. The physiology of *Ascidia atria* Lesueur. III. The blood system. *Am. J. Physiol.,* **45,** 157–87 (1918)
42. Hess, A. The sarcoplasmic reticulum, the T system, and the motor terminals of slow and twitch muscle fibers in the garter snake. *J. Cell Biol.,* **26,** 467–76 (1965)
43. Hinks, C. F. The dorsal vessel and associated structures in some Heteroptera. *Trans. Roy. Entomol. Soc. London,* **118** (12), 375–92 (1966)
44. Hodgkin, A. L., Huxley, A. F. A quantitative description of membrane current and its application to conduction and excitation in nerve. *J. Physiol.,* **117,** 500–44 (1952)
45. Hoffman, B. F., Cranefield, P. F. *Electrophysiology of the Heart.* (McGraw-Hill, New York, 323 pp., 1960)
46. Hoffman, B. F., Paes de Carvalho, A., De Mello, W. C., Cranefield, P. F. Electrical activity of single fibers of the atrioventricular node. *Circulation Res.,* **7,** 11–18 (1959)
47. Hoffman, J. A. Contribution a l'etude de la cellule pericardiale de *Locusta*

migratoria (Orthoptere). *Bull. Biol. France, Belg.,* **101,** 3–12 (1967)

48. Hoyle, G., McAlear, J. H., Selverston, A. Mechanism of contraction in a striated muscle. *J. Cell Biol.,* **26,** 621–40 (1965)
49. Huddart, H. The effect of potassium ions on resting and action potentials in lepidopteran muscle. *Comp. Biochem. Physiol.,* **18,** 131–40 (1966)
50. Huddart, H. The effect of sodium ions on resting and action potentials in skeletal muscle fibres of *Bombyx mori* (L.). *Arch. Intern. Physiol. Biochim.,* **74,** 592–602 (1966)
51. Huddart, H. Ionic composition of haemolymph and myoplasm in Lepidoptera in relation to their membrane potentials. *Arch. Intern. Physiol. Biochim.,* **74,** 603–13 (1966)
52. Huddart, H. Generation of membrane potentials in lepidopteran muscle I. Analysis of a mixed electrode system in the skeletal muscle fibres of *Sphinx ligustri* (L.). *Arch. Intern. Physiol. Biochim.,* **75,** 245–60 (1967)
53. Huddart, H. Generation of membrane potentials in lepidopteran muscle II. The effect of ionic depletion on resting and action potentials. *Arch. Intern. Physiol. Biochim.,* **76,** 519–32 (1968)
54. Huddart, H., Wood, D. W. The effect of DNP on the resting potential and ionic content of some insect skeletal muscle fibres. *Comp. Biochem. Physiol.,* **18,** 681–88 (1966)
55. Hutter, O. F., Trautwein, W. Vagal and sympathetic effects on the pacemaker fibers in the sinus venosus of the heart. *J. Gen. Physiol.,* **39,** 715–33 (1956)
56. Huxley, H. E. Evidence for continuity between the central elements of the triads and extracellular space in frog sartorius muscle. *Nature,* **202,** 1067–71 (1964)
57. Huxley, H. E., Hanson, J. The molecular basis of contraction in cross-striated muscles. In *The Structure and Function of Muscle I,* Chap. VII, 183–228. (Bourne, G. H., Ed., Academic Press, New York, 472 pp., 1960)
58. Irisawa, H., Irisawa, A. F., Kadotani, T. Findings on the electrograms of the cicada's heart (*Cryptotympana japonensis* Kato). *Japan. J. Physiol.,* **6,** 150–61 (1956)
59. Ishikawa, S. Membrane potentials of dorsal vessel of the silkworm, *Bombyx mori,* by intracellular method. *J. Sericult. Sci. Japan,* **28,** 295–97 (1959)
60. Jahn, T. L., Crescitelli, F., Taylor, A. B. The electrogram of the grasshopper (*Melanoplus differentialis*). *J. Cellular Comp. Physiol.,* **10,** 439–60 (1937)
61. Johnson, B. Fine structure of the lateral cardiac nerves of the cockroach *P. americana* (L.). *J. Insect Physiol.,* **12,** 645–55 (1966)
62. Johnson, E. A., Sommer, J. R. A strand of cardiac muscle. Its ultrastructure and the electrophysiological implications of its geometry. *J. Cell Biol.,* **33,** 103–29 (1967)
63. Jones, J. C. The heart and associated tissues of *Anopheles quadrimaculatus* Lay. *J. Morphol.,* **94,** 71–123 (1954)
64. Jones, J. C. The circulatory system of insects. In *The Physiology of Insecta,* **3,** Chap. 1, 1–109. (Rockstein, M., Ed., Academic Press, New York, 692 pp., 1964)
65. Kater, S. B. Cardioaccelerator release in *Periplaneta americana* (L.). *Science,* **160,** 765–67 (1968)
66. Kessel, R. G. Light and electron microscope studies on the pericardial cells of the nymphal and adult grasshoppers *Melanoplus differentialis* (Thomas). *J. Morphol.,* **110,** 79–104 (1963)
67. Kriebel, M. Conduction velocity and intracellular action potentials of the tunicate heart. *J. Gen. Physiol.,* **50,** 2097–2107 (1967)
68. Kriebel, M. Studies on cardiovascular physiology of tunicates. *Biol. Bull.,* **134,** 434–55 (1968)
69. Kriebel, M. Pacemaker properties of tunicate heart cells. *Biol. Bull.,* **135,** 166–73 (1968)
70. Kriebel, M. Electrical characteristics of tunicate heart cell membranes and nexuses. *J. Gen. Physiol.,* **52,** 46–59 (1968)
71. Krijgsman, B. J., Dresden, D., Berger, N. E. The action of rotenone and tetraethyl pyrophosphate on the isolated heart of the cockroach. *Bull. Entomol. Res.,* **41,** 141–51 (1950)

72. Kuhl, W. Der feinere bau des zirkulationssystems von *Dytiscus marginalis.* Ruckengefafs, pericardialseptum und pericardialgewebe. *Zool. Jahrb. Abt. Anat. Ontog. Tiere,* **46,** 75–198 (1924)
73. Kuwana, F. Morphological studies of the nervous system of the silkworm *Bombyx mori* Linne. 1. The innervation of the dorsal vessel. *Bull. Imperial Sericult. Sta. Japan,* **8,** 109–20 (1932)
74. Lasch, W. Einige beobachtungen am herzen der hirschkaferlarve. *Z. Allgem. Physiol.,* **14,** 312–19 (1913)
75. Lebrun, H. L'appareil circulatoire de *Corethra plumicornis. La Cellule,* **37,** 183–200 (1926)
76. Ledbetter, M. C., Porter, K. R. A "microtubule" in plant cell fine structure. *J. Cell Biol.,* **19,** 239–50 (1963)
77. Lehmkuhl, D., Sperelakis, N. Electrical activity of cultured heart cells. In *Factors Influencing Myocardial Contractility,* 245–78. (Tanz, R. D., Kavaler, F., Roberts, J., Eds., Academic Press, New York, 693 pp., 1967)
78. Ling, G., Gerard, R. W. The membrane potential and metabolism of muscle fibers. *J. Cellular Comp. Physiol.,* **34,** 413–38 (1949)
79. Loewenstein, W. R., Kanno, Y. Studies on an epithelial (gland) cell junction. I. Modifications of surface membrane permeability. *J. Cell Biol.,* **22,** 565–86 (1964)
80. Loewenstein, W. R., Nakas, M., Socolar, S. J. Junctional membrane uncoupling: permeability transformations at cell membrane junction. *J. Gen. Physiol.,* **50,** 1865–91 (1967)
81. McCann, F. V. Electrophysiology of an insect heart. *J. Gen. Physiol.,* **46,** 803–21 (1963)
82. McCann, F. V. Conduction in the moth myocardium. *Comp. Biochem. Physiol.,* **12,** 117–23 (1964)
83. McCann, F. V. The effect of anion substitution on bioelectric potentials in the moth heart. *Comp. Biochem. Physiol.,* **13,** 179–88 (1964)
84. McCann, F. V. Unique properties of the moth myocardium. *Ann. N. Y. Acad. Sci.,* **127,** 84–99 (1965)
85. McCann, F. V. The effect of intracellular current pulses on membrane potentials in the moth heart. *Comp. Biochem. Physiol.,* **17,** 599–608 (1966)
86. McCann, F. V. The effect of metabolic inhibitors on the moth heart. *Comp. Biochem. Physiol.,* **20,** 399–409 (1967)
87. McCann, F. V. The insect heart as a model for electrophysiological studies. In *Experiments in Physiology and Biochemistry.* (Kerkut, G., Ed., Academic Press, New York, in press)
88. McCann, F. V., Sanger, J. W. Ultrastructure and function in an insect heart. In *Comparative Physiology of the Heart: Current Trends. Experientia Suppl. 15.* (McCann, F. V., Ed., Birkhauser, Basel, in press)
89. McCann, F. V., Wira, C. L. Transmembrane ionic gradients in Lepidoptera as related to cardiac electrical activity. *Comp. Biochem. Physiol.,* **22,** 611–15 (1967)
90. McIndoo, N. E. Innervation of insect hearts. *J. Comp. Neurol.,* **83,** 141–55 (1945)
91. Marshall, J. M. Action of iodoacetic acid, 2, 4 dinitrophenol, and L-triiodothyronine on the electrical responses of the myocardium. *Am. J. Physiol.,* **180,** 350–56 (1955)
92. Maunsbach, A. B. Absorption of ferritin by rat kidney proximal tubule cells. *J. Ultrastruct. Res.,* **16,** 1–12 (1966)
93. Metcalf, R. L., Winton, M. Y., Fukuto, T. R. The effects of cholinergic substances upon the isolated heart of *Periplaneta americana. J. Insect Physiol.,* **10,** 353–61 (1964)
94. Miller, T. Role of cardiac neurons in the cockroach heartbeat. *J. Insect Physiol.,* **14,** 1265–75 (1968)
95. Miller, T. Initiation of activity in the cockroach heart. In *Comparative Physiology of the Heart: Current Trends. Experientia Suppl. 15.* (McCann, F. V., Ed., Birkhauser, Basel, in press)
96. Miller, T., Metcalf, R. L. A device for insect mechanocardiogram and some physiological measurements on the heart of *Periplaneta americana* L. (In preparation)
97. Miller, T., Metcalf, R. L. Site of action of pharmacologically active compounds on the heart of *Periplaneta americana* L. *J. Insect Physiol.,* **14,** 383–94 (1968)

98. Miller, T., Thomson, W. W. Ultrastructure of cockroach cardiac innervation. *J. Insect Physiol.,* **14,** 1099–1104 (1968)
99. Mills, R. P., King, R. C. The pericardial cells of *Drosophila melanogaster. Quart. J. Microscop. Sci.,* **106,** 261–68 (1965)
100. Mislin, H. Zur theorie der reversion der herzschlags bei den Tunikaten (*Ciona intestinalis* L.). *Rev. Suisse Zool.,* **72,** 865–73 (1965)
101. Mislin, H. Patterns of reversal in the heart of *Ciona intestinalis* L. In *Comparative Physiology of the Heart: Current Trends. Experientia Suppl. 15.* (McCann, F. V., Ed., Birkhauser, Basel, in press)
102. Mislin, H., Krause, R. Die schrittmachereigenschaften des herzschlauchs von *Ciona intestinalis* L. und ihre beziehung zur reversion des herzschlags. *Rev. Suisse Zool.,* **71,** 610–26 (1964)
103. Moe, G. K. Computer simulation of atrial fibrillation. In *Computers in Biomedical Research.* **II,** Chap. 9, 217–38. (Stacey, R. W., Waxman, B. D., Eds., Academic Press, New York, 363 pp., 1965)
104. Needham, A. E. The neurogenic heart and ether anesthesia. *Nature,* **166,** 9–11 (1950)
105. North, R. J. The fine structure of the myofibers in the heart of the snail *Helix aspersa. J. Ultrastruct. Res.,* **8,** 206–18 (1963)
106. Paes de Carvalho, A., De Mello, W. C., Hoffman, B. F. Electrophysiological evidence for specialized fiber types in rabbit atrium. *Am. J. Physiol.,* **196,** 483–88 (1959)
107. Page, S. G. A comparison of the fine structures of frog slow and twitch muscle fibres. *J. Cell Biol.,* **27,** 477–97 (1965)
108. Peachey, L. D., Huxley, A. F. Structural identification of twitch and slow striated muscle fibres of the frog. *J. Cell Biol.,* **13,** 177–80 (1962)
109. Pouzat, J., Pouzat, M.-H., Brocas, J. Etude electrophysiologique des organes pulsatiles chez un coleoptere carabique: *Anthia sexmaculata. Ann. Soc. Entomol. France* (N.S.), **3, 4,** 1129–32 (1967)
110. Prosser, C. L., Brown, F. A. *Comparative Animal Physiology,* 2nd ed. (W. B. Saunders Co., Philadelphia, 681 pp., 1961)
111. Robb, J. S. *Comparative Basic Cardiology.* (Grune & Stratton, New York, 602 pp., 1965)
112. Roth, T. F., Porter, K. R. Yolk protein uptake in the oocyte of the mosquito *Aedes aegypti* L. *J. Cell Biol.,* **20,** 313–32 (1964)
113. Ruska, H. Electron microscopy of the heart. In *Electrophysiology of the Heart,* 1–19. (Taccardi, B., Marchetti, G., Eds., Pergamon Press, New York, 344 pp., 1965)
114. Sanger, J. W. *The Ultrastructure of the Moth Myocardium and its Adjunctive Tissues.* (Doctoral thesis, Dartmouth College, Hanover, New Hampshire, 1967)
115. Sanger, J. W., McCann, F. V. Ultrastructure of the myocardium of the moth, *Hyalophora cecropia. J. Insect Physiol.,* **14,** 1105–11 (1968)
116. Sanger, J. W., McCann, F. V. Ultrastructure of moth alary muscles and their attachment to the heart wall. *J. Insect Physiol.,* **14,** 1539–44 (1968)
117. Sanger, J. W., McCann, F. V. Fine structure of pericardial cells of the moth *H. cecropia* and their role in protein uptake. *J. Insect Physiol.,* **14,** 1839–45 (1968)
118. Scher, A. M., Rodriguez, M. I., Liikane, J., Young, A. C. The mechanism of atrio-ventricular conduction. *Circulation Res.,* **7,** 54–61 (1959)
119. Senff, R. E. The electrophysiology of the adult heart of *Periplaneta americana:* evidence for a neurogenic heart. *Dissertation Abstr.,* Sect. B, **27**(7), 2499B–2500B (1967)
120. Slautterback, J. Cytoplasmic microtubules. I. Hydra. *J. Cell Biol.,* **18,** 367–88 (1963)
121. Smith, D. S., Gupta, B. L., Smith, U. The organization and myofilament array of insect visceral muscle. *J. Cell Sci.,* **1,** 49–57 (1966)
122. Smith, N. A. Observations on the neural rhythmicity in the cockroach cardiac ganglion. In *Comparative Physiology of the Heart: Current Trends. Experientia Suppl. 15.* (McCann, F. V., Ed., Birkhauser, Basel, in press)
123. Sperelakis, N., Lehmkuhl, D. Effect of current on transmembrane potentials in cultured chick heart cells. *J. Gen. Physiol.,* **47,** 895–927 (1964)

124. Sperelakis, N., Lehmkuhl, D. Ionic interconversion of pacemaker and nonpacemaker cultured chick heart cells. *J. Gen. Physiol.,* **49,** 867–95 (1966)
125. Stay, B. Protein uptake in the oocytes of the cecropia moth. *J. Cell Biol.,* **26,** 49–62 (1965)
126. Sutcliffe, D. W. The chemical composition of haemolymph in insects and some other arthropods, in relation to their phylogeny. *Comp. Biochem. Physiol.,* **9,** 121–35 (1963)
127. Swammerdam, J. *Bybel der Natuure.* (Severinus, Leyden, 1737–1738)
128. Tenney, S. M. Observations on the physiology of the lepidopteran heart with special reference to reversal of the beat. *Physiol. Comp. Oecologia,* **3,** 286–306 (1953)
129. Tilney, L. G., Hiramoto, Y., Marsland, D. Studies on the microtubules in Heliozoa. III. A pressure analysis of the role of these structures in the formation and maintenance of the Axopodia of *Actinosphaerium nucleofilum* (Barrett). *J. Cell Biol.,* **29,** 77–95 (1966)
130. Tilney, L. G., Porter, K. R. Studies on microtubules in Heliozoa I. The fine structure of *Actinosphaerium nucleofilum* (Barrett), with particular reference to the axial rod structure. *Protoplasma,* **60,** 317–44 (1965)
131. Toselli, P. A., Pepe, F. A. The fine structure of the ventral intersegmental abdominal muscles of the insect *Rhodnius prolixus* during the molting cycle: I. Muscle structure at molting. *J. Cell Biol.,* **37,** 445–61 (1968)
132. Truex, R. C. Comparative anatomy and functional considerations of the cardiac conduction system. In *Specialized Tissues of the Heart,* 22–43. (Paes de Carvalho, A., De Mello, W., Hoffman, B. F., Eds., Elsevier, Amsterdam, 218 pp., 1961)
133. Tzonis, V. K. Beitrag zur kenntnis des herzens der *Corethra plumicornis Larve Fabr.* (*Chaoborus crystallinus* Geer). *Zool. Anz.,* **116,** 81–90 (1936)
134. Verloren, M. Sur la circulation dans les insectes. *Mem. Cour. Acad. Roy. Sci. Belgique,* **19,** 1–96 (1844)
135. Whitten, J. M. Connective tissue membranes and their apparent role in transporting neurosecretory and other secretory products in insects. *Gen. Comp. Endocrinol.,* **4,** 176–92 (1964)
136. Wiener, J., Spiro, D., Loewenstein, W. R. Studies on an epithelial (gland) cell junction II. Surface structure. *J. Cell Biol.,* **22,** 587–98 (1964)
137. Wigglesworth, V. B. *The Principles of Insect Physiology,* 5th ed. (Methuen, London, 546 pp., 1953)
138. Williams, G. T., Ballard, R. C., Hall, S. C. Mechanical movement of the insect heart recorded with a continuous laser beam. *Nature,* **220,** 1241–42 (1968)
139. Wood, D. W. The sodium and potassium composition of some insect skeletal muscle fibres in relation to their membrane potentials. *Comp. Biochem. Physiol.,* **9,** 151–59 (1963)
140. Yaguzhinskaya, L. V. New data on the physiology and anatomy of the dipteran heart. (In Russian). *Biul. Mosk. Obsch. Ispyt. Prir., Otdel Biol.,* **59,** 41–50 (1954)
141. Yeager, J. F. Electrical stimulation of isolated heart preparations from *Periplaneta americana. J. Agr. Res.,* **59,** 121–37 (1939)
142. Zawarzin, A. Histologische studien über insekten I. Das herz der Aeschnalarven. *Z. Wiss. Zool.,* **97,** 481–510 (1911)

BIOLOGICAL RHYTHMS IN TERRESTRIAL ARTHROPODS[1]

By A. S. Danilevsky,[2] N. I. Goryshin, and V. P. Tyshchenko

Department of Entomology, Leningrad State University, Leningrad, U.S.S.R.

The term "biological rhythms" involves all cyclic processes in living organisms. The frequency spectrum of biological rhythms is very wide. The cycles can range from some milliseconds to some years. The rapid oscillations are especially characteristic of physiological rhythms which ensure the synchronization of all the internal processes in an organism. Such rhythms may be exemplified by the rhythmic events in the "enzyme-substrate" system or by the pulsations of the heart muscle. Oscillations with daily or longer periods rank as a rule among ecological rhythms which provide for adaptation of plants and animals to fluctuations of environmental factors. Internal oscillators are assumed to control such rhythms; they are synchronized by external timing elements (*Zeitgeber* = time-givers) which can be also termed as synchronizers or entraining agents. Photoperiod is especially important as this agent. The possession of internal oscillators enables organisms to prepare in good time for the changes of ecological conditions during each day (daily rhythms), month (lunar rhythms), or year (seasonal rhythms).

Adaptive implication and, correspondingly, mode of displaying the three types of rhythms varies considerably. That's why, until recently, circadian and seasonal rhythms in arthropods have been investigated quite independently. Daily rhythms were studied as biological oscillations related to time measuring. As to investigations into seasonal cyclic phenomena, most attention was paid to systems of neurosecretory control of development and diapause as well as to ecological aspects of the problem. The two points of view have been reflected in reviews by Harker (90) and de Wilde (207), published in previous volumes of *Annual Review of Entomology.*

The material accumulated shows that some general principles are the basis of all the ecological rhythms. It displays especially clearly the synchronization of the biological processes with rhythmic factors of the environment. The adaptations to lunar and seasonal periodicity evolve by way of

[1] The following abbreviations are used in the text: PhPR (photoperiodic reaction); τ (period of a free-running rhythm); LL (constant light); DD (constant darkness); LD (light-dark cycles which can be further specified as LD12:12 to denote a cycle of 12 light hours and 12 dark hours; LD20:4 to denote a cycle of 20 light hours and 4 dark hours).

[2] It is with deep regret that we record the death of Professor Danilevsky on June 27, 1969. *The Editors.*

time measuring within a daily cycle. Hence, the mechanisms of all ecological rhythms must involve circadian components. This is a conclusion many authors have arrived at; relations of different ecological rhythms have been reviewed recently in a most complete way by Beck (16). However, up to now there is no unified interpretation of concrete mechanisms connecting circadian and seasonal rhythms, and the proposed hypotheses hardly may be coordinated. This paper will stress the main principles in the display of daily and seasonal rhythms, with emphasis on their internal connection and community of synchronization mechanisms.

SEASONAL RHYTHMS

Alternation of Development and Diapause as the Essence of the Seasonal Rhythm

Regular variations in environmental conditions during the course of the year are the initial reason for the seasonal rhythm of development. The latter is especially pronounced and more completely investigated in temperate climates, where it is caused mainly by the yearly rhythm of the temperature, which drops in winter below the level necessary for development and reproduction of plants and poikilothermic animals. The inability of insects to stand such low temperatures during growth and morphogenesis gave rise to a specific resting stage in their life cycle—diapause, for which such conditions are the ecological norm. The state of rest may also arise to assure survival through other unfavourable seasonal conditions, i.e., too high temperature, long drought, excessive moisture, or recurrent absence of food. That is why seasonal rhythms often occur also under conditions of subtropical and tropical climates. Great biological significance and physiological specificity of diapause have led to detailed research into it both from ecological and physiological points of view. A great amount of literature has been summarized in the reviews of Andrewartha (3), Harvey (92), Lees (106) and Schneiderman (176); Müller (134) has attempted to classify different diapause forms.

Diapause was shown to be a consequence of changes in the activity of the neurohumoral system, controlling growth and metamorphosis. This explains the great evolutional lability of diapause and the possibility of its occurrence at any ontogenetic stage ranging from egg to imago. Though hormonal factors involved in the various types of diapause are different, the brain in every case serves as the primary regulating center, able to perceive both intrinsic and extrinsic stimuli influencing diapause.

This dual regulation results in different degrees of dependence of an insect's seasonal cycle on environmental conditions. Diapause becomes obligatory when the internal factors are dominant in the regulation of the cycle. When the exogenous factors play the major role, diapause is facultative and appears only under certain conditions.

In spite of a vast variety of the phenology patterns of different species, the basis of any seasonal rhythm is always an alternation of physiological activity and diapause, which ensures the life of an organism under the con-

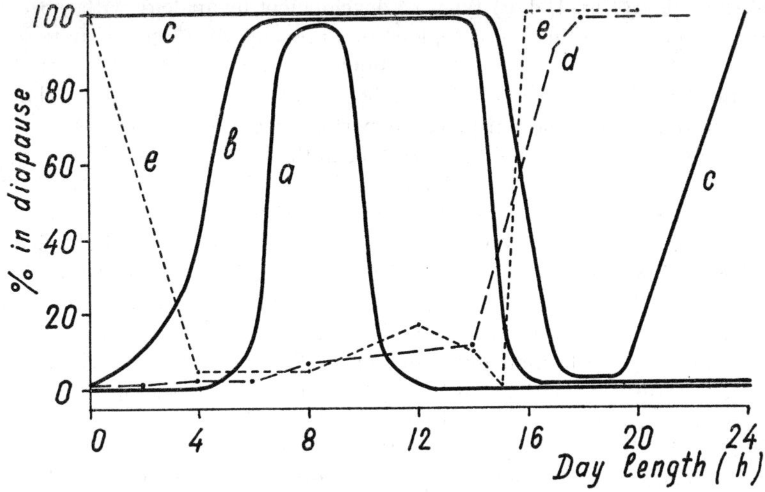

FIG. 1. The types of the photoperiodic reaction. a, b—the long-day type; d, e—the short-day type; c—the intermediate type. a—*Pieris brassicae* from Sukhumi; b—*P. brassicae* from Leningrad; c—*Euproctis similis* (67); d—*Stenocranus minutus* (133); e—*Neodiprion sertifer* (188).

ditions of climatic seasonal variations. Obviously, these adaptations are effective only when there is a reliable synchronization of seasonal rhythm phases with the dynamics of external conditions. Hence, the investigation of the mechanisms of seasonal rhythm synchronization presents an important ecological problem. The elucidation of the factors controlling diapause initiation and elimination is one of the major aspects of this problem.

The External Conditions Regulating Diapause

Various factors that regularly change during the course of the season may serve as signals for diapause initiation. They include temperature, humidity, and the chemical composition of food vegetation. But the most reliable signal is provided by the seasonal change in the day-length, the course of which during the year is not subject to chance fluctuations and is, in its essence, the initial reason for seasonal climatic variations. The other factors are additional regulators of the seasonal cycle. As all of them act through the common neurohumoral mechanism, their effect is seen only in the form of changes in photoperiodic reaction (PhPR) parameters. The problem of photoperiodic regulation of development in arthropods was investigated thoroughly enough and was dealt with in reviews by Lees (106, 111), Danilevsky (51), de Wilde (207), and Beck (16).

Reaction to the photoperiod is displayed in very different forms (Fig. 1). Two types are usually distinguished. They are called long-day (Fig. 1, a, b) and short-day (Fig. 1, d, c) types of development by analogy with those observed in plants. In the first case, long-day prevents the diapause, in the second it facilitates its initiation. The examples given in the figure make it clear that there are some transitional types between these two (Fig. 1, c). Therefore, all of them may be considered as particular expressions of the common regulating system. It should be noted that in the given curves only the right-hand part (between 10 and 20 hours) reflects the response to natural light rhythms. The effect of shorter photoperiods is a nonadaptive consequence of the physiological mechanism, on which the reaction to natural light rhythm is based. Of most ecological importance is the threshold region of the PhPR, in particular the critical photoperiod (causing diapause in 50 per cent of individuals), as it determines the calendar time of diapause initiation in nature. The day-length, inducing diapause in 100 per cent of individuals and preventing further development, is also significant. The photoperiodic reaction is most sensitive to the additional regulating factors when the photoperiod is in the threshold region (Fig. 2).

The photoperiod undoubtedly plays the main part in diapause induction, but there are diverse opinions as to the factors terminating diapause and synchronizing spring development. Well known is the specific influence of temperature on the process of the breaking up of diapause, i.e., reactivation. As numerous investigations have shown (3, 49, 106), the highest speed of reactivation in species of the temperate climate with winter diapause usually occurs in temperature between 0° and 12° C, i.e., below the level necessary for active life processes. Higher temperatures as well as temperatures below 0° hamper or completely prevent reactivation. As a result of such dependence on temperature, reactivation in nature takes place during wintering and the time at which resumption of activity occurs is determined by the rise of temperature over the threshold for development.

Many facts have recently been obtained showing the maintenance of sensibility to the photoperiod during the winter diapause (62, 115, 116, 169). In connection with this, some authors (1, 19, 114) suppose that in nature the time of the spring resumption of development is determined not by the temperature conditions but by day-length. However, investigations of the relative roles of cold and photoperiod in more than 20 insect species have not confirmed this point of view (54, 95, 214). It turned out that even if the PhPR is marked at the induction of the diapause, after the initiation of the latter it is often sharply weakened (e.g., in *Pieris brassicae, Antheraea pernyi*) or disappears completely (*Acronycta rumicis, Barathra brassicae,* and others). In contrast with that, the reactivating effect of low temperature (0° to 10°) is clearly marked in all these species. In the cases of maintenance of the PhPR during diapause as in *Laspeyresia pomonella, Grapholitha funebrana,* or *Chilocorus* species, the reactivating effect of day-length was apparent only at temperatures favourable for morphogenesis (54). Photoperiodic reaction disappears completely after sufficient chilling, development may then be renewed at any day-length. It follows from these data

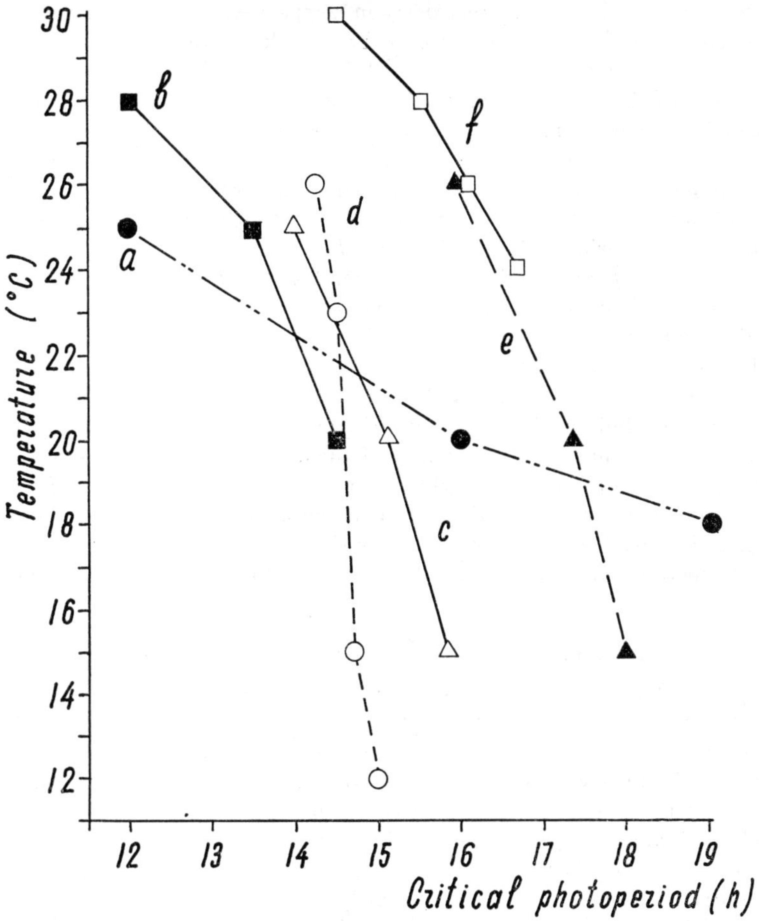

FIG. 2. The critical photoperiod variations depending on temperature. a—*Leucoma salicis* from Armenia; b—*Chloridea obsoleta* from the northern Caucasus; c—*Acronycta rumicis* from Sukhumi; d—*Pieris brassicae* from Leningrad; e—*A.* rumicis from Belgorod; f—*Leptinotarsa decemlineata* from Berlin.

that cold reactivation has originated in insects of temperate climate as a specific adaptation, replacing the photoperiodic one. The function of the PhPR during diapause is limited only to the maintainance of rest during autumn or during occasional warm weather in winter. It may control spring phenology only in species which resist reactivation by cold. In temperate climates such instances seem to be rare, but they are known in *Nemobius yezoensis* (122) and in *Ixodes ricinus* (22).

The reason for the prevalence of cold reactivation of insects in climatic regions with a cold winter may be explained ecologically. The yearly course of temperature displays a great delay in comparison with the change in the day-length. Thus, the temperature in spring is lower than in the autumn at the same day-length. Therefore, the PhPR only can hardly ensure autumn cessation of development and its spring resumption in the time limits favourable enough from ecological point of view. Besides, the photoperiodic reactivation would make impossible hibernation in soil and in other places inaccessible for light but soundly protected from frosts.

Other dependencies are observed with the regulation of the summer diapause. The experimental data for *Dasychira pudibunda* (72), *Hadena sordida* (25), *Exapate congelatella* (54), *Parthenolecanium corni* and its parasite *Blastothrix confusa* (54, 187), have shown that both onset and cessation of the summer diapause are determined mainly by photoperiod. Temperature only intensifies this effect.

In general, the reactions of the diapausing stages always reflect the peculiarities of light and temperature rhythms of the species range. Sharp differences in temperature optimums of the active and diapausing stages of the life cycle in the temperate climate species and the requirement for prolonged chilling for the reaction, serve as one of the reasons restricting the penetration of boreal elements of entomofauna into the tropical zone where the temperature rhythm during the course of the year is only slight (48, 180).

Types of Seasonal Cycles and Their Regulation

The system of seasonal adaptations, determining the yearly rhythm of insect development is not connected with the ontogenetic cycle. This system may embrace (*a*) several generations—in polyvoltine species; (*b*) one generation—in monovoltine ones; (*c*) only a part of ontogenesis—in species with perennial development. Every type of yearly rhythm has its own ecological aims and, correspondingly, characteristic features of the synchronizing mechanisms.

Polyvoltinism.—This is an extremely widespread and ecologically resourceful type of seasonal rhythm, in which the whole period of the year, provided there is food and favourable climatic conditions, is most fully used for reproduction. The generation number mainly depends on the sum of effective temperatures. Diapause here is always facultative and in the temperate climate is adapted for survival during winter.

The ecological regulation of diapause in polyvoltine species is well studied and has been shown to be monotypic. Photoperiodic reaction inducing diapause always is of a long-day type [for examples covering more than 100 species see (51)]. This ensures a continuous development of several generations during summer and the onset of diapause by the end of the vegetative period. High temperature as well as a long day prevent the diapause onset. Low temperature, on the contrary, stimulates diapause initiation and can fully suppress the photoperiodic effect. As the decrease in temperature usually causes the rise in the critical day-length (Fig. 2), the time of the diapause onset in nature and the generation number may vary depending on the weather conditions during the year. Thermoperiodicity, i.e., the daily temperature rhythm, can be of importance as well as the general temperature level (80). Changes in the chemical composition of food often provide an additional source of seasonal information (208). Feeding on senescent plants intensifies the tendency to diapause (84, 104, 184) and increases the critical day-length (76, 77, 102). It has been shown that the diapause stimulation in *Pectinophora gossypiella* is influenced by content of lipids in the food (6, 182, 183). An effect of water content on diapause onset has not yet been shown but it is not excluded.

The mode of reactivation is always based on the autumn-winter chilling effect. Therefore, spring development usually starts as soon as temperature exceeds the threshold for the morphogenesis. But in some species, such as *Chloridea obsoleta* (77) and *Antheraea pernyi* (215) feeding on late-developing plants, the resting state continues until the token stimulus is received, this being the temperature significantly exceeding the level necessary for further development.

In general, the synchronizing system in case of polyvoltinism is susceptible to parallel effects of various external factors, which makes it highly reliable and sufficiently mobile at the same time. Species differ only in the ontogenetic stage at which diapause occurs, in the threshold values of the regulating reactions, and in other particular characteristics.

Complications in polyvoltine type.—Simple alternation of active and resting stages in a polyvoltine cycle is often subject to considerable complications. One of them is seasonal di- or polymorphism, which can involve differences in colour, body size, the degree of development of wings and other structures and, in aphids, the alternation of parthenogenetic and sexual reproduction.

Most of these seasonal morphoses exhibit a close relation to the diapause state and are regulated mainly by photoperiod. The adaptive significance of such morphoses is not sufficiently clear. These phenomena could be treated as an indirect consequence of differences in metabolic processes in the diapause and active developmental stages in those cases in which differences in colour or in other features are observed between the active and diapausing individuals [*Polygonia c-aureum* (94); *Acleris fimbriana* (51); *Psylla pyri* (29)] or between individuals emerging from the diapause and active pupae [*Araschnia levana* (47, 130); *Pieris napi* (125); *Ascia monuste* (146); *Hylophila prasinana* (51)]. However, polymorphism in colour, body size, and even in the genital structure was found to be controlled by photoperiod in the leafhopper *Euscelis plebejus* (129, 131). It does not seem to have a direct relation to diapause. Seasonal colour and diapause in *Lycaena phleas daimio,* though both regulated by photoperiod, are probably also independent of this factor (171). The experiments of Fukuda & Endo (64) on *Polygonia c-aureum* have shown that the factor determining the light summer colour in nondiapausing butterflies, passes from medial neurosecretory cells through axons to the corpora cardiaca. The extirpation of corpora allata from larvae in the last instar did not influence the wing colouring. It seems that seasonal morphosis and diapause are, to a certain degree, independent phenomena, controlled by nonidentical hormonal factors.

Seasonal cycles become most complicated in aphids. The mechanisms controlling alternation of parthenogenetic and sexual reproduction, formation of winged parthenogenetic migrants and the change of host plants between seasons were discussed in special reviews by Bonnemaison (27), Shaposhnikov (178), and in a recent work by Lees (110). In general, the factors most important in the transition to sexual reproduction and the consequent deposition of diapausing eggs are photoperiod and low temperature. Migration and change of hosts are mainly controlled by overcrowding (group effect) and by seasonal changes in the chemical composition of the food plant.

Monovoltinism.—This system was usually considered as a simple and monotonous type of the seasonal rhythm in which the diapause appears at a certain stage of ontogenesis independent of external factors, and terminates under the influence of temperature. But such strictly obligatory diapause seems to be not the only, and probably not the major, way to maintain monovoltinism. An unexpected diversity of reactions determining development of the only generation a year has been found recently.

In some species with winter diapause, previously considered to be obligatory, there is an intermediate (Fig. 1, c) or long-day photoperiodic reaction with a very high critical threshold but it becomes apparent only at a sufficiently high temperature (above 20° C): *Leucoma salicis* (the northern race); *Euproctis chrysorrhoea; E. similis* (67); *Hyalophora cecropia* and *Antheraea polyphemus* (116); *Aelia sibirica* and *A. acuminata* (39). In nature, this potential ability to develop without diapause is not usually realized owing to the too short day-length in the southern part of the range. The diapause terminates during the hibernation under the influence of low temperature. Undoubtedly, this way of maintaining a single generation has originated from the polyvoltine system, but all transitory stages towards the obligatory diapause may be traced in it.

Monovoltinism is often gained in quite a different way, i.e., with the help of the short-day PhPR (Fig. I, d, e), causing various forms of summer cessation or delay in development. In this case, high temperature intensifies the retardation effect of the long day. These factors determine summer delay in the gonad maturation in beetles *Psylliodes chrysocephala* and *Ceuthorrhynchus pleurestigma* (4, 5), *Hypera postica* (97) and *H. variabilis* (172), in the leafhopper *Stenocranus minutus* (132, 133, 186), and probably in many other monovoltine species with an imaginal diapause. A change in photoperiodic conditions was found to be necesssary for the termination of the ovarian diapause in carabid beetles *Pterostichus nigrita* and *Agonum assimile,* which lay eggs in the spring (195). A short day is necessary for the final egg formation. In *A. assimile,* the process is complicated by the necessity of chilling for the male maturation.

Two separate periods of rest (summer and winter) are observed in other variations of this type. They occur at different ontogenetic stages. The summer rest is controlled by the short-day PhPR; the winter one arises under the influence of the long-day PhPR or spontaneously, and is terminated by cold reactivation. In this way very specific phenology is conditioned in *Dasychira pudibunda* (72), in *Hadena sordida* (25), in *Exapate congelatella* (54), in *Neodiprion sertifer* (188, 189), in *Parthenolecanium corni* (181). Doubtless, further investigations will also reveal other variations in the regulation of seasonal development which are based on the same principle.

Perennial cycles.—Such cycles are characteristic of many species living in water as well as in soil and in wood. In species residing on the open plant and soil surface such development is comparatively rare. The total duration of the generation is not firmly fixed and depends on temperature, as illustrated by the acceleration of development in the southern parts of the range, well known in *Melolontha melolontha,* in many elaterids, and in ticks. At the same time in every species a great constancy in the time of pupation and of adult emergence is observed, which indicates the ability to synchronize the development rhythm with the variations in external conditions. The synchronization mechanism has been investigated only in two species of lasiocampid moths—*Dendrolimus pini* and *D. sibiricus,* having a 1- or 2-year generation (70) and in the tick *Ixodes ricinus,* developing in nature for 3 to 5 years (22). This mechanism is, as in other cases, based on a diapause controlled by photoperiod and temperature. It is characteristic that in these species the susceptibility to photoperiod is retained for the most part of development and in connection with this, the diapause and wintering may appear in the ontogenesis more than once. Specific reactions to photoperiod and temperature make it possible to prolong the development until the next year if metamorphosis has not started at a favourable time.

The factors which control the seasonal rhythm of development in larvae living with absence of light in the soil, such as Scarabaeidae and Elateridae, have not been exposed. The main part here might be played by the yearly temperature rhythm.

Endogenous Variation in Photoperiodic Reaction During Season

It is known (27, 117, 210) that the fundatrices and the first parthenogenetic generations in aphids do not usually react to photoperiod. This reaction determining the appearance of the sexual generation, develops as the clone ages. Lees (107) showed that the appearance of the PhPR does not depend on the number of the past generations and is determined only by a particular time interval, which diminishes with increased temperature.

Very similar results have been recently obtained in spider mites in the photoperiodic regulation of imaginal diapause *Tetranychus urticae* (59), and of embryonic diapause in *Panonychus ulmi* and in *Schizotetranychus schizopus* (71, 157). The females of the first spring generation of *S. schizopus,* developing at 25°C, lay with any day-length, only nondiapausing eggs. But at 15° the PhPR is clear and the diapause occurs in response to a short-day. In the second natural generation the PhPR is well pronounced within a wide temperature range (from 15° to 25°C), its critical day-length, however, varying greatly. Further, the tendency to diapause increases to such a degree that the 4th generation females lay almost exclusively wintering eggs regardless of day-length and temperature conditions. The change in the tendency to diapause is endogenous, but this tendency is deeper when the previous generation has been reared under a higher temperature and a short day. In nature, the return to the initial weak susceptibility to photoperiod takes place as a result of the winter reactivation of the diapausing stages. However, there are also possible spontaneous rhythmic change of the response to photoperiod. Thus, in *Tetranychus crataegi* reared under long-day conditions and tested thereafter during several generations with a 14-hour photoperiod at 18° C the percentage of the diapause individuals increases only up to the 8th to 9th generation; after that it decreases and comes to zero by the 14th generation. Diapause, and the conditions required for its inception in *Nasonia vitripennis* (173), *Coeloides brunneri* (170), and *Lucilia caesar* (165), are very close in essence to these phenomena.

The ecological significance of such variations in the PhPR is obvious. On the one hand, it ensures the absence of the diapause in spring generations whose development occurs within a period of rather short days and low temperature and, on the other hand, it secures an obligatory onset of the winter diapause by the end of the season even in the case of a warm autumn. In aphids the appearance of the sexual generation in spring is excluded in this manner. It is, thus, one of the general mechanisms participating in the synchronization of the biological rhythms. The physiological part of this mechanism has not become clear. Lees (107) has supposed that the information communication transfer through generations, in this case, is carried out through cytoplasm and not through nuclear apparatus.

The endogenous rhythm with long periods might be characteristic not only of the reactions including the diapause. Dubynina (59) has discovered a spontaneous rhythm of cessation and resumption of diapause with a period of about 40 days in wintering females of spider mites *Tetranychus urticae* while keeping them at constant temperatures of 5° and 10°C. Periodicity of pupation (in 180 to 200, 350 to 400, and 750 to 800 days) was observed as well in the long diapausing individuals of the pine sawfly *Neodiprion sertifer* at 10° C (189).

Geographical Variation in Reactions Controlling Seasonal Rhythms

Photoperiodic regulation of seasonal development is based on the parallelism and stable correlations between seasonal course of day-length and of temperature. These correlations, however, may vary sharply, particularly depending on the geographical latitude. With increase in latitude the temperature decreases, causing a restriction in the number of generations and necessitating an earlier diapause onset. The length of the summer day, on the other hand, increases considerably; under the unvariable PhPR it should cause a disharmony between a life cycle and the seasonal conditions.

Many attempts have been made to explain the adjustments ensuring the correspondence of a species phenology with the alternation of climatic conditions within its area. These investigations have led to important conclusions.

It was shown that the influence of the geographical variations in the day-length may be partly compensated by the shift of the PhPR critical threshold under the influence of the environmental temperature (53, 75). However, the major means of adaptations are inherited intraspecific changes in the PhPR parameters and in temperature reactions regulating the diapause. At the same time the temperature relations of the active development stages are, as a rule, constant specific characteristics (50, 51).

The information on the geographical variation in seasonal cycle reactions was considered in the reviews by Masaki (119) and Danilevsky (51). Further investigations showed the existence of similar phenomena in many terrestrial arthropods: in the codling moth *Laspeyresia pomonella* (179, 180) and in some other Lepidoptera (6, 18, 53, 72, 125); among Diptera—in anopheline mosquitoes (56, 99, 198); among Hymenoptera—in the sawfly *Neodiprion sertifer* (200) and in parasitic species (123); among Coleoptera (145)—in Coccinellidae; in some aphids (10, 124), Orthoptera (140), in mites (59, 71), and ticks (22).

In all polycyclic species with the long-day PhPR and winter diapause, the variation depending on the latitude was found to have a uniform trend.

a. The critical day-length of the PhPR in local populations within the range of the species greatly increases from the south to the north (Fig. 3). The temperature threshold causing the diapause also shows a parallel increase. This explains an earlier date of diapause onset in the north than in the south.

b. A certain number of individuals in the northern parts of the range displays an inherited tendency to the obligatory diapause which is not regulated by the photoperiod. A partly homodynamic development may be, on the contrary, observed in the extreme south.

c. The hybrids between the geographical strains exhibit a stable intermediate type of the PhPR with a very slight splitting in F_2 and in further generations. This explains their better adaptation to the conditions of the intermediate regions as compared to those of both parent forms. It is one of the main reasons for the continuity of the species range.

In general, these rules may seem to be applied to all polyvoltine species independent of the ontogenetic stage at which the diapause occurs. But variations with the geographical latitude are not the same in different species.

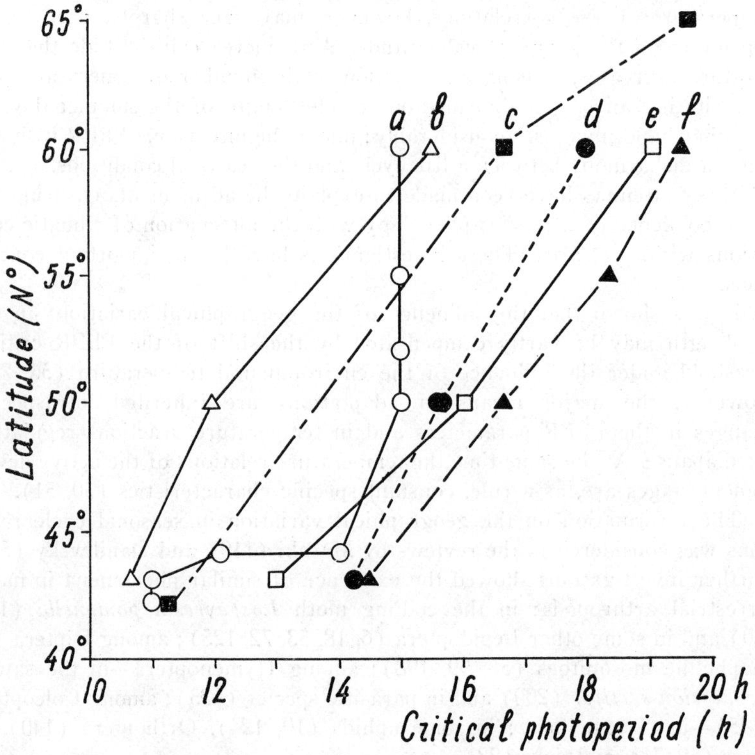

FIG. 3. Geographical variations of the critical photoperiod. a—*Pieris brassicae* (reared at 23° C); b—*P. rapae* (23° C); c—*Tetranychus urticae* (18° C); d—*Barathra brassicae* (25° C); e—*P. napi* (18° C); f—*Acronycta rumicis* (23° C). Populations from Hibiny, Leningrad, Vitebsk, Belgorod, northern Caucasus, and the Black Sea Coast of Caucasus were investigated.

More characteristic is a gradual clinal-type latitudinal variation in the critical day-length and in other parameters of the PhPR (Fig. 3, b-f). However, cases of interrupted variability of reaction are not rare. Those in *Pieris brassicae* (Fig. 3, a) and in *Leucoma salicis* strains with the invariable type of the PhPR are found over the larger part of the eastern European range, and only in the far south are there distinctly different strains with another type of reaction.

The data on the geographical variation of the monocyclic species with summer diapause regulated by the short-day PhPR, are scarce. Observations on the populations of *Dasychira pudibunda* (72) coming from Leningrad (60°N) and Sukhumi (43°N), and on the sawfly *Neodiprion sertifer* (200) coming from Canada (44°N), from Bulgaria (42°N and from Finland (62°N) show that the PhPR threshold in the northern forms is higher than in the southern ones. Due to that fact, the summer delays in development induced by a long-day cease earlier in the north than in the south.

Considerable inherited variations in local populations were found in the winter diapause duration as well as in the temperature level necessary for reactivation. Masaki (119) pointed out two opposite directions of geographical variations of these characteristics. In some species such as in *Chilo suppressalis* (63), Emma field cricket *Teleogryllus sp.* (120, 121) and *Lymantria dispar* (73, 74), the southern forms have a longer diapause and usually need a longer period of chilling than those from the north. This type also includes *Anopheles maculipennis messeae* (20, 21). Another type may be observed in the cricket *Pteronomobius fascipes* (119), the plant-hoppers *Sogata furcifera* and *Nilaparveta lugens* (135), and probably in *Ostrinia nubilalis* (7), in the sawfly *Gilpinia polytoma* (155), the butterfly *Pieris napi* (147). The northern forms of these species possess a deeper diapause and require more intensive chilling for its termination. The second type seems to be comprehensive as the duration and the requirements of the diapausing stages agree well with the duration and severity of the winter period. However, the first type may also have adaptive value, for a more stable diapause in the south prevents resumption of development under the conditions of a warm, changeable winter. Different types of variations may probably be due to some peculiarities of hibernation in various species and possibly to the participation of the PhPR in reactivation. Further investigations in this direction are necessary.

In subtropical and tropical species geographical variation in seasonal cycles and diapause is exhibited in peculiar forms. Thus, *Barathra brassicae* in the temperate European climate and in the north of Japan (Hokkaido) possess only winter facultative diapause, induced by a short-day and requiring chilling for reactivation. Further south in Japan, besides the winter diapause there may also be a summer diapause induced by a long-day and high temperature. Here, the summer diapause becomes so intensive that in the subtropical regions of Japan *B. brassicae* has only two generations a year—spring and autumn ones—separated by two periods of summer and winter rest (118).

Ankersmit & Adkisson (6) studied the influence of photoperiod and temperature on the diapause induction in six geographical populations of *Pectinophora gossypiella,* coming from latitudes ranging from Texas to Argentina. In all the strains the critical photoperiod remained stable, the maximum numbers of individuals entering diapause was observed at 10 to 12 hours photoperiod and the development became continuous with a 13-hour day. At the same time, the intensity of the PhPR obviously varied with latitude. It was quite pronounced and similar in Texas (32° and 28°N) and in Argentina (27°S) strains. But in the strains coming from the tropical areas (Venezuela, 10°N and Columbia, 3°N), the PhPR was extremely weak and almost completely disappeared at a high temperature (31°C). Thus, the adaptation to the tropical climate in this species is realized not by a change in the critical day-length but by reduction of the PhPR and by the adoption of homodynamic development.

According to Norris (140), in *Nomadacris septemfasciata* from Tanganyika (7° to 9°S), where the day-length changes by only 1 hour within a year, the critical photoperiod for the imaginal diapause lies between 12 and 13 hours. The population from the delta of the Central Niger, where the day changes by 2 hours within a year, has a higher photoperiodic threshold, between 13 and 14 hours.

Variations in the photoperiodic reaction under the climatic changes at the same latitude.—It is known that at the same latitude, i.e., with the same yearly course of day-length, the seasonal temperature dynamics and duration of the vegetation season may vary considerably. In consequence, the species phenology and the dates of diapause onset also change. Such facts were observed in *Anopheles maculipennis* (21) and in many pest insects. It was shown (69, 71), that the critical day-length in *Tetranychus urticae* populations coming from the Black Sea coast of the Caucasus, is notably lower than in populations from Kirghizia (Frunze) and South Kazakhstan (Alma-Ata), in spite of their inhabiting places with nearly the same latitude. This agrees well with an early autumn in the continental regions of Middle Asia.

Of particular interest are the variations in the PhPR in local strains *Pieris rapae* from the Black Sea coast of the Caucasus (Abkhazia) and from the Vladivostok district situated at an identical latitude (53). The vegetation period in Abkhazia is much longer and warmer, therefore *P. rapae* passes through six generations before diapause begins at the end of October-November. In the Far East this species has only three generations annually and diapause commences late in August or in September, i.e., with a considerably longer day. The experimentally found critical day-length for the Abkhazian form is less than 12 hours and for the Far East form it is 14.5 hours. The Far East strain of *P. rapae* was found to be similar in its PhPR to the European population of this species distributed 10° northwards. Thus, variations in the temperature conditions of the vegetation period observed at an identical latitude may considerably alter the regularity of the latitudinal geographical variation in reaction to photoperiod.

Regulation of seasonal development of insects in the mountains.—Climatic variations with altitude in the mountains cause biological zonation analogous to that on the plain in the northern direction. Correspondingly, considerable changes in phenology and an earlier seasonal onset of diapause are observed in insects as the altitude increases. There may arise, therefore, intraspecific inherited variations in the synchronizing reactions, similar to those described above. However, experimental data testify to a less important role of inherited variations in seasonal cyclic reactions in the mountains as compared to that on the plains; and they show a more important role of direct influence of climate.

The analysis in phenology of the leafroller *Polychrosis botrana* in the Transcaucasus (101) and of codling moth *Laspeyresia pomonella* in Austria (169), has proved that the decrease in the generation number and earlier incidence of diapause at higher localities are explained by delayed development through temperature shifting of the stage which is senstive to the photoperiod. In Armenia the highlands population (2000 m) of *Leucoma salicis,* always developing in one generation, brought to the plains, gave two to three generations like the local population (53).

In this respect it is interesting to compare the zonal adaptation in *Pieris napi*—the species which is common in Palearctic regions and ascends up to 2000 m in the Caucasus. On the East European Plain, the number of generations in this species varies from six on the Black Sea coast of the Caucasus (43°N) to two in the Leningrad district (60°N). The experimentally found critical day-length of the PhPR (at 18°C) in local populations increases gradually from 13 to 18 hours (Fig. 3, e), characterizing their adaptive inherited norms. In the mountains, the differentiation is much less pronounced. From the coast to a height of 1600 m there is a genetically uniform race—ssp. *meridionalis* with a 13-hour critical photoperiod. Still higher (1500 to 2000 m) it is replaced completely by a monocyclic form—ssp. *bryoniae* with a rudimentary PhPR in evidence only at a high temperature. This form bears a close ecological resemblance to a monocyclic subarctic subspecies *adalivinda* from Northern Scandinavia, possessing an almost obligatory diapause (147). The special character of ssp. *bryoniae* concerns not only the PhPR but also the reactivation temperature norms. This explains the fact that in the contact zone with ssp. *medirionalis* (1500–1600 m) these subspecies are phenologically separate and do not produce any transition forms.

The above data show that in the mountains the inherited ecological and physiological differentiation of populations is more hampered than on the plains, probably due to a less special isolation.

The Possibility of Spread and Acclimatization of Local Strains

Intraspecific differentiation and perfect adaptation of every geographical race to local light and temperature rhythm peculiarities restrict free-spreading even within a species' range. One of the factors first hindering the acclimatization is the lack of correlation between the PhPR of the introduced form and the new climatic conditions.

This suggestion was confirmed by the first experiments in which the indigenous and the southern races of *Acronycta rumicis* were reared under natural conditions in Leningrad district (50). The southern races with the low PhPR threshold failed to enter diapause here and died with first autumn frosts. Analogous results have been obtained for the monovoltine species *Dendrolimus pini* (68), *Dasychira pudibunda* (72), and for some other Lepidoptera (51). The results of numerous further experiments have been summarized by Danilevsky & Kuznetsova (53). They arrived at the following conclusions about the introduction of the southern race to the northern part of the range.

(*a*) The diapausing stages of all southern races normally hibernate in the north of the range and resume spring development simultaneously with the native ones. The latter fact proves that the spring development dates are controlled not by photoperiod but by temperature.

(*b*) The introduced southern races have not shown any marked differences in development time during the summer, or any growth suppression as compared to the indigenous strains.

(*c*) The diapause in all southern races commences much later than in indigenous ones. As a rule, the further southward the place of origin of a strain lies, the longer is the delay. Some southern forms (*Pieris brassicae, P. rapae,* and *Acronycta rumicis* from Sukhumi) do not form wintering stages at all in Leningrad province, whereas others have time to produce there just a small number of diapausing individuals.

(*d*) The possibility of acclimatization depends more on the degree of the geographical variation of the PhPR and on its thermal lability than on the latitudinal distance. The northern populations of *P. brassicae* and of *Leucoma salicis,* maintaining a stable PhPR (Fig. 3, a) over the most part of the European territory of the USSR, behave in Leningrad similarly to the local strain. High thermal lability of the PhPR (Fig. 2, a) makes it possible for the Caucasus forms of *Barathra brassicae* and *Leucoma salicis* to diapause in the North regions.

A few experiments carried out so far on the introduction of the northern race to more southerly locations (53) have revealed a sharp discrepancy of the cycle with the new seasonal conditions. The populations of *P. brassicae, P. napi, L. salicis, Polia oleracea,* and *Tetranychus crataegi,* coming from Leningrad district and having a high PhPR threshold, in some regions of the Caucasus and Middle Asia (40° to 44°N) produced only diapausing individuals, even when reared at a high temperature and long photoperiod. Thus, the effect of increased temperatures does not compensate for the decrease in the day-length. It must be noted that the southern forms of the same species developed without diapause.

On the whole the investigations have shown that the experimental analysis of the reaction system regulating the life cycle, gives a firm basis for making a prognosis in the acclimatization results.

Change in Photoperiodic Adjustments as a Result of Selection

There is no doubt now that the reactions controlling diapause can be easily changed by selection. Thus, by means of the selection of the nondiapausing individuals it has become possible to obtain continuously developing lines in *Gilpinia polytoma* (155), *Ephestia elutella* (201), *Locusta migratoria gallica* (103), *Choristoneura fumiferana* (91), *Gryllus* (23), *Pectinophora gossypiella* (13), and *Pseudosarcophaga affinis* (96). Most of these experiments were carried out without controlled photoperiodic conditions or under one regime of light and temperature. Therefore, those changes of the PhPR and its critical threshold, that appear at such selection, are difficult to judge. Only recently some information on the subject has been received.

Helle (93) has found a genetic heterogeneity in the reaction to photoperiod of a natural (Holland) population of the spider mite *Tetranychus urticae*. Some selected inbred lines were notable for the low percentage of diapause at any short-day photoperiods. The threshold of the PhPR in all these lines was virtually the same. Crossing of lines with high and with low percentage of diapause showed the first characteristic to dominate partly, the threshold remaining stable in the hybrids as well. But Geyspitz (71) discovered with the same *T. urticae* the possibility of prompt change of the PhPR threshold by means of strict selection of individuals not entering diapause at the critical photoperiod. She managed to decrease the critical day-length of the Leningrad population of the mites from 17 to 16 hours after selection within two generations. Further selection at the new level of the threshold allowed her to obtain a line entering diapause with only 10 to 12-hour photoperiods. This method of sibselection has helped to reveal a possibility for a decrease in the critical day-length. The newly selected lines showed a stable maintenance of this character and were in correspondence with natural geographic populations when the investigations were carried out in nature. Similar results were obtained by Maslennikova (123) in chalcid *Pteromalus puparum*. In the Leningrad population of this species the critical day-length was decreased by more than 2 hours by means of selection of nondiapausing individuals during four generations.

As with the spread of insects to new localities, there is an inevitable elimination of the individuals that do not fit new conditions as a result of the critical threshold of their PhPR; this reaction must soon assume an adaptive norm. This has been confirmed by field observations and by the analysis of local strains in the newly populated areas. Judging by phenological data of the initiation of hibernation, populations of *Laspeyresia pomonella* from Canada and New England have a higher photoperiodic threshold than strains of that species in the southern parts of the United States. The Australian population from Canberra and those from California and Middle Asia distributed at the similar latitude of the northern hemisphere bore a great resemblance in their PhPR threshold (66, 180). Adaptive differentiation also takes place in the European corn borer in northern America (18, 30), but apparently it still has not been completed.

Another type of geographical adaptation was found in *Pectinophora gossypiella* introduced several decades ago into subtropical and tropical zones of America (6). In this case adaptation to the tropical climate arises by a weakening and then a complete disappearance of the PhPR, its threshold remaining constant, i.e., just as in *Tetranychus urticae* in selection experiments by Helle (93).

Adaptation to the Rhythm of the Opposite Hemisphere

There are numerous well-known instances of the successful acclimatization of temperate climate insects brought accidentally or intentionally to countries located in the opposite hemisphere. From special observations by Geier (65, 66) on the phenology of the codling moth *Laspeyresia pomonella* naturalized in Australia, and from experiments on the Australian population (180), it has been found that the same photoperiodic and temperature mechanism of the original range is followed. But very little is known of how the life cycle, previously adapted to the seasonal rhythm of one hemisphere, switches over to the inverted rhythm of the opposite hemisphere. Such switching over must be accompanied even in the first stages of acclimitization either by a half-year prolongation or by shortening of the diapause period.

One way would be for the species to be reactivated merely by low temperature. Duclaux (60) described such a situation for the silkworm, *Bombyx mori.* Diapausing eggs of this species transported from South America to France, or in the opposite direction and then returned, did not experience chilling at the appropriate time and remained in diapause for a long time. Egg reactivation occurred only after wintering in the new locality so that the larvae emerged in the spring of the corresponding hemisphere. Few species are capable of such an extended diapause state. This must hamper considerably the acclimatization of many holarctic insects in the southern hemisphere.

The process of switching over is greatly simplified in those species whose development is reactivated by photoperiod. For instance, wintering larvae of *Laspeyresia pomonella* are inevitably exposed to the long-day and high temperature of summer when they are brought to the southern hemisphere. These conditions cause an abrupt elimination of diapause, which then opens the way for an immediate transition to a new rhythm of seasonal development (54). The same may take place in species with a short-term diapause, requiring no chilling, as, for instance, in *Melitaria junotilinella* introduced into Australia from Texas (86), and in species hibernating in a state of quiescence, e.g., *Plutella maculipennis.*

DAILY RHYTHMS

Daily rhythms of Arthropoda reflect their adaptation to the 24-hour periodicity of environmental conditions. Adaptive implication of daily rhythms in arthropods is less evident than in seasonal ones where it is clearly seen as an adjustment to seasonal change of climatic factors. To all appearances, the main aim of daily rhythm is the synchronization of intraspecific or interspecific relations. It concerns especially facilitation of contacts between the sexes during the reproductive period as well as trophic relations. It must be borne in mind that many daily rhythmic processes themselves have no ecological significance but only reflect the working of a mechanism controlling some other adaptive rhythms.

Daily rhythms can be observed as periodical changes in populations, organisms, organs, or cells. There are several types of such rhythms in arthropods: locomotor (general mobility, flight activity, taxes), reproductive (copulation, oviposition), ontogenetic (emergence of larvae and imago, moults), metabolic (feeding, respiration, and excretion), morphological (volume of the cell, nucleus and nucleolus, deposition of chemical substance layers in the cuticle, state of chromatophores), biochemical (enzymatic and secretory activity of cells, contents of different substances in the hemolymph, narcotic sensitivity), and biophysical (electrical activity of neurons). However, the biological clocks controlling all these rhythms behave similarly in many situations. That is why we will consider the characteristics of daily rhythms independently of their types.

Rhythm Pattern

When a daily rhythm has been synchronized, its free-running period (τ) becomes equal to 24 hours. But, in fact, it cannot always be measured as a distance between two neighbouring maxima because sometimes the curve of a daily rhythm has 2, 3, or more peaks within 24 hours.

In insects and other arthropods, the single-phase rhythm with one peak has been studied best of all. As a rule, its maximum coincides with the beginning of light or darkness. The maxima are more rarely observed at noon or midnight. Well-known examples of the single-phase rhythm are the daily rhythmic activity of cockroaches (85, 88, 126, 166, 167), crickets (45, 142), beetles (113, 202, 204), and arachnids (46, 213). The daily rhythms controlling the time of moulting and emergence in insects are also single-phase in most cases (160).

The two-phase rhythms seem to be widespread among arthropods. The curve of such rhythms has two daily peaks, usually corresponding to "lights-on" and "lights-off." Thus, there are two maxima in the emergence pattern in the tortricid *Sparganothis pilleriana* (83) and in some chironomids (185), or in the daily rhythms of locomotor activity in the beetle *Trogoderma glabrum* (41), in the woodlouse *Porcellio scaber* (55), and in the pill millipede *Arthrosphaera delayi* (144). In some cases, the two-phase rhythms are characteristic of many physiological functions of the same organism. For instance, in *Drosophila* at a photoperiod with LD 12:12, very similar two-peak curves were obtained for the daily rhythmic changes in locomotor activity, oxygen uptake, nucleus volume of neurosecretory cells in pars intercerebralis and in corpus allatum, that of cells in the prothoracic glands, and in the fat body (161).

The pattern of a daily rhythm may vary under the influence of environmental physical factors or during the season. In these cases, a single-phase rhythm often is converted into a two-phase one, or vice versa. In the crab, *Carcinus,* a temperature decrease of 5 to 10°C in the time between two daily maxima causes some additional peaks of locomotor activity (136). The beetles *Calandra granaria* reared under an LD 12:12 regime show only one peak of locomotor activity in the middle of the day; but with constant illumination or with any abnormal light regime they show a rhythmic process with two peaks (24). In Uzbekistan, in spring and autumn, the diurnal variation in flight activity of many synanthropic flies can be expressed in the form of a single-peak curve, whereas in summer the curve has morning and evening maxima (191). In Australia, similar changes of daily activity are observed in *Calliphora stygia* and *Microcalliphora varipes* when passing from colder to warmer periods of season (141). In some Palestinian Coleoptera, the general pattern of daily rhythms of behaviour is uniform within the whole year, but the maximum position in relation to the time of transition from light to dark (or vice versa) changes in the course of the season (26). Considerable seasonal variations in the daily rhythm of activity have been found in the beetles *Carabus* (204, 205). For example, *C. cancellatus* reared at room temperature is active at night in winter and in the day in May.

Three-phase rhythm seems to be rare. Several rhythms with $\tau = 8$ hours have been recently described in Lepidoptera. Such is the feeding rhythm in silkworm (112), and the daily variation of oxygen uptake and nucleus volume of the neurosecretory cells in *Ostrinia nubilalis* (15). In the caterpillars *Dendrolimus pini* daily rhythmic fluctuations of electrical activity of the nervous system show similar patterns (197). But, in this case, two additional peaks of activity, repeated every 8 and 16 hours after the first maximum, are extremely small and can be recognized only by using a logarithmic ordinate.

Exogenous and Endogenous Rhythms

Many authors have observed varying dependence of daily rhythm on the environment. Schneirla (177) concludes that in some insects activity rhythms are mainly controlled by external and in others by internal factors. Harker (89) suggests subdividing all rhythms into exogenous and endogenous. More complex classifications of daily rhythms based on different correlations of external and internal regulators have been also suggested (40).

In an exogenous rhythm (Fig. 4A) extrinsic conditions cause the rhythm and an insect is only a high-sensitive indicator of a daily variation of microclimate. In most cases exogenous rhythms are caused by the changes of illumination, e.g., rhythm of flashes in fireflies (156), daily vertical migrations of *Chaoborus flavicans* larvae (194), and rhythm of locomotor activity in the locust *Schistocerca gregaria* (143). Nevertheless, in the millipedes *Blaniulus guttulatus* and *Oxidus gracilis,* nocturnal activity is stimulated by a drop of temperature at night (42).

In an endogenous rhythm (Fig. 4B, C), extrinsic conditions only synchronize the phase of oscillations which arise inside the organism. Any rhythm may be rightfully called endogenous if it displays at least one of the following properties: (*a*) The rhythm can be initiated by a single, even short

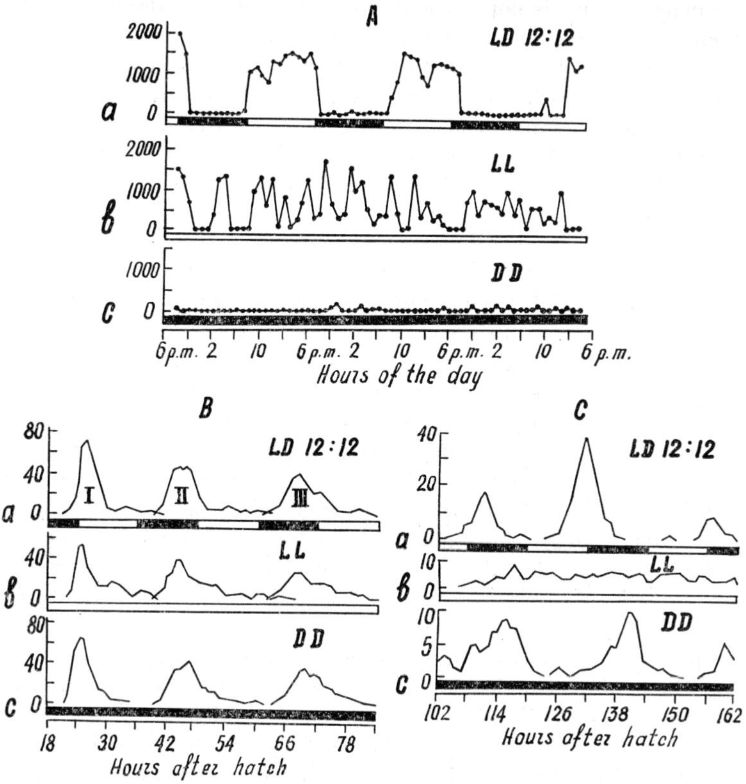

FIG. 4. Daily rhythms in insects. a—at 12-hour photoperiod; b—at constant light; c—at constant darkness. A—the exogenous rhythm of locomotor activity in the locust *Schistocerca gregaria* (143). Ordinate—the number of movings per 1 hour. B and C—the endogenous rhythm of larvae moults and of pupation in the mosquito *Aedes taeniorynchus* (137). Ordinate—the number of moulted or pupated larvae per 1 hour. I, II, III—the first, second, and third larvae moult.

stimulus. Thus, in *Drosophila* the rhythm of imago emergence does not occur in constant darkness or light. But a single transition from light to darkness, or vice versa, does initiate the rhythm. (*b*) The phase of the rhythm may be in advance of external synchronizing signals. Thus, in experiments on *Galeodes arabs* and many beetles of the genus *Carabus* with nocturnal activity, the insects emerged from cover shortly before the onset of darkness; the advance reaches 30 to 40 min (at LD 12:12) or even 4.5 hours (at LD 21:3) (46, 205). (*c*) The rhythm persists for some time in the absence of rhythmic light and temperature fluctuations, but may change to free-running oscillations gradually or suddenly. The behaviour of different daily rhythms under constant conditions will be discussed below.

It must be emphasized that a precise distinction between exogenous and endogenous rhythms is impossible. "Pure" exogenous or absolutely independent endogenous rhythms seems to be unlikely in living organisms. The daily rhythm appears to be always conditioned by endogenous signals or internal oscillators which can be suppressed by exogenous responses to favorable environmental conditions. In such case, any difference between exogenous and endogenous rhythms may be caused by the instability of oscillators. Some authors have already expressed similar opinions about the origin of biological rhythms in insects (40, 44).

Circadian Rhythms in Constant Conditions

In conditions which are constant in respect to light and temperature and in the absence of any other entraining signals, endogenous rhythms often persist. In such cases, biological oscillators have free-running nonsynchronized oscillations. In terrestrial arthopods the free-running period (τ) is close to 24 hours. Generally it is not less than 21.5 nor more than 27.5 hours.

The value of the free-running period in constant darkness is different from the value in constant light. According to Aschoff's rule (9), the length of period in diurnal animals measured in LL is decreased relative to the value in DD, whereas for nocturnal animals it is generally increased. It is true for vertebrates but this rule is absolutely inapplicable for arthropods (163). In arthropods the τ is always slightly increased in the light as compared to the value in darkness[3]), but the period changes can range from 0.5 to 4.5 hours. In many cases, the rhythm persists only in darkness and it disappears in continuous light (in *Schistocerca gregaria, Trogoderma glabrum, Crambus trisectus, Drosophila melanogaster,* and others).

In contrast to plants and vertebrates (9) in arthropods the rhythm is practically independent of light intensity. If a rhythm persists in continuous light, in passing from DD to LL, the τ increase takes place when the intensity of illumination is very small (about 1–2, 6 lux), and a further increase in intensity (to 600 to 700 lux) does not affect the free-running period (159, 213). But if a rhythm disappears in continuous light, the desynchronization of oscillators is more rapid when the intensity of illumination is higher (139, 211).

The effect of temperature on circadian rhythms is discussed in many works published recently (149, 190, 209). As in other living organisms, in arthropods τ depends only slightly on temperature. In most cases, the value of the free-running period measured at different temperatures from 6 to 7°C to 30 to 35°C is just the same. The Q_{10} (ratio of the period at one temperature to the period at a temperature 10° higher) is equal to 1 in the cases of retinal pigment migration in the crayfish *Cambarus* (206), daily colour rhythm in the fiddler crab *Uca* (31), the memory for time in bees *Apis* (98, 199), the rhythm of pupation in the mosquito *Aedes taeniorhynchus*

[3] *Pseudosmittia arenaria,* having τ in light more than in darkness, is the only known exception (158).

(137), and that of locomotor activity in the crab *Carcinus* (136) and in the beetle *Tenebrio* (113). The eclosion rhythm in *Drosophilia* has Q_{10} of more than 1.0. Bünning (33) obtained $Q_{10} = 1.1$ to 1.25 and Pittendrigh (148), 1.02 for this rhythm. In the cockroach *Periplaneta,* Q_{10} is less than 1.0 for the rhythm of locomotor activity (36). Persistent values of τ are, however, observed only within the ecological range of temperature, and when the latter is beyond this scope, considerable changes of the free-running period and the shape of the rhythm may take place. But, even within the limits between 6–7°C and 35°C, the temperature has an important influence upon the metabolic component of the rhythmic process, thus changing the oscillation amplitude, the total mobility, and activity time in animals (163). For example, temperature changes from 15° to 25°C give $Q_{10} = 5$ for the total amount of activity per unit-time in the beetle *Tenebrio* (113), and $Q_{10} = 3$ for intensity of oxygen uptake in *Drosophila* (162).

Entrainment of Circadian Rhythms

The phase of a circadian rhythm is synchronized by daily periodic changes of environmental factors. The main entraining signal is photoperiod, or a periodic repetition of light and darkness in a daily cycle (8). When an animal after experiencing constant conditions is exposed to a light-dark cycle, the latter causes a persistent rhythm with the same period as an entraining cycle. One may say that the rhythm is entrained to the external signal. The time of converting the rhythm is about 1 to 3 new LD cycles, more exactly from 48 to 72 hours in Carabidae (100), about 24 hours in the cockroach *Blaberus craniifer* (212), and about 38 hours in the leaf beetle *Aspidomorpha quinquefasciata* (168), but sometimes is greater (24, 87).

The "limits of entrainability" of the circadian rhythms to light-dark cycles are quite different for different arthropods, but as a rule, with the cycle length shorter than 18 hours, or longer than 28 hours, entrainment no longer occurs (45, 158, 167, 203). The *Drosophila* eclosion rhythm can be entrained by cycles from 13.8 to 34.9 hours, but if the cycle length is very different from 24 hours, the character (one-phase, two-phase, or three-phase) of the rhythm may change (see 32).

In artificial light-dark cycles the maximum of an entrained rhythm correlates with dawn (in the light-oscillator) or with the dusk (in the dark-oscillator). Different rhythms in one and the same organism have been estimated to have different relations with the beginning of "subjective" night or day. That is why it is very difficult to use the terms "nocturnal" or "diurnal" species for arthropods. For instance, in *Drosophila,* which seems to be regarded as a "day-active" species (58), oviposition and pupation occur mainly in the evening or at night (164).

The phase correlations of a circadian rhythm with dawn or dusk are perfect in LD 12:12, but they often break down when either the light interval or the dark interval is short. In such cases there is a definite phase-angle difference between the peak of the rhythm and the entraining signal. A positive phase shift takes place if the rhythm maximum is advanced; a negative phase shift occurs when the maximum is delayed in relation to the signal. Some experiments carried out on arthropods have shown that, independently of the oscillator type, a decrease in the dark period increases negative phase shift with respect to dawn, or positive phase shift with respect to dusk. While the L:D ratio is changed, such shifting has been observed in different rhythms of arthropods: pupation of *Aedes* (137), eclosion in *Pseudosmittia* (158) and in *Ephestia* (128), locomotor activity in *Blatta* (126), *Ecdyonurus* (87) and *Carabus* (205), as well as pupation, eclosion, oxygen uptake, and oviposition in *Drosophila* (149, 162, 164).

The position of the rhythm phase does not depend only on the L:D ratio (163). Intensity of illumination in LL and temperature are also important. In *Tenebrio molitor* increase of illumination intensity causes a decrease of positive phase shift in relation to the light-off (113). On the other hand, an increase of positive phase shift is observed when the temperature is increased from 20° to 30°C. A reversal of phase position at different temperatures takes place in some insects. The bear caterpillar *Halisidota argentata* is active at 5°C only in the daytime and at 23°C at night (61). In LD 12:12, the peaks of pupation rhythm in the mosquito *Aedes taeniorhynchus* coincide with the middle of light period at a temperature of 32°C, and with the beginning or with the middle of night at 24°C (137). But with the same photoperiod, position of the peaks in the rhythms of pupation and eclosion in *Drosophila* does not depend on temperature in the range 16° to 26°C (148, 164).

Thermoperiod entrains circadian rhythms of many arthropods as the photoperiod does: temperature increase in the daily cycle acting as light, and its decrease, like darkness (209). The entraining role of a temperature cycle is best manifested in either constant darkness or light (43, 100, 128). With few exceptions where there is an interaction of light and temperature cycles, light entrains more strongly than temperature does (137, 152, 167). However, the daily growth of resilin layers in the cuticle of *Schistocerca gregaria* (139), and the locomotor activity of the caterpillars *Hepytia phantasmaria* (61) are mainly regulated by temperature cycles.

Temperature changes seem also to play the main part in the rhythm synchronization in West African millipedes *Oxydesmus platycercus* and *Ophistreptus* sp., inhabiting tropical forests where light hardly penetrates through the canopy (42).

The above data testify undoubtedly to the leading role of the photoperiod in the circadian rhythm entrainment. However, they do not elucidate completely the entrainment mechanism. The results obtained in experiments involving application of short light impulses (from several minutes up to 1 to 2 hours) are of great importance in understanding the mechanism. It appeared that for the circadian rhythm synchronization, a so-called "skeleton photoperiod" can be used, in which a normal photoperiod is substituted by two short light impulses ("dawn" and "sunset") imitating the beginning and the end of the day. The entrainment of circadian rhythm by such skeleton photoperiods was thoroughly investigated by Pittendrigh and his co-workers (127, 150, 151, 154). In those experiments, the phase position of the eclosion rhythm in *Drosophila* corresponded closely to the position with the regular photoperiods up to the "day-length" of 12 hours. A cycle with two light impulses may be considered as any of the two complete photoperiods composing it (for example, the skeleton photoperiod LD 8:16 effectively simulates the complete photoperiods, LD 8:16 and LD 16:8). However, the rhythm is always phased with the least of two intervals of the skeleton cycle and therefore it appears impossible to imitate photoperiods longer than 12 hours. Similar results were obtained (154) for the circadian rhythm of oviposition in *Pectinophora*.

The rhythm entrainment by the "asymmetrical skeleton photoperiods" (a main photoperiod with a short light interruption during the night in each LD cycle) has much in common with the entrainment by symmetrical ones. It has been shown (150) that (*a*) day-length of a longer duration can be imitated by using the asymmetrical skeleton photoperiods; (*b*) a short light impulse can play the role of "dawn" or "sunset" depending on its position at the beginning or at the end of the night; (*c*) longer photoperiods are imitated more successfully if a night interruption acts as "dawn."

In order to understand the mechanism of the circadian rhythm entrainment, investigations of the rhythm response to a single light impulse were of principal importance. Pittendrigh with collaborators carried out detailed investigations of this kind on arthropods (32, 149–151, 154).

A period of free-running rhythm (τ) characterizes the specific scale of circadian time, during which sensitivity of an organism to the external factors changes regularly. These cyclic changes of sensitivity are most distinctly revealed in the experiments in which the single light impulse influences a "pre-entrained" circadian rhythm. Induced by the light impulse the phase shift of the free-running rhythm depends on the randomly chosen moment of circadian periodicity when the signal is applied. This shift (phase delay or advance) is described by a response curve given in the coordinates of circadian time scale (CT). The response curve is considered by Pittendrigh to be a fundamental characteristic of the circadian rhythm. The entrainment process is realized by the time correlation of the rhythm period (τ) with that of environment (T) and by setting a definite relation between the circadian oscillation and LD cycle; it ensures adaptive phase occurrence with the appropriate part of the day.

A harmonious theory of entrainment of a circadian rhythm to environmental periodicity was postulated by Pittendrigh and his co-workers (149–151, 153) as a result of experiments in which circadian rhythms were investigated under various LD cycles. The essence of the theory may be expressed as follows. Every circadian rhythm is controlled at least by two coupled oscillators; one of them plays the leading role and responds immediately to synchronizing signals "dawn" and "sunset" within the limits of each LD cycle. The phase of the leading oscillator is entrained in the leap manner. The other (subordinate) oscillator controls the entrainment of the circadian rhythm through "commands" of the leading oscillator. In Pittendrigh's opinion, the properties of the leading oscillator may be indirectly judged by the curve of the phase responses to single impulses. Only these properties are important for the mechanism of the circadian rhythm entrainment.

The entrainment theory allows the calculation of the position of a circadian rhythm phase under various LD cycles. The calculated data always agreed well with the experimental results.

For instance, in cycles with 15-min light impulse (one impulse per cycle) the circadian rhythm entrainment was shown to proceed in accordance with the curve of phase responses to a single signal (150, 154). If $\tau \neq T$ (cycles of the duration from 20 hr 40 min to 25 hr 00 min were used), the circadian rhythm was entrained to the external periodicity by means of a correction of its own period τ to equal the duration of T. Increase or decrease in the τ is achieved by a phase shift sufficient for the signal to reach the appropriate section of circadian time scale. It must be stressed that the phase position actually observed for the investigated rhythms of *Drosophila* and *Pectinophora* appeared to be close to the calculated ones.

The final position of the rhythm phase at a skeleton symmetric photoperiod may be considered as the result of phase shiftings of the leading oscillator $\Delta\varphi_1$ and $\Delta\varphi_2$ under the influence of each of two light impulses forming the skeleton (150). The algebraic sum of these shifts ensures the rhythm synchronization with the LD cycle in accordance with the equation

$$\Delta\varphi_1 + \Delta\varphi_2 = \tau - T$$

If the skeleton photoperiod is assymetric, the main photoperiod comes to have a summed $\Delta\varphi$ of its own that interacts with the $\Delta\varphi$ of an additional impulse. The effect of each of two light impulses as well as the summed effect may be evaluated by the corresponding curve of phase responses for a single impulse.

The entrainment theory makes it possible to calculate not only the rhythm phasing under various conditions but also extreme possibilities of its synchronization with environmental periodicity of any arbitrary duration. The range of synchronization of a circadian rhythm with the nondiurnal LD cycle is determined by maximal values of $\Delta\varphi$ on the phase responses curve ($\Delta\varphi_{\text{limit}}$) according to the formula:

$$T_{\text{limit}} = \tau + (\Delta\varphi_{\text{limit}})$$

where T_{limit} is the extreme value of the LD cycle period. The experimentally established limits of synchronization in *Drosophila* and in the hamster *Mesocricetus auratus* are in a good agreement with the theory (150).

If T of the entraining cycle exceeds the range of T_{limit}, disturbance of synchronization may be expected, which is often apparent as a transition to a free-running rhythm with the period τ.

THE OSCILLATORY CONCEPTION OF PHOTOPERIODISM

The Common Features of Circadian and Seasonal Rhythms

Resemblance between seasonal and circadian rhythmic phenomena is seen in many of their important characteristics. Both groups of phenomena are undoubtedly based on the capacity of living organisms for time measuring, and both are connected with certain cyclic factors, the photoperiod being the main one of these. It is neither the energy of light nor even the number of light hours in the cycle that is important for the circadian rhythm entrainment or for the photoperiodic control of development, but rather the time structure of the LD cycle.

Both groups of rhythms are characterized by the similar response to alternative aperiodic regimes. For instance, a free-running circadian rhythm may be usually observed at the regimes LL and DD. Thus, continuous light or darkness cause the same effect. The similar trends take place in photoperiodic regulation of development. Dependence of circadian and seasonal rhythms on temperature is also similar; that is true for temperature rhythms (51, 79, 80) or for short temperature rises and drops (52, 75, 174, 175) and other related effects. The common feature for both rhythms is a loose dependence of some their parameters (τ of circadian rhythms and the critical photoperiod of PhPR) on temperature. Q_{10} is usually 1 to 1.2 in both cases.

Biological rhythms are subjected to adaptive seasonal and geographical changes (see pages 210 and 219). A parallel is also observed in the nature of the inheritance of the critical day-length of PhPR and the position of peaks of daily rhythms with lunar periodicity. In both cases, the responses of hybrids between geographical races are intermediate in respect to their parents' responses (51, 138).

Some supplementary facts may serve as proof of the oscillatory conception of photoperiodism. Like the circadian rhythms, the PhPR takes place only in the range of certain deviations of the LD cycle period from the diurnal norm, although this lability of PhPR varies in different species.

Cycles of various durations with a dark phase greater than 12 hours induce diapause in the highest number of individuals at the following durations of the LD cycle: in *Acronycta rumicis*—from 16 to 72 hours (78, 79); in *Metatetranychus ulmi*—from 16 to 48 hours (105); in *Megoura viciae,* about 18 to 60 hours and more (108); in *Ostrinia nubilalis* from 16 to 36 hours (14). The narrower range of PhPR entrainment is known in *Laspeyresia molesta* (57). *Pieris rapae* (12), and *Pectinophora gossypiella* (1).

Thus, the limits of the PhPR synchronization with LD cycles of various durations vary considerably in different species of arthropods. These variations are in general rather close to the established limits of circadian rhythms entrainment. However, there are almost no data to confirm directly the limits of the PhPR entrainment and of circadian rhythms in the same species. We may note only in *Pectinophora gossypiella,* that LD cycles in narrow limits (*circa* 24 hours) are effective both in accomplishing the PhPR (1) and in entrainment of oviposition rhythm, corresponding to the phase response curve (127, 154). As was shown in our laboratory (A. G. Azarjan), the wider limits of oviposition entrainment in *Drosophila phalerata* correspond to the same wider limits of diapause photoperiodic induction.

The disturbance in PhPR with the ineffective cycle duration is observed during the transition to the normal reaction characteristic for aperiodic conditions (LL or DD); it may presumably be analogous to the "free-running" circadian rhythms.

One of the most important features of the circadian rhythms is their ability to entrain to symmetric skeleton photoperiods (p. 224). The PhPR has similar features as experiments in *Acronycta rumicis* have shown. The experimental larvae were placed after a week's entrainment by the complete photoperiods LD 8:16, 12:2, and 18:6 into corresponding skeleton photoperiods (two light impulses of one hour each). One hundred per cent of pupae were in diapause at the short-day skeleton photoperiods imitating LD 8:16 and 12:12 as well as at the complete photoperiods of the same duration (control). No doubt, entrainment to the skeleton photoperiod took place in these experiments, for in continuous darkness (after the entrainment by the short-day photoperiod during a week) the percentage of diapausing pupae did not exceed 50. Naturally, imitation of the long-day complete photoperiod 18:6 by means of a skeleton scheme failed: 82.5 per cent of experimental pupae *A. rumicis* were in diapause (whereas only 8.3 per cent of control pupae at LD 18:6), which is close to the number of pupael *A. rumicis* in diapause at the short photoperiod LD 6:18. Thus, the mechanism of seasonal cyclic reaction as well as the circadian rhythm is entrained to the shorter of two alternatives allowed by every skeleton.

There are also some analogous parallels in the actions of asymmetric skeleton photoperiods (p. 225). Influence of the latter photoperiod on insect diapause induction was investigated first by Bünning & Joerrens (37 (in preventing diapause), 38) in *Pieris brassicae.* Those well-known works showed that efficiency of a light impulse breaking the dark part of a short-day LD cycle varied a great deal during the night. Maximal efficiency coincided with the value of critical day-length calculated from the beginning of the main photoperiod. The authors interpreted the result as a confirmation of the mono-oscillatory hypothesis of photoperiodism proposed by Bünning. However, later it was demonstrated (82) that in *P. brassicae* there may be two (not just one) maxima of light efficiency during the night.

From our data, the bimodal character of light-sensibility dynamics during a dark part of the photoperiod is also not a rule. Comparable investigations of the question in various species and geographical races of Lepidoptera (82) indicated a wide variety of results obtained at the asymmetric skeleton photoperiods in spite of very distinct PhPR under the normal conditions.

We managed to obtain all the intermediates ranging from the cases in which light impulses were wholly ineffective at any moment of darkness, to complete prevention of diapause. These results seem to call for further experimental data with consideration of ecological diversity of the phenomena investigated.

Detailed investigation of *Pectinophora gossypiella* (1) has established the fact that the first and the second maxima of light sensibility in the dark period are equally remote (approximately by 14 hours) from the end of the following main photoperiod or from the beginning of the preceding one. So photoperiodic measuring of the critical day-length for *P. gossypiella* (13 hr 40 min) took place at the asymmetric skeleton photoperiods as well as at normal ones. The night break functions as "dawn" or "sunset" depending on its position on the Zeitgeber scale, i.e., PhPR entrainment at asymmetric skeleton photoperiods is analogous to that of circadian rhythms.

Photoperiodic efficiency of light interruptions of darkness was demonstrated by Bonnemaison (28) for the aphid *Dysaphis plantaginea,* by Barker (11) for *Pieris rapae,* etc. The latter author has found out that the short-day induction of diapause can be prevented by very short flashes of an impulse lamp (0.001 sec), if these break darkness at the moments of maximal sensibility. The great energy of the flashes might play a certain role; their effectiveness towards photoperiodic reaction is comparable with the synchronizing action of such flashes on circadian rhythms (32).

The experiments by Lees (109) in which "the cycle period (T)" in the aphid *Megoura viciae* varied due to the change in the main photoperiod (25.5; 13.5; 5.5 hours are of a great interest. The duration of the dark period was constant (10.5 hours); it was interrupted at different times by a 1-hour light impulse. Equally situated interruptions of darkness corresponded to asymmetric skeleton photoperiods of various duration. Light impulses at the beginning and at the end of a dark period prevented the diapause. The author confirmed the bimodal form of the curve expressing the dynamics of the efficiency of interruptions of night darkness and revealed independence of the curve on the duration of the main photoperiod. In Lees' opinion this and other facts established from his works contradict the oscillatory theory of photoperiodism; they prove the "biological clock" to measure a dark component of the LD cycle in the photoperiodic induction of diapause and to work on the principle of an "interval timer." However, the conclusion that the dark component is measured allows the possibility of another interpretation of the PhPR measurement and does not at all exclude its oscillatory nature (1, 2, 82).

The exact experiments by Lees on *Megoura viciae* and by Adkisson on *Pectinophora gossypiella* permit the emphasis of a new aspect of the problem of biological rhythms: the participation of processes with stable durations in the mechanisms of the rhythms. Such an aspect did not arise from the research of circadian rhythms but is unavoidable in investigations of seasonal phenomena. Many workers (51, 57, 105, 106, 192, 193, 207) pointed out an

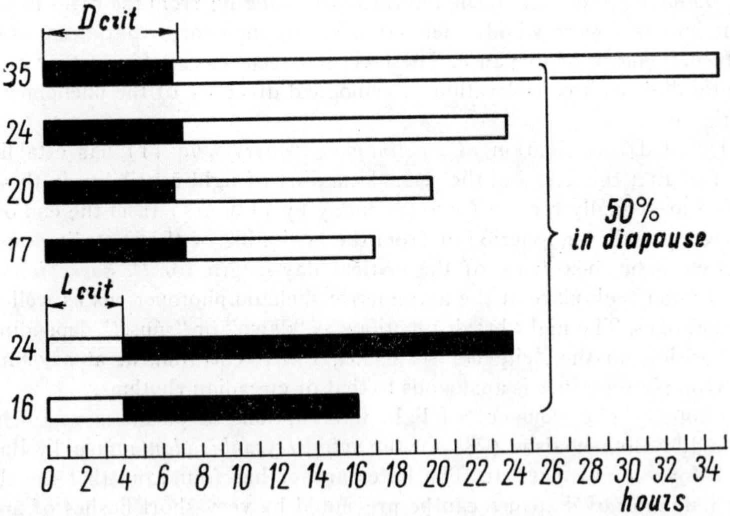

FIG. 5. Stable periods of darkness (D_{crit}) and of light (L_{crit}) characterizing the critical situation in the PhPR of *Acronycta rumicis* (50% in diapause). Values D_{crit} and L_{crit} are constant, independently of duration of LD cycles in the range from 17 to 35 hours. Figures on the left—duration of LD cycles.

important role of the dark period in the PhPR. The role of the light period in LD cycle remains less clear. Lees (109), as was mentioned above, arrived at the conclusion that it played no part in the PhPR. Adkisson (1, 2), basing his judgment on experiments with skeleton photoperiod, stresses the role of light in the indication of "dawn" and "sunset."

Goryshin & Tyshchenko (81) showed in detailed investigations in *Acronycta rumicis* a normal critical dark period (50 per cent in diapause) was the constant value that characterized a break in PhPR curves at LD cycles from 17 to 35 hours. Duration of light period did not matter in these experiments. On the other hand, a decrease in the light phase of the cycle (at a sufficient duration of its dark phase) established the existence of a "critical light period" in *Acronycta rumicis* (Fig. 5). The latter characterizes the break, if any, of the PhPR curve in the region of very short photoperiods. The data in *Pieris rapae* by Barker & Cohen (12) confirm this (82).

Thus, both LD cycle components are measured in the PhPR mechanism independently. Measuring the dark period has immediate ecological significance. Measuring the light period that does not coincide with the generally accepted idea of "critical day-length" is not adaptive but at least confirms the bimodal nature of the PhPR mechanism. The constant and independent dark and light periods characterizing the parameters of photoperiodic reaction in arthropods clearly separate this reaction from circadian rhythms;

however, similarity between the entraining primary mechanisms is not excluded.

Hypotheses of Photoperiodic Reaction Mechanisms

Bünning (34) was the first to have suggested an oscillatory hypothesis of PhPR physiological mechanism. In his opinion it was possible to explain all the photoperiodic effects on the basis of a rhythmic process controlled by only one oscillator. However, this supposition was rather unsuitable for analysis of many experimental data, especially in the field of photoperiodism ecology. Minis (127) and Pittendrigh & Minis (154) derived a bimodal modification as an improvement of Bünning's hypothesis. Not only an endogenous circadian rhythm of a substrate is responsible in their version for diapause induction but also an additional factor free from being controlled by the oscillator—a photosensitive enzyme. Beck (15, 16), on the other hand, completely rejected Bünning's mono-oscillatory model and proposed his own, explaining the PhPR mechanism as based on a principle of phase interaction of two oscillatory processes with the period approaching 8 hours.

All of the above-mentioned conceptions of PhPR mechanism have the same drawback: they do not account for experimentally determined definite durations of phases of oscillatory processes involved in the insect development regulation. Dickson (57) and Lees (108, 109) ascribe a determinant role in photoperiodic time measuring to the particular duration of a dark process; the authors consider it necessary to reject the idea of the oscillators taking part in insect PhPR and they deny utterly an active role of the processes which take place during the light part of photoperiod. None of the oscillatory hypotheses suggested is capable of explaining change of critical day-length and of PhPR type depending on temperature, and Bünning (35) even denies the existence of such change.

All these hypotheses are confined to attempts to explain only the essence of the critical day-length of PhPR and do not contain any analysis of the general form of PhPR curve and of its modifications. A model of PhPR proposed by Tyshchenko (196) seems to be more universal. According to this model, two unequivalent oscillators (*A* and *B*) are involved in the PhPR and the reaction is the result of the phase relations of the oscillators, the usual circadian periodicity of the hypothetical oscillators being assumed in contrast with Beck's proposal. The phase of the *A* oscillator is assumed to switch on at "dawn" and the phase of the *B* oscillator, in darkness. Rhythmic processes which are controlled by the oscillators are thus synchronized in the first case only with light rhythm; in the second case, only with darkness rhythm. Temporal coincidence, if only partial, of the oscillators' active phases is supposed to promote active development of insects, whereas their absolute separating—to induce diapause.

The assumed synchronization with the photoperiodic cycle means that interaction of or between the oscillators must depend on day- and night-length. Mutual coincidence of the oscillators' active phases (zone c on Figure 6) can

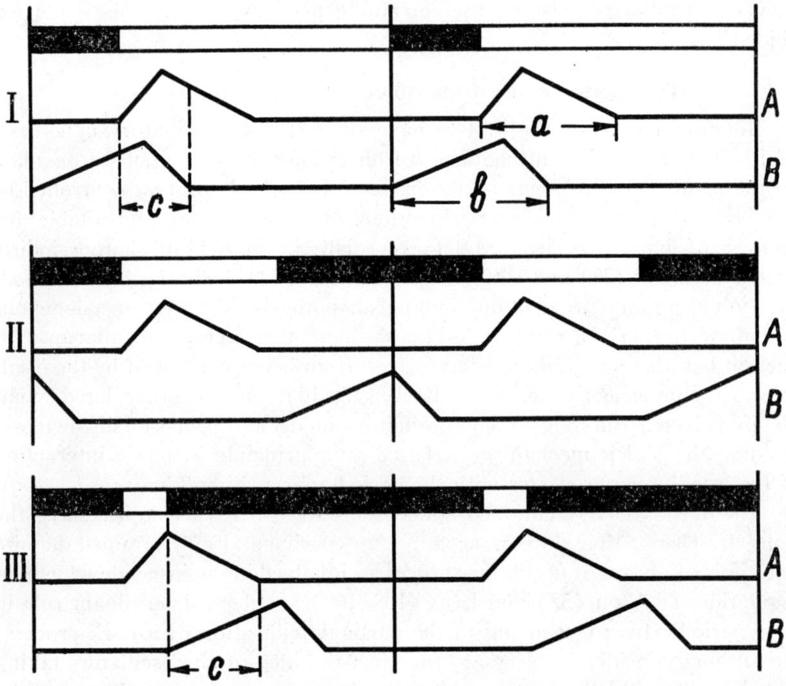

FIG. 6. Interaction of the hypothetic oscillators in the two-oscillatory model of the PhPR mechanism under various day lengths (I, II, III). A—the oscillator phasing with light; B—the oscillator phasing with darkness; a,b—their active phases; c—zone of active phases mutual coincidence which is necessary for development without diapause.

be achieved only in two cases: when phase *a* exceeds duration of the light portion of photoperiod (Fig. 6, III) or when phase *b* exceeds the dark portion (Fig. 6, I). There is a field between these two photoperiods of extreme duration where the active phases of the oscillators are completely separated in time and the c zone necessary for active development is not pronounced.

The described cases of interaction correspond to different portions of PhPR curve (Fig. 6). The I and II versions of interactions take place at the short- and long-day in the range of its natural variability during the vegetation season (the ecological part of PhPR curve); the version III conveys the interaction of the active phases under the very short photoperiods devoid of ecological significance. The model may thus be in one of three conditions. I and III being similar in principle because they both promote insect development without diapause, and condition II (separating the time of active phases of both oscillators) results in diapause. This condition seems to arise in "long-day" species under the influence of sufficiently short photoperiods (usually from 8 up to 15 hours) and increasing or decreasing of the day brings about the conditions I or III. In that way, a separating or partial matching of active phases of the oscillators A and B may be a base for the precise synchronization of insect development with the rhythm of climatic factors. Such synchronizing is expressed by a typical long-day curve of PhPR.

Change of the critical day-length in PhPR depending on temperature in terms of this model requires an additional assumption that the B oscillator is sensitive to temperature and that the amplitude of B oscillations (hence, duration of the active phase) depends on temperature conditions. In that case, a certain latent period must take place, counted from the beginning of darkness to the emergence of the active phase of B oscillator. In such a way, the PhPR threshold (50 per cent in diapause) corresponding to the field of contact of the oscillators' active phases is determined by three main parameters A: duration of the A oscillator active phase; B: duration of the B oscillator active phase; b: duration of the B oscillator latent period.

The classical curve of the PhPR of the long-day type (Fig. 1-a) has two thresholds: P_1—left-hand or "physiological" one corresponding to the situation when the end of the A oscillator active phase is contiguous to the beginning of the B oscillator active phase; P_2—right-hand or "ecological" threshold, when the beginning of the A oscillator active phase is contiguous to the end of the B oscillator active phase. In indexes of our model (A, B, and b) these thresholds are described by the formulas:

$$P_1 = A - b \qquad 1.$$

$$P_2 = 24 - (B + b) \qquad 2.$$

where all the values are expressed in hours. The value $A-b$ in Figure 1 is measured from the beginning of the photoperiodic scale up to P_1, and the value $B+b$ from P_2 up to the end of the scale.

Let us examine how the PhPR would change while changing A, B, and b. Obviously, when $b < A$, both thresholds indeed exist and are positive values; for example, the case of the PhPR curve of *Pieris brassicae* type (Fig. 1a). If $b=A$, then $P_1=0$, i.e., the PhPR curve has the only threshold P_2, which is characteristic, for instance, in *Leptinotarsa decemlineata*.

If $b > A$, then P_1 is a negative value. When the curve answering this demand is constructed it is easy to see that the threshold P_1 in this case passes to the right-hand part of the photoperiodic scale and the PhPR curve acquires a form characteristic for the intermediate type (*Euproctis similis*, Fig. Ic). When $B + b = 24$ hours (if $b > A$), the PhPR curve turns into the short-day one, peculiar, for instance, to *Stenocranus minutus* (Fig. 1d).

Thus, all the main types of the PhPR curves may have a common origin and may be considered as particular cases of a united physiological mechanism that works on a principle of phase interrelations of two oscillators. The principle, hence, easily accommodates a change in the form and the type of the PhPR curve when changing the main parameters of our model, i.e., duration of the photoperiodic oscillators' active phases and of the B oscillator

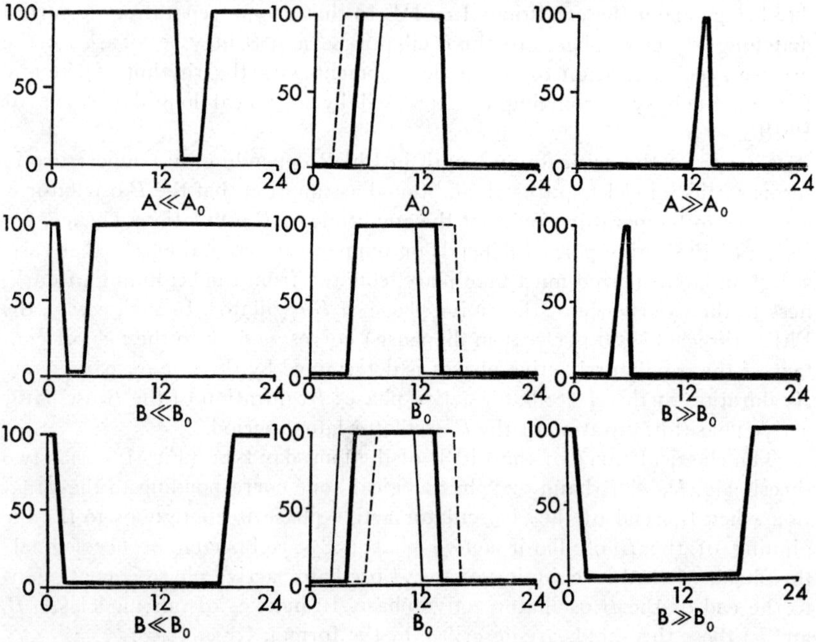

FIG. 7. Changes of photoperiodic curves as a result of increasing or decreasing A (upper series of curves); B (middle curves); C (lower curves). The curves situated in the middle of each series (A_o, B_o, C_o) show threshold changes with increasing (thin solid line) or decreasing (broken line) PhPR parameters. On the right and on the left, curves obtained when the decrease or increase of each parameter is more pronounced. Contact of the active phases of photoperiodic oscillators is assumed to conform with 50% of diapausing insects.

latent period (Fig. 7). Quite clearly, such lability of the photoperiodic mechanism serves as a good base for seasonal, geographical, and temperature variations of the PhPR to develop.

Formula 2 proved to be suitable for calculating the threshold of the long-day PhPR at nondiurnal cycles. For that purpose it should be expressed in more general form:

$$P_2 = T - (B + b) \qquad 3.$$

where T = duration of light-dark cycle. The PhPR thresholds in *Acronycta rumicis* calculated by Formula 3 for LD cycles of duration from 17 up to 35 hours coincided well with the threshold values which have been established experimentally by Goryshin and G. F. Tyshchenko (Table 1).

Formula 3 reflects stability of a critical dark period at duration of the LD cycle investigated (p. 230). On the other hand, Formula 1 conforms to

TABLE I

Calculated and Experimental Values of the PhPR Threshold in Geographical Strains of *Acronycta rumicis* at Non-Diurnal Light-Dark Cycles

Strain	Duration of LD cycle in hours	Threshold in hours	
		Calculated	Obtained in the experiments
Belgorod	24	—	16.6
	17	9.6	9.5
	20	12.6	13.3
	35	27.6	28.5
Sukhumi	24	—	
	17	8.5	8.1
	20	11.5	12.0

the constancy of the critical light period and to its independence on cycle length since the value of the cycle duration is absent in the formula. In such a way, the data obtained may be considered as supporting the correctness of our interpretation of the PhPR mechanism.

According to the two-oscillatory model, the process of perception of the photoperiodic signal consists in the phase entrainment of free oscillations of the *A* and *B* oscillators whose mutual synchronization determines active or nonactive condition of neuroendocrine organs regulating insect development. However, the problem of the photoperiodic oscillators localization in an organism is still far from being solved. Beck & Alexander (17) consider one such oscillator in larvae of *Ostrinia nubilalis* to be situated in the proctodeum and to be represented by endocrine cells secreting a specific hormone —proctodon. The secretion of proctodon begins at darkness and is then repeated each 8 hours. Phase interrelations of the hormone secretion periods with the rhythmic changes of the activity of brain neurosecretory cells may cause photoperiodic reactivation in a long-day or keep up the diapause in a short-day (15).

The main role in photoperiodic induction of diapause in larvae of *Dendrolimus pini* is played, in Tyshchenko's opinion (197), by "generator" nervous cells of the pars intercerebralis. These neurons possess clear daily rhythms of spontaneous electrical activity and are activated at the transition from darkness to light. They seem to be connected by synapses with the neurosecretory cells. Experiments with ligaturing showed that *D. pini,* as well as *O. nubilalis,* seem to have some endocrine organ localized in the abdomen. Cells of this organ are activated after the onset of darkness and secret a hormone that might influence the transmission of photoperiodic information from the generator neurons to the neurosecretory cells. The brain neurons and the cells of the abdomen endocrine organ are assumed to interact like the light and dark oscillators of the PhPR model (p. 236). Analysis of circadian rhythms of electrical activity in the nervous system permits modelling the PhPR curve that coincided closely in its parameters with the real curve in *Dendrolimus pini.*

In conclusion, difficulties in using this hypothesis (as well as other oscillatory ones) to interpret the PhPR physiological mechanism should be stressed: the photoperiodic oscillators still do not have a place in the modern theory of the neurohumoral mechanism of diapause regulation.

LITERATURE CITED

1. Adkisson, P. L. Action of the photoperiod in controlling insect diapause. *Am. Naturalist,* **98,** 357–74 (1964)
2. Adkisson, P. L. Internal clocks and insect diapause. *Science,* **154,** 234–41 (1966)
3. Andrewartha, H. G. Diapause in relation to the ecology of insects. *Biol. Rev.,* **27,** 50–107 (1952)
4. Ankersmit, G. W. On the influence of photoperiod and temperature on the life cycle of some univoltine insects. *Proc. Intern. Symp. Ontog. Insects,* 277–82 (1959)
5. Ankersmit, G. W. Voltinism and its determination in some beetles of cruciferous crops. *Med. Landbouwhg. Wagenigen,* **64,** 1–60 (1964)
6. Ankersmit, G. W., Adkisson, P. L. Photoperiodic responses of certain geographical strains of *Pectinophora gossypiella. J. Insect Physiol.,* **13,** 553–64 (1967)
7. Arbuthnot, K. D. Strains of the European corn borer in the United States. *U. S. Dept. Agr. Bull.,* **869,** 1–30 (1944)
8. Aschoff, J. Zeitgeber der tierischen Tagesperiodik. *Naturwissenschaften,* **41,** 49–56 (1954)
9. Aschoff, J. Exogenous and endogenous components in circadian rhythms. *Cold Spring Harbor Symp. Quant. Biol.,* **25,** 11–28 (1960)
10. Azarjan, A. G. On some peculiarities of intraspecific geographical adaptations in *Dysaphis anthrisci* C. B. *Entomol. Obozenie,* **45,** 500–8 (1966) (In Russian)
11. Barker, R. I. Inhibition of diapause in *Pieris rapae* L. by brief supplementary photophases. *Experientia,* **19,** 185 (1963)
12. Barker, R. I., Cohen, C. F. Light-dark cycles and diapause induction in *Pieris rapae* L. *Entomol. Exptl. Appl.,* **8,** 28–32 (1965)
13. Barry, B. D., Adkisson, P. L. Certain aspects of the genetic factors involved in the control of the larval diapause of the pink bollworm. *Ann. Entomol. Soc. Am.,* **59,** 122–25 (1966)
14. Beck, S. D. Photoperiodic induction of diapause in an insect. *Biol. Bull.,* **122,** 1–12 (1962)
15. Beck, S. D. Time-measurement in insect photoperiodism. *Am. Naturalist,* **98,** 329–46 (1964)
16. Beck, S. D. *Insect photoperiodism,* 1–288. (Academic Press, New York-London, 1968)
17. Beck, S. D., Alexander, N. Proctodone as insect developmental hormone. *Biol. Bull.,* **126,** 185–98 (1964)
18. Beck, S. D., Apple, J. W. Effect of temperature and photoperiod on voltinism of geographical population of the European corn borer, *Pyrausta nubilalis. J. Econ. Entomol.,* **54,** 550–58 (1961)
19. Beck, S. D., Cloutier, E. J., McLeod, D. G. R. Photoperiod and insect development. In *Insect Physiology,* 43–64. (Proc. 23, Biol. Coll. Oregon State Univ., 1963)
20. Beklemishev, W. N. Ecology of the anopheline mosquito. (Medgiz, Moscow, 1944) (In Russian)
21. Beklemishev, W. N., Shipitsina, N. K. Eds. Seasonal events in the life of the anopheline mosquitoes in the Soviet Union. (Moscow, 1957) (In Russian)

22. Belozerov, V. N. Photoperiodic control of seasonal development in ticks. In *Photoperiodic Adaptations in Insects and Acari,* 100–28. (Danilevsky, A. S., Ed., Leningrad State Univ., Leningrad, 1968) (In Russian)
23. Bigelow, R. S. Factors affecting developmental rates and diapause in field crickets. *Evolution,* **16,** 396–406 (1962)
24. Birukow, G. Aktivitäts- und Orientierungsrhythmik beim Kornfäfer (*Calandra granaria* D.). *Z. Tierpsychol.,* **21,** 279–301 (1964)
25. Bobinskaja, S. G. Effect of ecological conditions on the life cycle of *Hadena sordida* Bkh. *Zool. Zhur.,* **45,** 1659–65 (1966) (In Russian)
26. Bodenheimer, F. S. Studies on the ecology of Palestinian *Coleoptera.* Seasonal and diurnal appearance and activity. *Bull. Soc. Entomol. Egypte,* **1–2,** 211–41 (1934)
27. Bonnemaison, L. Contribution a l'etude des facteurs provoquant l'apparition des formes ailees et sexuees chez les *Aphindinae. Ann. Epiphyties,* **2,** 1–380 (1951)
28. Bonnemaison, L. Action de l'alternance de scotophase et de photophases dans un cycle de 24h sur la production des sexupares de *Dysaphis plantaginea* Pass. *Compt. Rend. Acad. Sci.,* **D262,** 2498–2501 (1966)
29. Bonnemaison, L., Missonier, J. Recherches sur le determinisme des formes estivales et hivernales et de la diapause chez le psylle du poirier (*Psylla pyri* L.). *Ann. Epiphyties,* **6,** 417–528 (1955)
30. Brindley, T. A., Dicke, F. F. Significant developments in European corn borer research. *Ann. Rev. Entomol.,* **8,** 155–76 (1963)
31. Brown, F. A., Webb, H. M. Temperature relations of an endogenous daily rhythmicity in the fiddler crab *Uca. Physiol. Zool.,* **21,** 371–81 (1948)
32. Bruce, V. G. Environmental entrainment of circadian rhythms. *Cold Spring Harbor Symp. Quant. Biol.,* **25,** 29–48 (1960)
33. Bünning, E. Zur Kenntnis der endogenen Tagesrhythmik bei Insekten und bei Pflanzen. *Ber. Deut. Botan. Ges.,* **53,** 594–623 (1935)
34. Bünning, E. Die endonome Tagesrhythmik als Grundlage der photoperiodischen Reaktion. *Ber. Deut. Botan. Ges.,* **54,** 590–607 (1936)
35. Bünning, E. *Die physiologische Uhr.* (Springer-Verlag, Berlin, 1958)
36. Bünning, E. Über den Temperatureinfluss auf die endogene Tagesrhythmik, besonders bei *Periplaneta americana. Biol. Zentr.,* **77,** 141–52 (1958)
37. Bünning, E. Joerrens, G. Versuche zur photoperiodischen Diapause-induction bei *Pieris brassicae. Naturwissenschaften,* **46,** 518–19 (1959)
38. Bünning, E., Joerrens, G. Tagesperiodische antagonistische Schwankungen der Blauviolett und Gelbrot-Empfindlichkeit als Grundlage der photoperiodischen Diapause-Induktion bei *Pieris brassicae* L. *Z. Naturforsch.,* **15,** 205–13 (1960)
39. Burov, V. N. Factors determining the dynamics of the number and injury of *Aelia* in virgin regions of northern Kazakhstan. *Entomol. Obozrenie,* **41,** 262–73 (1962) (In Russian)
40. Chernyshov, V. B. Problem of diurnal rhythms in insects. *Zhur. Obshch. Biol.,* **21,** 455–60 (1960) (In Russian)
41. Chernyshov, V. B. Analysis of the diurnal cycle exemplified by the behaviour of *Trogoderma glabrum* Herbst. *Zhur. Obshch. Biol.,* **26,** 577–84 (1965) (In Russian)
42. Cloudsley-Thompson, J. L. Studies in diurnal rhythms, I. Rhythmic behaviour in millipedes. *J. Exptl. Biol.,* **31,** 165–72 (1951)
43. Cloudsley-Thompson, J. L. Studies in diurnal rhythms. III. Photoperiodism in the cockroach *Periplaneta americana* (L.). *Ann. Mag. Nat. Hist., Ser. 12,* **6,** 705–12 (1953)
44. Cloudsley-Thompson, J. L. Diurnal rhythms of activity in terrestrial arthropods. *Nature,* **178,** 215 (1956)
45. Cloudsley-Thompson, J. L. Studies in diurnal rhythms. VIII. The endogenous chronometer in *Gryllus campestris* L. *J. Insect Physiol.,* **2,** 275–80 (1958)
46. Cloudsley-Thompson, J. L. Some aspects of the physiology and behaviour of *Galeodes arabs. Entomol. Exptl. Appl.,* **4,** 257–63 (1961)
47. Danilevsky, A. S. The photoperiodic reaction of insects in conditions

of artificial light. *Dokl. Akad. Nauk S.S.S.R.*, **60**, 481–84 (1948) (In Russian)
48. Danilevsky, A. S. The dependence of the geographical distribution of insects on ecological characteristics of their life cycles. *Entomol. Obozrenie*, **30**, 194–207 (1949) (In Russian)
49. Danilevsky, A. S. The temperature conditions of reactivation of diapausing stages of insects. *Tr. Leningr. Obshchestva Estestvoisp.*, **70**, 90–107 (1950) (In Russian)
50. Danilevsky, A. S. Photoperiodism as a factor in the formation of geographical races of insects. *Entomol. Obozrenie*, **36**, 5–27 (1957) (In Russian)
51. Danilevsky, A. S. *Photoperiodism and seasonal development of insects*, 1–242. (Leningrad State Univ., Leningrad, 1961) (In Russian. English translation of this book was published in 1965 by Oliver & Boyd, London)
52. Danilevsky, A. S., Goryshin, N. I. The relation between temperature and light conditions in regulating diapause in insects. *Tr. Peterhof. Biol. Inst.*, **18**, 147–68 (1960) (In Russian)
53. Danilevsky, A. S., Kuznetsova, I. A. The intraspecific adaptations of insects to the climatic zonation, 5–51. (See Ref. 22)
54. Danilevsky, A. S., Sheldeshova, G. G. The adaptive significance of the photoperiodic and cold reactivation, 80–99. (See Ref. 22)
55. Den Boer, P. J. The ecological significance of activity patterns in the woodlouse *Porcellio scaber* Latr. *Arch. Neerl. Zool.*, **14**, 283–409 (1961)
56. Depner, K. D., Harwood, R. F. Photoperiodic responses of two latitudinally diverse groups of *Anopheles freeborni*. *Ann. Entomol. Soc. Am.*, **59**, 7–11 (1966)
57. Dickson, R. C. Factors governing the induction of diapause in the oriental fruit moth. *Ann. Entomol. Soc. Am.*, **42**, 511–37 (1949)
58. Dobzhansky, T., Epling, C. Taxonomy, geographic distribution and ecology of *Drosophila pseudoobscura* and its relatives. *Carnegie Inst. Wash. Publ.*, 1–554 (1944)
59. Dubynina, T. S. The peculiarities of diapause induction and reactivation in the spider mite *Tetranychus urticae* Koch. *Entomol. Obozrenie*, **44**, 287–92 (1965) (In Russian)
60. Duclaux, M. E. Etudes physiologique sur la graine de ver a soie. *Ann. Chem. Phys., Ser. IV*, **24**, 290–306 (1871)
61. Edwards, D. K. Activity rhythms of depidopterous defoliators. II. *Halisidota argentata* Pack. and *Nepyria phantasmaria* Strr. *Can. J. Zool.*, **42**, 939–58 (1964)
62. Engelmann, W., Shapiro, D. G. Photoperiodic control of the maintenance and termination of larve diapause in *Chironomus tentans*. *Nature*, **207**, 548–49 (1965)
63. Fukaya, M. On the theoretical bases for predicting the occurrence of the rice stem borer in the first generation. *Ber. Ohara Inst. Landwirsch. Forsch.*, **9**, 357–76 (1951)
64. Fukuda, S., Endo, K. Hormonal control of the development of seasonal forms in the butterfly, *Polygonia c-aureum* L. *Proc. Japan Acad.*, **42**, 1082–87 (1966)
65. Geier, P. W. The life history of codling moth *Cydia pomonella* (L.) in the Australian Capital territory. *Australian J. Zool.*, **11**, 323–67 (1963)
66. Geier, P. W. Wintering and spring emergence of codling moth, *Cydia pomonella* L., in Southeastern Australia. *Australian J. Zool.*, **11**, 431–45 (1963)
67. Geyspitz, K. F. The reaction of univoltine *Lepidoptera* to daylength. *Entomol. Obozrenie*, **33**, 17–31 (1953)
68. Geyspitz, K. F. The adaptational significance of the photoperiodic reaction and its role in the ecology of the pine moth (*Dendrolimus pini* L.). *Uch. Zap. Leningr. Gos. Univ.*, **240**, 21–33 (1958) (in Russian)
69. Geyspitz, K. F. The effect of the conditions in which preceding generations were reared on the photoperiodic reaction of geographical forms of the cotton spider mite (*Tetranychus urticae* Koch). *Tr. Petergofsk. Biol. Inst., Leningr. Gos. Univ.*, **18**, 169–77 (1960) (In Russian)
70. Geyspitz, K. F. Photoperiodical and temperature reactions affecting the seasonal development of pine moths *Dendrolimus pini* L. and

D. sibiricus Tschetw. *Entomol. Obozrenie,* **44,** 538–53 (1965) (In Russian)

71. Geyspitz, K. F. Genetic aspects of variation of photoperiodic adaptations, 52–79. (See Ref. 22)
72. Geyspitz, K. F., Zarankina, A. I. Some peculiarities of photoperiodic reaction of *Dasychira pudibunda* L. *Entomol. Obozrenie,* **42,** 29–38 (1963) (In Russian)
73. Goldschmidt, R. Untersuchungen zur Genetik der geographischen Variation. V. Analyse der Ueberwinterungzeit als Anpassungscharakter. *Arch. Entw. Mech. Org.,* **126,** 671–768 (1932)
74. Goldschmidt, R. A note concerning the adaptation of geographic races of *Lymantria dispar* L. to the seasonal cycle in Japan. *Am. Naturalist,* **72,** 385–86 (1938)
75. Goryshin, N. I. The relation between light and temperature factors in the photoperiodic reaction in insects. *Entomol. Obozrenie,* **34,** 9–14 (1955) (In Russian)
76. Goryshin, N. I. Effect of day length on diapause formation in Colorado potato beetle, *Leptinotarsa decemlineata* Say. In *Colorado Potato Beetle,* **2,** 136–49. (Acad. Sci. Publ., USSR, Moscow, 1958)
77. Goryshin, N. I. The ecological analysis of the seasonal cycle of development of the Cotton-boll worm (*Chloridea obsoleta* F.) in the northern areas of its range. *Uch. Zap. Leningrad Gos. Univ.,* **240,** 3–20 (1958) (In Russian)
78. Goryshin, N. I. Light-dark periodicity and photoperiodic reaction of insects. *Entomol. Obozrenie,* **42,** (1963) (In Russian)
79. Goryshin, N. I. The effect of light and temperature rhythms upon the diapause formation in *Lepidoptera. Entomol. Obozrenie,* **43,** 86–93 (1964) (in Russian)
80. Goryshin, N. I., Kozlova, R. N. Thermoperiodism as a factor of insect development. *Zhur. Obshch. Biol.,* **28,** 278–88 (1967) (in Russian)
81. Goryshin, N. I., Tyshchenko, G. F. On the significance of the absolute longevity of the day and night in insect photoperiodic reaction. *Dokl. Akad. Nauk S.S.S.R.,* **171,** 754–57 (1966) (in Russian)
82. Goryshin, N. I., Tyshchenko, V. P. Physiological mechanism of photoperiodic reaction and the problem of endogenous rhythms, 192–269. (See Ref. 22)
83. Götz, B. Der Einfluβ von Tageszeit und Witterung auf Ausschlüpfe, Begattung und Eiblage der Springwurmwicklers *Sparganothis pilleriana* Schiff. *Z. Angew. Entomol.,* **31,** 261–74 (1949)
84. Grison, P. Developments and diapause des chenilles de *Euproctis phaeorrhaea. Compt. Rend. Acad. Sci.,* **225,** 1089–90 (1947)
85. Gunn, D. L. The daily rhythm of activity of the cockroach, *Blatta orientalis.* I. Actograph experiments, especially in relation to light. *J. Exptl. Biol.,* **30,** 525–33 (1940)
86. Hamlin, J. C. Seasonal adaptation of a Northern Hemisphere insect to the Southern Hemisphere. *J. Econ. Entomol.,* **16,** 420–23 (1923)
87. Harker, J. E. The diurnal rhythm of activity of mayfly nymphs. *J. Exptl. Biol.,* **30,** 525–33 (1953)
88. Harker, J. E. Factors controlling the diurnal rhythm of activity in *Periplaneta americana* L. *J. Exptl. Biol.,* **33,** 224–34 (1956)
89. Harker, J. E. Diurnal rhythms in the animal kingdom. *Biol. Rev.,* **33,** I–52 (1956)
90. Harker, J. E. Daily rhythms. *Ann. Rev. Entomol.,* **6,** 131–46 (1961)
91. Harvey, G. T. The occurrence and nature of diapause-free development in the spruce budworm, *Choristoneura fumiferana* Clem. *Can. J. Zool.,* **35,** 549–72 (1957)
92. Harvey, W. R. Metabolic aspects of insect diapause. *Ann. Rev. Entomol.,* **7,** 57–80 (1962)
93. Helle, W. Genetic variability of photoperiodic response in an arrhenotokous mite (*Tetranychus urticae*). *Entomol. Exptl. Appl.,* **11,** 101–13 (1968)
94. Hidaka, T., Aida, S. Day-length as the main factor of the seasonal form determination in *Polygonia c-aureum. Zool. Mag. (Tokyo).* **72,** 77–83 (1963)
95. Hodek, J. Diapause in females of *Pyrrhocoris apterus* L. *Acta Entomol. Bohemoslov.,* **65,** 422–35 (1968)
96. House, H. L. The decreasing occurrence of diapause in the fly *Pseudosarcophaga affinis* through laboratory-reared generations. *Can. J. Zool.,* **45,** 149–53 (1967)

97. Huggens, J. L., Blickenstaff, C. C. Effect of photoperiod on sexual development in the alfalfa weevil. *J. Econ. Entomol.*, **57**, 167–68 (1967)
98. Kalmus, H. Über die Nature des Zeitgedächtnisses des Bienen. *Z. Vergleich. Physiol.*, **20**, 405–19 (1934)
99. Kappus, K. D., Vernard, C. E. The effect of photoperiod and temperature on the induction of diapause in *Aedes trisseriatus* Say. *J. Insect Physiol.*, **13**, 1007–19 (1967)
100. Kirchner, H. Tageszeitliche Aktivitätsperiodik bei Carabiden. *Z. Vergleich. Physiol.*, **48**, 385–99 (1964)
101. Komarova, O. S. The life cycle and conditions of development of the grape berry moth (*Polychrosis botrana* Schiff.). *Zool. Zhur.*, **33**, 102–13 (1954) (In Russian)
102. Kuznetsova, I. A. Factors causing the coming of diapause in *Pectinophora malvella* Hb. *Entomol. Obosrenie*, **41**, 510–15 (1962) (In Russian)
103. Le Beree, J.-R. Contribution a l'etude biologique du criquet migrateur des landes (*Locusta migratoria gallica* Remandiere). *Bull. Biol.*, **87**, 227 (1953)
104. Lees, A. D. Environmental factors controlling the evocation and temination of diapause in the fruit tree red spider mites *Metatetranychus ulmi* Koch. *Ann. Appl. Biol.*, **40**, 449–86 (1953)
105. Lees, A. D. The significance of the light and dark phases in the photoperiodic control of diapause in *Metatetranychus ulmi* Koch. *Ann. Appl. Biol.*, **40**, 487–97 (1953)
106. Lees, A. D. *The physiology of diapause in Arthropods*, 1–151. (Cambridge Univ. Press, Cambridge, Engl., 1955)
107. Lees, A. D. The role of photoperiod and temperature in the determination of parthenogenetic and sexual forms in the aphid *Megoura viciae* Buckton. II. The operation of the "interval timer" in young clones. *J. Insect Physiol.*, **4**, 154–75 (1960)
108. Lees, A. D. Is there a circadian component in the *Megoura* photoperiodic clock? In *Circadian Clocks*, 351–56. (Aschoff, J., Ed., North-Holland Publ. Co., Amsterdam, 1965)
109. Lees, A. D. Photoperiodic timing mechanisms in insects. *Nature*, **210**, 986–89 (1966)
110. Lees, A. D. The control of polymorphism in aphids. *Advan. Insect physiol.*, **3**, 207–77 (1966)
111. Lees, A. D. Photoperiodism in insects. In *Photophysiology*, **IV**, 47–137. (Academic Press, New York-London, 1968)
112. Legay, J. M. Recent advances in silkworm nutrition. *Ann. Rev. Entomol.*, **3**, 75–86 (1958)
113. Lohmann, M. Der Einfluss von Beleuchtungsstärke und Temperatur auf die tagesperiodische Laufaktivität des Mehlkäfers, *Tenebrio molitor* L. *Z. Vergleich. Physiol.*, **49**, 341–89 (1964)
114. MacLeod, E. Experimental induction and elimination of adult diapause and autumnal coloration in *Chrysopa carnea*. *J. Insect Physiol.*, **13**, 1343–49 (1967)
115. MacLeod, D. G. R., Beck, S. D. Photoperiodic termination of diapause in an insect. *Biol Bull.*, **124**, 84–96 (1963)
116. Mansingh, A., Smallman, B. N. Effect of photoperiod on the incidence and physiology of diapause in two saturniids. *J. Insect Physiol.*, **13**, 1147–62 (1967)
117. Marcovitch, S. The migration of the *Aphididae* and the appearance of the sexual forms as affected by the relative length of daily light exposure. *J. Agr. Res.*, **27**, 513–22 (1924)
118. Masaki, S. The local variation in the diapause pattern of the cabbage moth, *Barathra brassicae L. Bull. Fac. Agr. Mie Univ.*, **13**, 29–46 (1956)
119. Masaki, S. Geographic variation of diapause in insects. *Bull. Fac. Agr. Hirosaki Univ.*, **7**, 66–98 (1961)
120. Masaki, S. Adaptation to local climatic conditions in the Emma field cricket. *Kontyu*, **31**, 249–60 (1963)
121. Masaki, S. Geographic variation in the intrinsic incubation period: a physiological cline in the Emma field cricket. *Bull. Fac. Agr. Hirosaki Univ.*, **11**, 59–90 (1965)
122. Masaki, S., Oyama, N. Photoperiodic control of growth and wing-form in *Nemobius yezoensis* Shiraki. *Kontyu*, **31**, 16–26 (1963)
123. Maslennikova, V. A. The regulation of seasonal development in parasitic insects, 129–52. (See Ref. 22)
124. Matinjan, T. K. On peculiarities of

seasonal cycle regulation in geographical forms of the aphid *Brevicorne brassicae* L. *Izvest. Akad. Nauk Armjanskoy S.S.R.*, **17**, 39–45 (1964) (In Russian)

125. Matinjan, T. K. Seasonal cycle adaptations to geographical variations of day-length and temperature in cruciferous butterflies (*Pieris napi* L. and *P. rapae* L.). *Izvest. Akad. Nauk Armjanskoy S.S.R.*, **18**, 81–98 (1965) (In Russian)

126. Mellanby, K. The daily rhythm of activity in the coackroach *Blatta orientalis*. *J. Exptl. Biol.*, **17**, 278–85 (1940)

127. Minis, D. H. Parallel peculiarities in the entrainment of a circadian rhythm and photoperiodic induction in the pink bollworm (*Pectinophora gossypiella*). In *Circadian Clocks*, 333–43. (Aschoff, J., Ed., North-Holland Publ. Co., Amsterdam, 1965)

128. Moriarty, F. The 24-hr rhythm of emergence of *Ephestia kühniella* Zell. from the pupa. *J. Insect Physiol.*, **3**, 357–66 (1959)

129. Müller, H. J. Der Saisondimorphismus bei Zikaden der Gattung *Euscelis* Brulle. *Beitr. Entomol.*, **4**, 1–54 (1954)

130. Müller, H. J. Die Saisonformenbildung von *Araschnia levana*, ein photoperiodisch gesteuerter Diapause-Effekt. *Naturwissenschaften*, **42**, 134–35 (1955)

131. Müller, H. J. Die Wirkung exogener Faktoren auf die zyklische Formenbildung der Insekten, insbesondere der Gattung *Euscelis*. *Zool. Jahrb., Abt., Syst., Ökol. Geogr. Tiere*, **85**, 317–430 (1957)

132. Müller, H. J. Über die Diapause von *Stenocranus minutus* F. *Beitr. Entomol.*, **7**, 203–26 (1957)

133. Müller, H. J. Über den Einfluβ der Photoperiode auf Diapause und Körpergröβe der Delphacide *Stenocranus minutus* Fabr. *Zool. Anz.*, **160**, 295–312 (1958)

134. Müller, H. J. Probleme der Insektendiapause. *Verhandel. Deut. Zool. Ges.*, **29**, 192–222 (1965)

135. Nasu, S. Problems in the ecological types and hibernation of planthoppers. *Nippon Oyo-Dobutsu-Kontyu Jakkai Taikai Simpo*, 3–8 (1960)

136. Naylor, E. Temperature relationships of the locomotor rhythm of *Carcinus*. *J. Exptl. Biol.*, **40**, 669–79 (1963)

137. Nayar, J. K. The pupation rhythm in *Aedes taeniorhynchus*. II. Ontogenetic timing, rate of development, and endogenous diurnal rhythm of pupation. *Ann. Entomol. Soc. Am.* **60**, 946–71 (1967)

138. Neumann, D. Die intraspezifische Variabilät der lunaren und täglichen Schlüpfzeiten von *Clunio marinus*. *Verhandel. Deut. Zool. Ges.*, **29**, 223–33 (1965)

139. Neville, A. C. Daily growth layers in animals and plants. *Biol. Rev.*, **42**, 421–41 (1967)

140. Norris, M. J. The influence of constant and changing photoperiods on imaginal diapause in the red locust (*Nomadacris septemfasciata* Serv.). *J. Insect Physiol.*, **11**, 1105–19 (1965)

141. Norris, K. R. Daily patterns of flight activity of blowflies in the Canberra district as indicated by trap catches. *Australian J. Zool.*, **14**, 835–54 (1966)

142. Nowosielski, J. W., Patton, R. L. Studies on circadian rhythm of the house cricket. *J. Insect Physiol.*, **9**, 401–10 (1963)

143. Odhiambo, T. R. The metabolic effects of the corpus allatum hormone in the male desert locust. II. Spontaneous locomotor activity. *J. Exptl. Biol.*, **45**, 51–63 (1966)

144. Paulpandian, A. A preliminary report on the diurnal rhythm in the locomotor activity of pill-millipede, *Arthrosphaera dalayi* Pocock. *Proc. Indian Acad. Sci., Ser. B*, **62**, 235–41 (1965)

154. Pantyukhov, G. A. On the photoperiodic reaction of *Chilocorus renipustulatus* Scr. *Entomol. Obozrenie*, **47**, 45–50 (1968) (In Russian)

146. Pease, R. W. Factors causing seasonal forms in *Ascia monuste*. *Science*, **137**, 987–88 (1962)

147. Petersen, B. Die geographische Variation einiger Fennoskandischer Lepidopteren. *Zool. Bidr. Uppsala*, **26**, 329–531 (1947)

148. Pittendrigh, C. S. On temperature independence in the clock system controlling emergence time in *Drosophila*. *Proc. Natl. Acad. Sci. U.S.*, **40**, 1018–29 (1954)

149. Pittendrigh, C. S. Circadian rhythms and the circadian organization of living systems. *Cold Spring Har-*

bor Symp. Quant. Biol., **25,** 159–84 (1960)

150. Pittendrigh, C. S. On the mechanism of the entrainment of a circadian rhythm by light cycles. In *Circadian Clocks,* 277–97. (Aschoff, J., Ed., North-Holland Publ. Co., Amsterdam, 1965)

151. Pittendrigh, C. S. The circadian oscillation in *Drosophila pseudoobscura* pupae: a model for the photoperiodic clock. *Z. Pflanzenphysiol.*, **54,** 275–307 (1966)

152. Pittendrigh, C. S., Bruce, V. G. An oscillator model for biological clocks. In *Rhythmic and Synthetic Processes in Growth,* 75–109. (Rudnick, D., Ed., Princeton Univ. Press, New Jersey, 1957)

153. Pittendrigh, C. S., Bruce, V. G. Daily rhythms as coupled oscillator systems; and their relation to thermo- and photoperiodism. In *Photoperiodism and Related Phenomena in Plants and Animals,* 475–505. (Withrow, R. B., Ed., Am. Assoc. Advan. Sci. Publ., Washington, D. C., 1959)

154. Pittendrigh, C. S., Minis, D. H. The entrainment of circadian oscillations by light and their role as photoperiodic clocks. *Am. Naturalist,* **98,** 261–94 (1964)

155. Prebble, M. L. The diapause and related phenomena in *Gilpinia polytoma* Hartig. II. Factors influencing the breaking of diapause. *Can. J. Res., Ser. D,* **19,** 322–46 (1941)

156. Rau, P. Rhythmic periodicity and synchronous flashing in the firefly *Photinus pyralis,* with notes on *Photorus pennsylvanicus. Ecology,* **13,** 7–12 (1932)

157. Razumova, A. P. Variability of photoperiodic reaction in a number of successive generations of spidermites. *Entomol. Obozrenie,* **46,** 268–72 (1967) (In Russian)

158. Remmert, H. Untersuchungen über das tageszeitliche gebundene Schlüpfen von *Pseudosmittia arenaria. Z. Vergleich. Physiol.,* **37,** 338–54 (1955)

159. Rensing, L. Aktivitätsperiodik des Wasserläufers *Velia currens* F. *Z. Vergleich. Physiol.,* **44,** 292–322 (1961)

160. Rensing, L. Ontogenetic timing and circadian rhythms in insects. In *Circadian Clocks,* 406–12. (Aschoff, J., Ed., North-Holland Publ. Co., Amsterdam, 1965)

161. Rensing, L. Die Bedeutung der Hormone bei der Steuerung circadianer Rhythmen. *Zool. Jahrb., Abt. 1,* **71,** 595–606 (1965)

162. Rensing, L. Zur circadian Rhythmik des Sauerstoffverbrauches von *Drosphila. Z. Vergleich. Physiol.,* **53,** 62–83 (1966)

163. Rensing, L., Brunken, W. Zur Frage der Allgemeingültigkeit circadianer Gesetzmäßigkeiten. *Biol. Zentr.,* **86,** 545–65 (1967)

164. Rensing, L., Hardeland, R. Zur Wirkung der circadianen Rhythmik auf die Entwicklung von *Drosophila. J. Insect Physiol.,* **13,** 1547–68 (1967)

165. Ring, R. A. Maternal induction of diapause in the larva of *Lucilia caesar* L. *J. Exptl. Biol.,* **46,** 123–36 (1967)

166. Roberts, S. K. Circadian activity rhythms in cockroaches. I. The free-running rhythm in steady-state. *J. Cellular Comp. Physiol.,* **55,** 99–110 (1960)

167. Roberts, S. K. Circadian activity rhythms in cockroaches. II. Entrainment and phase shifting. *J. Cellular Comp. Physiol.,* **59,** 175–86 (1962)

168. Roth, M. Sur les facteurs conditionnant les horaires de deplacement chez *Aspidomorpha quinquefasciata* Roheman. *Bull. Soc. Entomol. France,* **67,** 50–53 (1962)

169. Russ, K. Der Einfluß der Photoperiodizität auf die Biologie der Apfelwicklers (*Carpocapsa pomonella* L.). *Planzenschutz Ber.,* **36,** 27–92 (1966)

170. Ryan, R. B. Maternal influence on diapause in parasitic insect *Coeloides brunneri* Vier. *J. Insect Physiol.,* **11,** 1331–36 (1965)

171. Sakai, T., Masaki, S. Photoperiod as a factor causing seasonal forms in *Lycaena phlaeas daimio* Seitz. *Kontyu,* **33,** 275–83 (1965)

172. Saringer, G., Deseö, K. V. Effect of photoperiod and temperature on the diapause of the alfalfa weevil (*Hypera variabilis* Herbst.). *Acta Phytopathol. (Budapest),* **1,** 353–64 (1966)

173. Saunders, D. S. Larval diapause induced by a maternally operating photoperiod. *Nature,* **206,** 739–40 (1965)

174. Saunders, D. S. Time measurement in insect photoperiodism: reversal

of a photoperiodic effect by chilling. *Science,* **156,** 1126–27 (1967)

175. Saunders, D. S. Photoperiodism and time measurement in the parasitic wasp, *Nasonia vitripennis. J. Insect Physiol.,* **14,** 433–50 (1968)
176. Schneiderman, H. A. Onset and termination of insect diapause. In *Physiological Triggers,* 46–59. (Bullock, T. H., Ed., Am. Physiol. Soc., Washington, 1957)
177. Schneirla, T. C. Insect behaviour in relation to its setting. In *Insect Physiology,* 685–722. (Roeder, K. D., Ed., Academic Press, New York, 1100 pp., 1953)
178. Shaposhnikov, G. Kh. Changes of host and diapause in aphids in the process of adaptation to the annual cycles on their food plants. *Entomol. Obozrenie,* **38,** 483–504 (1959) (In Russian)
179. Sheldeshova, G. G. Geographical variability of photoperiodic reaction and seasonal development of codling moth, *Laspeyresia pomonella* L. *Tr. Zool. Inst. Akad. Nauk S.S.S.R.,* **36,** 5–25 (1965) (In Russian)
180. Sheldeshova, G. G. Ecological factors determining the distribution of codling moth, *Laspeyresia pomonella* L. in northern and southern hemispheres. *Entomol. Obozrenie,* **46,** 583–601 (1967) (In Russian)
181. Sheldeshova, G. G., Stekolnikov, A. A. On the type change in photoperiodic reaction during the development of *Parthenolecanium corni* Bouche. *Tr. Zool. Inst. Akad. Nauk S.S.S.R.,* **36,** 26–30 (1965)
182. Squire, F. A. Observations on the larval diapause of the pink bollworm *Platyedra gossypiella* Saund. *Bull. Entomol. Res.,* **30,** 475–81 (1939)
183. Squire, F. A. On the nature and origin of the diapause in *Platyedra gossypiella* Saund. *Bull. Entomol. Res.,* **31,** 1–6 (1940)
184. Steinberg, D. M., Kamensky, S. Les premisses oecologiques de la diapause de *Loxostege sticticalis* L. *Bull. Biol. France Belg.,* **70,** 145–83 (1936)
185. Strenzke, K. Die systematische und ökologische Differenzierung der Gattung *Chironomus. Ann. Entomol. Fennici,* **26,** 111–38 (1960)
186. Strübing, H. Zum Diapuseproblem in der Gattung *Stenocranus. Zool. Beitr.,* **9,** 1–119 (1963)
187. Sugonjaev, E. C. On the seasonal-cyclic adaptations of the parasite *Blastothrix confusa* Erd. to its host *Parthenolecanium corni Bouche. Izvest. Akad. Nauk S.S.S.R., Ser. Biol.,* **5,** 754–66 (1963) (In Russian)
188. Sullivan, C. R., Wallace, D. R. Photoperiodism in the development of the European pine sawfly, *Neodiprion sertifer* (Geoff.). *Can. J. Zool.,* **43,** 233–45 (1965)
189. Sullivan, C. R., Wallace, D. R. Interaction of temperature and photoperiod in the induction of prolonged diapause in *Neodiprion sertifer. Can. Entomologist,* **99,** 834–50 (1967)
190. Sweeney, B., Hastings, W. Effects of temperature upon diurnal rhythms. *Cold Spring Harbor Symp. Quant. Biol.,* **25,** 87–104 (1960)
191. Sytshevskaya, V. I. On the daily dynamics of the specific composition of synanthropic flies within a season. *Entomol. Obozrenie,* **41,** 545–53 (1962) (In Russian)
192. Tanaka, J. Studies on hibernation with special reference to photoperiodicity and breeding of the Chinese tussar-silkworm. III. *Nippon Sanshigaku Zasshi,* **19,** 580–90 (1950)
193. Tanaka, J. Studies on hibernation with special reference to photoperiodicity and breeding of the Chinese tussar-silkworm. V. *Nippon Sanshigaku Zasshi,* **20,** 132–38 (1951)
194. Teraguchi, M., Northcote, T. G. Vertical distribution and migration of *Chaoborus flavicans* larvae in Corbett lake, British Columbia. *Limnol. Oceanog.,* **11,** 164–76 (1966)
195. Thiele, H. U. Einflüsse der Photoperiode auf die Diapause von Carabiden. *Z. Angew. Entomol.,* **58,** 143–49 (1966)
196. Tyshchenko, V. P. Two-oscillatory model of the physiological mechanism of insect photoperiodic reaction. *Zhur. Obshch. Biol.,* **27,** 209–22 (1966) (In Russian)
197. Tyshchenko, V. P. The analysis of mechanism of the photoperiodic induction of diapause in caterpillars of the *Dendrolimus pini* L. on the basis of electrophysiologi-

cal study of the nervous system. *Entomol. Obozrenie,* **46,** 551–68 (1967) (In Russian)

198. Vinogradova, E. B. Experimental investigation of ecological factors inducing imaginal diapause in blood-sucking mosquitoes. *Entomol. Obozrenie,* **39,** 327–40 (1960) (In Russian)

199. Wahl, O. Neue Untersuchungen über das Zeitgedächtniss der Bienen. *Z. Vergleich. Physiol.,* **16,** 529–89 (1932)

200. Wallace, D. R., Sullivan, C. R. Geographic variation in photoperiodic reaction of *Neodiprion sertifer* (Geoff.). *Can. J. Zool.,* **44,** 147 (1966)

201. Waloff, N. Observations on larvae of *Ephestia elutella* Hb. during diapause. *Trans. Roy. Entomol. Soc.,* **100,** 147 (1949)

202. Warnecke, H. Vergleichende Untersuchungen zur tagesperiodischen Aktivität von 3 *Geotrupes*-Arten. *Z. Tierpsychol.,* **23,** 513–36 (1966)

203. Webb, H. M. Diurnal variation of response to light in the fiddler crab, *Uca. Physiol. Zool.,* **23,** 316–37 (1950)

204. Weber, F. Zur Tagaktivität von *Carabus*-Arten. *Zool. Anz.,* **175,** 354–60 (1965)

205. Weber, F. Zur tageszeitlichen Aktivitätsverteilung der *Carabus*-Arten. *Zool. Jahrb., Abt. 1,* **72,** 136–56 (1966)

206. Welsh, J. H. The sinus glands and twenty-four hour cyles of retinal pigment migration in crayfish. *J. Exptl. Zool.,* **86,** 35–49 (1941)

207. de Wilde, J. Photoperiodism in insects and mites. *Ann. Rev. Entomol.,* **7,** 1–26 (1962)

208. de Wilde, J., Ferket P. The host-plant as a source of seasonal information. *Med. Rijksfac. Landbouw.,* **32,** 387–92 (1967)

209. Wilkens, M. B. The influence of temperature and temperature changes on biological clocks. In *Circadian Clocks,* 146–63. (Aschoff, J., Ed., North-Holland Publ. Co., Amsterdam, 1965)

210. Wilson, F. Some experiments on the influence of environment upon the forms of *Aphis chloris* Koch. *Trans. Roy. Entomol. Soc.,* **87,** 165–80 (1938)

211. Wobus, U. Der Einfluß der Lichtintensität auf die circadiane Laufaktivität der Schabe *Blaberus craniifer* Burm. *Biol. Zentr.,* **85,** 305–23 (1966)

212. Wobus, U. Der Einfluss der Lichtintensität auf die Resynchronisation der circadianen Laufaktivität der Schabe *Blaberus craniifer* Burm. *Z. Vergleich. Physiol.,* **52,** 276–89 (1966)

213. Wuttke, W. Untersuchungen zur Aktivitätsperiodik bei *Euscorpius carpathicus* L. *Z. Vergleich. Physiol.,* **53,** 405–48 (1966)

214. Zaslavskij, V. A., Bogdanova, T. P. The peculiarities of imaginal diapause in two species of *Chilocorus. Tr. Zool. Inst. Akad. Nauk S.S.S.R.,* **36,** 89–95 (1965) (In Russian)

215. Zolotarev, E. Ch. On some peculiarities of development of Chinese tussar-silkworm in relation of diapause presence in its ontogenesis. *Vestn. Mosk. Univ., serija biol. i pochvovedenija,* **6,** 93–100 (1950) (In Russian)

RECENT ADVANCES IN INSECT PATHOLOGY

By Jaroslav Weiser

Laboratory of Insect Pathology, Institute of Entomology, Academy of Sciences, Praha, Czechoslovakia

Today's situation in insect pathology is characterized by a shift of workers from research laboratories to university chairs. In this way the field gains more attention, but many experienced scientists who formed the basis of research work with knowledge and experience are more and more involved in administration and lose much personal, direct contact. Nevertheless, the increase of published papers is reflected by the fact that the *Journal of Invertebrate Pathology* had to publish, in 1968, an additional volume to accommodate a backlog of manuscripts it had accepted for publication. On the other hand, long-term field projects, diagnostics, taxonomy, and morphology of insect pathogens are not as common pursuits as before, and many centers concentrate on a set of useful laboratory models of infections. With the increased amount of research in biochemistry and other experimental areas of insect pathology, including pathophysiology, the lack of information concerning normal, "classical" insect pathology is more and more resented by certain segments of the field.

Recently, research on virus infections has concentrated on some "modern" groups of newly discovered noninclusion and spindle inclusion viruses, whereas the classical nuclear polyhedroses or granuloses of the fat body are being neglected relatively. An exception to this trend are some reports on viruses with abnormal development such as a nuclear polyhedrosis of *Wiseana cervinata* (23). The arched rods of this virus are of exceptional length and their intimate membranes are most unusual; they shrink during dissolution into spherical sacs and cause the virus rod to bend under pressure into a spiral. After rupture of the membrane the virus rods straighten out into their original shape. Great masses of free rods have been observed in the nucleus beside the polyhedra.

Another unusual nuclear polyhedrosis appears to be the infection of the gut epithelium in the mountain ash sawfly, *Pristiphora geniculata* (75). Here, the elongate, ovoidal inclusion bodies are melted in rosettelike clusters. The virus etiology of this infection has not been demonstrated by isolating the virus particles.

Two species of laboratory-reared insects, *Plodia interpunctella* (4, 5) and *Carpocapsa pomonella* (80), have been found most suitable for studies of the ultrastructure of insects infected with granulosis viruses. In the first case, the deposition of capsule protein around the virus rods and the formation of compound capsules offer some ideas regarding the capsule formation. The codling moth granulosis virus presents stromata of coiled filamentous structures and an important demonstration of the passage of the virus into the gut via the Malpighian tubules, a fact not previously noted.

Special attention has been paid to spindle-inclusion viruses of Lepidoptera and Coleoptera, and new infections by these viruses have been described (83, 93). In some infections aberrant sterile protein threads are formed from transformed polyhedra (92, 93). The fine structure of vagoiaviruses (8, 9, 82) show close relation to poxlike viruses at the early stages of development, during the formation of the oval empty membranes. The fully formed virus particle in its ovoid shape with a corrugated surface and the irregular internal structure compared with the helix of the other viruses differs in many respects from the smallpox virus particle. A much closer resemblance to the poxvirus has been found in particles demonstrated in "refreshed" material from *Camptochironomus tentans,* studied in ultrathin sections 20 years after its first description (90, 93). In contrast to *Vagoiavirus* infections, the nuclei of infected cells are involved, and cushion-shaped virus particles are present in oval inclusions. The virion is covered with a dense smooth membrane containing two sandwiched plates.

Nonoccluded viruses are a new and interesting group. A new iridescent virus (MIV) has been described from black flies (90, 91), and other strains have been studied in mosquitoes (20, 26, 97). Another group of non-inclusion viruses with spherical particles from saturniid moths (56, 67) may be related to the flacherie group of viruses which cause flaccidity of the silkworm. Although the results of cross-infection experiments with three strains showed only a slight relationship, preliminary serological tests show a distinct relationship. Another virus of this group, the cause of lethargy of *Melolontha melolontha* grubs (34), although causing similar symptoms, differs in the size of its particles. Here, as everywhere, the morphological and taxonomic classification are the necessary basis for the differentiation of strains or species. We must require in all these instances a detailed description of their morphology, of their development, their localization in host cells and tissues, and an evaluation of participation of cell components in virus synthesis. Perhaps an obligatory part of the description should be to prepare a differential diagnosis of all similar infections, using cross-infections and serology, as well as other criteria, for differentiation. In this sense, a binomial nomenclature of viruses based on rules used everywhere in microbiology, zoology, and botany, with all possible rules for changes, emendations, synonymizations, and transfers to other groups would be a real benefit for virology.

The polyhedrosis viruses, causing the classical type of virus infections of insects, have been infections appearing only in insects of one group, not able to infect other host groups (vertebrates). This is not so certain in the case of noninclusion viruses where morphological similarities with viruses of plants and vertebrates are quite numerous. This is the case mainly with insect vectors of plant or animal disease agents. The same hesitations may be in recent evaluations of *Mycoplasma*-like organisms (29).

Another way of obtaining information on the complicated relations among the various virus groups is the use of serological tests. At present, sucrose gradient centrifugation methods have been designed for almost all types of insect viruses, for polyhedra as well as for free particles (18, 31, 51, 52, 57). It has been confirmed (18, 19) that serological methods are suitable for differentiation in strains of nuclear polyhedrosis and granulosis viruses in the various hosts, while cytoplasmic polyhedrosis viruses are more closely related. Serological differences have been found also in the tipula iridescent virus and sericesthis iridescent virus complex on the one hand and the mosquito iridescent virus (MIV) (20) on the other hand. The same procedures show the identity or close relation of the known MIV strains in Europe and America. Serology of insect viruses, being still in its initial stage, is as yet unable to solve a number of questions mainly in relation to old taxonomic descriptions, "old" viruses about which information on their serology is missing. Also missing is information on the relation of the viruses to their host tissues, their development in the nucleus or cytoplasm of cells, and the details of their RNA and DNA synthesis. In the future, the resin-inbedded blocks with virus-infected tissues may be used as an analogy for the type specimens in other descriptions.

Light radioautographic studies of protein synthesis in nuclear and cytoplasmic polyhedroses (63, 64, 87, 88) reveal a steady synthesis of labeled protein in virogenic stromata, which increases before massive polyhedra formation. Distinct evidence of infectivity has been demonstrated in the RNA isolated from cytoplasmic polyhedrosis infected tissues (42) which was lost by RNase treatment and not changed after DNase application or use of homotypic serum (89). Evaluation of changes of tissue proteins of gut epithelia with cytoplasmic polyhedrosis by the method of disc-electrophoresis (89) reveals constant zones where changes occur. Here, as in all other cases where material from whole insects and many insects is used for preparation of samples for evaluation, the danger of combination of the studied virus with other, noninclusion, viruses of inapparent infections is critical. A further complication may result from the activation by a foreign polyhedral material (55).

Intensive rearing of infected insects in the laboratory brought further improvement in techniques of insect infection and in evaluation of the results. Hemocyte counts in polyhedrosis-infected *Galleria mellonella* (71) showed a reduction of hemocytes, mainly of spherical and fusiform plasmatocytes, analogous to changes during starvation of caterpillars. In virus-infected animals there was a relative increase in number of spherule cells. A similar situation was found in *Pseudaletia unipuncta* (96) infected with granulosis and nuclear polyhedrosis, at least during the first instar. But from the 8th day onwards, the number of spherule cells decreased in specimens infected with granulosis, while the plasmatocytes regained their original numbers.

The evaluation of virus infection experiments (65) resulted in the preparation of a computing program for this type of work. The most important factors in this type of experiment are the uniformity of infectious material and the absolute health of the host insects. Aging of different types of viruses produces different degrees of deterioration and loss of activity (65,

72) as shown with granulosis viruses. The antiviral effect on nuclear polyhedrosis of sodium hypochloride (84) and formalin (37, 84) was demonstrated by direct action, as well as by incorporation into infected food. Irradiation by X rays is able to destroy free virus particles on leaves (39, 72), but there is no increased susceptibility to infection with polyhedrosis or granulosis viruses in irradiated insects. Decreased metabolic activity in either irradiated or starving insects most probably offers no suitable conditions for virus development. Only in chronic infections with low mortality, such as in the cytoplasmic polyhedrosis of *Calophasia lunula* (12), does irradiation cause higher mortality compared with nonirradiated insects.

Information yet to be obtained on fine structure may locate some recently discovered infectious agents between viruses and rickettsiae. This may be the case with the "rickettsia-like organism" in *Evarthrus alternatus,* a pathogen very similar to the known *Rhabdionvirus* group although it occurs in the cytoplasm of infected cells (78). It differs from known rickettsiae in its internal arrangement, in its special outer membrane and in the absence of dividing stages or developmental spherules in ultrathin sections. These differences are much more striking when the *Evarthrus* pathogen is compared with typical rickettsial organisms in new hosts. What is probably a new group of rickettsiae has been described (22) in cavities in the intestinal cells of saturniid moths. The short rods of this pathogen are covered with thin pili, enlarging their surface. Another case, a rickettsia which was first described as an insect pathogen as early as 25 years ago, *Rickettsiella chironomi* (92, 94) was refixed and cut from old paraffin materials and its rickettsial nature was demonstrated together with the fact that it is a flat bean-shaped organism with a surface grid and a thread-like structure coiled directly under the surface of its body. It divides by lateral splitting and forms a pile of particles in the infected cell plasm. The intestinal rickettsiae of saturniid moths (22) produced higher mortality after peroral infection than did *Rickettsiella melolonthae* in grubs of *Melolontha melolontha* (35, 36) at different temperatures or in combinations with bacteria and protozoa.

Insect bacteriology achieved very satisfactory results in studies of the structure of the most important bacterial toxins active in insects. The comprehensive study of the taxonomy of *Bacillus thuringiensis* strains (1, 6, 47) offered a good base for identifying known and new strains (30, 48, 49). The crystal-like endotoxin (53, 66), when treated with proteases in digestive fluids of *Pieris brassicae,* splits into protein and peptide groups up to $M = 5000$ which retain their toxic activity. According to recent results (53, 66), the substance of the crystal may be a protoxin, transformed by the enzymatic activity of the host into the true toxin (3). The different host specificity of the toxin may reflect the ability of its proteases to split the molecule at certain peptide bonds. A model of this type of toxic substances for further study may be the comparison with velinomycin (2). In another connection, it has been demonstrated that delta endotoxin of *B. thuringiensis* forms a complex with proteins having an alkaline isoelectric pH (25).

The second important constituent of *B. thuringiensis* toxicity, the thermostable exotoxin (68), was identified as a substance with molecular weight

M = 850 and an analogue of the adenosine triphosphoric acid. It contains a phosphorylated allaric acid bond on adenosyl glucose (24). Every rupture of the original molecule brings a loss of its toxicity. Its inhibitory action on the *de novo* synthesis of RNA and the DNA-dependent RNA-polymerase has been studied. The enzymatically dephosphorylated toxin was nontoxic. This change may occur in part during the process of absorption of the substance when ingested and this may explain the lower toxicity *per os* compared with injection (69). An unfavorable reaction (74) on five varieties of *Bacillus thuringiensis* by volatile substances of some plants was revealed during field control with bacteria. In two other instances, *B. thuringiensis* was used under special conditions. In one case, the two wax moths (14) were brought under control by addition of *B. thuringiensis* to wax in bee hives. In the second case, predatory moths in colonies of the lack scale were treated with bacterial dusts with good results (60).

Contrary to all expectations, the mass production (as was the research) of *Bacillus popilliae*—the cause of milky disease in the Japanese beetle, *Popillia japonica*—seems to be slow in progress. The repeatedly demonstrated increased activity and germination in heated spore preparations (41) raises the question whether such heat activation is a phenomenon likely to occur in the field. Disc-electrophoresis investigation of *Popillia* hemolymph, after *Bacillus popilliae* infection (11), revealed the disappearance of a single protein line, containing in the normal hemolymph lipid and carbohydrate moieties (11). In the group of bacteria participating in septicemias of insects, *Pseudomonas aeruginosa* is the bacterium best known to produce a virulent exotoxin. Studies of this pathogen (50, 58) demonstrated that the proteinase forming the toxin is composed of 4 fractions, which could be separated on DEAE cellulose. Their toxicity for *Galleria mellonella* was found to vary with the LD_{50} ranging from 36×10^{-3} to 53×10^{-3} toxic units.

As with bacteria, the main interest in research on entomophytic fungi has been concentrated on the production of toxic substances. After long efforts by many workers, a definite mycotoxin has been demonstrated in the white muscardine fungus *Beauveria bassiana* (95) by injecting filtrates of cultures, and a similar activity has been demonstrated in the case of *Aspergillus* in experimental applications (7, 70). *Beauveria bassiana* was also used for the experimental larviciding of mosquitoes (17), and in three mosquito genera the larvae suffered infections and high mortality when spores of the fungus were dusted on the water surface. Another important fungus, *Metarrhizium anisopliae,* was studied as one of the few factors reducing the populations of *Schistocerca gregaria* (86), mainly from the standpoint of its infectivity and development in the insect host. Aschersonia-fungi were used in large-scale field applications (21).

Production of toxic substances has also been studied in the host-specific group Entomophthoraceae. Experiments revealed that *Entomophthora coronata* and *E. apiculata* (99) released, in liquid media, substances toxic for *Galleria mellonella* and *Musca domestica.* These toxins may be identical with alkaline proteinases studied in 10 different strains of *Entomophthora* in submerse cultures (40). The quantity of toxins produced depends on the strain and on the composition of the medium. The toxins act only after injection (or released by fungi in the body of the host) into the insect hemolymph. In addition to proteases it has been possible to demonstrate also chitinases and lipases in some *Entomophthora* species (27, 28) and changes in fatty acids (81). Although the physiology of fungi of this group may be essentially the same, it is interesting that on artificial media some strains are more active in toxin production than are others. The most active in all tests was *Entomophthora coronata* which is also the most tolerant species in experiments with humidity, temperature, and integrated use with insecticides (98) when compared with *E. apiculata* or *E. virulenta*. These differences speak for its isolation in the separate genus *Conidiobolus*.

Infections of insects by protozoa differ in many respects from other infections. The pathogens are not able to grow on artificial media and their manipulation is possible only with use of natural or secondary hosts. A considerable number of papers have been published describing new species. The newness of these species may be valid, but the general presentation of descriptions in the last years has been rather poor, with morphological data based only on dry, Giemsa-stained smears. There has not been a study of smears and sections and, hence, an elucidation of the manner and tissue succession of the infection in the host and a true differential diagnosis has not been presented. The variability of spores is largely underestimated and schematic drawings or photographs of dry smears do not allow a reliable comparison. Of course, "author's licence" gives the liberty to describe the pathogens in their own way. But they risk a synonymization of their species in the future, and they present taxonomists with a very difficult task if they have to make determinations in such groups as Microsporidia of mosquitoes and of some Lepidoptera.

Some descriptions bring perspectives of possible use of the new pathogens in natural control 15, 16). It is definitely interesting to know that the Colorado potato beetle acquired, during its expansion in the Ukraine, a microsporidian (54) although information will be helpful as to which organs are infected and what is the pathogenicity of the infection. Another important pest, the sawfly *Pristiphora erichsoni,* is infected with a microsporidian, *Thelohania pristiphorae.* The protozoan was laboratory transmitted to the tent caterpillars, *Malacosoma disstria* and *M. americanum* (76), and it is not clear which host is the primary seat of the infection. But the spores do not lose infectivity for the other hosts when produced in one, and both hosts can be used for the production of infectious material. Another pathogen, useful mainly for laboratory experiments, is *Nosema plodiae,* the cause of an infection in *Plodia interpunctella* (44). A third model object, in some cases complicating evaluations of physiological experiments, is the schizogregarine *Sporomyxa tenebrionis* (33) of the yellow mealworm, *Tenebrio molitor.*

The infectivity as well as the development of the infection in host insects is dependent on the rearing temperature. This was demonstrated again with *Nosema necatrix* (59) whose reproduction in host tissues is optional at a temperature of 30° C. In the same way in infections caused by *Octosporea muscaedomesticae* (45, 46) the optimum was between 25 and 30° C and the development slowed down at temperatures lower or higher than the optimum. This optimum may depend also on the usual temperature optimum of the regular host. It has been demonstrated in *Nosema bombycis* that this microsporidian is able to develop in primary cell cultures of mammalian and chicken embryo (38) when the cultures were kept at normal temperatures of insect tissues, suitable also for vertebrate cells, e.g., 28° C. At a temperature of 37° C there was no growth of microsporidia in the cultures. Experiments with simultaneous infections of grubs with *Nosema melolonthae* or *Adelina melolonthae* brought similar results, with increased mortality in the 20 to 25° C temperature range, compared with chronic long-lasting infections at lower temperatures (34).

Studies of microsporidian infections in black flies were concentrated on identification by normal histology (61, 62) of cellular reactions to parasites in tissues, hypertrophy of infected cells, and on ultrastructures of the most important parasites (85). The different tissues may have some infections which may replace each other under certain conditions. *Haplosporidium simulii* in *Simulium venustum* in Pennsylvania may be a such case (10). The infection deserves further morphological evaluation of the parasite for an easier taxonomic diagnosis.

Very important are some recent observations concerning the relations of microsporidia to the sex of their host. In some hosts not only hypertrophies of infected cells but also direct selection of host sex may occur. In an invertebrate host, the amphipod, *Gammarus duebeni* that occurs on the Helgoland shore, the microsporidian *Octosporea effeminans* infects only the ovary of females (13), males are not infected. The microsporidian invades the egg follicles and changes the infected eggs to female. The progeny of infected amphipods are only females. In rare intersexes not only the ovary, but also the testicular tissues are infected. The infected area produces less and less male animals and is "sterilized" by the pathogen. A hormonal change in eggs due to the activity of the pathogen may be anticipated.

Another similar situation in mosquitoes was described in *Anopheles maculipennis* and *A. quadrimaculatus* infected with *Thelohania legeri* (32). In nature, usually 1 per cent of the female mosquitoes is infected with this microsporidian in middle Europe, about the same as in Florida. When *A. quadrimaculatus* was colonized in the laboratory the infected females produced in the progeny from their eggs a 50 per cent infection with white cysts obvious in the larvae. This group of larvae died before pupation. The other, apparently healthy larvae, did not show any cysts, but during pupation the microsporidian developed in oenocytes, lymphocytes, and in the ovary, and also infected the egg follicles. Only females hatch from the apparently healthy larvae. The infection produces the same situation in the next generation and colonies can be maintained only with healthy males from other colonies. The elimination of males in infected colonies may be a good model for biological control in isolated populations of the mosquitoes. In other infections in mosquitoes or other Diptera (e.g., in *Chaoborus*), this selective activity does not occur and is not so strict (73).

Teratology of insects was a common object of study of insect physiologists in connection with disorders in nutrition, hormonal troubles, regeneration, etc. In some instances, abnormal individuals were produced by larval insects infected with some protozoan diseases where the supplies of fat in the fat body were absorbed by Microsporidia or Schizogregarina. A comprehensive study was devoted to the teratology of *Tenebrio molitor* without any connection with infections, and a series of changes in the formation of wings, pupae, and adult bodies were described (77, 100). Studies of experimental responses of insects to different pathogens is a promising expansion of insect pathology.

In ending this brief evaluation we would like to emphasize that all approaches to new methods and techniques in biochemistry, immunology, or other fields of research on insect pathogens, must recognize and use as basis the "old," traditional morphological methods first in order to maintain continuity with past descriptions and studies, and second, so as not to lose results of intensive laboratory work with new methods because of an unidentified contamination or mixed infection or confusion in the determination of the pathogen. The journals in which the main publications in this field appear have an obligation to care for adequate taxonomic descriptions or identifications. More and more, it is apparent how the lack of cooperation between taxonomists, ecologists, and physiologists on insect diseases wastes a great amount of important material, information and knowledge. Both sides may profit a great deal from a close cooperation.

LITERATURE CITED

1. Angus, T. A. The use of *Bacillus thuringiensis* as a microbial insecticide. *World Rev. Pest Control,* **7,** 11–26 (1968)
2. Angus, T. A. Similarity of effect of valinomycin and *Bacillus thuringiensis* parasporal protein in larvae of *Bombyx mori. J. Invertebrate Pathol.,* **11,** 145–46 (1968)
3. Angus, T. A., Norris, J. R. A comparison of the toxicity of some varieties of *Bacillus thuringiensis* Berliner for silkworm larvae. *J. Invertebrate Pathol.,* **11,** 289–95 (1968)
4. Arnott, H. J., Smith, K. M. An ultrastructural study of the development of a granulosis virus in the cells of the moth *Plodia interpunctella* (Hbn.). *J. Ultrastruct. Res.,* **21,** 251–68 (1968)
5. Arnott, H. J., Smith, K. M. Ultrastructure and formation of abnormal capsules in a granulosis virus of the moth *Plodia interpunctella* (Hbn.) *J. Ultrastruct. Res.,* **22,** 136–58 (1968)
6. deBarjac, H., Bonnefoi, A. A classification of strains of *Bacillus thuringiensis* Berliner with a key to their differentiation. *J. Invertebrate Pathol.,* **11,** 335–47 (1968)
7. Beard, R. L. Mycotoxin: a cause of death in milkweed bugs. *J. Invertebrate Pathol.,* **10,** 438–39 (1968)
8. Bergoin, M., Devauchelle, G., Duthoit, J. L., Vago, C. Étude au microscope electronique des inclusions de la virose à fuseaux des Coleopteres. *Compt. Rend. Acad. Sci.,* **266,** 2126–28 (1968)
9. Bergoin, M., Devauchelle, G., Vago, C. Observations au microscope électronique sur le développement du virus de la maladie à fuseaux du coleoptere *Melolontha melolontha* L. *Compt. Rend. Acad. Sci.,* **267,** 382–85 (1968)
10. Beaudoin, R. L., Wills, W. *Haplo-*

sporidium simulii sp.n. parasitic in larvae of *Simulium venustum* Say. *J. Invert. Pathol.*, **10,** 374–78 (1968)

11. Bennett, G. A., Shotwell, O. L., Hall, H. H., Hearn, W. R. Hemolymph proteins of healthy and diseased larvae of japanese beetle, *Popillia japonica*. *J. Invertebrate Pathol.*, **11,** 112–18 (1968)
12. Bucher, G. B., Harris, P. Virus diseases and their interaction with food stress in *Calophasia lunula*. *J. Invertebrate Pathol.*, **10,** 235-44 (1968)
13. Bulnheim, H. P., Vávra, J. Infections by the microsporidian *Octosporea effeminans* sp.n. and its sex determining influence in the amphipod *Gammarus duebeni*. *J. Parasitol.*, **54,** 241–48 (1968)
14. Burges, H. D., Bailey, L. Control of the greater and lesser wax moth [*Galleria mellonella* and *Achoria grisella*] with *Bacillus thuringiensis*. *J. Invertebrate Pathol.*, **11,** 184–95 (1968)
15. Cali, A., Briggs, J. D. The biology and life history of *Nosema tracheophila* sp.n. [Protozoa:Cnidospora:Microsporidia] found in *Coccinella septempunctata* L. *J. Invertebrate Pathol.*, **9,** 515–22 (1968)
16. Chapman, H. C., Kellen, R. *Plistophora caecorum* sp.n., a microsporidian of *Culiseta inornata* [Dipt., Culicidae] from Louisiana. *J. Invertebrate Pathol.*, **9,** 500–2 (1967)
17. Clark, T. B., Kellen, W. R., Fukuda, T., Lindegren, J. E. Field and laboratory studies on the pathogenicity of the fungus *Beauveria bassiana* to three genera of mosquitoes. *J. Invertebrate Pathol.*, **11,** 1–7 (1968)
18. Cunningham, J. C. Serological and morphological identification of some nuclear polyhedrosis and granulosis viruses. *J. Invertebrate Pathol.*, **11,** 132–41 (1968)
19. Cunningham, J. C., Longworth, J. F. The identification of some cytoplasmic polyhedrosis viruses. *J. Invertebrate Pathol.*, **11,** 196–202 (1968)
20. Cunningham, J. C., Tinsley, T. W. A serological comparison of some iridescent non-occluded insect viruses. *J. Gen. Virol.*, **3,** 1–8 (1968)
21. Dolidze, M. D., Timofeeva, T. V. Use of the Aschersonia-fungus in control of the citrus whitefly, *Dialeurodes citri* Riley et How. *Sbornik po karantinu rastenij*, **19,** 131–46 (1967)
22. Entwistle, P. F., Robertson, J. S. Rickettsiae pathogenic to two saturniid moths. *J. Invertebrate Pathol.*, **10,** 345–54 (1968)
23. Entwistle, P. F., Robertson, J. S. An unusual nuclear polyhedrosis virus from larvae of a hepialid moth. *J. Invertebrate Pathol.*, **11,** 487–95 (1968)
24. Farkaš, J., Šebesta, K., Horská, K., Samek, Z., Dolejš, L., Šorm, F. The structure of exotoxin from *Bacillus thuringiensis* v. *gelechiae*. *Coll. Czechoslov. Chem. Commun.*, Sect. 34, 1118–20 (1969)
25. Faust, R. M. *In vitro* chemical reaction of the δ endotoxin produced by *Bacillus thuringiensis var. dendrolimus* with other proteins *J. Invertebrate Pathol.*, **11,** 465–75 (1968)
26. Faust, R. M., Dougherty, E. M., Adams, J. R. Nucleic acid in the blue-green and orange mosquito iridescent viruses (MIV) isolated from larvae of *Aedes taeniorhynchus*. *J. Invertebrate Pathol.*, **10,** 160 (1968)
27. Gabriel, B. P. Enzymatic activities of some entomophthorous fungi. *J. Invertebrate Pathol.*, **11,** 70–81 (1968)
28. Gabriel, B. P. Histochemical study of the insect cuticle infected by the fungus *Entomophthora coronata*. *J. Invertebrate Pathol.*, **11,** 82–89 (1968)
29. Granados, R. R., Maramorosch, K., Shikata, E. *Mycoplasma:* suspected etiologic agent of corn stunt. *Proc. Natl. Acad. Sci. U.S.*, **60,** 841–44 (1968)
30. Gukasjan, A. B. Taxonomy of the crystalliferous germ *Bacillus insectus* Guk. *Kristallonos. mikroorganismy i perspektivy ich ispolzovanija v lesnom chozjajstve.* Nauka, Moskva, 5–20 (1967)
31. Hayashi, Y., Bird, F. T. The use of sucrose gradient in the isolation of cytoplasmic polyhedrosis virus particles. *J. Invertebrate Pathol.*, **11,** 40–44 (1968)
32. Hazard, E. I., Weiser, J. Spores of *Thelohania* in adult female *Anopheles:* Development and transovarial transmission, and redescription of *T. legeri* Hesse and

T. obesa Kudo. *J. Protozool.*, **15**(4), 817–23 (1968)

33. Huger, A. Epizootics among populations of the yellow mealworm, *Tenebrio molitor*. *J. Invertebrate Pathol.*, **9**, 572–74 (1967)
34. Hurpin, B. La léthargie, nouvelle virose des larves de *Melolontha melolontha* L. [Col., Scarab.]. *Entomophaga*, **12**, 311–15 (1967)
35. Hurpin, B. The influence of temperature and larval stage on certain diseases of *Melolontha melolontha*. *J. Invertebrate Pathol.*, **10**, 252-62 (1968)
36. Hurpin, B., Robert, P. Experiments on simultaneous infections of the common cockchafer, *Melolontha melolontha*. *J. Invertebrate Pathol.*, **11**, 203–13 (1968)
37. Ignoffo, C. M., Garcia, C. Formalin inactivation of nuclear polyhedrosis virus. *J. Invertebrate Pathol.*, **10**, 430–32 (1968)
38. Ishihara, R. Growth of *Nosema bombycis* in primary cell cultures of mammalian and chicken embryo. *J. Invertebrate Pathol.*, **11**, 328–30 (1968)
39. Jafri, R. H. The susceptibility of irradiated larvae of *Galleria mellonella* to VND-virus. *J. Invertebrate Pathol.*, **10**, 355–60 (1968)
40. Jönsson, A. G. Protease production by species of entomophthora. *Appl. Microbiol.*, **16**, 450–57 (1968)
41. Julian, St. G., Hall, H. H. Infection of *Popillia japonica* larvae with heat-activated spores of *Bacillus popilliae*. *J. Invertebrate Pathol.*, **10**, 48–53 (1968)
42. Kawase, S., Hukuhara, T. Amino acid composition of the *Tipula* Iridescent Virus. *J. Invertebrate Pathol.*, **9**, 273–74 (1967)
43. Kawase, S., Miyajima, S. Infectious ribonucleic acid of the cytoplasmic polyhedrosis virus of the silkworm, *Bombyx mori*. *J. Invertebrate Pathol.*, **11**, 63–69 (1968)
44. Kellen, W. R., Lindegren, J. E. Biology of *Nosema plodiae* sp.n., a microsporidian pathogen of the Indian meal moth, *Plodia interpunctella* (Hbn.) *J. Invertebrate Pathol.*, **11**, 104–11 (1968)
45. Kramer, J. P. An octosporeosis of the black blowfly, *Phormia regina*: Incidence rates of host and parasite. *Z. Parasitenk.*, **30**, 33–39 (1968)
46. Kramer, J. P. An octosporeosis of the black blowfly, *Phormia regina*: Effect of temperature on the longevity of diseased adults. *Texas Repts. Biol. Med.*, **26**, 199–204 (1968)
47. Krieg, A. Ueber das Vorkommen verschiedener Varietäten von *Bacillus thuringiensis* in Deutschland. *Zentr. Bakteriol. Parasitenk. Abt. I, Orig.*, **207**, 83–90 (1968)
48. Krieg, A. Neues über *Bacillus thuringiensis* und seine Anwendung. *Mitt. Biol. Bundesanstalt Land-Forstwirtsch*, **125**, 106 pp. (1967)
49. Krieg, A., deBarjac, H., Bonnefoi, A. A new serotype of *Bacillus thuringiensis* isolated in Germany: *Bacillus thuringiensis* var. *darmstadtiensis*. *J. Invertebrate Pathol.*, **10**, 428–29 (1968)
50. Kučera, M., Lysenko, O. The mechanism of pathogenicity of *Pseudomonas aeruginosa*, V. Isolation and properties of the proteinases toxic to larvae of the greater wax moth *Galleria mellonella* L. *Folia Microbiol.*, **13**, 288–94 (1968)
51. Kurstak, E., Cote, J. R. Morphologie du virus de la densonucléose purifié par centrifugation en gradient. *Rev. Can. Biol.*, **27**, 83–87 (1968)
52. Kurstak, E., Goring, I., Garzon, S., Coté, J. R. Étude de la densonucléose de *Galleria mellonella* L. par les techniques de fluorescence. *Naturaliste can.*, **95**, 773–83 (1968)
53. Lecadet, M. M., Martouret, D. Enzymatic hydrolysis of the crystals of *Bacillus thuringiensis* by the proteases of *Pieris brassicae*. I. Preparation and fractionation of the lysates. *J. Invertebrate Pathol.*, **9**, 310–21 (1967); II. Toxicity of the different fractions of the hydrolysate for larvae of *Pieris brassicae*. *J. Invertebrate Pathol.*, **9**, 322–28 (1967)
54. Lipa, J. J. *Nosema leptinotarsae* sp.n., a microsporidian parasite of the Colorado potato beetle, *Leptinotarsa decemlineata* [Say]. *J. Invertebrate Pathol.*, **10**, 111–15 (1968)
55. Longworth, J. F., Cunningham, J. C. The activation of occult nuclear polyhedrosis by foreign nuclear polyhedrosis. *J. Invertebrate Pathol.*, **10**, 361–67 (1968)
56. Longworth, J. F., Harrap, K. A. A nonoccluded virus isolated from

four saturniid species. *J. Invertebrate Pathol.*, **10**, 139–45 (1968)

57. Longworth, J. F., Tinsley, T. W., Barwise, A. H., Walker, I. O. Purification of a non-occluded virus of *Galleria mellonella*. *J. Gen. Virol.*, **3**, 167–74 (1968)
58. Lysenko, O., Kučera, M. The mechanism of pathogenicity of *Pseudomonas aeruginosa*, VI. The toxicity of proteinases for larvae of the greater wax moth, *Galleria mellonella* L. *Folia Microbiol.*, **13**, 295–99 (1968)
59. Maddox, J. V. Generation time of the microsporidian *Nosema necatrix* in larvae of the armyworm, *Pseudaletia unipuncta*. *J. Invertebrate Pathol.*, **11**, 90–96 (1968)
60. Malhotra, C. P., Choudhary, S. G. Control of *Eublema amabilis* Moore and *Holcocera pulverea* Meyr, predators of the Lac Insect *Kerria lacca* [Kerr] by *Bacillus thuringiensis* Berl. *J. Invertebrate Pathol.*, **11**, 429–39 (1968)
61. Maurand, J. *Plistophora simulii* (Lutz, Splendore 1904) microsporidie parasite des larves de *Simulium:* cycle, ultrastructure, ses rapports avec *Thelohania bracteata* (Strickl. 1913). *Bull. Soc. Zool. France*, **91**, 621–30 (1968)
62. Maurand, J., Manier, J. F. Actions histopathologiques comparées de parasites coelomiques des larves de Simulies. *Ann. Parasitol.*, **43**, 79–85 (1968)
63. Morris, O. N. Metabolic changes in diseased insects. II. Radioautographic studies on DNA and RNA synthesis in nuclear polyhedrosis and cytoplasmic polyhedrosis virus infections. *J. Invertebrate Pathol.*, **11**, 475–86 (1968)
64. Morris, O. N. Metabolic changes in diseased insects, I. Autoradiographic studies of DNA synthesis in normal and in polyhedrosis virus infected Lepidoptera. *J. Invertebrate Pathol.*, **10**, 28–38 (1968)
65. Paschke, J. D., Lowe, R. E., Giese, R. L. Bioassay of the nucleopolyhedrosis and granulosis viruses of *Trichoplusia ni*. *J. Invertebrate Pathol.*, **10**, 327–34 (1968)
66. Pendleton, I. R. Toxic subunits of the crystal of *Bacillus thuringiensis*. *J. Appl. Bacteriol.*, **31**, 208–12 (1968)
67. Peters, D., Staal, G. B. Viruslike particles from *Hyalophora cecropia* larvae succumbing from an intestinal disease. *J. Invertebrate Pathol.*, **11**, 330–34 (1968)
68. Šebesta, K., Horská, K., Vaňková, J. Isolation and properties of the insecticide exotoxin of *Bacillus thuringiensis* var. *gelechiae*. *Collection Czechoslov. Chem. Commun.*, **34**, 891–900 (1969) ; Inhibition of *de novo* RNA synthesis by the insecticide exotoxin of *Bacillus thuringiensis* var. *gelechiae*. *Collection Czechoslov. Chem. Commun.*, **34**, 1786–91 (1969)
69. Šebesta, K., Horská, K., Vaňková, J. Mechanism of action of the insecticide exotoxin of *Bac. gellechiae:* Inhibition of de novo RNA—synthesis. *5th Meeting of Fed. Europ. Biochem. Soc., Prague*, 250 (1968)
70. Shah, V. K., Matsumura, F., Knight, S. G. Fungal toxins against larvae of the yellow fever mosquito, *Aedes aegypti*. *J. Invertebrate Pathol.*, **11**, 146–48 (1968)
71. Shapiro, M. Pathogenic changes in the blood of the greater wax moth, *Galleria mellonella*, during the course of nucleopolyhedrosis and starvation, II. Differential hemocyte count. *J. Invertebrate Pathol.*, **10**, 230–34 (1968)
72. Sidor, C. The granulosis of *Pygaera anastomosis* L. caterpillars. *Zbornik za prirod. nauke*, **34**, 105–19 (1968)
73. Sikorowski, P. P., Madison, C. H. Host-parasite relationships of *Thelohania corethrae* from *Chaoborus stictopus*. *J. Invertebrate Pathol.*, **11**, 390–97 (1968)
74. Smirnoff, W. A. Effect of volatile substances released by foliage of various plants on the entomopathogenic *Bacillus cereus* group. *J. Invertebrate Pathol.*, **11**, 513–15 (1968)
75. Smirnoff, W. A. A nuclear polyhedrosis of the mountain ash sawfly, *Pristophora geniculata*. *J. Invertebrate Pathol.*, **10**, 436–38 (1968)
76. Smirnoff, W. A. Adaptation of the microsporidian *Thelohania pristiphorae* to the tent caterpillars *Malacosoma disstria* and *M. americanum*. *J. Invertebrate Pathol.*, **11**, 321–25 (1968)
77. Steinhaus, E. A., Zeikus, R. D. Teratology of the beetle *Tenebrio molitor*. I. Gross morphology of cer-

tain abnormality types. *J. Invertebrate Pathol.,* **10,** 190-210 (1968); III. Ultrastructural alterations in the flight musculature of the pupal winged adult. *J. Invertebrate Pathol.,* **12,** 40-52 (1968)

78. Sutter, G. R., Kirk, V. M. Rickettsia-like particles in fat-body cells of carabid beetles. *J. Invertebrate Pathol.,* **10,** 445–49 (1968)
79. Tanada, Y., Chang, G. Y. Resistance of the Alfalfa caterpillar, *Colias eurytheme,* at high temperature to a cytoplasmic polyhedrosis virus and thermal inactivation point of the virus. *J. Invertebrate Pathol.,* **10,** 79–83 (1968)
80. Tanada, Y., Leutenegger, R. Histopathology of a granulosis virus disease of the codling moth, *Carpocapsa pomonella. J. Invertebrate Pathol.,* **10,** 39–47 (1968)
81. Tyrrell, D. The fatty acid composition of some entomophthoraceae, II. The occurrence of branched chain fatty acids in *Conidiobolus denaesporus* Drechsl. *Lipids,* **3,** 368–72 (1968)
82. Vago, C., Bergoin, M. Viruses of invertebrates. *Advan. Virus Res.,* **13,** 247–303 (1968)
83. Vago, C., Monsarrat, P., Duthoit, J.-L., Amargier, A., Meynadier, G., Waerebeke, D. Nouvelle virose a fuseaux observée chez un Lucanide de Madagascar. *Compt. Rend. Acad. Sci.,* **266,** 1621.3 (1968)
84. Vail, P. V., Henneberry, T. J., Kishaba, A. N., Arakawa, K. Y. Sodium hypochlorite and formalin as antiviral agents against nuclear polyhedrosis virus in larvae of the cabbage looper. *J. Invertebrate* Pathol., **10,** 84–93 (1968)
85. Vávra, J. Ultrastructural features of *Caudospora simulii* Weiser. *Folia Parasitol.,* **15,** 1–9 (1968)
86. Veen, K. H. Recherches sur la maladie due a *Metarrhizium anisopliae* chez le criquet pélerin. *Mededel. Landbouwhogeschool Wageningen,* **68** (5), 1–77 (1968)
87. Watanabe, H. Site of viral RNA synthesis within the midgut cells of the silkworm *Bombyx mori* infected with cytoplasmic polyhedrosis virus. *J. Invertebrate Pathol.,* **9,** 480–87 (1967)
88. Watanabe, H. Light radioautographic study of protein synthesis in the midgut epithelium of the silkworm, *Bombyx mori,* infected with a cytoplasmic polyhedrosis virus. *J. Invertebrate Pathol.,* **11,** 310–15 (1968)
89. Watanabe, H. Disc electrophoretic patterns of larval midgut proteins in normal and virus-infected silkworms, *Bombyx mori L. Appl. Entomol. Zool.,* **3,** 74–80 (1968)
90. Weiser, J. *Nemoci hmyzu (Insect Diseases Manual).* (Academia, Praha, 554 pp., 1966)
91. Weiser, J. Iridescent virus from the blackfly *Simulium ornatum* Meig. in Czechoslovakia. *J. Invertebrate Pathol.,* **12,** 36–39 (1968)
92. Weiser, J. *An Atlas of Insect Diseases.* (Academia, Praha, 300 pp., 1969)
93. Weiser, J., Žižka, Z. Elektronenmikroskopische Untersuchung alter Virus-Materialien. *Mikroskopie,* **22,** 336–40 (1968)
94. Weiser, J., Žižka, Z. Electron microscope studies of *Rickettsiella chironomi* in the midge *Camptochironomus tentans. J. Invertebrate Pathol.,* **12,** 222–30 (1968)
95. West, E. J., Briggs, J. D. *In vitro* toxin production by the fungus *Beauveria bassiana* and bioassay in greater wax moth larvae. *J. Econ. Entomol.,* **61,** 684–87 (1968)
96. Wittig, G. Phagocytosis by blood cells in healthy and diseased caterpillars, III. Some observations concerning virus inclusion bodies. *J. Invertebrate Pathol.,* **10,** 211–29 (1968)
97. Woodard, D. B., Chapman, H. C. Laboratory studies with the mosquito iridescent virus [MIV]. *J. Invertebrate Pathol.,* **11,** 296–301 (1968)
98. Yendol, W. G. Factors affecting germination of *Entomophthora* conidia. *J. Invertebrate Pathol.,* **10,** 116–21 (1968)
99. Yendol, W. G., Miller, E. M., Behnke, C. N. Toxic substances from entomophthoraceous fungi. *J. Invertebrate Pathol.,* **10,** 313–19 (1968)
100. Zeikus, R. D., Steinhaus, E. A. Teratology of the beetle *Tenebrio molitor.* II. The development and gross description of the pupal winged adult. *J. Invertebrate Pathol.,* **11,** 8–24 (1968)

MODE OF ACTION OF PYRETHROIDS, NICOTINOIDS, AND ROTENOIDS

BY IZURU YAMAMOTO[1]

Department of Agricultural Chemistry, Tokyo University of Agriculture, Setagaya, Tokyo, Japan

Rotenone, pyrethrins, and nicotine were reviewed by LaForge & Markwood (37) in the *Annual Review of Biochemistry* in 1938. Since then great progress has been made in understanding their chemistry and mode of action (82). These three natural insecticides occupy special positions in insect toxicology, because they interact with different sites of action, which again differ from the sites of action of organophosphorus and carbamate insecticides. In brief, pyrethrins act on the nerve axon, nicotine on the synapse of insect central nervous system, and rotenone on the respiratory chain of tissues including the nerve and the muscle.

PYRETHROIDS

Within this *Annual Review* series alone, a variety of descriptions have appeared on the chemistry and mode of action of pyrethroids including the synergists (9, 35, 40, 45, 64, 80). Pyrethrins cause a very rapid paralytic action in insects, even when given in small dosages, but the effect is usually temporary. The dosage required to kill insects is much higher and a longer time is also needed. Both the paralytic and killing effects can be potentiated several times when mixed with synergists. Pyrethrins also show very low toxicity to mammals when applied dermally or orally. These unique characteristics are closely associated with the interaction of pyrethrins with the nervous system and the detoxication system. Metcalf (44) and Brown (4) summarized in detail the symptomatology. Although the nervous system as the site of action seems indisputable, the exact part of the system (central versus peripheral) has been a matter of controversy (35). Pyrethrins act on the sensory nerve endings causing temporary paralysis, and further poisoning effects may occur from localized action along the nerve. However, the central effect of pyrethrins is also evident; first they stimulate the nerve cells and nerve fibers to discharge repetitively and later they paralyze them.

The study of the mechanism of action of pyrethroids on the nerve at the cellular level has been improved greatly with the aid of an intracellular microelectrode inserted into the giant axon of the cockroach, permitting the

[1] Grateful acknowledgment is made to Professor Ryo Yamamoto, Professor John E. Casida, Dr. Toshio Narahashi, Dr. Jun-ichi Fukami, Dr. Toshio Fujita, and Dr. Michihiko Sakai for their advice, and Mrs. Jane Clarkin for assistance in the preparation of manuscript.

analysis of the excitation phenomena in terms of physicochemistry. Narahashi (49, 50, 53, 57) made a major contribution in this field. Allethrin exerts three major actions on nerve, namely, repetitive discharge, increase in negative after-potential, and conduction block. With low concentrations of allethrin (say, 3×10^{-7} g/ml) the negative after-potential is induced in the poisoned axon of the cockroach. This increase is regarded as one of the causes of repetitive excitation which initiates abnormal excitation and convulsion of insects. Upon repetitive stimuli the negative after-potentials are built up remarkably. With high concentrations (say, 10^{-6} g/ml) the negative after-potential is first increased and then decreased, and a conduction block follows in a short period of time.

The mechanism whereby the negative after-potential is increased by allethrin appears to be different from that by DDT. In the DDT-poisoned axon, the potassium activation and the sodium inactivation mechanisms are inhibited, thereby producing a large negative after-potential (54, 55, 57). In contrast, it is suggested that an accumulation of some substance inside or outside the nerve membrane is responsible for the large negative after-potential caused by allethrin (50). The substance is neither potassium nor sodium ion, but perhaps chloride ion or some metabolic substance produced by the excitation. The nerve paralysis caused by allethrin may be due to the block of both sodium and potassium conductances of the nerve membrane (51). The voltage clamp study applied to lobster axons supports this conclusion (53). The situation rather resembles that of cocaine and procaine, but is in contrast to that of tetrodotoxin which inhibits only the sodium-activation mechanism (56). Narahashi (52) suggested that the allethrin molecule directly plugs the sodium and potassium channels themselves at their gates or that the allethrin molecule penetrates the nerve membrane at interchannel regions, thereby affecting the channels through intermolecular forces. Thus, the action of pyrethrins is primarily a physicochemical process on the nerve membrane rather than biochemical interactions. Pyrethrins do not affect muscle respiration (16) and cholinesterase (25), and the *in vitro* inhibition of cytochrome oxidase (48) seems to be an artifact. It is interesting that a neurotoxin (not a pyrethroid metabolite) is induced by pyrethrins or other stress and released into the haemolymph of the cockroach (2), as observed with DDT poisoning (74), seeming secondarily responsible for further stimulation and eventual paralysis of the nerve. Such neurotoxin as released by DDT poisoning appeared to be an attractive subject to be studied but the chemical identity of the substance is not clear, an aromatic amine containing an ester group being suggested (27); the same might be conceived with neurotoxin from pyrethrins.

The structure-activity relationship suggests the nature of the receptor and mechanism of action at the receptor. The appearance of a series of new pyrethroids, alkenylbenzyl esters (12, 33), furylmethyl esters (11, 34), and imidomethyl esters (32) having increased insecticidal activity opened a new era of pyrethroid insecticides and it is a remarkable feature that all active pyrethroids contain a planar or pseudoplanar ring system in the alcohol moiety having a hydroxyl function not coplanar with the ring. The planar ring might fit the planar part of the receptor. As for the chrysanthemumic acid moiety, there are four stereoisomers, but the natural acid has the *d-trans* configuration and this is the most insecticidal among isomers when esterified with the alcohols. Among eight possible isomers of allethrin, *d-trans-d*-isomer(*d-trans*-chrysanthemumoyl *d*-allethrolone), having the same configuration as natural pyrethrin I, is most active, thus being provided with the most suitable arrangement of indispensable moieties to fit on the receptor. By analyzing the variation of the insecticidal activity depending on the configuration changes (10, 82), an understanding was made by postulating that (*a*) the absolute configuration of C_1, on which the alcohol moiety connects through ester linkage, is the most important factor determining the arrangement of *gem*-dimethyl grouping relative to the alcohol portion, and (*b*) the contribution of the isobutenyl side chain to the intrinsic toxicity is minor.

Matsui & Kitahara (42) prepared a shortened vinylog of chrysanthemumic acid, 2,2,3,3-tetramethylcyclopropanecarboxylic acid, the esters of which were highly active. The *gem*-dimethyl grouping is essential for insecticidal activity and the presence of two *gem*-dimethyl groupings may serve to increase the chance to fit the hypothetical receptor. As described later, the isobutenyl side chain of the chrysanthemumic acid is the site of detoxication. The higher insecticidal and the increased mammalian toxicity, as well as good knock-down activity of the ester might be due partly to the lack of this site. The significance of the cyclopropane skeleton is conceived as a supporter for the *gem*-dimethyl. Berteau, Casida & Narahashi (1) examined an ester of an open-chain carbamic acid bearing the *gem*-dimethyl groupings. Despite the rather drastic changes in chemical structure and potency involved (about 1/40 of allethrin to house flies, alone or synergized), the ester is synergized with piperonyl butoxide and blocks nerve excitability of the giant axon of the crayfish in the same manner as allethrin does. They also challenged the unessentiality of the ester linkage. The allethrin type compound having a methylene instead of an oxygen of the ester group acts on the nerve in the same manner as allethrin and is synergized with piperonyl butoxide. The compound, a ketone, is not expected to be broken at the center of molecule. Thus, chrysanthemumoylation of the nerve component by pyrethroids as the mechanism of action in the analogous manner to phosphorylation or carbamoylation has proved to be improbable. The nature of the receptor and the manner in which it is affected by pyrethroids, however, remains to be explored. Matsumura & Hayashi (43) tried to study the binding of the radioactive insecticides, phthalthrin, nicotine, and several chlorinated hydrocarbons with the nerve components of rat brain and of axonic and ganglionic portion of cockroach nerve cord. Though the specific distribution pattern was observed for each of the insecticides, and attention was given to the separation of DDT-nerve component complex of macromolecular type by gel filtration and electrophoresis, whether this binding represents one controlling the conductance change of the nerve or merely an unspecific binding remains to be solved.

The most recent advance in our understanding of pyrethroids is the elucidation of the pathway of degradation. The critical history of the earlier investigation was well written by O'Brien (62). Early studies on pyrethroid metabolism in insects led to the suggestion that hydrolysis of the ester linkage is the major detoxication mechanism. This was based upon the scant production of unspecified acid from incubation of pyrethrins *in vitro* with the acetone powder of cockroach or house fly [Chamberlain (6)] and upon the incomplete identification of chrysanthemumic acid and the ketoalcohols from metabolism of ^{14}C-pyrethrins *in vivo* by cockroach [Zeid et al. (91)] (unfortunately, the sample was contaminated with much-labeled nonpyrethroids). Later work with ^{14}C-labeled allethrin and pyrethrum constituents failed to support this hypothesis, but neither could they positively deny the hypothesis, in spite of the investigator's claims. In the most complete study of this type, Chang & Kearns (7) interpreted their results obtained with the metabolism of randomly labeled pyrethrin I and cinerin I in house flies, as indicating that hydrolysis of the ester linkage is not a major detoxication mechanism, but that the detoxication process is apparently initiated on the keto-alcohol moiety while the acid moiety and the ester linkage are still intact. Three out of five metabolites plus chrysanthemumic acid gave positive Denigés color reaction on saponification, thus being assumed to be chrysanthemumic acid ester. However, the specificity of the test has not been established for various chrysanthemumic acid derivatives. Whether the other two including the major metabolite were esters or not was not known. Only a little chrysanthemumic acid, never exceeding 2.6 per cent of the applied dose was produced after good metabolism of the pyrethrins. Trace production of the acid or no production whatsoever [Hopkins & Robbins (28); Bridges (3)] cannot necessarily deny the possibility of hydrolysis, however, if followed by the further degradation or conjugation of the free acid. Some of the metabolites bound with body constituents and were not extracted with polar solvent like ethanol [Winteringham et al. (79)]. It was also noted in reviewing the paper that the polar metabolite(s) chromatographing like keto-alcohol, using the paper chromatographic techniques developed by Winteringham (78, 79), were often assumed or cited to be the keto-alcohol. In fact, the Rf value of the keto-alcohol was almost 0 or 1 depending on the solvent systems, and the low resolution did not allow any reliable identification. The *in vitro* metabolism of the acid-labeled allethrin by house flies gave the radioactive metabolite with the same Rf value as allethrolone [Hopkins & Robbins (28)]; it was certainly not allethrolone. Thus, the previous studies could not give evidences for the claimed pathway of metabolism.

Recently Yamamoto and Casida (83, 85) first succeeded in defining the metabolic pathway of pyrethroids in insect. First they prepared the acid moiety-labeled pyrethrin I, allethrin, dimethrin, and phthalthrin, stereochemically pure and of high specific activity, from *d-trans*-chrysanthemumic acid -1-^{14}C, and the alcohol-moiety labeled allethrin and phthalthrin. Problems of limitations in earlier preparations of radio-labeled pyrethroids were reviewed by Nishizawa & Casida (60) and Yamamoto & Casida (84). Secondly, the *in vitro* metabolizing system from resistant house fly abdomen homogenate fortified with NADPH was provided, which effectively metabolized all pyrethroids tested. The details of the system were studied by Tsukamoto & Casida (76). With this metabolizing system and thin layer chromatographic procedures to resolve closely related metabolites or degradation products, an experimental approach to give unequivocal evidence for the intact ester nature of all metabolites was devised. Allethrin-^{14}C was converted to more than 10 metabolites and the same products were detected on thin-layer radioautograms when the acid-labeled and alcohol-labeled compounds were used as substrates, indicating that each metabolite has both the acid and the alcohol moieties, and thus ester. One major metabolite representing about half of the total radioactivity was characterized as allethrin-ω_t-oic acid or O-demethylallethrin II as it gave allethrin II by methylation. Pyrethrin I, dimethrin, and phthalthrin-^{14}C also gave the same type of pyrethroid-oic acids as evidenced by degradation and synthesis studies. Among other metabolites, two were characterized as allethrin-ω_t-ol and allethrin-ω_t-al. It is noticeable that all the metabolites investigated have the intact allethrolone moiety as well as the intact ester linkage, and there is very little, if any, modification of the allethrolone moiety of the metabolite mixture as a whole. This is in sharp contrast to the prediction made by earlier investigators (3, 6, 28, 47, 91). Therefore, the major pathway is that one of the methyl groups in the isobutenyl side chain on a cyclopropane carboxylic acid moiety is hydroxylated and further converted, via an aldehyde, to a carboxylic compound. The metabolism of allethrin in living house flies also gave

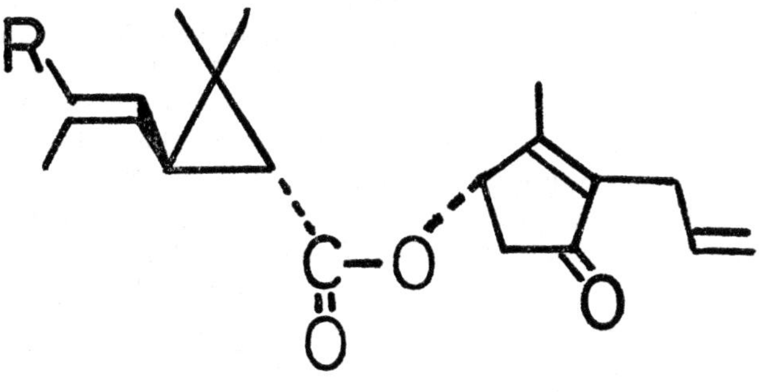

FIG. A. Metabolites of allethrin (R = H_3C—).
R = HOH_2C— allethrin-ω_t-ol
OHC— allethrin-ω_t-al
HOOC— allethrin-ω_t-oic acid

the same type of the metabolites without any indication of hydrolysis, but the hydroxylated allethrin is a major metabolite, which appears also as its conjugate. Though other metabolites of polar nature are formed probably by further oxidation of the acid and alcohol moieties, their amounts are not comparable to the major ones. As a whole, no hydrolysis occurs, no chrysanthemumic acid is produced, and there is little, if any, modification of the alcohol moiety. This metabolism leading to allethrin-oic acid or the conjugate of the alcohol intermediate involves a detoxication, because allethrin is more than 30 times more toxic than allethrin-oic acid when injected into house flies. Such hydroxylation as an initial step of the metabolism is a very common phenomenon in the biological system, which is regarded as a function of microsomal cell fraction. For mammals, pyrethroids are low in toxicity, but the intravenous toxicity of allethrin, pyrethrin, and phthalthrin is very high; for example, a lethal dose of pyrethrins for a dog is 6 to 8 mg/kg (58). Therefore, it is suggested that pyrethroids are potent in intrinsic toxicity and the remarkable selectivity is due to the metabolic differences in mammals and insects. When barthrin and dimethrin are fed to rabbits, they give chrysanthemumic acid and the corresponding alcohols or the acids derived from the alcohols in the urine (41), though the major part of the metabolite is not known. ^{14}C-Phthalthrin (labeled on the ring in the alcohol moiety) orally administered to rats is metabolized rapidly and the alcohol moiety (N-hydroxymethyl-3,4,5,6-tetrahydrophthalimide) is further metabolized. The major metabolite excreted is probably 3-hydroxycyclohexane-1,2-dicarboximide, as investigated by Miyamoto et al. (46). These data indicate that hydrolysis may take part in the case of mammalian metabolism.

ROTENOIDS

Rotenone has decisive lethal action on insects and fishes. It depresses respiration as well as heart beat in insects; the paralysis occurs from the initial stage of intoxication and these symptoms differ greatly from those caused by pyrethrins, nicotine, DDT, and organophosphorus insecticides. Detailed insect toxicological data were reviewed by Metcalf (44), Brown (4) and also appeared in the *Handbook of Toxicology* (58). On the primary action of rotenone, Tischler (75) suggested the inhibition of the ability to utilize oxygen in insects, and Harvey & Brown (26) gave additional evidence for this hypothesis. Fukami and his colleagues accumulated biochemical data on the mode of action of rotenone (15, 16). Their studies were partly reviewed in the *Annual Review of Entomology* by Winteringham & Lewis (80) and by Roan & Hopkins (64), and in more complete form by O'Brien (61, 62). The primary action of rotenone was shown to be the inhibition of cell respiratory metabolism in insect tissues, where rotenone acts first on the nerve, then on the muscle. The biochemical lesion was considered to be responsible for the block of nerve conduction (19, 89), though Tischler (75) described no specific action on either peripheral or central nerves. The important finding was that rotenone inhibits the respiration of NADH-linked substrate on isolated mitochondria, but leaves that of succinate unaffected and is not primarily concerned with oxidative phosphorylation (13, 16, 21, 38, 39). Though Fukami & Tomizawa (21) made the first finding that the *in vitro* oxidation of L-glutamate in the mitochondrial fraction of insect muscle is inhibited by rotenone, they diagnosed the result as the glutamic dehydrogenase inhibition. This is, in fact, the glutamate-supported respiration of mitochondria and so care should be taken to read their papers or the citations. Lindahl & Öberg (38, 39), while studying the inhibition of the oxygen uptake of gill filament by rotenone, first indicated while using rat liver mitochondria that rotenone blocks a link in the electron transport of the respiratory chain, which is situated at the diaphorase level. The same conclusion was obtained with American cockroach muscle (17). Inhibitors which interfere with specific electron transfer reaction, such as rotenone, Amytal (sodium amobarbital) and antimycin A, provide an important tool for elucidating this vital biochemical process. Therefore, many related investigations followed or preceded this finding on rotenone. The reviews were made in part by Pullman & Schatz (63) and by Klingenberg (36), but the present view on rotenone action will be outlined here. Though rotenone and Amytal selectively inhibit the oxidation of NADH-linked substrate, the sites of their action in the respiratory chain have been variously debated as to whether they act between NADH and NADH-dehydrogenase or the oxygen side of the dehydrogenase. These controversies were reviewed in the paper by Horgan, Singer & Casida (30). Their critical studies and the availability of rotenone-6a-^{14}C (2.36 m/mmole) (59) made it possible to define the site of action as the oxygen side of the NADH-dehydrogenase (30). Rotenone is tightly bound not only at the specific site in the dehydrogenase segment of the respiratory chain to cause the inhibition, but also at the other sites in submitochondrial particles. Bovine serum albumin was used to minimize the unspecific binding and the specific binding was shown to be reversible in contrast to the previous assumption (13). Rotenone binds without any modification of the structure, but the binding does not occur on the dehydrogenase or on soluble NADH-coenzyme Q reductase in its natural condition. It is interesting that piericidin A, Amytal, and some rotenoids including dihydrorotenone and 6aβ,12aβ-rotenolone compete at the same specific site with rotenone according to their inhibitory power, but not to the concentrations. Piericidin A is a more specific and more tightly bound inhibitor than rotenone or Amytal (29). However, identity of the respiratory chain component binding the above inhibitors remains undetermined.

These investigations revealed the primary biochemical lesion by rotenone, but as it is common to insects, fishes, and mammals, the selective toxic action to them awaits further studies. Santi & Tóth (68), using the solubilized rotenone, showed its rather high degree of mammalian toxicity (LD_{50} mg/kg, rat): 0.2, i.v.; 1.6, i.p; 60, or. The practical lack of poisoning to man and the low toxic data often reported (44, 58) were ascribed by them to the extremely low solubility in water or to the instability in the intestinal juice. Papaverine-like myolytic effect of rotenone was also reported (67). The cause of the selective toxicity at the biochemical level was considered from various aspects. The inhibitory effect of rotenone on mitochondria was overcome by adding vitamin K_3 which activated a bypass of the rotenone-sensitive site (13), and similar antagonism was observed in isolated guinea pig atria as well as in the intact animal (68). Such effect of K_3, however, was not observed in mitochondria of roach muscle and midgut (18). Detoxication mechanism is usually the main cause of physiological selective toxicity of insecticides. However, no report had been available for the metabolic pathway of rotenone, until Fukami, Yamamoto & Casida (22), using rotenone-6a-^{14}C (59), studied the *in vitro* metabolism using homogenates of mice or rat livers or house fly abdomens fortified with NADPH and the *in vivo* metabolism in mice and house flies. Hydroxylation occurs to give 6aβ,12aβ-rotenolone (A), 6aβ,12aα-rotenolone (B), 8′-hydroxyrotenone (C), 6′,7′-dihydro-6′,7′-dihydroxyrotenone (F), two rotenolones of each of the last mentioned two compounds (D,E and G,H), obtained as the ether-soluble metabolites. The characterization was made by Yamamoto et al. (86). The five major metabolites, A,C,D,F, and G always occur but the minor ones, B,E, and H are produced more in the fly system. Their production is the function of microsomal mixed function oxidase system. More polar metabolites not recovered in ether and probably the conjugates, are present in larger amounts in the mixture of microsome and the soluble fractions of liver than in that from the house fly abdomen. The function of the soluble fraction and the probable presence of various kinds of natural inhibitors in insect tissues might be involved in such metabolic differences (20, 76).

The structure-biological activity relationships contributed to the examination of the validity of the proposed biochemical mechanism of action and the essential molecular features for the activity. It should be mentioned here that although the chemistry of rotenoids has a long history, the stereochemical aspects of some key compounds, including rotenone, have only recently been established (8). Rotenone is 6aβ,12aβ-compound, rotenolone I is a mixture of 6aβ,12aβ- and 6aα,12aα-compounds, and rotenolone II is a mixture of 6aβ,12aα- and 6aα,12aβ-compounds, all of them having 5′β-side chain on the E ring. Fukami, Nakatsugawa & Narahashi (19) found that the *in vitro* inhibition of the glutamate oxidase system correlated approximately to the block of nerve conduction and insecticidal activity, and Burgos & Redfearn (5) pursued the structural feature pertinent to the inhibitory potency. As summarized by Yamamoto (82), the A,B,C, and D rings with *cis*-6aβ,12aβ fusion of B/C rings as in rotenone is essential, though not sufficient, for the inhibitory potency. Slight modifications at B and C rings, as far as they do not affect the conformation of B/C fusion, and the introduction of hydroxyl group at 11 position are permissible. Various types of modifications of the E ring occur without serious loss of potency. Two

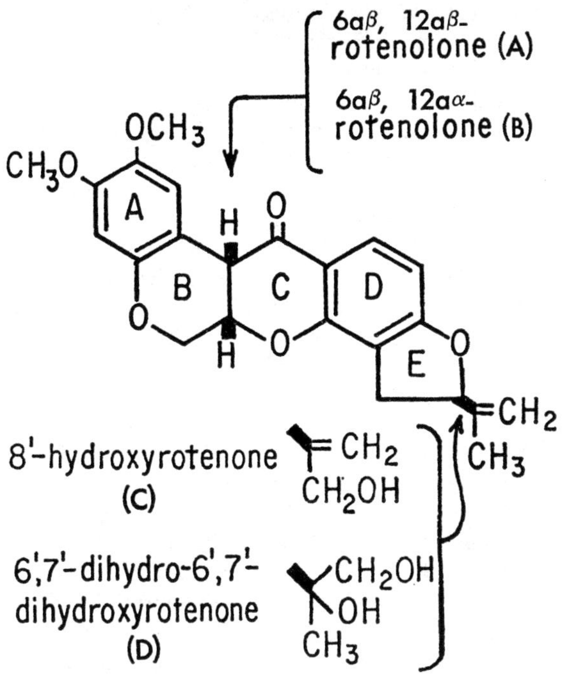

FIG. B. Metabolites of rotenone.
Combination of C and A: metabolite D
of C and B: metabolite E
of F and A: metabolite G
of F and B: metabolite H

anomalies to the essentiality of *cis*-6aβ,12aβ fusion of B/C rings were noted: high potency of acetylrotenone (19) and the $NaBH_4$ reduction product of epirotenone (15). An *ad hoc* interpretation is that acetylrotenone might be incorrectly named, probably one reduced at the enol double bond, and epirotenone might epimerize in part to rotenone type under alkaline conditions. Due to the many factors involved in toxicity, high inhibitory potency does not necessarily give high toxicity, but there has been no case of toxicity without respiratory inhibition. It was noted that rotenone inhibition does not develop immediately, but shows a pronounced time lag, which is minimized by preincubation with NADH (5, 30). Such an effect should be considered for structure-activity correlation, as it differs among different rotenoids.

NICOTINOIDS

Action of nicotinoids on insects was summarized by Brown (4), Metcalf (44), and in the *Handbook of Toxicology* (58). Recently, O'Brien (62) reviewed the action of nicotine based on the general background of neuropharmacology of insects and vertebrates. The transmission of the insect central nervous system is cholinergic, which is the target of nicotine and anticholinesterase, but is protected from the penetration of the ionic substance by the so-called ion-impermeable barrier (62). This makes nicotine, an ionizable substance (roughly 90 per cent ionized), inefficient to intoxicate most insect species except aphids. On the other hand, the neuromuscular junction which is cholinergic and not protected from the ionic poison in mammals, is not cholinergic in insects and is insensitive to nicotine and anticholinesterase. This makeup of the nervous system and the structure-insecticidal activity correlation (31, 72, 87) among nicotinoids prompted Yamamoto (81, 87) to propose that nicotine mimics acetylcholine by interacting with the receptor at the synapse of the insect central nervous system and this is a result of the binding of nicotine on the anionic site at the postsynaptic membrane with the cationized nitrogen of the pyrrolidine moiety. Guthrie's group (23, 24, 69, 70) has investigated the mechanism of insensitivity to nicotine among some insects. Each of these insects has developed its own manner of resistance to nicotine by rapid detoxication, rapid excretion, or by avoiding its uptake. Though so far only cotinine has been characterized among the several metabolites in insects, many metabolites were identified from the metabolism by mammalian systems (77). The hydroxylation at the three vicinal positions of the pyrrolidyl nitrogen which is followed by appropriate reactions may explain the pathway to most metabolites. Yang & Guthrie (90) compared various physiological parameters in nicotine-insensitive tobacco hornworm and in nicotine-sensitive silkworm. The penetration rate of ^{14}C-nicotine into the isolated nerve cord was higher at higher pH, indicating the preferential penetration of the free base form and was higher in the silkworm than in the hornworm, suggesting a lower ion-impermeable barrier in the silkworm. However, nicotine was shown to have no interaction with insect nerve homogenate or with haemolymph. When the mixture of ^{14}C-nicotine and the nerve homogenate was centrifuged in the subcellular fractions, most of the nicotine was present in the soluble protein fraction in both insect preparations. Matsumura & Hayashi (43) also observed that nicotine did not show a particularly high affinity toward particulate nerve components of roach or rat brain. The weak interaction of nicotine with the receptor probably does not allow the demonstration of the insecticide-nerve component complex in these cases. More elaborated experimental procedures are needed for the demonstration.

Yamamoto et al. (71, 88) studied the interaction of nicotinoids with house fly head cholinesterase, which is experimentally feasible by observing the enzyme inhibition. The concentration required for the inhibition is quite high and the inhibition is not the mechanism of action per se. The significant correlation among the inhibition, the toxicity to house flies, and the structural features, however, as well as the competitive nature of the inhibition were shown; these indicate some similarity between the receptor site and the active center of cholinesterase for combining the nicotinoid molecules. 3-Pyridylmethylamine moiety with fairly high basicity is the preferential structural feature for both activities. Such amines can penetrate the ion barrier as the free base form, but once penetrated, they can interact with the receptor as the ionized form. Therefore, if we assume the percentages of the ionized form and the free base form as parameters of intrinsic toxicity and penetrability, respectively, the apparent maximum toxicity is given to the nicotinoids having the same pKa value as the insect body pH. Dihydronicotyrine (pKa, 7.45 at 25° C) almost idealized the requirement, giving the highest toxicity. The other extreme are nicotyrine (in almost free form) and nicotine mono-methiodide (100 per cent ionized). The effect on fly nerve activity also correlated with the proposed structural requirement. Though the active center of the cholinesterase is only a receptor model of first approximation, its detailed physicochemical analysis might reflect the mechanism of interaction actually happening on the receptor. Fujita, Yamamoto & Nakajima (14) have tried such an analysis by using the substituent constants relating electronic and hydrophobic natures. Most biologically active substances are stereospecific for the actions. However, natural *l*-nicotine and its antipode have the same toxicity to house flies, American cockroaches, and rice weevils, while *l*-isomer is about two times more toxic than *d*-isomer to rice stem borers and mealy plum aphids (73). The difference is more pronounced in *Aphis rumicis* (44). Soeda & Yamamoto (73), by reviewing various biological activities of both isomers of nicotine and nornicotine, assumed that the interaction with the receptor is the same for both isomers, but one may be more metabolized or excreted than the other in some cases, causing the difference in toxicity.

Nereistoxin, isolated from a marine annelid, *Lumbrineris heteropoda,* is 4-N,N-dimethylamino-1,2-dithiolane and paralyzes insects, particularly lepidopterous larvae, without excitation. It blocks synaptic transmission without any depolarization by interfering with the postsynaptic acetylcholine receptor in the central nervous system of insects (65). Detailed citation and the discussion of this are beyond the scope of this review, but the substance is interesting because of its unique mode of action as it relates to acetylcholine and nicotine, the similarity of the structural feature to the above two compounds, and the finding of a practical insecticide, cartap [1,3-bis(carbamolythio)-2-(N,N-dimethylamino)propane hydrochloride]. This latter compound was synthetically developed from the nereistoxin (66).

CONCLUDING REMARKS

Modern insecticides would seem to have expelled the old-fashioned natural insecticides. The interest in nicotine, rotenone, and pyrethrins, however, has recently revived. Several things have stimulated further investigation of their mode of action. Natural insecticides persist for a shorter length of time; this is an important goal for synthetic pesticides and is very likely the reason for the rare occurrences of insecticide resistance to natural insecticides. They are also selective in their toxic action. In order to find nonpersistent and selectively toxic insecticides knowledge of the mechanism of biochemical and physicochemical instability and the physiological barriers in biological systems for insecticides must be increased.

There is at least one known insecticide which attacks a certain site of action, causing the death of the insect; therefore, we can expect others that attack the same site. This requires a greater depth of understanding in our search to define the site of action. Nicotine, rotenone, and pyrethrins interact with different sites in completely different ways. However, our understanding of the nerve axon and synapse and the respiratory chain component which provide the receptor site of these insecticides is still limited. The unique biological activities and the complex structural (electronic, hydrophobic, and conformational) features of these insecticides furnish an excellent tool for elucidating these vital points of life. Having these tools, we should then further rely on the sophisticated experimental approaches available from various segments of biology, chemistry, and physics to solve these problems. The recent finding of new insecticidally active compounds, particularly in the pyrethroid group, further stimulates us to be involved with greater vigor in the aforementioned investigations.

LITERATURE CITED

1. Berteau, P. E., Casida, J. E., Narahashi, T. Pyrethroid-like biological activity of compounds lacking cyclopropane and ester groupings. *Science,* **161,** 1151–53 (1968)
2. Blum, M. S., Kearns, C. W. Temperature and the action of pyrethrum in the American cockroach. *J. Econ. Entomol.,* **49,** 862–65 (1956)
3. Bridges, P. M. Absorption and metabolism of ^{14}C-allethrin by the adult housefly, *Musca domestica* L. *Biochem. J.,* **66,** 316–20 (1957)
4. Brown, A. W. A. Chemical injuries. In *Insect Pathology,* **I,** Chap. 3, 65–131. (Steinhaus, E. A., Ed., Academic Press, New York, London, 661 pp., 1963)
5. Burgos, J., Redfearn, E. R. The inhibition of mitochondrial-reduced nicotinamide-adenine dinucleotide oxidation by rotenoids. *Biochim. Biophys. Acta,* **110,** 475–83 (1965)
6. Chamberlain, R. W. An investigation on the action of piperonyl butoxide with pyrethrum. *Am. J. Hyg.,* **52,** 153–83 (1950)
7. Chang, S. C., Kearns, C. W. Metabolism *in vivo* of C^{14}-labeled pyrethrin I and cinerin I by house flies with special reference to the synergistic mechanism, *J. Econ. Entomol.,* **57,** 397–404 (1964)
8. Crombie, L. Chemistry of the natural rotenoids. In *Progress in the Chemistry of Organic Natural Products,* **XXI,** 275–325. (Zechmeister, L., Ed., Springer Verlag, Wien, 325 pp., 1963)
9. Dahm, P. A. The mode of action of insecticides exclusive of organic phosphorus compounds. *Ann. Rev. Entomol.,* **2,** 247–60 (1957)
10. Elliott, M. Allethrin, *J. Sci. Food Agr.,* **5,** 505–14 (1954)
11. Elliott, M., Farnham, A. W., Janes, N. F., Needham, P. H., Pearson, B. C. 5-Benzyl-3-furylmethyl chrysanthemate: a new potent insecticide. *Nature,* **213,** 493–94 (1967)
12. Elliott, M., Janes, N. F., Jeffs, K. A., Needham, P. H., Sawicki, R. M. New pyrethrin-like esters with high insecticidal activity. *Nature,* **207,** 938–40 (1965)
13. Ernster, L., Dallner, G., Azzone, G. F. Differential effects of rotenone and Amytal on mitochondrial electron and energy transfer. *J. Biol. Chem.,* **238,** 1124–31 (1963)
14. Fujita, T., Yamamoto, I., Nakajima, M. The analysis of the physiological activities of nicotine-related compounds with substituent constants. *43rd Annual Meeting Agr. Chem. Soc. Japan, April, 1968,* 323–24 (Abstract of papers)
15. Fukami, J. Effects of rotenone on the succinoxidase system in the muscle of the cockroach. *Japan. J. Appl. Zool.,* **19,** 29–37 (1954)
16. Fukami, J. Effect of some insecticides on the respiration of insect organs, with special reference to the effects of rotenone. *Botyu-Kagaku,* **21,** 122–28 (1956)
17. Fukami, J. Effect of rotenone on respiratory enzyme system of insect muscle. *Bull. Natl. Inst. Agr. Sci. Japan, Ser. C,* **13,** 33–45 (1961)
18. Fukami, J. (Private communication, 1969)
19. Fukami, J., Nakatsugawa, T., Narahashi, T. The relation between chemical structure and toxicity in rotenone derivatives. *Japan. J. Appl. Entomol. Zool.,* **3,** 259–65 (1959)
20. Fukami, J., Shishido, T., Fukunaga, K., Casida, J. E. Oxidative metabolism of rotenone in mammals, fish, and insects and its relation to selective toxicity. *J. Agr. Food, Chem.* (Submitted for publication)
21. Fukami, J., Tomizawa, C. Effects of rotenone on the *l*-glutamic oxidase system in insect. *Botyu-Kagaku,* **21,** 129–33 (1956)
22. Fukami, J., Yamamoto, I., Casida, J. E. Metabolism of rotenone *in vitro* by tissue homogenates from mammals and insects. *Science,* **155,** 713–16 (1967)
23. Guthrie, F. E., Campbell, W. V., Baron, R. L. Feeding sites of the green peach aphid with respect to its adaptation to tobacco. *Ann. Entomol. Soc. Am.,* **55,** 42–46 (1962)
24. Guthrie, F. E., Ringler, R. L., Bowery, T. B. Chromatographic separation and identification of some alkaloid metabolites of nicotine in certain insects. *J. Econ. Entomol.,* **50,** 821–25 (1957)
25. Hartley, J. B., Brown, A. W. A. The effect of certain insecticides on the cholinesterase of the American

cockroach. *J. Econ. Entomol.*, **48,** 265–69 (1955)

26. Harvey, G. T., Brown, A. W. A. The effect of insecticides on the rate of oxygen consumption in Blatella. *Can. J. Zool.*, **29,** 42–53 (1951)
27. Hawkins, W. B., Sternburg, J. Some chemical characteristics of a DDT-induced neuroactive substance from cockroaches and crayfish. *J. Econ. Entomol.*, **57,** 241–47 (1964)
28. Hopkins, T. L., Robbins, W. E. The absorption, metabolism, and excretion of C^{14}-labeled allethrin by house flies. *J. Econ. Entomol.*, **50,** 684–87 (1957)
29. Horgan, D. J., Ohno, H., Singer, T. P. Interactions of piericidin with the mitochondrial respiratory chain. *J. Biol. Chem.*, **243,** 5967–76 (1968)
30. Horgan, D. J., Singer, T. P., Casida, J. E. Binding sites of rotenone, piericidin A, and Amytal in the repiratory chain. *J. Biol. Chem.*, **243,** 834–43 (1968)
31. Kamimura, H., Matsumoto, A., Miyazaki, Y., Yamamoto, I. Studies on nicotinoids as an insecticide. IV. Relation of structure to toxicity of pyridylmethylamines. *Agr. Biol. Chem.*, **27,** 684–88 (1963)
32. Kato, T., Ueda, K., Fujimoto, K. New insecticidally active chrysanthemates. *Agr. Biol. Chem.*, **28,** 914–15 (1964)
33. Katsuda, Y., Ogami, H. Studies on the substituted benzyl esters of chrysanthemic acid. *Botyu-Kagaku*, **31,** 30–33 (1965)
34. Katsuda, Y., Ogami, H., Kunishige, Y. Novel active chrysanthemic esters. *Agr. Biol. Chem.*, **31,** 259–60 (1967)
35. Kearns, C. W. The mode of action of insecticides. *Ann. Rev. Entomol.*, **1,** 123–47 (1956)
36. Klingenberg, M. The respiratory chain. In *Biological Oxidations*, 3–54. (Singer, T. P., Ed., Intersci. Publ., New York, London, Sydney, 722 pp., 1968)
37. LaForge, F. B., Markwood, L. N. Organic insecticides. *Ann. Rev. Biochem.*, **7,** 473–90 (1938)
38. Lindahl, P. E., Öberg, K. E. Mechanism of the physiological action of rotenone. *Nature*, **187,** 784 (1960)
39. Lindahl, P. E., Öberg, K. E. The effect of rotenone on respiration and its point of attack. *Exptl. Cell Res.*, **23,** 228–37 (1961)
40. Martin, H. The chemistry of insecticides. *Ann. Rev. Entomol.*, **1,** 140–66 (1956)
41. Masri, M. S., Jones, F. T., Lundin, R. E., Bailey, G. F., DeEds, F. Metabolic fate of two chrysanthemumic acid esters, barthrin and dimethrin. *Toxicol. Appl. Pharmacol.*, **6,** 711–15 (1964)
42. Matsui, M., Kitahara, T. Studies on chrysanthemic acid. XVIII. A new biologically active acid component related to chrysanthemic acid. *Agr. Biol. Chem.*, **31,** 1143–50 (1967)
43. Matsumura, F., Hayashi, M. Comparative mechanisms of insecticide binding with nerve components of insects and mammals. In *Experimental Approaches to Pesticide Metabolism, Degradation and Mode of Action*, 231–38. (U.S.-Japan Seminar, Nikko, 295 pp., 1967)
44. Metcalf, R. L. *Organic Insecticides.* (Intersci. Publ., New York, 392 pp., 1955)
45. Metcalf, R. L. Mode of action of insecticide synergists. *Ann. Rev. Entomol.*, **12,** 229–56 (1967)
46. Miyamoto, J., Sato, Y., Yamamoto, K., Endo, M., Suzuki, S. Biochemical studies on the mode of action of pyrethroidal insecticides. I. Metabolic fate of phthalthrin in mammals. *Agr. Biol. Chem.*, **32,** 628–40 (1968)
47. Moore, J. B. Chemistry and biochemistry of pyrethrins. In *Natural Pest Control Agents*, 39–50. (Am. Chem. Soc., Washington, D.C., 146 pp., 1966)
48. Morrison, P. E., Brown, A. W. A. The effects of insecticides on cytochrome oxidase obtained from the American cockroach. *J. Econ. Entomol.*, **47,** 723–30 (1954)
49. Narahashi, T. Effect of the insecticide allethrin on membrane potentials of cockroach giant axons. *J. Cellular Comp. Physiol.*, **59,** 61–76 (1962)
50. Narahashi, T. Nature of the negative after-potential increased by the insecticide allethrin in cockroach giant axons. *J. Cellular Comp. Physiol.*, **59,** 67–76 (1962)
51. Narahashi, T. The physiology of insect axons. In *The Physiology of the Insect Central Nervous System.*, 1–20. (Treherne, J. E., Beament, J. W. L., Eds., Academic Press, London, New York, 277 pp., 1965)
52. Narahashi, T. Mode of action of DDT

and allethrin on nerve: cellular and molecular mechanisms. In *Experimental Approaches to Pesticide Metabolism, Degradation and Mode of Action,* 239–52. (U.S.-Japan Seminar, Nikko, 295 pp., 1967)
53. Narahashi, T., Anderson, N. C. Mechanism of excitation block by the insecticide allethrin applied externally and internally to squid giant axons. *Toxicol. Appl. Pharmacol.,* **10,** 529–47 (1967)
54. Narahashi, T., Haas, H. G. DDT: interaction with nerve membrane conductance changes. *Science,* **157,** 1438–40 (1962)
55. Narahashi, T., Haas, H. G. Interaction of DDT with the components of lobster nerve membrane conductance. *J. Gen. Physiol.,* **51,** 177–98 (1968)
56. Narahashi, T., Moore, J. W., Scott, W. R. Tetrodotoxin blockage of sodium conductance increase in lobster giant axons. *J. Gen. Physiol.,* **47,** 965–74 (1964)
57. Narahashi, T., Yamasaki, T. Mechanism of increase in negative after-potential by dicophanum (DDT) in the giant axons of the cockroach. *J. Physiol.,* **152,** 122–40 (1960)
58. Negherbon, W. O. *Handbook of Toxicology,* **III,** 508–19, 636–51 661–73. (E. B. Saunders, Philadelphia, Pa., 854 pp., 1959)
59. Nishizawa, Y., Casida, J. E. Synthesis of rotenone-6a-C^{14} on a semimicro scale. *J. Agr. Food Chem.,* **13,** 522–24 (1965)
60. Nishizawa, Y., Casida, J. E. Synthesis of d-trans-chrysanthemumic acid-1-C^{14} and its antipode on a semimicro scale. *J. Agr. Food Chem.,* **13,** 525–27 (1965)
61. O'Brien, R. D. Mode of action of insecticides. *Ann. Rev. Entomol.,* **11,** 369–402 (1966)
62. O'Brien, R. D. *Insecticides. Action and Metabolism,* Chaps. 8, 9, 10, 148–72; Chap. 16, 253–90. (Academic Press, New York, London, 332 pp., 1967)
63. Pullman, M. E., Schatz, G. Mitochondrial oxidations and energy coupling. *Ann. Rev. Biochem.,* **36,** 539–610 (1967)
64. Roan, C. C., Hopkins, T. L. Mode of action of insecticides. *Ann. Rev. Entomol.,* **6,** 333–46 (1961)
65. Sakai, M. Studies on the insecticidal action of nereistoxin, 4-N, N-dimethylamino-1, 2-dithiolane. V. Blocking action on the cockroach ganglion. *Botyu-Kagaku,* **32,** 21–33 (1967)
66. Sakai, M., Sato, Y., Kato, M. Insecticidal activity of 1,3-bis (carbamoylthio)-2-(N,N-dimethylamino) propane hydrochloride, cartap, with special reference to the effectiveness for controlling the rice stem borer. *Japan. J. Appl. Entomol. Zool.,* **11,** 125–34 (1967)
67. Santi, R., Ferrari, M., Tóth, E. Pharmacological properties of rotenone. *Farmaco (Pavia) Ed. Sci.,* **21,** 689–703 (1966)
68. Santi, R., Tóth, C. E. Toxicology of rotenone. *Farmaco (Pavia), Ed. Sci.,* **20,** 270–79 (1965)
69. Self, L. S., Guthrie, F. E., Hodgson, E. Metabolism of nicotine by tobacco-feeding insects. *Nature,* **204,** 300–1 (1964)
70. Self, L. S., Guthrie, F. E., Hodgson, E. Adaptation of tobacco hornworms to the ingestion of nicotine. *J. Insect Physiol.,* **10,** 907–14 (1964)
71. Soeda, Y., Yamamoto, I. Studies on nicotinoids as an insecticide. V. Inhibition of house fly head cholinesterase by nicotine. *Agr. Biol. Chem.,* **32,** 568–73 (1968)
72. Soeda, Y., Yamamoto, I. Studies on nicotinoids as an insecticide. VI. Relation of structure of toxicity of pyridylalkylamines *Agr. Biol. Chem.,* **32,** 747–52 (1968)
73. Soeda, Y., Yamamoto, I. Physiological activities of the optical isomers of nicotinoids. *Botyu-Kagaku,* **34,** 57–62 (1969)
74. Sternburg, J. G., Kearns, C. W. The presence of toxins other than DDT in the blood of DDT-poisoned roaches. *Science,* **116,** 114–17 (1952)
75. Tischler, N. Studies on how derris kills insects. *J. Econ. Entomol.,* **28,** 215–20 (1936)
76. Tsukamoto, M., Casida, J. E. Metabolism of methylcarbamate insecticides by the $NADPH_2$-requiring enzyme system from houseflies. *Nature,* **213,** 49–51 (1967)
77. Von Euler, U. S., Ed. *Tobacco Alkaloids and Related Compounds,* 53–74, 87–99, 101–104. (Pergamon Press, Oxford, London, Edinburgh, New York, Paris, Frankfurt, 346 pp., 1965)

78. Winteringham, F. P. W. Separation and detection of the pyrethrin-type insecticides and their derivatives by reversed phase paper chromatography. *Science,* **116,** 452–33 (1952)
79. Winteringham, F. P. W., Harrison, A., Bridges, P. M. Absorption and metabolism of [^{14}C] pyrethroids by the adult housefly, *Musca domestica* L., *in vivo. Biochem. J.,* **61,** 359–67 (1955)
80. Winteringham, F. P. W., Lewis, S. E. On the mode of action of insecticides. *Ann. Rev. Entomol.,* **4,** 303–18 (1959)
81. Yamamoto, I. Nicotinoids as insecticides. *Advan. Pest Control Res.,* **6,** 231–60 (1965)
82. Yamamoto, I. Mode of action of natural insecticides. In *Experimental Approaches to Pesticide Metabolism, Degradation and Mode of Action,* 145–59. (U.S.-Japan Seminar, Nikko, 295 pp., 1967)
83. Yamamoto, I., Casida, J. E. O-Demethyl pyrethrin II analogs from oxidation of pyrethrin I, allethrin, dimethrin, and phthalthrin by a house fly enzyme system. *J. Econ. Entomol.,* **59,** 1542–43 (1966)
84. Yamamoto, I., Casida, J .E. Synthesis of ^{14}C-labeled pyrethrin I, allethrin, phthalthrin, and dimethrin on a submillimole scale. *Agr. Biol. Chem.,* **32,** 1382–91 (1968)
85. Yamamoto, I., Kimmel, E. C., Casida, J. E. Pyrethroid metabolism in house flies: specificity for oxidation of *trans*-methyl group of isobutenyl moiety. *J. Agr. Food Chem.* (Submitted for publication)
86. Yamamoto, I., Ezashi, I., Fukami, J., Casida, J. E. (In preparation)
87. Yamamoto, I., Kamimura, H., Yamamoto, R., Sakai, S., Goda, M. Studies on nicotinoids as an insecticide. I. Relation of structure to toxicity. *Agr. Biol. Chem.,* **26,** 709–16 (1962)
88. Yamamoto, I., Soeda, Y., Kamimura, H., Yamamoto, R. Studies on nicotinoids as an insecticide. VII. Cholinesterase inhibition by nicotinoids and pyridylalkylamines—its significance to mode of action. *Agr. Biol. Chem.,* **32,** 1341–48 (1968)
89. Yamasaki, T., Narahashi, T. Effects of metabolic inhibitors, potassium ions and DDT on some electrical properties of insects nerve. *Botyu-Kagaku,* **22,** 354–67 (1957)
90. Yang, R. S. H., Guthrie, F. E. Physiological responses of insects to nicotine. *Ann. Entomol. Soc. Am.,* **62,** 141–46 (1969)
91. Zeid, M. M. I., Dahm, P. A., Hein, R. E., McFarland, R. H. Tissue distribution, excretion of $C^{14}O_2$ and degradation of radioactive pyrethrins administered to the American cockroach. *J. Econ. Entomol.,* **46,** 324–36 (1953)

ENTOMOLOGY OF THE COCOA FARM

By Dennis Leston
University of Ghana, Accra, Ghana

Cocoa[1] is grown in South and Central America and the Caribbean (24 per cent of world production, 1963–67) (2), West Africa, Cameroun and Madagascar (74 per cent), south and southeast Asia and Melanesia (2 per cent). Although areas under cocoa are increasing in the first and last named, West Africa and Cameroun are likely to dominate production at least for the next two decades. Using the most rapid varieties, it is three to four years from planting before the first crop can be harvested. This review is therefore largely of entomological problems encountered in Africa where the facts "are well-documented, thanks mainly to the outstanding work of scientists associated with the West African Cocoa Research Institute" (19).

In Cameroun and Côte d'Ivoire, cocoa is often a plantation crop but in Ghana and Nigeria, the two largest producers, it is confined to peasant-owned farms (8, 71–73). On these the forest cover is thinned and larger or less valuable timber left to provide cover—lightmeter readings suggest this type of cultivation gives conditions on the floor differing little from those of primary forest, and the term "cocoa forest" is a preferred one (96, 122). It has been said that cocoa is the subdominant in the West African secondary forest (140) but the plant cover under cocoa is generally similar to that of primary rather than secondary forest, and the dense vegetation, climbing towers, etc., of the secondary succession are uncommon. The importance of this for the ant fauna is discussed below. Although a tree crop with a potential yield life of at least sixty years—the oldest datable trees in Ghana, at Aburi, Eastern Region, now exceed this—"the historical picture of the successive development of the cocoa-growing areas in the world is a glorified picture of shifting cultivation on a global scale" (52). This is exemplified in Ghana: one-time centres of production in Eastern Region such as Bisa and Nsawam are now insignificant in yield due to drought or virus disease (with its cutting-out method of control), or both, that occurred between the late 'twenties and early 'forties (6, 64, 131, 134), whereas production in Brong-Ahafo, roughly 150 miles to the northwest, was negligible in the period mentioned but now surpasses that of the entire Eastern Region (10).

Reviews of the entomological problems of cocoa are many: on cocoa-

[1] The practice of writing *cocoa* for the crop and *cacao* for the tree, is needless pedantry. The West African trade has ever used *cocoa* for both. And *cocoa-capsids* has been a vernacular term for some fifty years; *mirids* in this context is an unnecessary innovation.

capsids (44, 45, 47, 81, 132), which also have a bibliography (21); on virus diseases and vectors (33); on pollination insects (112); on stored cocoa pests (87); on minor pests in West Africa (41); cocoa Lepidoptera (121), etc. But, in view of the rapid turnover in research personnel and the lack of museums in the producing countries, regional lists of insects associated with the cocoa farm, with diagnoses and descriptions of damage, form a more useful part of the literature. Notable in this respect is the work of Alibert (1) for francophone West Africa, Boulard (11) for Republique Centrafricaine and Szent-Ivanyi (128, 129) for Papuasia. Similar works exist for Latin America, e.g., for Bahia (119), while Szent-Ivanyi's work has been extended (120).

Work in Nigeria in the late 'fifties led to development of a generalised "cocoa degeneration" approach (82, 84, 138, 140). This was brought about by the discovery that insecticide treatment for cocoa-capsids also alleviated damage attributed to swollen-shoot virus (SSV); the cocoa degeneration syndrome was an amalgam of such factors as light, water-stress, cocoa-capsids, fungal diseases and SSV. This work has been reviewed and given a wider audience by Clark et al. (19), who regard it as a good demonstration of the complexity of some crop damage situations. They give an operational analysis of cocoa degeneration but some of their facts are erroneous. It has seemed to research workers in the past five years (or so their publications would suggest), that the cocoa degeneration approach, while employing for the first time wholesome ecological thinking, must fail insofar as too little is known of the components, the biology, phenology and population dynamics of the insects involved. There is thus a lack of knowledge regarding the timing of improvement operations, chemical or cultivational. Nor does it take fully into account the most important sector of the background entomofauna, ants.

A number of the nonentomological contributors to degeneration are studied individually—*Calonectria* fungus, SSV, water-stress, etc.; all, however, have associations with insects. Water-stress, for instance, is important, the effects of thrips (51) and cocoa-capsids (61) varying with the water status of the host. Above all, the direct effect of shade removal, in addition to its secondary effect through contributing to water loss, is paramount. It has been shown that removal of shade can, optimally, increase yields tenfold if insects are controlled (31); in open conditions, nutrients are more gainfully utilised and fertilisers can usefully be applied. But cocoa is subject in most areas to a dry season varying in length and severity from year to year (95). Drought, it has been claimed, was the direct cause of the death of cocoa in parts of Ghana (134). And even with a good soil water balance the effect of low atmospheric humidities is considerable, causing wilt and premature leaf-fall, and leaving trees more susceptible to capsid damage (103); much of the cocoa area of Ghana and Nigeria is in a belt subject, in most years, to a day or two with mid-afternoon RH of 20 per cent or less (92).

The present review considers the principal entomological associations with cocoa under five separate heads; not because each is independent but because it is impossible to integrate non-numerical data—and most cocoa-lore is still only that.

Pollination

Both self- and cross-compatible strains of cocoa exist. The phenomenon is reviewed by Cope (23) from the standpoint of the plant morphologist. Until about 1940, wind pollination or ants were the invoked agents. Subsequently, it was shown that female Ceratopogonidae account for almost all pollination (9, 110, 112, 127). Pollination by ants is now discounted but one worker is still of the opinion that wind plays a part, if only through directing the flights of the biting-midges (63).

Although only a relatively low proportion of flowers is successfully pollinated, this is sufficient. Cocoa has a self-limiting mechanism—wilting of the surplus young pods (cherelles). This ensures that only a few dozen pods ever reach maturity; generally less than 50 per tree in the entire year (100, 101), but on West African farms only half this number mature (138). The ceratopogonids breed in wet vegetable trash, requiring three or four weeks to develop from egg to adult. Females feed in part on material from the cocoa flower but also need blood, perhaps insectan, for successful egg maturation (127). The flies have morning and afternoon activity peaks, the latter being the smaller. At Tafo, Ghana, their numbers increase from April to October—the wet season—stay high until January, then collapse (112). A chemical control programme in cocoa is unlikely to have any appreciable effect on pollination; this is born out by there being no report of reduced yield following spraying (although, of course, this may be due to the inefficiency of most spray routines in cocoa).

One problem remains, that of clarification of the identity of the ceratopogonids. Most of the six or so species so far found to be concerned are *Forcipomyia* sens. str., and there is evidence that West Africa and Trinidad have forms in common (35, 58, 98). But Saunders now believes that all are members of *Proforcipomyia;* his work may be taken as authoritative (115a).

Cocoa-Capsids

Throughout the tropics some members of the Bryocorinae (Hemiptera, Miridae) feed upon parts of cocoa (47, 90) (Table I). This practice can be seen as a segment of the overall feeding pattern of many bryocorines. Man's sources of tropical stimulants—tea, coffee, cocoa, cola, guarana-nut, maté tea—form a botanically diverse group but each contains methyl purines (caffeine, theophylline, theobromine, etc)., and each harbours bryocorines. Little is known of the metabolic pathways of methyl purines in plants or of their function and it could well be that they have independently evolved as protection against the attacks of some groups of insects. Within bryocorines there subsequently arose the ability to exploit such otherwise

TABLE I

List of Cocoa-Capsids of the World (Hemiptera, Miridae, Bryocorinae)

Group	Species	Distribution on cocoa
I	*Sahlbergella singularis* Haglund	Sierra Leone to Rep. Centrafricaine and Congo (K.)
	Distantiella theobroma (Distant)	Côte d'Ivoire to Rep. Centrafricaine and Cameroun
	Bryocoropsis laticollis Schumacher	Côte d'Ivoire to Congo (K.)
	Odoniella reuteri Haglund	Congo (K.)
	Boxiopsis madagascariensis Lavabre	N. E. Madagascar
	Platyngomiriodes apiformis Ghauri	Sabah
	Pseudodoniella pacifica China & Carvalho	New Britain
	P. duni (China & Carvalho)	New Britain
	P. typicus (China & Carvalho)	New Britain, Papua
	P. cheesmanae (China & Carvalho)	Papua
	P. laensis Miller	Papua, New Guinea
II	*Helopeltis* about 12 spp.	All Old World cocoa areas
III	*Monalonion* about 12 spp.	South & Central American cocoa areas

protected tissues—the present pan-tropical but fragmented distribution of the mirids concerned indicates this must have taken place in the earliest days of mirid radiation. It is notable that almost all the species listed in Table I, except for *Helopeltis,* were first discovered on cocoa, an introduced crop everywhere except in America. This indicates that the world's tropical forests harbour bryocorines on plants which, for the most part, are undetected as mirid hosts. It is therefore desirable before establishing cocoa that a survey of forest mirids be attempted. It would also be of value to elucidate the role of methyl purines as attractants (unlikely because of chemical structure, but if true, opening up a possible method of control through baiting), phagostimulants or essential substances for bryocorines.

The subfamily Bryocorinae is at least diphyletic but the current classification does not reflect the situation (17, 18). The cocoa-associated genera have eggs with a stalked process which is lacking in *Bryocoris* (20, 74); the guts of many species show unusual features (66, 67); there is diversity in testis follicle numbers (91)—consequently it is probable that none of the forms found on cocoa is consubfamilial with *Bryocoris* (20). Recent systematic work has been concerned with the description of new species (59, 107) but recognition of a "swollen scutellum group" (Group 1 in the Table)—due originally to Schouteden (116)—promises to lead to a more useful taxonomy, one which will permit predictions to be made. The systematics of *Helopeltis* is confused and no reliance can be placed on any determinations therein; the situation in *Monalonion* is but little better.

Eggs, larvae and adults of cocoa-capsids are reasonably well known and their life histories worked out (28, 32, 37, 44, 47, 102). Most species develop from egg to adult within five to six weeks; interspecific differences are less than intraspecific variation [but little constant temperature rearing has been attempted, and laboratory breeding has so far proven difficult (114, 115, 149)].

Since de Jong's (34) study of *Helopeltis*, only Goodchild (66, 67) has investigated internal anatomy. He has looked at the gut and salivary glands from a functional viewpoint, the latter in relation to feeding and lesion production. The two dominant West African species, *Sahlbergella singularis* and *Distantiella theobroma*, feed on the superficial parenchymatous tissue of pods or on vegetative growth (30, 62, 66) and it is this ability to feed on fruits or green stems that makes them so important. There is evidence that some *Monalonion* and *Pseudodoniella* species, at first noted as pod-feeders, have changed subsequently to vegetative feeding, that is, once the capsid has switched from its autochthonous host it can begin rapidly to exploit leaf petioles and basal shoots (44, 83, 102, 120). The evidence that *Helopeltis* is able to affect both pods and vegetative material is contradictory but this is probably due to differences between the species (54, 65). *Bryocoropsis* feeding is apparently confined to pods (29). In *Distantiella* there is a marked association of rapid development with a good water level in the plant. With low water, there is survival because of enhanced plant regeneration but low water content causes death if the bugs cannot migrate (61).

Feeding by cocoa-capsids on pods causes distortion, bean decay and pod wilt but these effects are seldom serious (104); but *Monalonion* and *Helopeltis* damage can sometimes be severe (47). Generally, serious pod damage is attributed to the additive action of pathogenic fungi—*Gleosporium* in New Guinea (37) and *Phytophthora* in Ceylon (54)—but the evidence for fungal attack following pod feeding by capsids in West Africa is conflicting (1, 47). Of far more importance is the damage to the tree or seedling subsequent to the bug feeding on vegetative tissue. A few feeding lesions are sufficient to cause death of an apical shoot, and phytotoxic effects are enhanced by pathogenic fungi, notably *Calonectria* in West Africa (30, 66, 83, 140, 146).

Three stages of damage are usually recognised: blast—small discolored lesions on young green shoots which later wilt, the dead leaves remaining on the trees; staghead—a thin crown with many leafless branches; capsid pockets—trees reduced to bare trunks with numerous lateral shoots (109, 146). In the last situation, regeneration attempts by the tree through development of lateral and basal shoots provide further optimal conditions for capsid increase. Heavy damage ensues when water balance is improved after a spell of desiccation (61). Attempts to evaluate crop loss by capsid or capsid-plus-fungus attack have not been very successful; estimates range as high as one third of potential yield for Ghana or "cannot be less than 20 per cent of total annual production" (12, 44, 143). On a plantation in New Britain, a 60 per cent loss in production in one year was attributed for the most part to capsid damage (37). Certainly, blasted or stagheaded trees are a conspicuous feature of the Ghana and Côte d'Ivoire scene at the end of the dry season (but severity of harmattan may contribute directly to this). However, recovery is usually rapid although the first, minor, crop may be reduced.

The behaviour of *Distantiella* and *Sahlbergella* has been studied, notably by Williams (146, 147), and much work currently is proceeding in this field. Basal shoots are the preferred egg-laying sites, the eggs being inserted fully into the plant tissue. However, egg totals in canopy shoots (fans) are greater because there are more of these available (40); pod stalks are also utilised (146). It would seem that feeding sites, in the main, condition oviposition sites and that light is at least of equal importance with nutrient status or palatability. Oviposition is more frequent in the canopy when this is good, lower on the tree when the canopy is poor. Egg production, it is claimed, is far higher in pod-reared than in shoot-reared bugs (75).

Feeding sites are also conditioned by harvesting. As this proceeds, fewer pods and pod-stalks are available, thus causing a mass movement to the canopy and general dispersal, which is more marked in *Sahlbergella* (62). On pods, feeding is concentrated around the stalk and attachment area; it could be that there is some pooling of saliva which encourages aggregation (76). But there remain a number of contradictions regarding the selection of sites, relative palatability and nutritional qualities of the selected tissues (146).

It is a matter of observation that some trees harbour capsids, others do not. Youdeowei (150) suggests there is true aggregation. It is likely that this is brought about by the bugs showing, at most times, a preference for dark places or, more probably, higher humidities (the dark angle between pod and stem is a favoured site) but some experiments have shown movements by females to well-lit areas (77). So far, experiments in which light and humidity are independently controlled are lacking. In *Distantiella*, water relations are critical (61): if the plant is under water-stress the bug is unable to combat low ambient humidities and dies; this is most marked in larvae, perhaps because of their relative immobility.

Distantiella can start to fly 20 hrs after the final moult (93). Early work on range was unsuccessful due to poor recovery of marked bugs, but from experiments in which the insects were flown to exhaustion and from speed trials in the open it is apparent that the range of a single flight can be over 2 km (and may, in fact, be double this) (93). Flight initiation at times is caused by attacks of predators; the bugs fall for about 1 to 1.2 metres than fly sunwards, to be intercepted by the canopy. Bugs thrown in the air in the open fly upward in a spiral and zig-zag manner if the sun is high (93).

Light is the single environmental factor to have been studied in any detail because the degree of damage is affected by this. Pockets tend to be found under shade of staghead cocoa in the open (109, 132, 146). Breaks in the cocoa canopy, Williams believes (146), usually precede capsid attack rather than result from it: thus, SSV or water-stress, damaging the canopy, allows more light to reach the low levels where increased basal shoot production permits capsids to flourish. The degree of overhead shade from forest trees is the dominant factor. Although negatively phototactic animals when walking, it has been established that capsids flourish best in more or less open conditions. In experiments in the Cameroun, populations of *Sahlbergella* were three and one-half times more numerous in the open than in medium or heavy shade (14). It is of note that all estimates of light regimes published so far have been based on arbitrary scoring; lightmeter readings show that this leads to inaccuracy (93).

Alternate host plants of cocoa-capsids have been listed (47, 81) but knowledge of them is meagre. *Sahlbergella* occurs not infrequently on *Cola* and other Sterculiaceae, *Distantiella* on *Ceiba;* other hosts listed have been tested by rearing larvae on them but it is not known if the adults are of normal fecundity. Only for *Bryocoropsis* have alternate hosts been clearly indicated—in this case some Annonaceae (81). The problems of *Helopeltis* and *Monalonion* hosts are confused by poor taxonomy.

Distantiella is subject to attack by a great range of predators as established by direct observation, inference or radiotracer studies (81, 99); the bug lacks the functional thoracic stinkglands of most Miridae. In the field, the predators—long-horned grasshoppers and crickets, mantids, reduviids, etc.—are themselves controlled by dominant ants. The predating ants fall into two groups: canopy nesting (*Oecophylla, Crematogaster* and *Macromischoides*) and ground or stump nesting (*Pheidole, Platythyrea conradti,* notably) (93); all of which are predatory on *Distantiella. Oecophylla longinoda, Crematogaster africana,* and *C. depressa* are themselves mutually exclusive and establish considerable realms in which few other predators (save, perhaps, *Pheidole* spp.) can survive. Experiments with *Oecophylla* show that on trees on which it nests, usually with good canopy but little top shade, *Distantiella* is unable to establish itself. Early doubts as to the efficacy of *Oecophylla* (122, 148) have been cleared up by experimentation and observation. It is less effective against *Sahlbergella* and is itself self-limiting insofar as each colony is surrounded by a no-man's-land in which, at times of capsid maxima such as January–February at Tafo, *Distantiella* still builds up (93). But a survey has shown that 10 to 15 per cent of Ghana's cocoa trees are more or less permanently protected by *Oecophylla* from *Distantiella* damage (22). Random surveys made in Brong-Ahafo and Western Region during 1969 suggest that natural protection may range up to 50 per cent in some areas. Only the use of insecticides or an upset in the ant balance can destroy this protection. The relationship between cocoa-capsids and *Crematogaster* spp. is more complex because, while the ant attacks capsids, they tend mealybugs. *Platythyrea conradti* and *P. frontalis* attack singly and thinly and, alone, are unlikely to be important while colonies of *Macromischoides* are too few and too feeble to be widely effective (93). Williams (148) has attempted to assess predation but in view of the patchy distribution of most predators, a mean total is likely to be misleading.

Parasitism by euphorine Braconidae is unknown for *Distantiella* but it is of major importance to *Sahlbergella,* of which over 20 per cent of fifth instar larvae may be infected. This suggests *Distantiella* is a relatively recent arrival in the cocoa area, but it has been known as a cocoa pest for about 60 years (81). That research has not been concentred on the *Sahlbergella* euphorine—probably *Leiophron sahlbergellae* (Wilkinson) in much of West Africa—is surprising but "there has been little quantitative work on the effect of parasites and predators" (81).

Early attempts to gauge population size and its annual variation in *Sahlbergella* and *Distantiella* were, for the most part, based on hand collecting, with or without ladders (148); but the bugs are camouflaged and difficult to spot, separation of larvae of the two species and identification of instars needs experience, and all insects above hand reach were usually ignored—no formula for correcting for this last has been devised. Egg counts (40), dusting, with sheets placed beneath the trees (89), ultraviolet light-trapping—only effective with *Sahlbergella* (62, 133)—have all been utilised but it was only by using four different methods simultaneously that Gibbs, Pickett & Leston (62) were able to arrive at a reasonably reliable picture of the changes in population taking place within a year. They used pyrethrum knockdown with a rigorous regime of sample rotation within a single circumscribed plot; weekly counts in a small isolated pocket of pollarded trees by an experienced technician; nightly trapping by a Robinson-type ultraviolet trap; and randomised hand-collecting (62). Indirectly, their work is a tribute to that of Williams (148) for the results are somewhat similar.

Populations in Ghana and Côte d'Ivoire build up in mature cocoa in July and August to reach a peak in September or October (62, 88, 89, 148). This may be followed by continuous high levels to December, when a drop can occur or, more usually, a second peak arises in January and February. The earlier peak is on pods and after harvesting *Sahlbergella* becomes widely dispersed, though with numbers higher in January and February than in March to July; *Distantiella* after harvest moves to fan or shoot tips in the same area and its numbers remain high (or go even higher with a new canopy-bred generation) to sometime in February. Late February sees a collapse in both species, which is possibly closely connected with spells of low humidity, but the effect of low RH, while it may kill larvae in exposed conditions, gives rise to a nutritional and regenerative state in the plant which is suitable for the survivors to build on again in localised areas (62). That low humidities may delay hatching is also possible. Above all, local differences are great, annual variation even greater. Late 1968 at Tafo saw almost complete replacement of *Distantiella* (for many years previously the dominant species) by *Sahlbergella* (22); possibly the excessive rains of 1968 upset the usual parasite balance in the latter species.

In an isolated capsid pocket in which immigration was negligible, *Distantiella* built up at each generation—September, late October, December—with a slight drop in the late January brood and collapse of the March generation (62). Dispersal by *Distantiella* seems to reflect population increase and goes on at an increasing rate from August to December, whereas it fluctuates at a low level in *Sahlbergella* until toward the end of harvesting, when a massive movement generally occurs (62). However, there is a moderate discrepancy between these data, obtained by random sampling and noting the presence or absence in the samples, and light trap captures (perhaps due to direct attraction by ultraviolet lamps) (62).

The results of all the sampling in West Africa, where reasonably well done, agree closely in the shape of the annual population curves. It is to be regretted, however, that work in Côte d'Ivoire has often lumped data on the two species (89). Absolute population size, it is now clear, was very much underrated in the past; estimates of the order of 2500 bugs per hectare as comprising a "normal maximum" (148) must be doubled or trebled at least, according to the figures reached by multimethod sampling (62). But damage done depends in part on population size, in part on the state of the tree in relation to age, water, and light regimes (62).

The phenology of cocoa and a wide range of its associated insects has been investigated by Gibbs and Leston (60, 95). By combining sunshine with rainfall data they were able to find climatological parameters which fitted biologically defined seasons. Cocoa leaf feeders and their predators had two peaks: March to April (first wet sunny season) and October to November (second wet sunny season). Cocoa-capsids, because of their ability to exploit two distinct categories of plant tissue, cut across the seasons, according to present data, in a unique fashion. Their decline coincides with the end of the dry sunny season but build-up takes place at the end of the first wet, dull season or in the short dry, dull season (60, 62, 95).

Although seemingly competing for the same resources and with one species at times replacing the other, there remain differences between *Sahlbergella* and *Distantiella,* some of which have already been mentioned. That *Sahlbergella* flies by night, *Distantiella* by day, is perhaps of note in indicating higher humidity preferences by the former, but in Nigeria, the driest cocoa area in West Africa, *Sahlbergella* is more important than *Distantiella* and the same applies to the wettest cocoa area, Cameroun (but does this reflect local population differences?). In general, there is a tendency for *Sahlbergella* to be found on mature trees, *Distantiella* on seedling and young trees, but this has been noted only in Ghana (147). It is claimed that feeding, especially on vegetative tissues, does so much damage as rapidly to make the tissue unusable by further bugs, that is, there is a density-dependent mortality factor (147); that this operates on larvae in small pockets is very likely.

Geographical distribution within West Africa of the two main species has been reviewed (148) and there is some recent information on Côte d'Ivoire (88, 89). It seems best to regard *Sahlbergella* as occupying the whole of the cocoa bloc (Sierra Leone to Cameroun) and *Distantiella* as a less permanent inhabitant of the central part of this area. Changes in status of the two are likely to be greatest in the overlap area, eastern Côte d'Ivoire to Western Nigeria; outside of this, *Distantiella* is rare and patchily distributed [for example, in a small part of Republique Centrafricaine (11)]. A final difference between the species is the presence of insecticide-resistant strains of *Distantiella* (38, 39) in Ghana, and of *Sahlbergella* in Nigeria (43).

Mealybugs

Pseudococcidae are found on cocoa everywhere. Thus, in Trinidad, *Planococcus citri* (Risso) is the dominant species; field colonies of this and other mealybugs are ant-attended (especially by *Wasmannia auropunctata* (Roger)) (50). *P. citri* and *P. lilacinus* (Cockerell) occur on cocoa in New Guinea together with other species—for which *Solenopsis* spp. are the more important associated ants (120, 128). In most cocoa areas outside West Africa it was thought that mealybugs do direct damage only through feeding but there is evidence that they are also vectors of virus diseases in Ceylon (16) and Trinidad (86) and, it is suspected, encourage pathogenic fungi in New Guinea and probably elsewhere (130). But it is in West Africa, through their association with virus diseases, that mealybugs are of major importance (33, 64). Indeed, the impact of mealybugs, SSV, and a government-enforced cutting-out programme for control played a leading part in the stormy politics of the Gold Coast in the decade before independence (6, 64).

Some 20 or more mealybug species can be found on West African cocoa, from Sierra Leone to Cameroun (5, 33, 125), and their systematics is now more or less stable (48, 145). Other coccoid families on cocoa include Stictococcidae, Lecaniidae, Diaspididae and Margarodidae (125). Swollen shoot was recognised as a plant disease by Steven in 1936 (124) and its virus nature elucidated by Posnette (108); Cotterell (29) was the first to show mealybugs were the insect vectors (33).

In laboratory tests almost all the mealybugs have been found to transmit strains of virus diseases (139) but *Planococcoides njalensis* (Laing) is now considered the most important vector, especially of the virulent forms of SSV. Some mealybugs, e.g., *Ferrisiana virgata* (Cockerell) and *Pseudococcus adonidum* (Linnaeus), are relatively specific in that they can transmit only a limited number of strains (111, 113). *P. njalensis* is not only the most effective vector of SSV but is also the most abundant and widespread cocoa mealybug in Ghana (125).

Cocoa mealybugs fall into two groups: members of the first are obligate associates of ants, have short legs and are generally ovoviviparous; the second are sometimes found with ants, have longer legs and are mostly oviparous. *P. njalensis* is one of the first group (125). "Eggs" hatch within minutes of laying and these and the three larval instars take about six weeks to reach the imaginal stage. The first instar larva, "crawler," is able to walk a few millimetres as is each later instar just after emergence. Females lay between 30 and 40 eggs but there is usually a delay of about three weeks before oviposition which spreads over about 20 days maximally (125). Second group species, for example, *Planococcus citri* and *Ferrisiana virgata,* produce true eggs and are far more active and mobile (27, 125).

P. njalensis is found on all parts of cocoa but most abundantly on new growth and petioles; they can also thickly encrust older pods. In Nigeria, *P. citri* and *Planococcus kenyae* (La Pelley) are almost as frequent as *P. njalensis:* they prefer mature trees but *F. virgata* is more often found on seedlings and young plants (33, 125).

It is due to Strickland (125, 126) that we have a considerable body of data on the ants attending mealybugs on cocoa. Unfortunately, the systematics of African ants is in a confused state—there has been little work of value (and much that is bad) since the days of Wheeler, and far too many forms have been identified without reference to types. When it is realised that "there are more than 70 species of ants associated with cacao and the cacao Coccids in the Gold Coast" (125) the problem of correct determination becomes vital. It is likely that the complete collapse of studies on the entomological aspects of SSV since about 1960 is due in large part to workers being scared of ant taxonomy.

In Ghana, Strickland (125) recognised the following ant groups in Crematogasterini: *Sphaerocrema*—about six species, all coccidophilic; *Crematogaster* s.str.—one species; *Atopogyne*—at least ten forms, of which *Crematogaster (A.) africana* Mayr was a major attendant of mealybugs; a group of five other *Crematogaster* spp. of little importance. Other than Crematogasterini, *Pheidole megacephala* (Fabricius) was another important mealybug associate; nearly 20 Myrmicinae tended mealybugs; five Dolichoderinae were important on seedling cocoa; a great range of Formicinae were generally found with coccoids other than mealybugs (125).

After a considerable amount of sampling, Strickland (126) reached a number of conclusions regarding *P. njalensis.* In the field it was almost invariably associated with Crematogasterini, many of which build carton tents over the mealybugs. Some of the arboreal nesting Crematogasterini were consistently associated with higher mealybug populations and tree-to-tree variation in mealybug populations reflected differences in the ant species present. *P. njalensis* "is maintained by ant protection at a density level considerably above that prevailing amongst closely related pseudococcid species which are not ant-attended to the same extent." Where the mealybugs carried SSV, the host tree usually became less able to support a high population and harvesting also removed a great number of mealybugs (a favorite site for tents is between pods and the main trunk): this kept mealybug populations at a "protected steady density" (Strickland's quotes). Above all, it was concluded that predation and parasitism maintained balance and the former was enhanced through competition between ant species, especially involving *Oecophylla longinoda* (strongly negatively associated with Crematogasterini). Using entirely different sampling methods, Leston (93) has confirmed the mutual exclusiveness existing between *Oecophylla* and various *Crematogaster* s. lat.

A survey of parasitism in *P. njalensis* showed that about 3 per cent were infected by Hymenoptera with over 10 per cent hyperparasitism, but marked local differences occur (36). The parasites covered a wide range of Encyrtidae, with evidence of some changes in the species balance over a few years (36, 126). Parasitism by encrytids is not confined to West Africa; it has been reported, for example, for cocoa mealybugs in Trinidad (85). Predators appeared to be unimportant except for the larvae of some Cecidomyidae (predators on "eggs" and larvae) and the coccinellid *Platynaspis higginsi* Crotch, of which all stages attack *P. njalensis* within crematogasterine tents; it is tolerated by the ants, whereas most potential predators of their mealybugs are not (36).

Swollen shoot virus has been shown to have two types of spread (26, 27, 136): radial, from the edges of outbreaks; and jump, new infections at a distance from existing outbreaks. There is apparently sufficient movement of the mealybugs through the canopy, only occasionally with ant assistance, to effect the first but the latter spread, it has been shown, is mediated by wind (24, 26, 27). Cornwell (27) did an elaborate set of measurements of wind speed and direction amid and above cocoa, using sticky traps and exposed bug-free seedlings. His work demonstrated that wind dispersal was highly effective but successful colonisation of a new tree depended on the presence of coccidophilic ants as well (but even so, success was less frequent than failure). Behavioural differences between mealybug species made some more likely to be wind-dispersed than others (27) but in view of the frequency of line squalls in the Ghana cocoa area this is unlikely to be very important.

Whereas *P. njalensis,* even within relatively uniform plots of cocoa, is patchily distributed (126) (probably because of the mosaic distribution of dominant ants) it was at first believed seasonal changes were minimal. Subsequently, it was shown that in "well cultivated" cocoa (the degree of top shade was not mentioned), populations remained low from January to July then trebled to a peak in October–November followed by a sharp decline in December. In "bush" cocoa, numbers declined steadily from January to a trough in July, to be followed by a steep build-up to an October–November peak (25). Sampling was spread over a 13-month period. The differences in the population within the bush plots in the two Januarys sampled are sufficient to suggest that little meaning can be deduced from this, although Cornwell believes the trends to have been brought about by changes in the abundance of predators and parasites. It could equally be explained by the bush plots having a more equable microclimate, the well-cultivated being subject to the full effect of the dry season, and hence in December–January the crash to a February low. The drop in population in the bush plots (January to July) was most marked in the more heavily infested subplots. Until more is known of the feeding preferences of mealybugs it is difficult to fit them into a seasonal framework; there is clearly no simple wet season/dry season pattern. But the work of Fennah with *P. citri* in Trinidad showed a negative correlation with light and a marked positive association with plants with a high nitrogen status (50).

Trees infected by severe strains of SSV die and also provide foci for the spread of the disease; with mild strains, the canopy is marred, enabling cocoa-capsids to build up. It is therefore impossible to obtain an estimate of damage directly attributable to mealybugs (135). But, during the years 1947 to 1957, some 70 million trees were cut out by control gangs in Ghana, under varying schemes of compensation or noncompensation, and moderate or severe compulsion (10, 33, 135). Certainly, the loss has run, in Ghana alone, into many millions of £ sterling. Although SSV is less virulent in Nigeria, an equally drastic scheme there has caused untold loss (137). But it could be claimed that the great rise in cocoa production in Ghana in the early 'sixties was in part made possible only through the cutting out programme curtailing the spread of SSV (10).

The Background Fauna

Cocoa supports a huge number of insects: 248 cocoa insects are reported in New Guinea (130); 147 Lepidoptera—excluding the infrequent species—are associated with the crop in Ghana (121); 56 genera of beetles, some, such as *Xyleborus* s.lat., with dozens of species, have been found on cocoa in francophone West Africa (1). Totals comprise a mixture of autochthonous forest forms with not necessarily native foodfarm, farmbush and savanna species: the second group includes some tropicopolitan tramps.

Orthoptera are in general unimportant though some Caelifera are useful capsid predators (60). Phasmida do minor damage in New Britain (129); Isoptera play an undetermined role, attacking and killing a few trees but seldom, if at all, as primary enemies [but cf. (128)]. Leafhoppers may be more important than hithertho supposed: Fernando's (53) study of *Empoasca devastans* Distant showed this jassid to cause marked hopperburn (not his term) to cocoa in Ceylon while similar damage has been found recently in Ghana (93) and Côte d'Ivoire (7), possibly due to the same species. Hopperburn has previously been confused with sun damage to young leaves.

In Sternorrhyncha, besides the coccoids and the aphid *Toxoptera theobromae* Schouteden, the cocoa-psyllid *Tyora (=Mesohomotoma) tessmanni* (Aulmann) is important, ranked third after capsids and mealybugs in West Africa by Gerard (56), but without numerical data. It attacks leafbuds, wherein it both oviposits and feeds, the excreted sugars providing a substrate for further fungal damage (1). Predators include syrphids, anthocorine cimicids and the mirid *Deraeocoris crigi* [Leston & Gibbs (94)]. The circumstance encouraging the psyllid is mainly removal of top shade: it is a strikingly seasonal insect with a population peak about May in Ghana, when up to 70 per cent of leaf buds may be affected (60, 94).

Within Hemiptera, many Pentatomidae and Coreidae have been associated with cocoa, for example, in Trinidad (15), and Bahia (118), but there is little precise information on damage. *Bathycoelia thalassina* (Herrich-Schaeffer) has been known to attack cocoa-pods in central and West Africa for over 50 years but symptoms are often confused with those of boron deficiency (3, 57).

Three major environmental alterations can be associated with the change from being a minor part of the background fauna to becoming a pest: water-stress, over-insolation, insecticide application. The cocoa-thrips, *Solenothrips rubrocinctus* (Giard), probably neotropical in origin but now found in cocoa almost everywhere, with a range of alternative hosts, is associated with unshaded cocoa (1, 56, 105). Fennah's (49, 51) detailed work suggests that water-stress is the most important concomitant of severe thrips damage or, as he puts it "... establishment of a cacao-thrips colony on cacao is obligately dependent on the prior occurrence of metabolic derangement in the host leaf leading to a retardation of protein synthesis ... (this) results from the effect on the tree of seasonally adverse factors in its physical environment" (51). Shade removal is the factor most conducive to damage by caterpillars. *Earias biplaga* Walker (Noctuidae) damages terminal buds in West Africa (41), has a similar periodicity to the cocoa-psyllid (60), and rates as a major pest only when cocoa is grown in the open (56); *Anomis leona* Schaus, another noctuid, is comparable in all three respects (41).

Anti-capsid spray programmes led, in Ghana, to big increases in a number of species, one being *Bathycoelia,* mentioned above. The pod huskminer, *Spulerina* (=*Marmara*) sp. (Gracillariidae), which covers cocoa-pods with a fine scribbling, increased in a trial plot to pest status following use of γ-BHC (41, 46, 55, 121). Similar increases took place in beetle and lepidopterous stem borers (41, 42, 46), but in the case of the cossid *Eulophonotus myrmeleon* Felder, increase in Nigeria was attributed to a severe dry season. New pests appearing after spray regimes are mainly borers or miners because, it is generally believed, their free-flying parasites get killed but the effects of spraying on predatory ants are little known.

Factors associated with increase in Scolytoidea are diverse. Of the numerous West African species, *Xyleborus morstatti* Hagedorn is the dominant; it attacks healthy wood of trees mostly under three years old and damage is caused by pathogenic fungi entering the borings. It is especially important where the dry season is long as in Nigeria and Sierra Leone, and plants under water-stress are the more severely damaged (41). In the Americas, it would seem that trees already infected with *Ceratostomella* fungus are more readily attacked, attacks being most frequent at ground level. To what extent wind dispersal of the spores is of more importance than the beetle in dissemination is not clear but Iton (79) now believes the beetles carry the pathogen directly (78–80).

Some beetles are of interest in that larvae may damage one part of cocoa, the adults another. Thus, certain eumolpids do minor damage to leaves and petioles in West Africa (1): the discovery that high larval populations build up on roots indicates a need for further investigation (60). But perhaps the chief group is the pachyrynchine weevil genus *Pantorhytes* in New Guinea; the larvae do major damage in burrowing through trunk and main lateral branches, adults feed on leaves and other tissues (128). Good plantation hygiene minimises *Pantorhytes* populations but as they have a range of alternate forest hosts, control is difficult; at least five species attack cocoa (128).

Ants

There seems little doubt that in entomological problems of tropical tree crops ants play the dominant role. This has been demonstrated in coconuts (13, 142, 144). All the work on cocoa farms in Ghana and Nigeria reviewed above leads to a similar conclusion. Ants can directly damage cocoa—leaf-cutters in Latin America and, in a minor fashion, *Macromischoides* in West Africa; they tend mealybugs and other Homoptera, thereby aiding spread of virus diseases; above all, it is as predators that they decide the make-up of the insect spectrum. By examining a range of weeds of the cocoa farm which carry extrafloral nectaries and, scoring for the ants on these, it was concluded that each farm carried a mosaic of dominant species: the mutually exclusive components were three *Crematogaster* species, *Oecophylla longinoda* and, less marked, were various species of *Pheidole*. The ground fauna was overrun by a range of ponerines of which only *Platythyrea* ascended trees (93). The ever-changing boundaries between species and, at least for *Oecophylla*, between colonies of a single species, give rise to gaps in which capsids may flourish: it has been shown that high light intensity minimises the effects of SSV in greenhouses (4) but to get shade removal on the farm is to remove the high forest trees which provide the nest sites of some *Crematogaster*, the chief attendants of SSV-potential mealybugs. Scattered throughout the cocoa insect literature are references to ant predation but, except for recent work in Ghana (93), there have been no numerically valid studies. Indonesia once saw attempts to control *Helopeltis* by introducing nests of a dolichoderine but, of course, this is likely to be difficult and time-consuming (45, 141).

The major environmental factors—water, light, insecticides—are also limiting for ants and it is well known in West Africa that shade removal leads to the build-up of dense populations of ground nesting *Camponotus* species. Shade thinning reduces *Oecophylla* (if the canopy is broken) and eliminates most *Crematogaster,* all of which are capsid predators, and encourages *Camponotus* spp. which are not capsid predators. A considerable number of cocoa-farm ants are primary forest species, again markedly associated with light regimes (93). Little work has been done in cocoa on the effects of spraying upon ants but it has been suggested that the build-up of *Spulerina* may be attributed to mortality among predatory ants rather than in encrytid parasites (55).

It seems to the writer that any attempts at chemical control of pests must first take into account the effect of insecticides on the ant populations. Manipulation of the ant spectrum, perhaps by cultivational methods, might lead to a less damaging insect fauna.

Some General Considerations

Capsid control by chemical sprays at first met with great success in Ghana but the development of resistance in *Distantiella* and the post-independence breakdown in extension services led, in less than ten years or so of spraying, to a virtual impasse (38, 45, 68, 81, 123). Since 1963, teams of scientists have studied resistance, candidate insecticides and capsid biology but much of their work is unpublished. The current situation is that farmers can buy mistblowers at a subsidised price but spraying is a matter for the individual, who is generally undercapitalised and an absentee, leaving farm management to illiterate and unskilled workers. With the present inputs, to use Schultz's (117) phraseology, increased yield is unlikely: socio-economic changes are the desiderata rather than more potent insecticides.

As for swollen-shoot virus, no insecticide programme has yet gone beyond the stage of field trials, wherein systemics have been used (69, 70, 106). But experience in Nigeria shows that capsid control, when done under effective management, also minimises SSV damage (138, 140). In fact, Nigerian practice is largely to ignore virus diseases and to concentrate on capsid control and restoration of the canopy (97).

Perhaps the most useful recent research has been in the field of phenology. Williams (148) worked out the population cycle for capsids in Ghana but only now have phenological data been published on other pests (60, 95). At least it should be possible to use chemical control more efficiently by timing it to coincide with vulnerable development stages in the pest; but, again, adequate extension service is required in order to reach the farmer.

As a tenfold increase in yield is possible under plantation conditions, given capital adequate to support machinery and chemicals, the future might seem to lie in big farming. But, against this, fluctuations in market price are so wide as to deter the investor even where national land policy or unstable regimes do not debar him. We are left, therefore, with no choice but to support the present small farms and to continue recommending this or that formulation to farm labourers who cannot read what's on the tin.

ACKNOWLEDGMENTS

Many colleagues working with the International Capsid Research Team made unpublished work available. The writer was sponsored for three years by the I.C.R.T. and now, while at University of Ghana, is presently in receipt of research grants from the International Office of Cocoa and Chocolate and Ghana Cocoa Growers Research Association. Above all, thanks are due to W. R. Feaver (Cadbury Brothers, Ltd.), H. D. Popplewell (Cadbury and Fry) and Professor D. W. Ewer (University of Ghana) for their continued support.

LITERATURE CITED

1. Alibert, H. Les insectes vivant sur les cacaoyers en Afrique occidentale. *Mem. Inst. Frnç Afrique noire, Dakar,* **15,** 1–174 (1951)
2. Anon. *Cocoa Statistics.* (Gill & Duffus, London, 1968)
3. Asomaning, E. J. A., Kwakwa, R. S. Boron deficiency symptoms in cocoa fruits. *Proc. Conf. Intern. Rech. Agron. Cacao, Abidjan,* 1965, 39–42 (1967)
4. Asomaning, E. J. A., Lockard, R. G. Studies on the physiology of Cocoa (*Theobroma cacao* L.) 1. Suppression of Swollen-Shoot virus symptoms by light. *Ann. Appl. Biol.,* **54,** 193–98 (1964)
5. Attafuah, A., Blencowe, J. W., Brunt, A. A. Swollen-shoot diseases of cocoa in Sierra Leone. *Trop. Agr.,* **40,** 229–32 (1963)
6. Austin, P. *Politics in Ghana.* (Oxford Univ. Press, London, 1964)
7. Ban, N. (Personal communication)
8. Beckett, W. H. Akokoaso: a survey of a Gold Coast village. *Monograph Social. Anthropol.,* **10,** 1–96 (1944)
9. Billes, D. J. Pollination of *Theobroma cacao* in Trinidad, B.W.I. *Trop. Agr.,* **18,** 151–56 (1941)
10. Birmingham, W. B., Neustadt, I., Omoboe, E. N. (Eds.) *A study of contemporary Ghana. 1. The economy of Ghana.* (Allen and Unwin, London, 1966)
11. Boulard, M. Hémiptéroides nuisibles ou associés aux cacaoyers en République Centrafricaine, Premiere Partie. *Café Cacao Theé, Paris,* **11,** 220–34 (1967)
12. Box, H. E. The *Sahlbergella* menace to Gold Coast cocoa. *Mem. Cocoa Res. Sta., Tafo,* **9,** 1–8 (1944)
13. Brown, E. S. Immature nutfall of coconuts in the Solomon Islands. 1. Distribution of nutfall in relation to that of *Amblypelta* and of certain species of ants. *Bull. Entomol. Res.,* **50,** 97–133 (1959)
14. Bruneau de Miré, P. Les recherches sur l'écologie des mirids du cacaoyer (*Sahlbergella singularis* Hagl. et *Distantiella theobroma* Dist.) en Republique du Cameroun. *Proc. Cong. Intern. Rech. Agron. Cacao, Abidjan,* 1965, 128–36 (1967)
15. Callan, E. McC. Cacao stinkbugs in Trinidad, B.W.I., *Rev. Entomol., Rio de Janeiro,* **15,** 321–24 (1944)
16. Carter, W. Notes on some mealybugs of economic importance in Ceylon, *Food Agr. Organ. Plant Protect. Bull., Rome,* **4,** 49–52 (1956)
17. Carvalho, J. C. de M. On the major classification of the Miridae (with keys to subfamilies and tribes and a catalogue of the world genera). *Ann. Acad. Brasil. Cienc.,* **24,** 31–110 (1952)
18. Carvalho, J. C. de M. Catalogo dos mirideos do Mundo, Parte 1, Subfamilies Cylapinae, Deraeocorinae, Bryocorinae. *Arquiv. Mus. Nac.* [*Rio*], **44,** 1–158 (1957)
19. Clark, L. R., Geier, P. W., Hughes, R. D., Morris, R. F. *The Ecology of Insect Populations in Theory and Practice.* (Methuen, London, 1967)
20. Cobben, R. H. *Evolutionary Trends*

in Heteroptera Part I. Eggs, architecture of the shell, gross embryology and eclosion. (Pudoc, Wageningen, 1968)
21. Cochrane, T. W., Entwistle, P. F. Preliminary world bibliography of mirids (= capsids) and other Heteroptera associated with cocoa (*Theobroma cacao* L.) *Cocoa Res. Inst. Tafo,* Lib. List No. 2, ii + 17 (1964)
22. Collingwood, C. A. (Personal communication)
23. Cope, F. W. Pollen incompatibility in Cuatrecasas, J. Cacao and its allies. *U.S. Natl. Museum Bull.,* 447–49 (1964)
24. Cornwell, P. B. Some aspects of mealybug behaviour in relation to the efficiency of measures for the control of virus diseases of cacao in the Gold Coast. *Bull. Entomol. Res.,* **47,** 137–66 (1956)
25. Cornwell, P. B. An investigation into the effect of cultural conditions on populations of the vectors of virus diseases of cacao in Ghana with evaluation of seasonal population trends. *Bull. Entomol. Res.,* **48,** 375–96 (1957)
26. Cornwell, P. B. Movements of the vectors of virus diseases of cacao in Ghana. I. Canopy movement in and between trees. *Bull. Entomol. Res.,* **49,** 613–30 (1958)
27. Cornwell, P. B. Movements of the vectors of virus diseases of cacao in Ghana. II. Wind movements and aerial dispersal. *Bull. Entomol. Res.,* **51,** 175–201 (1960)
28. Cotterell, G. S. Life history and habits, etc. of *Sahlbergella singularis* Hagl. and *Sahlbergella theobroma* Dist. *Bull. Dept. Agr. Gold Coast,* **7,** 40–43 (1927)
29. Cotterell, G. S. Entomology. *Rept. Central Cocoa Res. Sta., Tafo, 1938–42,* 46–51 (1943)
30. Crowdy, S. H. Observations on the pathogenicity of *Calonectria rigidiuscula* (Berk. and Br.) Sacc. on *Theobroma cacao* L. *Ann. Appl. Biol.,* **34,** 45–59 (1947)
31. Cunningham, R. K., Lamb, J. Cocoa shade and manurial experiment in Ghana. *Nature,* **182,** 119 (1958)
32. da Costa Lima, A. Hemipteros, in *Insetos do Brasil,* **2,** 1–351 Rio de Janeiro (1940)
33. Dale, W. T. Diseases and pests of cocoa. A. Virus diseases. In *Agriculture and land use in Ghana,* 286–316. (Wills, J. B., Ed., Oxford Univ. Press, London, 1962)
34. De Jong, J. K. Anatomische waarnemingen bij *Helopeltis* (*Helopeltis antonii* Sigh.), *Arch. Theecult. Ned.-Ind. 1934,* 38–57 (1934)
35. Dessart, P. Contribution a l'etude des Ceratopogonidae, VII. *Proc. Inst. Roy. Sci. Nat. Belg.,* **72,** 1–151 (1963)
36. Donald, R. G. The natural enemies of some Pseudococcidae in the Gold Coast. *J. West Afr. Sci. Assoc.,* **2,** 48–60 (1956)
37. Dun, G. S. Notes on cacao capsids in New Guinea. *Papua New Guinea Agr. Gaz.,* **8,** 7–11 (1954)
38. Dunn, J. A. Insecticide resistance in the cocoa capsid, *Distantiella theobroma* (Dist.). *Nature,* **199,** 1207 (1963)
39. Dunn, J. A. The resistance pattern in a strain of the cocoa capsid (*Distaniella theobroma* (Dist.)) resistant to BHC. *Entomol. Exptl. Appl.,* **6,** 304–8 (1963)
40. Entwistle, P. F. Population studies. *Rept. W. African Cocoa Res. Inst., Tafo, 1957–58,* 43–45 (1958)
41. Entwistle, P. F. Diseases and pests of cocoa. E. Minor insect pests. In *Agriculture and land use in Ghana,* 342–48 (Wills, J. B., Ed., Oxford Univ. Press, London, 1962)
42. Entwistle, P. F. A note on *Eulophonotus myrmeleon* Fldr., a stem borer of cocoa in West Africa. *Bull. Entomol. Res.,* **54,** 1–3 (1963)
43. Entwistle, P. F. The distribution of mirid species and of resistant mirids in Nigeria. *Proc. Conf. Mirids Pests Cacao, Ibadan, 1964,* 9–17 (1964)
44. Entwistle, P. F. Cocoa mirids. Part I. A world review of biology and ecology. *Cocoa Growers' Bull.,* **5,** 16–20 (1965)
45. Entwistle, P. F. Cocoa mirids. Part 2. Their control. *Cocoa Growers' Bull.,* **6,** 17–22 (1966)
46. Entwistle, P. F., Johnson, C. G., Dunn, E. New pests of cocoa (*Theobroma cacao* L.) in Ghana following applications of insecticides. *Nature,* **184,** 2040 (1959)
47. Entwistle, P. F., Youdeowei, A. A preliminary world review of cacao mirids. *Proc. Conf. Mirids Pests Cacao, Ibadan, 1964,* 71–78 (1964)

48. Ezzat, Y. M., McConnell, H. S. A classification of the mealybug tribe Planococcini. *Bull. Md. Agr. Expt. Sta., A 84,* 1–108 (1956)
49. Fennah, R. G. The epidemiology of cacao-thrips on cacao in Trinidad. *Rept. Cacao Res., Trinidad, 1954,* 7–26 (1955)
50. Fennah, R. G. Nutritional factors associated with the development of mealybugs on cacao. *Rept. Cacao Res., Trinidad, 1957–58,* 18–28 (1959)
51. Fennah, R. G. The influence of environmental stress on the cacao tree in predetermining the feeding sites of cacao-thrips, *Selenothrips rubrocinctus* (Giard), on leaves and pods. *Bull. Entomol. Res.,* **56,** 333–49 (1965)
52. Fennah, R. G., Murray, D. B. The cocoa tree in relation to its environment. *Proc. Cocoa Conf., London, 1957,* 222–27 (1958)
53. Fernando, H. E. Studies of *Empoasca devastans* Dist., a new pest of cacao causing defoliation, and its control. *Trop. Agriculturist,* **115,** 121–44 (1960)
54. Fernando, H. E., Manickvasgar, R. Economic damage and control of the cacao capsid *Helopeltis* sp. in Ceylon. *Trop. Agriculturist,* **112,** 23–36 (1956)
55. Gerard, B. M. Side effects from the use of insecticides. II. Effects on insects other than mirids. *Proc. Cacao Mirid Control Conf., Tafo, 1963,* 28–31 (1964)
56. Gerard, B. M. Insects associated with unshaded *Theobroma cacao* L. in Ghana. *Proc. Conf. Mirids Pests Cacao, Ibadan, 1964,* 101–11 (1964)
57. Gerard, B. M. *Bathycoelia thalassina* (Herrich-Schaeffer); a pest of *Theobroma cacao* L. *Nature,* **207,** 881 (1965)
58. Gerard, B. M. Pollination studies. *Rept. Cocoa Res. Sta., Tafo, 1963–65,* 46–47 (1966)
59. Ghauri, M. S. K. A new bryocorine genus and species associated with cocoa in North Borneo. *Ann. Mag. Nat. Hist.,* (13) **6,** 235–40 (1963)
60. Gibbs, D. G., Leston, D. Insect phenology in a forest cocoa-farm locality in West Africa. *J. Animal Ecol.* (In press, 1969)
61. Gibbs, D. G., Pickett, A. D. Feeding by *Distantiella theobroma* (Dist.) on cocoa. I. The effects of water stress in the plant. *Bull. Entomol. Res.,* **57,** 159–70 (1966)
62. Gibbs, D. G., Pickett, A. D., Leston, D. Seasonal population changes in cocoa capsids in Ghana. *Bull. Entomol. Res.,* **58,** 279–93 (1968)
63. Glendinning, D. R. Natural pollination of cocoa. *Nature,* **193,** 1305 (1962)
64. Gold Coast Colony. *Interim report of the committee of enquiry to review legislation for the treatment of the swollen shoot disease of cocoa.* (Accra, Government printing Dept., 19 pp., 1948)
65. Golding, F. D. Capsid pests of cacao in Nigeria. *Bull. Entomol. Res.,* **32,** 83–89 (1941)
66. Goodchild, A. J. P. A study of the digestive tract of the West African cacao capsid bugs. *Proc. Zool. Soc. London,* **122,** 543–72 (1952)
67. Goodchild, A. J. P. Studies on the functional anatomy of the intestines of Heteroptera. *Proc. Zool. Soc. London,* **141,** 851–910 (1963)
68. Hammond, P. S. Notes on the progress of pest and disease control in Ghana. *New Gold Coast Fmr.,* **1,** 109–15 (1958)
69. Hanna, A. D. Heatherington, W., Mapother, H. R., Wickens, R. An investigation into the possible control of the mealybug vectors of cacao swollen-shoot virus by trunk implantation with dimefox. *Bull. Entomol. Res.,* **50,** 209–25 (1959)
70. Hanna, A. D., Judenko, E., Heatherington, W. Systemic insecticide for the control of insects transmitting swollen-shoot virus disease of cocoa in the Gold Coast. *Bull. Entomol. Res.,* **46,** 669–710 (1955)
71. Hill, P. *The Gold Coast cocoa farmer: a preliminary survey.* (Oxford Univ. Press, London, 1957)
72. Hill, P. Social factors in cocoa farming. In *Agriculture and land use in Ghana.* (Wills, J. B., Ed., Oxford Univ. Press, London, 1962)
73. Hill, P. *Migrant cocoa-farmers of southern Ghana.* (Cambridge Univ. Press, Cambridge, 1963)
74. Hinton, H. E. The structure of the shell and respiratory system of the eggs of *Helopeltis* and related genera. *Proc. Zool. Soc. London,* **139,** 483–88 (1962)
75. Houillier, M. Regime alimentaire et disponibilité de ponte des mirides dissimules du cacaoyer. *Rev.*

Pathol. Vegetale Entomol. Agr. France, **43,** 195–200 (1965)

76. Houillier, M. Étude experimentale de la repartition des piqures de mirides (*Sahlbergella singularis* Hagl. et *Distantiella theobromae* Distant) sur cabosse de cacaoyer. *Rev. Pathol. Vegetale Entomol. Agr. France,* **43,** 201–8 (1965)
77. Houillier, M. Influence de l'éclairement du cacaoyer sur le comportement de ponte du miride *Sahlbergella singularis* Hagl. *Rev. Pathol. Vegetale Entomol. Agr. France,* **43,** 209–11 (1965)
78. Iton, E. F. Ceratostomella wilt in cacao in Trinidad. *Proc. Inter-Amer. Cacao Conf., 8th, Trinidad, 1959,* 201–7 (1960)
79. Iton, E. F. Unpublished Doctoral thesis. University of London (1966)
80. Iton, E. F., Conway, G. R. Studies on a wilt disease of cacao at River Estate. III. Some aspects of the biology and habits of *Xyleborus* spp. and their relation to disease transmission. *Rept. Cacao Res., Trinidad, 1959–60,* 59–65 (1961)
81. Johnson, C. G. Diseases and pests of cocoa. B. Capsids: a review of present knowledge. In *Agriculture and land use in Ghana,* 316–31. (Wills, J. B., Ed., Oxford Univ. Press, London, 1962)
82. Johnson, C. G. Diseases and pests of cocoa. F. The ecological approach to cocoa diseases and health. In *Agriculture and land use in Ghana,* 348–52. (Wills, J. B., Ed., Oxford Univ. Press, London, 1962)
83. Kay, D. Die-back of cocoa. *Tech. Bull. West Afr. Cocoa Res. Inst.,* **8,** 5–20 (1961)
84. Kay, D., Longworth, J. F., Thresh, J. M. The interaction between swollen shoot disease and mirids on cocoa in Nigeria. *Rept. Proc. Inter-Amer. Cacao Conf., 8th, Trinidad, 1959,* 224–35 (1961)
85. Kerrich, B. J. Report on Encyrtidae associated with mealybugs on cacao in Trinidad and on some other species related thereto. *Bull. Entomol. Res.,* **44,** 789–810 (1953)
86. Kirkpatrick, T. W. Insect pests of cacao and insect vectors of cacao virus disease. *Rept. Cacao Res., Trinidad, 1945–51,* 122–25 (1953)
87. Lavabre, E. M. Les insectes des stocks de cacao et de café. *Café Cacao Thé (Paris),* **9,** 193–205 (1965)
88. Lavabre, E. M., Decelle, J., Debord. P. Recherches sur les variations des populations de Mirides en Cote d'Ivoire. *Café Cacao Thé (Paris),* **6,** 287–95 (1962)
89. Lavabre, E. M., Decelle, J., Debord, P. Étude de l'évolution régionale et saisonniére des populations de Mirides en Côte d'Ivoire. *Café Cacao Thé (Paris),* **7,** 267–89 (1963)
90. Leston, D. Testis follicle number and the higher systematics of Miridae. *Proc. Zool. Soc. London,* **137,** 89–106 (1961)
91. Leston, D. Systematics of Bryocorinae. *Rept. Cocoa Res. Inst. Tafo 1965–66,* 73–74 (1968)
92. Leston, D. Harmattan in the West African forest zone. *Ghana J. Sci.* (In press)
93. Leston, D. (Unpublished data)
94. Leston, D., Gibbs, D. G. A new deraeocorine predacious on *Mesohomotoma tessmanni* (Aulmann) on cocoa. *Proc. Roy. Entomol. Soc. London, Ser. B,* **37,** 73–79 (1968)
95. Leston, D., Gibbs, D. G. Climate and phenology in a tropical high-forest area. *Oikos, Lund* (In press)
96. Leston, D., Hughes, B. The snakes of Tafo, a forest cocoa-farm locality in Ghana. *Bull. Inst. Fond. Afr. noire, (A),* **30,** 737–70 **(1968)**
97. Longworth, J. F. The effect of swollen-shoot disease on mature cocoa in Nigeria. *Trop. Agr. (London),* **40,** 275–83 (1963)
98. Macfie, J. W. S. Ceratopogonidae collected in Trinidad from cacao flowers. *Bull. Entomol. Res.,* **35,** 297–300 (1944)
99. Marchart, H., Leston, D. Radioisotope tagging experiments. *Rept. Cocoa Res. Inst., Tafo, 1965–66,* 52–54 (1966)
100. McKelvie, A. D. Cherelle wilt of cacao: I. Pod development and its relation to wilt. *J. Exptl. Botany,* **7,** 252–63 (1956)
101. McKelvie, A. D., Cocoa, B. Physiology. In *Agriculture and land use in Ghana,* 256–60. (Wills, J. B., Ed., Oxford Univ. Press, London, 1962)
102. Morales, E. M., Matarrita, A. A. El capsido del cacao y su importancia

en el cultivo del cacao en Costa Rica. *El Cacaotero,* **3,** 11–14 (1961)

103. Murray, D. B. Soil moisture regimes. *Ann. Rept. Cacao Res. Trinidad, 1965,* 34–39 (1966)
104. Nicol, J. Cocoa capsids and their control, 1–6. *Rept. Cocoa Conf. London, 1948,* 54–58 (1948)
105. Nicol, J. Les Insectes du Cacaoyer. *Bull. Agr. Congo Belg., Bruxelles,* **39,** 779–802 (1948)
106. Nicol, J. Systemic insecticides and the mealybug vectors of swollen-shoot virus of cacao. *Nature,* **169,** 120 (1952)
107. Odhiambo, T.-R. Review of some genera of the subfamily Bryocorinae. *Bull. Brit. Museum (Entomol.),* **11,** 245–331 (1962)
108. Posnette, A. F. Transmission of swollen-shoot. *Trop. Agr., Trinidad,* **17,** 98 (1940)
109. Posnette, A. F. Botany. *Rept. Centr. Cocoa Res. Sta. Tafo, 1938–42,* 19–30 (1943)
110. Posnette, A. F. The pollination of cacao in the Gold Coast. *J. Hort. Sci.,* **25,** 155–63 (1950)
111. Posnette, A. F. Virus diseases of cacao in West Africa VII. Virus transmission by different vector species. *Ann. Appl. Biol.,* **37,** 378–84 (1950)
112. Posnette, A. F., Entwistle, H. M. The pollination of cocoa flowers. *Proc. Cocoa Conf., London, 1957,* 66–68 (1958)
113. Posnette, A. F., Robertson, N. F. Virus diseases of cacao in West Africa. VI. Vector investigations. *Ann. Appl. Biol.,* **37,** 363–77 (1950)
114. Prins. G. A laboratory rearing method for the cocoa mirid *Distantiella theobroma* (Dist.). *Bull. Entomol. Res.,* **55,** 615–16 (1965)
115. Raw, F. An insectary method for rearing cacao mirids, *Distantiella theobroma* (Dist.) and *Sahlbergella singularis* Hagl. *Bull. Entomol. Res.,* **50,** 11–12 (1959)

115a. Saunders, L. G. *Can. J. Zool.* **34,** 657–705 (1956); **37,** 33–51 (1959); *Tech. Rept. Cocoa Res. Inst., Tafo,* 6 pp. (1963)

116. Schouteden, H. Sahlbergella nouveaux du Congo Belge. *Rev. Zool. Botan. Afric.,* **26,** 473–76 (1935)
117. Schultz, T. W. *Transforming Traditional Agriculture.* (Yale Univ. Press, New Haven, 1964)
118. Silva, P. Insect pests of cacao in the State of Bahia, Brazil. *Trop. Agr. Trinidad,* **21,** 8–14 (1944)
119. Silva, P. Problemas entomologicas do Cacaueiro com referencia especial à Bahia. *Reun. IV. Com. Tec. Interameric. Cacau,* 60–72 (1957)
120. Smee, L. Insect pests of *Theobroma cacao* in the Territory of Papua and New Guinea: their habits and control. *Papua N. Guinea Agr. J.,* **16,** 1–19 (1963)
121. Smith, M. R. A list of Lepidoptera associated with cocoa in West Africa with notes on identification and biology of species in Ghana. *Tech. Bull. Cocoa Res. Inst., Tafo,* **9,** 1–68 (1965)
122. Squire, F. A. On the economic importance of the Capsidae in the Guinea region. *Rev. Entomol. Rio de Janeiro,* **18,** 219–47 (1947)
123. Stapley, J. H., Hammond, P. S. Large-scale trials with insecticides against capsids on cacao in Ghana. *Empire J. Exptl. Agr.,* **27,** 343–53 (1959)
124. Steven, W. F. A new disease of cocoa in the Gold Coast. *Gold Coast Farmer,* **5,** 122–44 (1936)
125. Strickland, A. H. The entomology of swollen shoot of cacao. I. The insect species involved, with notes on their biology. *Bull. Entomol. Res.,* **41,** 725–48 (1951)
126. Strickland, A. H. The entomology of swollen shoot of cacao. II. The bionomics and ecology of the species involved. *Bull. Entomol. Res.,* **42,** 65–103 (1951)
127. Sumner, H. M. Pollination. In *Agriculture and land use in Ghana* 260–61. (Wills, J. B., Ed., Oxford Univ. Press, London, 1962)
128. Szent-Ivany, J. J. H. Insect pests of *Theobroma cacao* in the Territory of Papua and New Guinea. *Papua N. Guinea Agr. J.,* **13,** 127–47 (1961)
129. Szent-Ivany, J. J. H. Further records of insect pests of *Theobroma cacao* in the Territory of Papua and New Guinea. *Papua Guinea Agr. J.,* **16,** 37–44 (1963)
130. Szent-Ivany, J. J. H. Insect pests of *Theobroma cacao* L. in the Territory of Papua and New Guinea. *Proc. Conf. Mirids Pests Cacao. Ibadan, 1964,* 85–89 (1964)
131. Szereszewski, R. *Structural changes in the economy of Ghana, 1891–*

1911 (Weidenfeld & Nicholson, London, 1965)

132. Taylor, D. J. A summary of the results of capsid research in the Gold Coast. *Tech. Bull. W. Afr. Cocoa Res. Inst.*, **1**, 1–20 (1954)

133. Taylor, D. J. Capsid studies: light trapping. *Rept. W. Afr. Cocoa Res. Inst. Tafo, 1955–56,* 54–55 (1957)

134. Thomas, A. S. The dry season in the Gold Coast and its relation to the cultivation of cacao. *J. Ecol.*, **20,** 263–69 (1932)

135. Thresh, J. M. The control of cacao swollen shoot disease in West Africa. *Tech. Bull. W. Afr. Cocoa Res. Inst.*, **4,** 1–36 (1958)

136. Thresh, J. M. The spread of virus disease in cacao. *Tech. Bull. W. Afr. Cocoa Res. Inst., Tafo,* **5,** 1–36 (1958)

137. Thresh, J. M. The control of cacao swollen shoot disease in Nigeria. *Tropical Agr. (London),* **36,** 35–44 (1959)

138. Thresh, J. M. Capsids as a factor influencing the effect of swollen-shoot disease on cacao in Nigeria. *Empire J. Exptl. Agr.*, **28,** 193–200 (1960)

139. Thresh, J. M., Tinsley, T. W. The viruses of cacao. *Tech. Bull. W. Afr. Cocoa Res. Inst.*, **7,** 1–32 (1959)

140. Tinsley, T. W. The ecological approach to pest and disease problems of cocoa in West Africa. *Trop. Sci.*, **6,** 38–46 (1964)

141. Van der Meer Mohr, J. C. Au sujet du rôle de certaines fourmis dans les plantations coloniales. *Bull. Agr. Congo Belge,* **18,** 97–106 (1927)

142. Vanderplank, F. L. The bionomics and ecology of the red tree ant, *Oecophylla* sp., and its relationship to the coconut bug, *Pseudotheretus wayi* Brown. *J. Animal Ecol.*, **29,** 15–33 (1960)

143. Vernon, A. J. Cocoa yield increase by control of capsids. *Chem. Ind. (London), 1964,* 320–21 (1964)

144. Way, M. J. The relationship between certain ant species with particular reference to biological control of the coreid *Theraptus* sp. *Bull. Entomol. Res.*, **44,** 669–91 (1953)

145. Williams, D. J. The mealybugs described by W. J. Hall, F. Laing, A. H. Strickland from the Ethiopian Region. *Bull. Brit. Museum (Entomol.),* **7,** 1–37 (1958)

146. Williams, G. Field observations on the cacao mirids, *Sahlbergella singularis* Hagl. and *Distantiella theobroma* (Dist.), in the Gold Coast, I. Mirid damage. *Bull. Entomol. Res.*, **44,** 101–19 (1953)

147. Williams, G. Field observations on the cacao mirids, *Sahlbergella singularis* Hagl. and *Distantiella theobroma* (Dist.), in the Gold Coast, II. Geographical and habitat distribution. *Bull. Entomol. Res.*, **44,** 427–37 (1953)

148. Williams, G. Field observations on the cacao mirids, *Sahlbergella singularis* Hagl. and *Distantiella theobroma* (Dist.), in the Gold Coast III. Population fluctuations. *Bull. Entomol. Res.*, **45,** 723–44 (1954)

149. Youdeowei, A. Progress in the laboratory rearing of cocoa mirids in Nigeria. *Proc. Conf. Mirids Pests Cacao, Ibadan,* 1964, 98–100 (1964)

150. Youdeowei, A. A note on the spatial distribution of the cocoa mirid *Sahlbergella singularis* Hagl. in a cocoa farm in Western Nigeria. *Nigerian Agr. J.*, **2,** 66–67 (1965)

POME FRUIT PESTS AND THEIR CONTROL

By Harold F. Madsen and C. V. G. Morgan[1]

Research Station, Research Branch, Canada Department of Agriculture, Summerland, British Columbia, Canada

Research on control of pome fruit pests since the last review of this subject by Barnes (13) has emphasized an integrated control approach, although practical control measures remain largely chemical. There has been considerable progress in developing control measures that are nonchemical in nature, but implementation of these new approaches is slow. The emphasis on integrated control has been stimulated by pest resistance to spray chemicals and legal restrictions on registration of new compounds. Any review of pest control practices over the past ten years would be remiss without mentioning the impact of Rachel Carson's book *Silent Spring* (44). Although the merits of the book have been vigorously debated, there is no question that it has influenced public attitudes towards pesticides and stimulated support of research on control measures other than chemical.

A review of the developments in pest control on pome fruits is best presented by discussing the principal insects and mites which attack apples and pears. The codling moth is still the key insect in pome fruits because it requires a schedule of preventative sprays in most fruit growing areas of the world.

CODLING MOTH

Chemical Control

DDT was the principal chemical used to control codling moth, *Carpocapsa pomonella,* during the 1950's, but the development of resistance brought a shift to other insecticides. DDT resistance in the field was confirmed by laboratory tests and reported from the United States (104), Canada (71), Australia (123), and other apple and pear growing regions of the world. The degree of resistance varied, but most laboratory tests indicated the adults were more tolerant to DDT than the larvae (10). There was no indication that any cross-resistance to the principal substitute insecticides azinphosmethyl and carbaryl was present (14).

Putnam (203), in his comprehensive review of the codling moth, pointed out that the difficulty of controlling the codling moth is proportional to the number of generations. In North America, the codling moth has two or

[1] The authors express their appreciation to Miss B. Jessica Angle and Miss Rossalynn Norman for assisting with the literature search and preparation of the manuscript.

more generations in most fruit growing areas and is univoltine only along the northern fringe of apple production. A similar situation exists in Europe and Asia, but multiple generations are common in most apple and pear growing regions of the world. Because of the necessity of maintaining an effective insecticide deposit on leaves and fruit, most investigations with new spray chemicals have been with persistent compounds. Although DDT has been supplanted by other insecticides, it is still an effective insecticide for codling moth control in several countries. Azinphosmethyl and carbaryl have given the best residual control of a wide range of insecticides tested against the codling moth (176). Other materials which have given adequate control are Zectran® (4-dimethylamino-3,5-xylyl methylcarbamate), fenthion (100), trichlorfon (161), carbophenothion (89), ethion, dioxathion (25), and phosalone (45). A number of organophosporous compounds, in addition to providing protection, kill larvae after they enter the fruit (61).

Studies on the intrinsic toxicity of insecticides to codling moth larvae (89), demonstrated that carbophenothion, parathion, and diazinon have the highest toxicity closely followed by azinphosmethyl. Malathion, DDT, and carbaryl were of low toxicity. When these same materials were evaluated for minimum days of effective deposit levels, DDT was effective for 24 days, carbaryl, 14, azinphosmethyl, 12, parathion and carbophenothion, 11, malathion, 7, and diazinon for 6.

In contrast to the general opinion that residual materials are needed for codling moth control, short-lived products were used effectively in an area where codling moth development was continuous (69). When materials were compared at treatment intervals of 11 to 16 days, dimetan, phosphamidon, dimethoate, and parathion all gave good control.

The side effects of codling moth sprays on nontarget organisms influence the choice of an insecticide for local recommendations. Mite increases have been reported following carbaryl applications (48), but not following sprays of azinphosmethyl or diazinon (224).

Chemical control of the codling moth is dependent upon correct timing of sprays, which must be adjusted to local conditions. In areas where two generations per season occur, two sprays for the first brood and two for the second are required (2). Attempts to control larvae seeking winter quarters have not been generally successful, but it has been reported (93) that spraying the trunks and scaffold branches with azinphosmethyl, diazinon, or parathion left insecticide deposits toxic to these larvae. In most current spray recommendations, azinphosmethyl is the preferred insecticide for codling moth control.

The current insecticides do not create residue problems on fruit when used according to label directions (239). There have been efforts to increase the persistence of materials used for codling moth control with conflicting results. In one instance (8) glyodin improved the control when added to azinphosmethyl, carbaryl, and ethion, but in another (25), the effectiveness of azinphosmethyl was reduced by the addition of glyodin. In Canada, the persistence of azinphosmethyl was reduced when combined with petroleum oil (147). The addition of urea (106) to DDT reduced its effectiveness against codling moth both in the laboratory and in the field.

One interesting observation was made after the introduction of organophosphorous compounds for codling moth control. Asquith (9) stated that during the lead arsenate era, the early segment of the first generation of codling moth was the critical one that had to be controlled. Since the organophosphorous compounds have been used, this segment of the first brood is rarely a problem and the last segment of the first generation has become of paramount importance. It was suggested that this may be a result of selection by the codling moth in response to a changed spray schedule.

Behavior studies and investigations on adult activity as reflected by attraction to various lures have contributed to improvements in control of the codling moth. In nonsprayed orchards, damage occurs every season where multiple generations exist (175). Although biology studies from different regions indicate variation in the emergence of the overwintered generation and the summer generations, there seems to be little difference within a given region (150). This has resulted in fairly uniform control recommendations from year to year within each area. Although both apples and pears are readily attacked by codling moth, there is a difference in the time of larval development in the two hosts (255), and this creates the need for separate spray schedules for second-brood codling moth in some regions.

Madsen (138) has reviewed the various lures that have been used to time spray applications for codling moth control and indicated their relative effectiveness. Investigations in the last ten years have shown that light traps are far more accurate than bait pans. Females attracted to molasses bait pans are reproductively old (78), and probably are not attracted until after eggs have been laid and the first eggs hatched. Blacklight traps, containing 15 watt ultraviolet tubes attracted both male and female codling moths, and 90 per cent of the captured females were reproductively young (77). Data on flight behavior of codling moths based upon blacklight traps showed no flights at temperatures below 12° C, and that heavy rains reduced or prevented flights (213). There has been good correlation between maximum flights and maximum oviposition by females (170) and the development of low intensity, battery-operated blacklight traps (15) has made these traps practical for orchard use. The most recent development in attractants is the use of sex pheromone traps (43) to lure the male codling moth, but it has not been determined how these traps can be used as a means of timing spray applications.

European authors have reported on warning services for growers in order to time the first spray for codling moth control correctly. The timing has been determined in two ways, one by emergence of adults from cages containing overwintered larvae and the other by the sum of mean daily temperatures. Results with the use of cages has been satisfactory in some cases (11) and too variable in others (67). When using temperature data, it was suggested that the first spray be applied when the sum of mean daily temperatures above a threshold development of 10° C reached 218° C (246). In some areas the sum of effective temperatures led to a variation of 6 to 18 days in the forecast date between localities (220). Better results were obtained from combining observations on adult emergence from cages and adult activity as determined by trapping.

Control by Means Other Than Chemical

Biological control of the codling moth has not been of practical value in most apple and pear orchards but parasites have contributed to a general reduction in codling moth populations. A number of parasites have been reared from overwintering larvae of codling moth. In one study (76) the larvae were parasitized by three species of Ichneumonidae, seven of Braconidae and three of Tachinidae. The principal larval parasites have been *Ephialtes caudatus, E. punctulatus, Pristomerus vulnerator* and *Ascogaster quadridentata* (254). Parasites were effective only in unsprayed orchards, and were in low numbers or absent in orchards sprayed with DDT or azinphosmethyl. There was an increase in parasites reared from codling moth larvae in orchards where selective sprays were employed (188). Encouraging results from egg parasitism by several species of *Trichogramma* have been obtained by European workers. The release of laboratory-reared *Trichogramma cacoeciae* resulted in a reduction of codling moth damage ranging from 35 to 65 per cent (228) in experimental plots. In seasons with high temperatures and where codling moth populations were present, egg parasitism was reduced and ranged from 10 to 45 per cent. These parasites were normally released from cages suspended in the trees but in one case were sprayed into the crown of the tree (219). Because of the interest in the use of *Trichogramma,* a method of mass rearing the parasites has been developed using codling moth eggs obtained from laboratory colonies (62). Associated with the use of parasites was the need for providing a favorable environment for optimum parasite development. In addition to the use of selective insecticides, it was reported that the orchard ground cover was important (130) and that parasitism of codling moths was five times greater in orchards with a source of wild flowers than in orchards with poor floral undergrowth.

Parasitic nematodes have been evaluated as biological control agents for several apple pests (114) but have not been effective against codling moth. This is probably due to the difficulty in contracting larvae and poor nematode survival in dry habitats.

Fungi have been evaluated in the laboratory for effects upon codling moth, and *Beauveria bassiana, B. lobulifera,* and *Metarrhizium anisopliae* were toxic to codling moth larvae (7). In nature, fungi attack the overwintered larvae and *B. bassiana* significantly reduced codling moth populations in the field (212). The fungi are often more effective when combined with a pesticide than when used alone (234), and a combination of *B. bassiana* and DDT gave 61 per cent reduction of the first generation and 71 per cent reduction of the second generation when sprayed on bark against larvae seeking cocooning sites. *Bacillus thuringiensis* effectively controlled the codling moth in laboratory tests (113), but was inadequate in field trials although 50 per cent reduction of fruit injury was obtained (171).

A naturally occurring granulosis virus of the codling moth was recently described (233) from larvae collected on apple and pear trees in Chihuahua, Mexico. When the virus was propagated in codling moth larvae and sprayed on apple trees in the field, a majority of the larvae perished soon after feeding upon the surface of sprayed fruit. Preliminary results indicate considerable potential for the virus if mass culture techniques can be developed (70).

Autocidal control of the codling moth has received considerable attention in recent years since this method has the potential of eradication. The subject of induced sterilization and control of insects including the codling moth has been reviewed by Proverbs (202). Since that review, the method has been tested in a commercial orchard and plans are now formulated for an area release trial (201). Along with the studies on autocidal control of the codling moth, an artificial medium for mass culture has been developed (205).

Integrated Control

The philosophy of integrated control of orchard pests and pest management has been discussed by a number of authors including Geier (79) and Beirne (17). Application of the principle to orchard entomology and implementation in integrated control programs has been slow because of the presence of key pests such as the codling moth which require a planned schedule of sprays that are not compatible with an integrated control program. On apples and pears, the broad-spectrum pesticides used for codling moth control precluded the widespread use of pest management practices (139). However, in areas where the codling moth is univoltine, integrated control programs have been developed and put into practice. Nova Scotia has pioneered research on integrated control in apple orchards and the method has been adopted by growers. In Nova Scotia, a number of natural enemies can reduce the overwintering larvae, and woodpeckers play a significant role (133). The key insecticide for codling moth control in this integrated control program was ryania, and when combined with a program of selective sprays for other pests, an environment was created which permitted the survival of arthropod predators (135). The use of either ryania or lead arsenate with glyodin has formed the basis of integrated control in several other areas where the codling moth is univoltine (253) or where populations are low (172). Predators and parasites of codling moth eggs have increased in integrated orchards and the predators included mirids, coccinellids, pentatomid bugs, clerid beetles, and thrips (134).

In areas where the codling moth has multiple generations, modified spray programs have allowed biological control of a few other apple pests. The use of ethion to control codling moth in France did not interfere with the release of parasites against woolly apple aphid, *Eriosoma lanigerum* (21). In the Pacific northwest apple region of the United States, low rates of azinphosmethyl have given adequate control of the codling moth and allowed biological control of the McDaniel spider mite, *Tetranychus mcdanieli* (109).

TORTRICIDS AND OTHER LEPIDOPTEROUS PESTS

The fruit-tree leaf roller, *Archips argyrospilus,* is a sporadic pest of apples and pears and is rarely encountered in orchards that receive a full spray program for codling moth control (103). It is listed among the pests that cause direct fruit damage in nonsprayed orchards (175) where the intensity of its attack varies from year to year. Some of the factors reported to cause population fluctuations of the fruit-tree leaf roller are egg dessication in winter, dispersal of larvae in spring, destruction of larvae by birds, parasitism of pupae, and a slight preponderance of males in the adult population (179). Some 25 species of parasites have been recovered from larvae and pupae of the fruit-tree leaf roller, and a seven-year study of the population dynamics of this pest in Quebec (181) showed that parasites which attack larvae are not as important in population regulation as parasites which attack pupae.

When sprays are necessary for the control of the fruit-tree leaf roller, DDT is generally used. In one instance, the use of DDT in an integrated orchard was followed by an outbreak of the eye-spotted bud moth, *Spilonota ocellana,* and the pistol casebearer, *Coleophora serratella* (132). In eastern Canada, the peak of fruit-tree leaf roller emergence coincides with flowering of McIntosh apple trees (178) and the best time to spray was at petal fall. Other materials which adequately controlled the fruit-tree leaf roller included carbaryl (63), TDE, Perthane® (1,1-dichloro-2,2-bis (*p*-ethylphenyl) ethane), and azinphosmethyl (180).

Recent progress on control of the codling moth by the sterility method has led to speculation that other pests may become troublesome when codling moth sprays are no longer applied. One of those potential pests is the fruit-tree leaf roller. In anticipation of a possible increase of fruit-tree leaf roller, it was demonstrated (141) that a prebloom spray of azinphosmethyl controlled the pest without disrupting the release program of sterile codling moths and without upsetting the biological control of orchard mites.

Resistance to TDE by the red-banded leaf roller, *Argyrotaenia velutiana* (92), stimulated studies on new chemicals and alternative methods of control. Carbaryl, azinphosmethyl (9), and diazinon (176) gave good control, and a systemic insecticide carbofuran controlled the insect on nursery trees (227). In Ontario, a recent outbreak of the red-banded leaf roller was traced to resistance to insecticides, improved orchard practices which created shoot growth favorable to the insect, and the use of air-blast sprayers which did not cover the ground vegetation (103).

A granulosis virus was tested in the laboratory and field (83), and although most of the larvae were destroyed, damage to the fruit was not prevented. *B. thuringiensis* was reported to control the insect (159), but has not been recommended in commercial orchards. The red-banded leaf roller is attacked by a number of parasites, and in one study (173) 22 species of Hymenoptera were reared from larvae and pupae. *Trichogramma minutum* has been reported to parasitize the eggs (85).

A continuous rearing method was developed for red-banded leaf roller (86) and an artificial diet (206) permitted mass culture. These developments plus the isolation of the female sex pheromone (211) may lead to control of the pest by biological rather than chemical means.

The eye-spotted bud moth, *Spilonota ocellana,* is particularly susceptible to natural mortality factors and therefore can be controlled by an integrated program. The larvae are attacked by a number of parasites, predators, and pathogens (128) including a nuclear-polyhedrosis virus disease (115). In Wisconsin (129), pupal mortality from undetermined causes was considered an important factor in natural reduction of the eye-spotted bud moth. A detailed study of the population dynamics of the insect in Quebec (131) showed that a number of mortality factors regulated its abundance. One of the more important factors was winter cold, and it was concluded that if winter temperature dropped to −21°F or lower, there was no need to apply sprays the following summer.

The eye-spotted bud moth is susceptible to *B. thuringiensis* (127), but sprays of spore preparations did not provide commercial control in orchards. Most of the chemicals used to control the pest fit into a program of selective or modified sprays, and integrated programs are reported to result in satisfactory control (188).

The winter moth, *Operophthera brumata,* a pest of apples in many areas of Europe, was accidentally introduced into North America, but it is not considered to be a serious pest as long as orchards receive a regular program of spray chemicals for other insects. It does, however, present a problem in areas where an integrated control program is in use. In Nova Scotia, a selective insecticide program resulted in a reduction of several major pests, but the winter moth required special insecticidal treatments (183). In forested areas, populations were reduced by the introduction of two parasites, *Cyzenis albicans* and *Agrypos flaveolatum* (68), but in sprayed orchards the parasites were rarely found (136). The insect is very cold-hardy, and climate is not expected to limit its spread over eastern North America. Although low rates of DDT and azinphosmethyl are effective against winter moth, two applications were required for good control and this program upset the natural balance in integrated orchards (218). Lead arsenate has been the preferred pesticide for integrated control of this pest, but restrictions on residues in foreign markets may restrict the use of this material. The winter moth illustrates how the introduction of a new pest may upset carefully planned integrated control programs.

APHIDS

The three species of aphids considered to be the most serious pests of apples are the apple aphid, *Aphis pomi,* the rosy apple aphid, *Dysaphis plantaginea,* and the wooly apple aphid, *Eriosoma lanigerum.*

The apple aphid can be controlled with insecticides directed against overwintering eggs, but reinfestation in the spring and summer necessitates foliage sprays. Most aphicides gave good initial kill of apple aphids, but reinfestation was rapid following sprays of diazinon, carbophenothion, endosulfan, or azinphosmethyl (191). Carbaryl gave residual action for 17 days (190), and the best control was obtained from the systemic compounds mevinphos and dimethoate (192).

Studies on economic levels of the apple aphid showed that the distribution of aphids on a growing shoot was more important than the actual number of aphids per leaf (143). Apple aphids prefer young leaves, and the offspring per female was 60 to 90 per cent lower on old leaves than new growth (251). When the population increased so that leaves near the center of a shoot were infested, honeydew damage to fruit was likely to occur. Evidence that the apple aphid can artifically transmit *Erwinia amylovera,* the organism causing fireblight (193), could lower the economic level for this aphid.

A number of mortality factors contribute to the natural control of the apple aphid. In one study (174), parasitism was less than 1 per cent, but a number of predators attack the aphids. Syrphids, lacewings, anthocorid nymphs, and coccinellids are effective predators and reduce the apple aphid in integrated control orchards. Other factors contributing to apple aphid mortality are extremes of weather in the fall when sexual forms are present and the reduction of overwintering eggs by pruning practices (250).

The rosy apple aphid is easily controlled by dormant sprays of oil plus dinitrocresol or dinitrocyclohexylphenol. These mixtures were injurious to cover crop plants, and combinations of oil plus ethion or carbophenothion gave satisfactory control with no damage to the cover crop (140), Bonnemaison (33) has conducted a thorough study of the bionomics of the rosy apple aphid in France and found that in the study area, migration to a secondary host was obligatory. Parasites were of little significance in controlling the insect and predators offered only limited control. Because aphid attack causes leaves to curl tightly, summer sprays of systemic insecticides or chemicals with a fumigant action provided the best control. The population of winter eggs was dependent upon time of leaf drop and rains during migration which killed many alates. In studies on fall sprays, parathion or endrin applied in November against egg-laying females gave better than 95 per cent control the following year (105).

The woolly apple aphid is a major pest of apples in most apple-growing regions of the world. It infests both roots and aerial portions of the tree, and dispersal of nymphs from roots creates reinfestation problems following chemical sprays for control (110). Control on nursery stock is difficult, and although high rates of BHC, lindane, and carbaryl were effective as soil drenches and dips (94), they were impractical because of cost. In field trials, systemic insecticides gave good control of woolly aphids, and the most effective materials were methyl-demeton, dimethoate, thiometon, and vamidothion (26). Although the systemic insecticides were less toxic to the principal parasite of the woolly aphid, *Aphelinus mali,* than contact insecticides (46), parasite mortality was still high. Conclusions were that most sprays directed against the woolly apple aphid were detrimental to *A. mali* unless applied before June (34).

The subject of host resistance to wooly apple aphid has been reviewed by Knight et al. (125). The authors list 18 apple varieties which are resistant to woolly apple aphid, 19 additional which are resistant in most localities and susceptible in a few, and 43 with conflicting data on resistance. Of the standard varieties, Northern Spy has maintained outstanding resistance wherever it is grown.

APPLE MAGGOT

The apple maggot, *Rhagoletis pomonella,* continues to be an important pest of apples in northeastern United States and southeastern Canada. It has not been reported outside the continent which indicates that the measures used to prevent its introduction have been successful. Rivard (210) has prepared a comprehensive synopsis and annotated bibliography of the apple maggot which summarizes the present knowledge of the biology and control of this pest.

PEAR PSYLLA

There are several species of psylla which attack pears in Europe and those of major importance are *Psylla pyricola, P. pyri, P. pyrisuga,* and *P. melanoneura* (32). Only *P. pyricola* will be considered in this review because of space limitations and the importance of this species in North America as well as in Europe.

The pear psylla is important not only because of the direct injury it causes to pear trees, but it has been identified as the vector of pear decline virus (116) and leaf curl, a related disease of pear trees (90).

The pear psylla is a prolific insect and in one study an average of 664 eggs were laid per female (40). The number of generations per year vary from four in cool areas (252) to seven in warmer climes (55). This potential for rapid increase makes multiple spray applications necessary to obtain control. The pear psylla will colonize and maintain a population on abandoned pear trees. If the trees have been abandoned for several years only low populations are present, but those abandoned one or two years support high populations capable of migrating to commercial orchards (249).

In eastern North America, the pear psylla is susceptible to a number of insecticides, but in the West, the pear psylla has developed resistance to a wide range of pesticides. Burts (39) has summarized the resistance pattern starting with parathion. The pear psylla first developed resistance to this compound with a cross-resistance to malathion, diazinon, and other related organophosphorous compounds. Dieldrin then failed to control the pest, and laboratory tests showed cross-resistance to endrin and toxaphene (97). Azinphosmethyl provided control for a short period until resistance developed to this pesticide with cross-resistance to BHC, mevinphos, and a number of experimental organophosphorous compounds. At the present time, Perthane®, Dilan® [2-nitro 1,1-bis(*p*-chlorophneyl) propane and butane mixture (1 to 2 ratio)], and Morestan® [6-methyl-2,3-quinoxaline dithial cyclic S, S-dithio carbonate] are the only effective commercially available insecticides for pear psylla control in the Pacific northwest.

The problem of resistance led to studies on the use of petroleum oil for pear psylla control. Dormant oil alone or in combination with dinitrocresol gave good control of overwintered adults and prevented oviposition on oil-sprayed twigs by surviving adults (142). Oil plus lime sulfur was less effective than oil alone. In the eastern United States, oils were toxic to psylla adults and nymphs, but eggs were unaffected by direct treatments with concentrations of 20 per cent or higher (223). In the West, oils were toxic to pear psylla eggs (145) which may have been due to higher temperatures and lower humidity in this region. Summer oil sprays with both supreme and superior-type oils gave good pear psylla control if treatments were properly timed and complete coverage was obtained (146). An area control program using Perthane® against overwintered adults resulted in improved seasonal control and reduced the need for multiple summer sprays (41).

The pear psylla is attacked by a number of predators, the more important species being *Anthocoris melanocerus* (137), *A. antevolens, Symphero-bius angustus* (148), *Chrysopa oculata, Campylomma verbasci,* and *Deraeo-coris brevis piceatus* (161). A parasite, *Trechnites insidiosus* has been reported from several areas, but hyperparasitism limited its effectiveness. Another parasite, *Prionomitus mitratus,* was introduced into British Columbia, but did not significantly reduce pear psylla populations (160). Natural control of the pear psylla by predators and parasites was reported from nonsprayed pear orchards, but control by natural enemies has not been demonstrated in commercial orchards (248). Other natural conditions which suppressed pear psylla populations were summer temperatures above 100°F, the absence of new growth, and poor condition of foliage (144).

Male pear psylla have been sterilized with tepa (119), and cage tests showed that fertile females mated to tepa-sterilized males laid infertile eggs. Tepa, however, was toxic to the male pear psylla, and the difficulty of developing a means of mass culture makes autocidal control of this insect improbable.

SCALES

Three monographs published during the past decade have made significant contributions to the knowledge of the ecology and control of the armored scale insects of apple and pear. One of the most important is a catalogue of world literature of the Diaspididae up to 1964 (36). The author, N. S. Borchsenius, has included a list of all the journals (and their abbreviated titles) in which the literature has been published; the distribution, food plants, and bibliography for each species; an index to the scientific names of the scales; a host index; and a list of the English and Russian common names of the insects. Kosztarab's review of the armored scale insects of Ohio is mainly a taxonomic treatise with a selected list of literature; an index to families, subfamilies, tribes, genera, and species; and an index to hosts plants (126). Applied entomologists will be interested in his notes on the biology and economic importance of each species and the parasitic and predacious insects, mites, and fungi reared from, or associated with, each species. The population studies of Samarasinghe & LeRoux (216, 217) are a milestone in the biological control of the oyster shell scale, *Lepidosaphes ulmi*. By constructing 81 life tables for three generations they were able to show that the key factors in the regulation of low-to-medium populations of the oyster shell scale on apple trees in Quebec are the mite predator *Hemiscarcoptes malus* and the chalcid parasite *Aphytis mytilaspidis*. The more important mortality factors of other scale insects could possibly be elucidated by similar studies. Many other taxonomic and biological papers contain valuable information for anyone interested in the control of scale insects (35, 37, 38, 74, 82, 165, 184, 197, 209, 236–238, 241).

About 45 to 50 species of armored scales occur throughout the world on trees of the genera *Malus* and *Pyrus* (36) and there may be as many as seven or more species in any one geographical area (126, 235, 240). Though many are of economic importance, the following review deals chiefly with the San Jose scale, *Aspidiotus* (=*Quadraspidiotus*=*Diaspidiotus*) *perniciosus,* as it is generally the most troublesome species throughout the world. Nevertheless, in some areas, the European fruit scale, *A. ostreaeformis,* is the dominant species (163, 208). The distribution of a number of species have been mapped (4). Reports are issued annually on the distribution, degree of infestation, and control measures of the San Jose scale in Europe and the Mediterranean basin (6).

During the last ten years, the San Jose scale has extended its range of distribution and several countries have become infested for the first time (3, 5, 245). As a result, more countries are imposing quarantine regulations and it is becoming increasingly difficult to export fruit. The regulations may be eased if infested apples are kept in standard cold storage or controlled atmosphere storage which kills the San Jose scale but not the European fruit scale (162). Eradication by insecticides has been attempted in several countries but has been successful only in a few orchards of France and Switzerland (16, 152, 185).

Comparatively little work has been done on the San Jose scale on apple and pear in North America during the past decade. The implications are that the pest has not been a serious problem on this continent, that it is not spreading, or that it is being effectively controlled. However, in the fruit-growing areas of the Pacific northwest, especially in British Columbia, control measures have had to be modified to cope with increased injury caused by San Jose scale and the European fruit scale (164). Current insecticides used for the control of the San Jose scale in the United States and Canada consist mainly of petroleum oil or wettable powder formulations of organophosphorous compounds such as parathion or diazinon. The oils are applied alone in the dormant stage and up to the tight cluster bud stage. Ethion, carbophenothion, lime sulphur, or dinitrocresol are often added to the oil if mites and aphids are also problems. Parathion or diazinon may be applied in the tight cluster bud stage but they are more commonly used for the summer control of the San Jose scale. Two summer sprays are often applied, the first when the crawlers begin to appear and the second about two weeks later. Though parathion has been used for many years in some areas there have been no authenticated reports of the development of resistant strains of the San Jose scale.

In other countries, substituted dinitrophenols are the insecticides most commonly used in the dormant period, but because of their toxicity they are gradually being replaced or supplemented by organophosphorus compounds (58, 120, 124, 194). The dinitro compounds are often used with oil as this combination gives simultaneous control of other pests (226). In some areas, this mixture is more effective than parathion plus oil (199, 200, 226). The relative effectiveness of the organophosphorous insecticides vary in different areas but generally it is recognized that parathion is the best and that it is most effective when used with oil (66, 238, 243). Certain types of oils when used alone in sufficient quantities are highly effective against scale insects (91, 182, 221, 222, 242). Many other insecticides, fungicides, and miticides are toxic to scales and some such as carbaryl, DDT, azinphosmethyl, demeton, and dimethoate are recommended for their control (27, 28, 98, 198).

Of all research done in the last ten years on the San Jose scale on apple and pear trees, more has been done on its biological control than on any other aspect. Numerous reports have been published especially in Europe and Asia. Probably one of the reasons why the San Jose scale has not been too troublesome in some parts of North America is the presence of various species of parasites and predators which have reduced the population levels of infestations and enhanced the effects of artificial control measures (72). Nevertheless, many parasitic and predacious species of insects, mites, and fungi attack scales in other countries (49, 56, 80, 158, 166, 204, 232); on the Black Sea coast alone, 17 species of aphelinids are known to parasitize the coccids that damage woody plants (52). However, only a few of the parasites and predators of scale insects have been studied in much detail (18, 81, 88, 169, 177, 231). Most of the work has been done on the aphelinid *Prospaltella perniciosi* as it is believed to be the most effective of all natural enemies in controlling the San Jose scale (31, 53, 157, 168). Both the parasite and the scale originated in the Soviet Far East, North China, and the Korean Peninsula and, though they have had parallel development, the parasite does not adapt very easily to conditions other than those of its homeland (51). Nevertheless, the parasite may be successfully established in other areas. This is most desirable because the Far Eastern strain of *Prospaltella* is much more fertile and active than other strains and has a greater potential for biological control. Various countries have exchanged a number of strains (19, 111, 207, 229), the most effective being those from the Far East and the United States (87, 156, 167, 215). The degree of parasitism can often be correlated with climatic conditions (24, 87, 111, 230). In northern areas, some strains tend to become unisexual which decreases their effectiveness because parthenogenetic strains lay fewer eggs than do bisexual forms (157, 168). The parasite is less tolerant than the host to temperature and humidity changes and thus these factors are most important in determining how much biological control can be achieved from specific strains. Various techniques have been developed for rearing the parasites, for distributing them in orchards on media such as scale-infested watermelon and squash, and for prolonging their life by providing supplementary food such as red clover flowers (22, 50, 170, 195). *P. perniciosi* has given excellent biological control of the San Jose scale in many experimental orchards but it has failed to give the control demanded by fruit growers (20, 23). Commercial control has been obtained where the parasite was supported by the judicial use of insecticides (153). Biological and chemical control methods for the San Jose scale are difficult to combine but such integrated programs are being developed as more is discovered about the effects of chemicals in general on parasites and predators (21, 29, 154, 155, 159).

MITES

Some of the more important species of tetranychid and eriophyid mites of apple and pear trees are the European red mite, *Panonychus ulmi*; the brown mite, *Bryobia rubrioculus;* the McDaniel spider mite, *Tetranychus mcdanieli*; the Pacific spider mite, *T. pacificus*; the two-spotted spider mite, *T. urticae*; the four-spotted spider mite, *T. canadensis*; the apple rust mite, *Aculus* (*Vasates*) *schlectendali;* the pear leaf blister mite, *Eriophyes pyri;* and the pear rust mite, *Epitrimerus pyri.* All species feed on the leaves, but the pear leaf blister mite injures the fruit buds before they open in late winter and the pear rust mite feeds on the fruit after the petals drop. The ecology of tetranychid mites has just been reviewed in detail (112). Included in the review is an exhaustive discussion of the predators of mites and how they and other factors affect the abundance and control of tetranychids. In addition, the authors summarize various aspects of basic biology and the effects of chemical control on pest and prey. Comparatively little is known about eriophyid mites; they are small, live in secluded niches, and are difficult to study, and so acarologists tend to avoid them. Recent work suggests that they may be key factors in the success of integrated control programs (101, 109).

The control of mites continues to be one of the most serious of all problems in the growing of apples and pears throughout the world chiefly because acarines possess the genetic ability to develop strains that are resistant to most of the modern acaricides, especially to the organophosphorous compounds. Helle reviewed resistance in the Acarina up to 1964 (99). Since then, resistance has developed to several new compounds and new aspects of the problem have been investigated. The following references represent only a few of the many papers that have been published on resistance in recent years (12, 42, 54, 60, 73, 102, 107, 118). Though the development of resistance in agricultural pests is a deplorable indirect result of the use of chemicals, biologists have, until recently, neglected the fact that resistance can occur in insect and acarine predators as well as in the pest. An integrated control program for the McDaniel spider mite is becoming a reality in some orchards of Washington and British Columbia partly because its mite predator *Typhlodromus occidentalis* has produced strains resistant to several organophosphorous compounds which are used for the control of insects (109). Also, the problems of resistance have forced acarologists to re-examine the practicability of using new formulations of older materials such as petroleum oils for the control of the European red mite (30, 47, 64, 149, 151). Petroleum oils have no history of resistance. They cost less than most acaricides and are very low in toxicity to man and other animals. Oils fit well into integrated control programs because they are commonly applied in the prebloom stage of tree development when the least harm can be done to predators.

The principles of chemical and integrated control of phytophagous mites have been outlined by Jeppson (117). The philosophy of control of arthropod pests of orchards, especially of mites, has changed considerably during the past decade. Formerly, acarologists were inclined to use broad-spectrum chemicals (13) whereas today the trend is to acaricides that kill the pests and do not harm the predators (65, 108, 131, 172, 189). Few chemicals possess this selectivity and for this and other reasons, mainly economic, comparatively few true acaricides have been introduced in recent years. Many pesticides have some acaricidal activity; over 80 are recognized in Canada (225) of which about 23 have been introduced since 1960. Less than a third of these, namely binapacryl, Omite® (2-(*p-tert*-butylphenoxy) cyclohexyl 2-propynyl sulfite), Chinomethionate, Lovozal® (dichlorophenoxycarbonyltrifluoromethylbenzimidazole), Neoron® (isopropyl 4,4′-

dibromobenzilate), Chlorphenamidine, and Dinobuton, are promising acaricides and only a few show selective activity. Sixty-five compounds are classified as acaricides in the United States (121) many of which are still in the experimental stage. Plictran® (tricyclohexyltin hydroxide), and phosalone are representative of some of these new compounds. Considering the number of the pesticides that affect mites in some way or another, it is not surprising that in the past decade a voluminous amount of literature has been published on their control. Martin's supplements of the World Review of Pest Control are an excellent guide to new compounds as are the technical data sheets and product manuals of chemical companies. Acaricides currently used in the United States, Canada, the West Indies, South Africa, India, and Australia have been listed by Hanna (95, 96). Over 20 different chemicals are used as acaricides on apple and pear trees alone, the most common to all countries being dicofol, chlorbenside, tetradifon, azinphosmethyl, dimethoate, and malathion. Many other chemicals have been tested as acaricides in these and other countries and a number, in addition to the above, have come into common use (1, 45, 57, 59, 75, 84, 118, 122, 186, 187, 196, 214, 244, 247).

LITERATURE CITED

1. Allison, W. E., Doty, A. E., Hardy, J. L., Kenaga, E. E., Whitney, W. K. Laboratory evaluations of Plictran® miticide against two-spotted spider mites. *J. Econ. Entomol.,* **61,** 1254–57 (1968)
2. Andriano, M., Maldovan, E., Guiria, M., Suta, V., Sapunaru, T. Contributions to the study of the biology, ecology, and control of the codling moth, *Carpocapsa pomonella* (L.). *Ann. Inst. Cercet. Agr.,* **27,** 213–32 (1960) (In Rumanian, French summary)
3. Anon. Outbreaks and new records. *Food Agr. Organ. U.N. FAO Plant Protect. Bull.,* **10,** 42–45 (1962)
4. Anon. Distribution maps of pests. *Commonwealth Inst. Entomol., Ser. A* (*Agr.*), 7 (revised) (1967)
5. Anon. First record of San Jose scale (*Quadraspidiotus perniciosus* (Comst.) on apple, pear and peach) in Greece. *Eur. Medit. Plant Protect. Organ. Repting. Serv.* ref 68/7–314, RSE, 1 (1968)
6. Anon. San Jose scale *Quadraspidiotus perniciosus* (Comst.) in Europe and the Mediterranean Basin in 1966. *Publ. OEPP Ser. B,* **64** (1968)
7. Arkhipova, V. D. Fungous diseases of the codling moth, *Carpocapsa pomonella* (L.). *Entomol. Rev.,* **44,** 48–54 (1965)
8. Asquith, D. Spray combinations for control of the codling moth and the red-banded leaf roller of apples. *J. Econ. Entomol.,* **51,** 378–79 (1958)
9. Asquith, D. Insecticides change the importance of segments of insect generations. *J. Econ. Entomol.,* **53,** 694–95 (1960)
10. Bailey, J. B., Madsen, H. F. A laboratory study of three strains of codling moth, *Carpocapsa pomonella* (Linnaeus), exhibiting tolerance to DDT in the field. *Hilgardia,* **35,** 185–210 (1964)
11. Balevski, A. D., Vaser, A. N., Ivanor, S. K., Lazarov, A. V., Tsvethova, T. T. The codling moth, (*Laspeyresia pomonella* (L.)), bionomics, ecology, control measures and possibilities for the introduction of a warning service. *Nauch. Tr. Nauchno-izsled. Inst. Zasht. Rast.* **1,** 105–58 (1958) (In Russian, German summary)
12. Ballantyne, G. H., Harrison, R. A. Genetic and biochemical comparisons of organophosphate resistance between strains of spider mites. *Entomol. Exptl. Appl.,* **10,** 231–39 (1967)
13. Barnes, M. M. Deciduous fruit in-

sects and their control. *Ann. Rev. Entomol.*, **4**, 343–62 (1959)

14. Barnes, M. M., Moffit, H. R. Resistance to DDT in the adult codling moth and reference curves for Guthion and carbaryl. *J. Econ. Entomol.*, **56**, 722–25 (1963)
15. Barnes, M. M., Wargo, M. J., Baldwin, R. L. New low intensity ultraviolet light trap for detection of codling moth activity. *Calif. Agr.*, **19**(10), 6–7 (1965)
16. Beauchard, J. L'eradication d'un foyer de pou de San Jose dans le sudouest de la France. *Phytoma*, **16**, 45–49 (1964)
17. Beirne, B. P. *Pest Management.* (Leonard Hill, London, 123 pp., 1967)
18. Benassy, C. Contribution to the study of the influence of some ecological factors on the limitation of outbreaks of diaspine coccids. *Ann. Epiphyties*, **12**, 1–157 (1961) (In French)
19. Benassy, C., Bianchi, H. Sur l'ecologie de *Prospaltella perniciosi* Tower, parasite specifique importe de *Quadraspidiotus perniciosus* Comst. *Entomophaga*, **5**, 165–81 (1960)
20. Benassy, C., Bianchi, H., Milaire, H. Observations sur la repercussion des traitements preconises en vergers, sur *Prospaltella perniciosi* Tower, parasite specifique de *Quadraspidiotus perniciosus* Comst. *Phytiat.-Phytopharm.*, **9**, 26–36 (1960)
21. Benassy, C., Bianchi, H., Milaire, H. Experimentation en matiere de lutte integree dans deux vergers francais. *Entomophaga*, **9**, 215–80 (1964)
22. Benassy, C., Bianchi, H., Milaire, H. Observations sur l'incidence de quelques produits insecticides et fongicides sur l'association pou de San Jose et parasite specifique. *Rev. Zool. Agr.*, **63**, 27–37 (1964)
23. Benassy, C., Bianchi, H., Milaire, H. Recherches sur l'utilization de *Prospaltella perniciosi* Tow. en France. *Ann. Epiphyties*, **15**, 457–72 (1965)
24. Benassy, C., Bianchi, H., Milaire, H. Note sur l'efficacite en France de *Prospaltella perniciosi* Tow. *Entomophaga*, **12**, 241–55 (1967)
25. Bengston, M. Studies on codling moth control in the Stanthorpe district, Queensland. *Queensland J. Agr. Animal Sci.*, **22**, 59–68 (1965)
26. Bengston, M. Control of woolly aphid (*Eriosoma lanigerum* (Hausm.)) in the Stanthorpe district, Queensland. *Queensland J. Agr. Animal Sci.*, **22**, 469–73 (1965)
27. Bengston, M. San Jose scale time for spraying. *Queensland Fruit Veg. News*, **34**, 128 (1968)
28. Beran, F. Remarkable side effect of Sevin against the San Jose scale. *Pflanzenarzt*, **15**, 135–36 (1962) (In German)
29. Biliotti, E., Benassy, C., Bianchi, H., Milaire, H. Premiers essais experimentaux d'acclimatation en France de *Prospaltella perniciosi* Tower, parasite specifique importe de *Quadraspidiotus perniciosus* Comst. *Compt. Rend. Hebd. Seanc. Acad. Agr. France*, **46**, 707–11 (1960)
30. Bobb, M. L. Pre-bloom oil sprays for European red mite control. *J. Econ. Entomol.*, **62**, 94–95 (1969)
31. Böhm, H. A further note on the rearing and establishment of the chalcidoid parasite of the San Jose scale, *P. perniciosi* Tow., in Austria. *Pflanzenarzt.*, **18**, 104 (1965) (In German)
32. Bollow, H. Die Blattsauger der Apfel und Birnbaume, Auftreten, Aussehex, lebensweire, Varaussage und Bekampfung. *Pflanzenschutz*, **12**, 159–66 (1960)
33. Bonnemaison, L. Le puceron cendre du pommier (*Dysapidis plantaginea* Pass.). Morphologie et biologie. Methodes de lutte. *Ann. Epiphyties*, **10**, 257–320 (1959)
34. Bonnemaison, L. Toxicite de divers insecticides de contact ou endotherapiques vis-a-vis des predateurs et parasites des pucerons. *Phytiat.-Phytopharm.*, **11**, 67–84 (1962)
35. Bonnemaison, L. Morphologie comparee du 'Pou de San Jose' (*Aonidiella perniciosa* Comst.) et de l'Aspidiotus des arbres fruitiers (*Aspidiotus ostreaeformis* Curtis). *Rev. Pathol. Vegetale. Entomol. Agr. France*, **23**, 230–43 (1963)
36. Borchsenius, N. S. A catalogue of the armoured scale insects (Diaspidoidea) of the world. *Izdat. "Nauka". Leningrad*, 449 pp. (1966) (In Russian)
37. Brimblecombe, A. R. Studies of the

Coccoidea. 14. The Genera *Aspidiotus, Diaspidiotus* and *Hemiberlesia* in Queensland. *Queensland J. Agr. Sci.,* **25,** 39–56 (1968)

38. Brown, S. W., McKenzie, H. L. Evolutionary patterns in the armored scale insects and their allies. *Hilgardia,* **33,** 133–70A (1962)
39. Burts, E. C. An evaluation of insecticides for the control of pear psylla. *Wash. State Univ. Agr. Expt. Sta., Sta. Circ.,* **438,** 1–11 (1964)
40. Burts, E. C., Fischer, W. R. Mating behavior, egg production, and egg fertility in the pear psylla. *J. Econ. Entomol.,* **60,** 1297–1300 (1967)
41. Burts, E. C. An area control program for the pear psylla. *J. Econ. Entomol.,* **61,** 261–63 (1968)
42. Busvine, J. R. Detection and measurement of insecticide resistance in arthropods of agricultural or veterinary importance. *World Rev. Pest Control,* **7,** 27–41 (1968)
43. Butt, B. A., Hathaway, D. O. Female sex pheromone as attractant for male codling moths. *J. Econ. Entomol.,* **59,** 476–77 (1966)
44. Carson, R. *Silent Spring.* (Houghton Mifflin Co., Boston, 355 pp., 1962)
45. Cessac, M., Burgaud, L. Efficacite pratique d'un nouvel insecticide-acaricide de contact: la phosalone (11.974 R.P.). *Phytiat.-Phytopharm.,* **13,** 45–54 (1964)
46. Chaboussou, F. Action de divers insecticides et notamment de certains produits endotherapiques vis-a-vis d'*Aphelinus mali* Hald. evaluant a l'interieur du puceron lanigere du pommier, *Eriosoma lanigerum* Hausm. *Rev. Pathol. Vegetale. Entomol. Agr. France,* **40,** 17–29 (1961)
47. Chapman, P. J., Lienk, S. E., Avens, A. W., White, R. W. Selection of a plant spray oil combining full pesticidal efficiency with minimum plant injury hazards. *J. Econ. Entomol.,* **55,** 737–44 (1962)
48. Chiswell, J. R. Field comparisons of insecticides for control of the codling moth, *Cydia pomonella* (L.) with observations on effects of treatments on mite populations. *J. Hort. Sci.,* **37,** 313–25 (1962)
49. Chumakova, B. M. The entomophages of the Californian or San Jose scale in the USSR and means to enhance their effectiveness. *Intern. Conf. Insect Pathol. Biol. Control, Prague, 1st,* 481–85 (1958) (In Russian)
50. Chumakova, B. M. Supplementary feeding as a factor increasing the activity of parasites of harmful insects. *Tr. Vses. Inst. Zashchity Rast.,* **15,** 57–70 (1960) (In Russian)
51. Chumakova, B. M. The San Jose scale, *Diaspidiotus perniciosus* Comst. and its parasites in the Soviet Far East. *Entomol. Rev. (USSR),* **43,** 272–79 (1964) (English transl.)
52. Chumakova, B. M. A review of the species of the family Aphelinidae parasitising Coccids that damage woody plants on the Black Sea coast of the RSFSR. *Tr. Vses. Inst. Zashchity Rast.,* **21,** 14–39 (1964) (In Russian)
53. Chumakova, B. M., Goryunova, Z. S. Development of males of *Prospaltella perniciosi* Tow., parasite of San Jose scale. *Entomol. Rev. (USSR),* **42,** 178–81 (1963) (English transl.)
54. Cranham, J. E. Laboratory determination of resistance to tetradifon in the fruit tree red spider mite, *Panonychus ulmi* (Koch). *Rept. E. Malling Res. Sta.,* **55,** 165–68 (1968)
55. Davatchi, A., Esmaili, M. *Psylla pyricola* Förster. *Entomol. Phytopathol. Appl.,* **24,** 14–30 (1966)
56. DeBach, P. Some Species of *Aphytis* Howard, in Greece. *Ann. Inst. Phytopathol., Benaki,* **7,** 5–18 (1964)
57. de Pietri-Tonelli, P., Corradini, V., Caracalli, N., Siddi, G. Winter control of the European red mite, *Panonychus ulmi,* and certain fruit insects by M-2060 (2-fluoroethyl (4-biphenylyl) acetate) in Italy. *J. Econ. Entomol.,* **62,** 107–12 (1969)
58. Dikov, I. Control of the Californian scale should not be abated. *Rastit. Zasht.,* **13,** 8–9 (1965) (In Bulgarian)
59. Dittrich, V. N-(2-methyl-4-chlorophenyl)-N′,N′-dimethylformamidine (C-8514/Schering 36268) evaluated as an acaricide. *J. Econ. Entomol.,* **59,** 889–93 (1966)
60. Dittrich, V. Chlorphenamidine nega-

tively correlated with OP resistance in a strain of two-spotted spider mite. *J. Econ. Entomol.,* **62,** 44–47 (1969)

61. Dobrokhotova, N. M. The comparative toxicity of some insecticides to larvae of the codling moth. *Zakhyst Rostyn,* **1,** 52–58 (1964) (In Russian, English summary)
62. Dolphin, R. E., Cleveland, M. L. *Trichogramma minutum* as a parasite of the codling moth and red-banded leaf roller. *J. Econ. Entomol.,* **59,** 1525–26 (1966)
63. Downing, R. S. Sevin as an orchard insecticide in British Columbia. *Proc. Entomol. Soc. Brit. Columbia,* **56,** 41–45 (1959)
64. Downing, R. S. Petroleum oils in orchard mite control. *J. Entomol. Soc. Brit. Columbia,* **64,** 10–13 (1967)
65. Downing, R. S., Arrand, J. C. Integrated control of orchard mites in British Columbia. *Brit. Columbia Dept. Agr., Entomol. Branch, Process. Publ.,* 7 pp. (1968)
66. Drgon, S., Huba, A., Beros, D. The possibilities of using organophosphorus insecticides in combination with mineral oils for the control of the San Jose scale. *Prace Lab. Ochr. Rast.,* 221–32 (1962) (In Czechoslovakian)
67. Ehrenhardt, H. Untersuchungen zur Prognose der Obstmadenbekämpfung im sudwestdeutschen Raum. *Proc. Intern. Congr. Crop Protect., 4th,* **1,** 247–51 (1961)
68. Embree, D. G. The population dynamics of the winter moth in Nova Scotia 1954–1962. *Mem. Entomol. Soc. Can.,* **46,** 1–57 (1965)
69. Emonnot, P., Ferand, G. Essais de nouveaux insecticides dans la lutte contre le carpocapse. *Phytiat.-Phytopharm.,* **11,** 201–5 (1963)
70. Falcon, L. A., Kane, W. R., Bethell, R. S. Preliminary evaluation of a granulosis virus for control of the codling moth. *J. Econ. Entomol.,* **61,** 1208–13 (1968)
71. Fisher, R. W. Note on resistance to DDT in the codling moth, *Carpocapsa pomonella* (L.), in Ontario. *Can. J. Plant. Sci.,* **40,** 580–82 (1960)
72. Flanders, S. E. The status of San Jose scale parasitization (including biological notes). *J. Econ. Entomol.,* **53,** 757–59 (1960)
73. Foott, W. H. A strain of *Tetranychus urticae* from the ground cover of an apple orchard resistant to chlorfenson. *Can. J. Plant. Sci.,* **45,** 226–28 (1965)
74. Freitas, A. The bio-ecological behaviour of the San Jose scale (*Quadraspidiotus perniciosus* in continental Portugal (Years 1960–61 and 1963–64) 1. Annual cycle on apple. *Agron. Lusitana,* **26** (1964), 289-335 (1966) (In Portugese)
75. Galbiati, F. Investigations of a new series of nitrogenous organophosphorus compounds. Characteristics and properties. *Proc. Brit. Insecticide Fungicide Conf., Brighton, Engl. (1961),* **2,** 507–18 (1962)
76. Gaprindashvili, N. K., Novitskaya, T. N. The natural enemies of the codling moth *Laspeyresia pomenella* (L.) and the effect of chemical treatments on their useful activity. *Entomol. Rev. (USSR),* **46,** 39–42 (1967) (English transl.)
77. Gehring, R. D., Madsen, H. F. Some aspects of the mating and oviposition behavior of the codling moth, *Carpocapsa pomonella. J. Econ. Entomol.,* **56,** 140–43 (1963)
78. Geier, P. W. Physiological age of codling moth females (*Cydia pomonella* (L.) caught in bait and light traps. *Nature,* **185,** 709 (1960)
79. Geier, P. W. Management of insect pests. *Ann. Rev. Entomol.,* **11,** 471–90 (1966)
80. Gerson, U. Observations on *Hemisarcoptes coccophagus* Meyer with a new synonym. *Acarologia,* **9,** 632–38 (1967)
81. Ghani, M. A., Rafiq Ahmad. Biology of *Pharoscymnus flexibilis* Muls. *Tech. Bull. Commonwealth Inst. Biol. Control,* **7,** 107–11 (1966)
82. Giliomee, J. H. Morphology and taxonomy of adult males of the family Coccidae. *Brit. Mus. Nat. Hist. Bull. Entomol., Suppl. 7,* 4–168 (1967)
83. Glass, E. H. Laboratory and field tests with the granulosis virus of the red-banded leaf roller. *J. Econ. Entomol.,* **51,** 454–57 (1958)
84. Glass, E. H. Some results obtained in New York State with a new systemic insecticide, vamidothion, on apple trees. *Mededel. Landbouwhogeschool Opzoekingssta. Staat Gent,* **27,** 869–72 (1962)

85. Glass, E. H. Parasitism of the red-banded leaf roller, *Argyrotaenia velutinana* by *Trichogramma minutum*. *Ann. Entomol. Soc. Am.*, **56**, 564 (1963)
86. Glass, E. H., Hervey, G. E. R. Continuous rearing of the red-banded leaf roller *Argyrotaenia velutinana*. *J. Econ. Entomol.*, **55**, 336–40 (1962)
87. Goryunova, Z. S. Intraspecific differentiation in *Prospaltella* (*Prospaltella perniciosi* Tow.) the parasite of the San Jose scale. *Tr. Vses. Inst. Zashchity Rast.*, **21**, 40–55 (1964) (In Russian)
88. Goryunova, Z. S. The bionomics of *Aphytis proclia* Wlk., a parasite of the San Jose scale, and ways of using it. *Tr. Vses. Inst. Zashchity Rast.*, **24**, 211–16 (1965) (In Russian)
89. Gratwich, M., Sillibourne, J. M., Tew, R. P. The toxicity of insecticides to larvae of the codling moth, *Cydia pomonella* (L.). 1. Intrinsic toxicity and persistence. *Bull. Entomol. Res.*, **56**, 367–76 (1965)
90. Griggs, W. H., Jensen, D. D., Iwakiri, B. T. Development of young pear trees with different rootstocks in relation to psylla infestation, pear decline and leaf curl. *Hilgardia*, **39**, 153–204 (1968)
91. Grunberg, A. Mode of action of mineral oils upon armoured scales in conjunction with low-volume application. *Plant Protect. Abstr.*, **4**, 81 (1968)
92. Haines, R. G. Results of field experiments with new insecticides and acaricides on Michigan fruit. *Mich. State Univ. Agri. Expt. Sta. Quart. Bull.*, **40**, 628–36 (1958)
93. Hamilton, D. W., Fahey, J. E. Effect of insecticides on codling moth larvae seeking cocooning quarters. *J. Econ. Entomol.*, **51**, 672–73 (1958)
94. Hamstead, E. O. Control of the woolly apple aphid on roots of apple nursery stock. *J. Econ. Entomol.*, **53**, 217–20 (1960)
95. Hanna, A. D. A review of the use of insecticides on the main agricultural crops of the world—1. *World Rev. Pest Control*, **7**, 97–114 (1968)
96. Hanna, A. D. A review of the use of insecticides on the main agricultural crops of the world—2. Section 2: Deciduous fruit trees. *World Rev. Pest Control*, **7**, 184–203 (1968)
97. Harries, F. H., Burts, E. Laboratory studies of pear psyllas resistant to dieldrin and some related compounds. *J. Econ. Entomol.*, **52**, 530 (1959)
98. Harrow, K. M. Control of San Jose scale. *Orchard. New Zealand*, **33**, 133–37 (1960)
99. Helle, W. Resistance in the Acarina: Mites. *Advan. Acarol.*, **2**, 71–93 (1965)
100. Hennequin, J. Analyse du mode d'action et valeur pratique de quelques insecticides sur le carpocapse des pommes (*Laspeyresia pomonella* (L.)) *Phytiat.-Phytopharm.*, **13**, 65–75 (1964)
101. Herbert, H. J., Sanford, K. H. The influence of spray programs on the fauna of apple orchards in Nova Scotia XIX. Apple rust mite, *Vasates schlechtendali*, a food source for predators. *Can. Entomologist*, **101**, 62–67 (1969)
102. Herne, D. H. C., Brown, A. W. A. Inheritance and biochemistry of OP-resistance in a New York strain of the two-spotted spider mite. *J. Econ. Entomol.*, **62**, 205–9 (1969)
103. Hikichi, A. Some factors influencing the control of the red-banded leaf roller, *Argyrotaenia velutinana* (Wlkr.) on apple in Norfolk County, Ontario. *Proc. Entomol. Soc. Ontario*, **92**, 182–88 (1962)
104. Hough, W. S. Resistance to insecticides by codling moth and red-banded leaf roller. *Virginia Agr. Expt. Sta. Tech. Bull.*, **166**, 32 pp. (1963)
105. Hough, W. S. Fall sprays to control rosy apple aphids the next year. *Virginia Fruit*, **51**, 16–18 (1963)
106. Hoyt, S. C. Effect of urea on the control of some apple insects with certain insecticides. *J. Econ. Entomol.*, **53**, 685 (1960)
107. Hoyt, S. C. Resistance to binapacryl and Union Carbide 19786 in the McDaniel spider mite, *Tetranychus mcdanieli*. *J. Econ. Entomol.*, **59**, 1278–79 (1966)
108. Hoyt, S. C. The theory of integrated chemical and biological control. *Better Fruit*, **61**, 15, 18 (1967)
109. Hoyt, S. C. Integrated chemical con-

trol of insects and biological control of mites on apple in Washington. *J. Econ. Entomol.*, **62**, 74–86 (1969)

110. Hoyt, S. C., Madsen, H. F. Dispersal behavior of the first instar nymphs of the woolly apple aphid. *Hilgardia*, **30**, 267–99 (1960)

111. Huba, A. Effectiveness of an introduction of parasites of the San Jose scale in Czechoslovakia. *Trans. Intern. Conf. Insect Pathol. Biol. Control, 1st, Prague,* 395–403 (1959) (In German)

112. Huffaker, C. B., van de Vrie, M., McMurtry, J. A. The ecology of tetranychid mites and their natural control. *Ann. Rev. Entomol.*, **14**, 125–74 (1969)

113. Jaques, R. P. Control of some lepidopterous pests of apple with commercial preparations of *Bacillus thuringiensis* Berliner. *J. Insect Pathol.*, **3**, 167–82 (1961)

114. Jaques, R. P. Mortality of five apple insects induced by the nematode DD136. *J. Econ. Entomol.*, **60**, 741–43 (1967)

115. Jaques, R. P., Stultz, H. T. The influence of a virus disease and parasites on *Spilonota ocellana* in apple orchards. *Can. Entomologist*, **98**, 1035–45 (1966)

116. Jensen, D. D., Griggs, W. H., Gonzales, C. Q. Schneider, H. Pear decline virus transmission by pear psylla. *Phytopathology*, **54**, 1346–51 (1964)

117. Jeppson, L. R. Principles of chemical control of phytophagous mites. *Advan. Acarol.*, **2**, 31–51 (1965)

118. Jeppson, L. R., Jesser, M. J., Complin, J. O. Chemical structure and toxicity of some carbamoyloxy phosphorodithioates to susceptible and organophosphorus-resistant strains of mites. *J. Econ. Entomol.*, **59**, 185–87 (1966)

119. Kaloostian, G. H. Chemosterilization of male pear psylla with Tepa. *J. Econ. Entomol.*, **61**, 573–74 (1968)

120. Karadzhov, S. The prospects of spraying fruit trees with DNC preparations in early spring. *Gradinar. Lozar. Nauka*, **2**, 743–52 (1965) (In Bulgarian)

121. Kenaga, E. Commercial and experimental organic insecticides (1966 Revision). *Bull. Entomol. Soc. Am.*, **12**, 161–217 (1966)

122. Kenaga, E. E., Whitney, W. K., Hardy, J. L. Doty, A. E. Laboratory tests with Dursban insecticide. *J. Econ. Entomol.*, **58**, 1043–50 (1965)

123. Kerr, R. W. Notes on arthropod resistance to chemicals used in their control in Australia. *J. Australian Inst. Agr. Sci.*, **30**, 33–38 (1964)

124. Kislyi, A. The yellow pear scale. *Zashchita Rast. ot Vreditelei i Boleznei*, **6**, 31–32 (1965) (In Russian)

125. Knight, R. I., Briggs, J. B., Massee, A. M., Tydeman, H. M. The inheritance of resistance to woolly aphid, *Eriosoma lanigerum* (Hasmn.) in the apple. *J. Hort. Sci.*, **37**, 207–18 (1962)

126. Kosztarab, M. The armored scale insects of Ohio. *Bull. Ohio Biol. Surv.* (N. S.), **11**(2) (1963)

127. Legner, E. F., Oatman, E. R. Effects of Thuricide on the eye-spotted bud moth *Spilonota ocellana*. *J. Econ. Entomol.*, **55**, 677–78 (1962)

128. Legner, E. F., Oatman, E. R. Natural biotic control factors of the eye-spotted bud moth, *Spilonota ocellana*, on apple in Wisconsin. J. *Econ. Entomol.*, **56**, 730–32 (1963)

129. Legner, E. F., Oatman, E. R. Limitation on the pupal development of the eye-spotted bud moth in northeastern Wisconsin. *J. Econ. Entomol.*, **58**, 359–60 (1965)

130. Leius, K. Influence of wild flowers on parasitism of tent caterpillar and codling moth. *Can. Entomologist*, **99**, 444–46 (1967)

131. LeRoux, E. J. The application of ecological principles to orchard entomology in Canada. *Can. Entomologist*, **96**, 348–56 (1964)

132. LeRoux, E. J., Reimer, C. Variation between samples of immature stages, and of mortalities from some factors of the eye-spotted bud moth, *Spilonota ocellana* (D. & S.), and the pistol casebearer, *Coleophora serratella* (L.) on apple in Quebec. *Can. Entomologist*, **91**, 428–49 (1959)

133. MacLellan, C. R. Woodpeckers as predators of the codling moth in Nova Scotia. *Can. Entomologist*, **91**, 673–80 (1959)

134. MacLellan, C. R. Predator populations and predation on the codling moth in an integrated control orchard—1961. *Mem. Entomol. Soc. Can.*, **32**, 41–54 (1963)

135. MacPhee, A. W., Sanford, K. H.

The influence of spray programs on the fauna of apple orchards in Nova Scotia. XII—Second supplement to VII. Effects on beneficial arthropods. *Can. Entomologist,* **93,** 671–73 (1961)

136. MacPhee, A. W. The winter moth, *Operophthera brumata,* a new pest attacking apple orchards in Nova Scotia and its cold hardiness. *Can. Entomologist,* **99,** 829–34 (1967)

137. Madsen, H. F. Notes on *Anthocoris melanocerus* Reuter as a predator of the pear psylla in British Columbia. *Can. Entomologist,* **93,** 660–62 (1961)

138. Madsen, H. F. Codling moth attractants. *Pesticide Abstr. Sect. A.,* **13,** 333–44 (1967)

139. Madsen, H. F. Integrated control of deciduous tree fruit pests. *World Crops,* **20,** 20–28 (1968)

140. Madsen, H. F., Bailey, J. B. Control of the apple aphid and the rosy apple aphid with new spray chemicals. *J. Econ. Entomol.,* **52,** 493–96 (1959)

141. Madsen, H. F., Downing, R. S. Integrated control of the fruit-tree leaf roller, *Archips argyrospilus* (Walker), and the eye-spotted bud moth, *Spilonota ocellana* (Denis & Schiffermuller). *J. Entomol. Soc. Brit. Columbia,* **65,** 19–21 (1968)

142. Madsen, H. F., Marshall, J. Dormant sprays for the control of the pear psylla, *Psylla pyricola,* in British Columbia. *J. Econ. Entomol.,* **54,** 1000–3 (1961)

143. Madsen, H. F., Westigard, P. H., Falcon, L. A. Evaluation of insecticides and sampling methods against the apple aphid, *Aphis pomi. J. Econ. Entomol.,* **54,** 892–94 (1961)

144. Madsen, H. F., Westigard, P. H., Sisson, R. L. Observations on the natural control of the pear psylla, *Psylla pyricola* Forster, in California. *Can. Entomologist,* **95,** 837–44 (1963)

145. Madsen, H. F., Williams, K. Control of the pear psylla with oils and oil-insecticide combinations. *J. Econ. Entomol.,* **60,** 121–24 (1967)

146. Madsen, H. F., Williams, K. The effects of petroleum oils on Bartlett pears and on pear psylla, *Psylla pyricola. Can. Entomologist,* **100,** 290–95 (1968)

147. Madsen, H. F., Williams, K. Effectiveness and persistence of low dosages of azinphosmethyl for control of the codling moth. *J. Econ. Entomol.,* **61,** 878–79 (1968)

148. Madsen, H. F., Wong, T. T. Y. Effects of predators on control of pear psylla. *Calif. Agr.,* **18**(2), 2–3 (1964)

149. Madsen, H. F., Wong, T. T. Y. European red mite control with petroleum oils. *Calif. Agr.,* **18,** 11 (1964)

150. Mailloux, N., LeRoux, E. J. Further observations on the life-history and habits of the codling moth, *Carpocapsa pomonella* (L.) in apple orchards of southwestern Quebec. *Rept. Pomol. Soc. Quebec, 1960,* 45–56 (1961)

151. Malbrunot, P. De l'interet des traitements d'hiver. *Phytoma,* **18,** 23–24 (1966)

152. Mathys, G. Vers un tournant dans la lutte contre le pou de San Jose? *Rev. Romande Agr. Viticult. Arboricult.,* **15,** 53–56 (1959)

153. Mathys, G., Possibilities de lutte contre le pou de San Jose par la methode biologique et integrée. In Proceedings of the FAO symposium on integrated pest control, October, 1965, Rome. *Food Agr. Organ. U.N.,* **3,** 53–64 (1966)

154. Mathys, G., Guignard, E. L'efficacité de *Prospaltella perniciosi* Tow., parasite du pou de San Jose (*Quadraspidiotus perniciosus* Comst.). Essai de conciliation de la lutte biologique et de la lutte chimique en vergers exploites commercialement. *Rev. Romande Agr. Viticult. Arboricult.,* **17,** 53–56 (1961)

155. Mathys, G., Guignard, E. Un important allie dans la lutte contre le pou de San Jose: *Prospaltella perniciosi* Tow. *Agr. Romande,* **1,** 59–61 (1962)

156. Mathys, G., Guignard, E. Enseignements recueillis au cours de neuf ans de travaux avec *Prospaltella perniciosi* Tow., parasite du pou de San Jose (*Quadraspidiotus perniciosus* Comst.). *Entomophaga,* **12,** 212–22 (1967)

157. Mathys, G., Guignard, E. Quelques aspects de la lutte biologique contre le pou de San Jose (*Quadraspidiotus perniciosus* Comst.) à l'aide de l'aphelinide *Prospaltella perniciosi* Tow. *Entomophaga,* **12,** 223–34 (1967)

158. Mathys, G., Guignard, E., Stahl, J. L'identification des cochenilles du genre *Quadraspidiotus* importantes en arboriculture (*Quadraspidiotus perniciosus, Q. ostreaeformis, Q. piri, Q. marani* et *Q. lenticularis*). *Agr. Romande,* **4** (*Ser. A*), 65–68 (1965)
159. Mathys, G., Guignard, R. C. E., Stahl, J. Peut-on concilier la lutte contre le pou de San Jose (*Quadraspidiotus perniciosus* Comst.) avec la lutte contre le carpocapse (*Carpocapsa pomonella* (L.))? *Agr. Romande,* **6,** 24–26 (1967)
160. McMullen, R. D. New records of chalcidoid parasites and hyperparasites of *Psylla pyricola* Forster in British Columbia. *Can. Entomologist,* **98,** 236–39 (1966)
161. McMullen, R. D., Jong, C. New records and discussion of predators of the pear psylla, *Psylla pyricola* Forster in British Columbia. *J. Entomol. Soc. Brit. Columbia,* **64,** 35–39 (1967)
162. Morgan, C. V. G. Fate of the San Jose scale and the European fruit scale on apples and prunes held in standard cold storage and controlled atmosphere storage. *Can. Entomologist,* **99,** 650–59 (1967)
163. Morgan, C. V. G., Angle, B. J. Notes on the habits of the San Jose scale and the European fruit scale on harvested apples in British Columbia. *Can. Entomologist,* **100,** 499–503 (1968)
164. Morgan, C. V. G., Arrand, J. C. San Jose scale and European fruit scale in interior British Columbia. *Brit. Columbia Dept. Agr. Entomol. Branch,* **67**(1), 1–12 (1966)
165. Morrison, H., Morrison, E. R. An annotated list of generic names of the scale insects. *U. S. Dept. Agr., ARS Misc. Publs., 1015,* 1–206 (1966)
166. Nagaraja, H., Hussainy, S. U. A study of six species of *Chilocorus* predaceous on San Jose and other scale insects. *Oriental Insects,* **1,** 249–56 (1968)
167. Neuffer, G. On release experiments with *Prospaltella perniciosi* Tower against the San Jose scale (*Quadraspidiotus perniciosus* Comstock) in Baden-Wurttemberg. *Entomophaga,* **9,** 131–36 (1964) (In German)
168. Neuffer, G. Remarks on the parasite fauna of *Quadraspidiotus perniciosus* Comst. and on the rearing of bisexual *Prospaltella perniciosi* Tow. in the insectary. *Z. Pflanzenkrankh. Pflanzenschutz,* **71,** 1–11 (1964) (In German)
169. Neuffer, G. On the parasite fauna of *Quadraspidiotus perniciosus* Comstock with particular reference to the imported *Prospaltella perniciosi* Tower. *Entomophaga,* **11,** 383–92 (1966) (In German)
170. Neuffer, G. An account of the mass rearing of *Prospaltella perniciosi* Tow. in a modified insectarium in Stuttgart. *Entomophaga,* **12,** 235–39 (1967) (In German)
171. Oatman, E. R. The effect of *Bacillus thuringiensis Berliner* on some lepidopterous larval pests, apple aphid and predators, and on phytophagous and predacious mites on young apple trees. *J. Econ. Entomol.,* **58,** 1144–47 (1965)
172. Oatman, E. R. Studies on integrated control of apple pests. *J. Econ. Entomol.,* **59,** 368–73 (1966)
173. Oatman, E. R., Jenkins, L. The biology of the red-banded leaf roller, *Argyrotaenia velutinana* (Wlkr.) in Missouri with notes on its natural control. *Missouri Univ. Agr. Expt. Sta. Res. Bull.,* **789,** 1–14 (1962)
174. Oatman, E. R., Legner, E. F. Bionomics of the apple aphid, *Aphis pomi,* on young nonbearing apple trees. *J. Econ. Entomol.,* **54,** 1034–37 (1961)
175. Oatman, E. R., Legner, E. F., Brooks, R. F. An ecological study of arthropod populations on apple in northeastern Wisconsin: Species affecting the fruit. *J. Econ. Entomol.,* **59,** 165–68 (1966)
176. Oatman, E. R., Libby, J. L. Progress on insecticidal control of apple insects. *J. Econ. Entomol.,* **58,** 766–70 (1965)
177. Pantyukhov, G. A. The effect of temperature and humidity on the development of *Chilocorus renipustulatus* Scriba. *Tr. Zool. Inst. Acad. Nauk SSSR,* **36,** 70–85 (1965) (In Russian)
178. Paradis, R. O. Repression de la tordeuse du pommier, *Archips argyrospilus* (Wlk.) dans le sud-ouest du Quebec. *Rept. Quebec Soc. Plant Protect., 41st, 1959,* 76–82 (1960)
179. Paradis, R. O. Essai d'analyse des

facteurs de mortalite chez *Archips argyrospilus* (Walk.). *Ann. Entomol. Soc. Quebec,* **6,** 59–69 (1961)
180. Paradis, R. O. Essais de traitements insecticides pour la repression simultanee du charancon de la prune, *Conotrachelus nenuphar* (Hbst.) et de la tordeuse du pommier, *Archips argyrospilus* (Wlk.). *Rept. Quebec Soc. Plant Protect., 42nd,* 1960, 14–19 (1962)
181. Paradis, R. O., LeRoux, E. J. Reserches sur la biologie et la dynamique des populations naturelles *d'Archips argyrospilus* (Wlk.) dans le sud-ouest du Quebec. *Mem. Entomol. Soc. Can.,* **43,** 1–77 (1965)
182. Parent, B., Pitre, D. Essais de lutte contre la cochenille virgule du pommier, *Lepidosaphes ulmi* (L.), dans les vergers du Quebec. *Phytoprotection,* **49,** 26–37 (1968)
183. Patterson, N. A. The influence of spray programs on the fauna of apple orchards in Nova Scotia. 12—The long term effect of mild pesticides on pests and their predators. *J. Econ. Entomol.,* **59,** 1430–35 (1966)
184. Phillips, J. H. H. Notes on species of *Lecanium* Burmeister in the Niagara Peninsula, Ontario, with a description of a new species. *Can. Entomologist,* **97,** 231–38 (1965)
185. Phillipp, W. Is the San Jose scale still a problem. *Gesunde Pfl.,* **17,** 218–19 (1965) (In German)
186. Pianka, M. Structures and pesticidal activities of derivatives of dinitrophenols. 1.—Structure and acaricidal activity of certain carbonates of dinitrophenols. *J. Sci. Food Agr.,* **17,** 45–56 (1966)
187. Pianka, M., Polton, D. J. Synthesis and insecticidal activity of N-methlenefluoroacetamide derivatives. *J. Sci. Food Agr.,* **16,** 330–41 (1965)
188. Pickett, A. D. Utilization of native parasites and predators. *J. Econ. Entomol.,* **52,** 1103–5 (1960)
189. Pickett, A. D. Pesticides and the biological control of arthropod pests. *World Rev. Pest Control,* **1**(2), 19–25 (1962)
190. Pielou, D. P., Williams, K. The effectiveness of residues of insecticides in preventing reinfestation of apple leaves by apple aphid, *Aphis pomi* DeG. 1. Diazinon and Sevin. *Can. Entomologist,* **93,** 93–101 (1961)
191. Pielou, D. P., Williams, K. The effectiveness of residues of insecticides in preventing reinfestation of apple leaves by apple aphid, *Aphis pomi* DeG. 2. Thiodan and Guthion. *Can. Entomologist,* **93,** 1036–41 (1961)
192. Pielou, D. P. The control of the apple aphid on dwarf apple trees with bark applications of dimethoate. *Can. J. Plant Sci.,* **41,** 407–12 (1961)
193. Plurad, S. B., Goodman, R. N., Enns, W. R. Persistence of *Erwinia amylovora* in the apple aphid, *Aphis pomi* DeGeer, a probable vector. *Nature,* **205,** 206 (1965)
194. Pokrovskii, E. A., Unterberger, V. K., Deniskina, G. P. Insecticides to control the San Jose scale. *Zashchita Rast. ot Vreditelei Boleznei* **5** (1960) (In Russian)
195. Popova, A. I. On a method for the large-scale breeding of a parasite of the San Jose scale. *Tr. Vses. Inst. Zashchity Rast.,* **20,** 61–64 (1964)
196. Powell, K. M., Linke, W. Thioquinox —a new specific acaricide of the quinoxaline group. *Proc. Brit. Insecticide Fungicide Conf., Brighton, Engl.,* **2,** 489–98 (1962)
197. Prints, E. Ya. On the susceptibility of different pear varieties to damage by the San Jose Scale. *Vredit. Bolez. Fauna Bespozvon. Moldavii,* **2,** 49–52 (1965) (In Russian)
198. Prints, E. Ya., Slonovskii, I. F. The effectiveness of merkaptofos for the control of the San Jose scale. In Problems of the ecology and economic importance of the terrestrial fauna. *Kishiven, Inst. Zool., Akad. Nauk Moldavsk. SSR.* (1961) (Prints, Ya., Ed. In Russian)
199. Priore, R. Experiments on the winter chemical control of the principal Diaspine Coccids of fruit trees in the Provinces of Naples and Caserta. Years 1961–64. *Boll. Lab. Entomol. Agr. Portici,* **22,** 179–204 (1964) (In Italian)
200. Priore, R. Experiments on the winter control of coccids in the Provinces of Naples and Caserta in 1965. *Boll. Lab. Entomol. Agr. Fillipo Silvestri,* **23,** 193–210 (1965) (In Italian)
201. Proverbs, M. D., Madsen, H. F.

Progress on the use of sterile insects for codling moth control in 1968. *Brit. Columbia Fruit Growers Assoc. Quart. Rept.*, **13**(3), 45–47 (1968)
202. Proverbs, M. D. Induced sterilization and control of insects. *Ann. Rev. Entomol.*, **14,** 81–102 (1969)
203. Putnam, W. L. The codling moth, *Carpocapsa* (*Cydia*) *pomonella* (L.): A review of its bionomics, ecology and control on apple with special reference to Ontario. *Proc. Entomol. Soc. Ontario,* **93,** 22–60 (1963)
204. Rafiq Ahmad, Ghani, M. A. Biology of *Chilocorus infernalis* Muls. *Tech. Bull. Commonwealth Inst. Biol. Control,* **7,** 101–6 (1966)
205. Redfern, R. E. Concentrate medium for rearing the codling moth. *J. Econ. Entomol.*, **57,** 607–8 (1964)
206. Redfern, R. E. Concentrate media for rearing red-banded leaf roller. *J. Econ. Entomol.,* **56,** 240–41 (1963)
207. Rehman, M. H., Ghani, M. A., Kazimi, S. K. Introduction of exotic natural enemies of San Jose scale into Pakistan. *Tech. Bull Commonwealth Inst. Biol. Control,* **1,** 165–77 (1961)
208. Richards, A. M. Scale Insect Survey on Apples 1950–60. *New Zealand J. Agr. Res.*, **3,** 693–98 (1960)
209. Richards, A. M. The oyster-shell scale, *Quadraspidiotus ostreaeformis* (Curtis), in the Christchurch district of New Zealand. *New Zealand J. Agr. Res.*, **5,** 95–100 (1962)
210. Rivard, I. Synopsis et bibliographie annottee sur la mouche de la pomme *Rhagoletis pomonella* (Walsh). *Mem. Entomol. Soc. Quebec,* **2,** 1–158 (1968)
211. Roelofs, W. L., Feng, K. Sex pheromone specificity tests in the Tortricidae, an introductory report. *Ann. Entomol. Soc. Am.*, **61,** 312–16 (1968)
212. Russ, K. Uber ein bemekenswertes Auftreten von *Beauveria bassiana* (Bals.) Vuill. an *Carpocapsa pomonella* (L.). *Pflanzenschutz Ber.*, **31,** 105–8 (1964)
213. Russ, K. Studie über die Abhangigkeit der Populationshdynmik des Apfelwicklers (*Carpocapsa pomonella* (L.)) vom Fruchtertrag der Wirtspflanze. *Pflanzenschutz Ber.*, **35,** 165–69 (1967)
214. Saggers, D. T., Clark, M. L. Trifluoromethyl-benzimidazoles—a new family of acaricides. *Nature,* **215,** 275–76 (1967)
215. Sahai, B., Joshi, L. D. Bionomics and biological control of San Jose Scale (*Quadraspidiotus perniciosus*) in U.P. *Punjab Hort. J.*, **5,** 37–43 (1965)
216. Samarasinghe, S., LeRoux, E. J. Preliminary results on the sampling of populations of the oystershell scale, *Lepidosaphes ulmi* (L.), on apple in Quebec. *Ann. Entomol. Soc. Quebec,* **9,** 104–20 (1964)
217. Samarasinghe, S., LeRoux, E. J. The biology and dynamics of the oystershell scale, *Lepidosaphes ulmi* (L.), on apple in Quebec. *Ann. Entomol. Soc. Quebec,* **11,** 206–92 (1966)
218. Sanford, K. H., Herbert, H. J. The influence of spray programs on the fauna of apple orchards in Nova Scotia. XV. Chemical controls for winter moth, *Operophthera brumata* (L.), and their effects on phytophagus mites and predator populations. *Can. Entomologist,* **98,** 991–99 (1966)
219. Schiitte, F., Franz, J. M. Investigations on the control of the codling moth, *Carpocapsa pomonella* (L.), with the help of *Trichogramma embryophagum* Hartig. *Entomophaga,* **6,** 237–74 (1962)
220. Sedivy, J., Kodys, E. Forecasting and warning, a basis for the control of the codling moth (*Cydia pomonella*). *Ved. Pr. vyzk. Ust. zivic. Uyroby Uhrinevsi,* **6,** 195–220 (1962) (In Czechoslovakian, German summary)
221. Sharma, P. L., Bhalla, O. P. Control schedule for San Jose scale. *Himachal Hort.*, **5,** 17–20 (1964)
222. Sharma, P. L., Bhalla, O. P. Studies on the control of San Jose scale *Quadraspidiotus perniciosus* (Comstock). *Indian J. Entomol.*, **27,** 323–30 (1965)
223. Smith, E. H. The susceptibility of life history stages of the pear psylla to oil treatment. *J. Econ. Entomol.*, **58,** 456–64 (1965)
224. Soueref, S. T., Komblas, K. N. On the possibility of substitution of synthetic organic insecticides for lead arsenate in the control of *Cydia pomonella* (L.). *Geoponika (Ser. B)*, **7,** 256–58 (1961) (In Greek, English summary)

225. Spencer, E. Y. *Guide to the chemicals used in crop protection,* 5th ed., 1093. (Canada Dept. Agr. Process. Publ., 1968)
226. Stanev, M. Ts. A study of the bionomics and measures for the control of the Californian scale *Quadraspidiotus perniciosus* Comst. in Bulgaria. *Izv. Inst. Zasht. Rast.,* **5,** 5–27 (1963) (In Bulgarian)
227. Stanley, W. W., Russell, W. G. Growth responses of flowering crabapple and control of red-banded leaf roller following applications of synthetic insecticides. *J. Econ. Entomol.,* **60,** 885 (1967)
228. Stein, W. Versuche zur biologischen Bekamfung des Apfelwicklers (*Carpocapsa pomonella* (L.)) durch Eiparasiten der Gattung. *Entomophaga,* **5,** 237–59 (1960)
229. Sudha Rao, V., Rao, V. P. Introduction of *Prospaltella perniciosi* for control of San Jose scale in India and Pakistan. *Food Agr. Organ. U.N., FAO Plant Protect. Bull.,* **8,** 120–23 (1960)
230. Sugonyaev, E. S. On the ecology, distribution and economic importance of the brown scale (*Parthenolecanium corni* Bouche) in the North Caucasus. *Trudy Zool. Inst., Akad. Nauk SSSR,* **36,** 180–90 (1965) (In Russian)
231. Sumaroka, A. F. Factors affecting the sex ratio of *Aphytis proclia* Wlk., external parasite of the San Jose scale. *Entomol. Rev. (USSR),* **46,** 179–85 (1967) (English transl.)
232. Tadic, M. D. Indigenous insect natural enemies of the San Jose scale in Yugoslavia. *Arh. Poljoprivredne Nauke,* **14,** 11–132 (1961) (In Serbo-Croat.)
233. Tanada, Y. A granulosis virus of the codling moth, *Carpocapsa pomonella* (Linnaeus). *J. Insect Pathol.,* **6,** 378–79 (1964)
234. Telenga, N. A. Die Anwendung der Muskardinenpilze im Verein mit Inisektiziden zu der Bekämfung der Schädlingsinsekten. *Trans. Intern. Conf. Insect Pathol. Biol. Control, 1st,* **1,** 155–68 (1958)
235. Tereznikova, E. M. Scale insects—injurious to fruit crops of the Transcarpathian region. *Dopovidi Akad. Nauk. Ukr. RSR,* **1,** 110–14 (1960) (In Ukrainian)
236. Tereznikova, E. M. On the bionomics of scale insects in the conditions of the Transcarpathian region. *Dopovidi Akad. Nauk. Ukr. RSR,* **4,** 536–39 (1960) (In Ukrainian)
237. Tereznykova, E. M. The economic importance of the Coccoidea of the forest zone of the Ukraine. *Dopovidi Akad. Nauk. Ukr. RSR,* **5,** 678–81 (1966) (In Ukrainian)
238. Terrosi, U. Further investigations on light mineral oil and some organophosphorous compounds used against coccids and as a winter ovicide. *Notiz. Mal Piante,* **52,** 65–74 (1960) (In Italian)
239. Tew, R. P., Sillibourne, J. M., Silva-Fernandes, A. M. Harvest residues of codling moth insecticides on apples. *J. Sci. Food Agr.,* **12,** 666–74 (1961)
240. Timlin, J. S. The distribution and relative importance of some armoured scale insects on pip fruit in the Nelson/Marlborough orchards during 1959–60. *New Zealand J. Agr. Res.,* **7,** 531–35 (1964)
241. Timlin, J. S. The biology, bionomics and control of *Parlatoria pittospori* Mask.: a pest on apples in New Zealand. *New Zealand J. Agr. Res.,* **7,** 536–50 (1964)
242. Unterberger, V. K. The penetration of petroleum oil under the shield of California Scale *Aspidiotus perniciosus* Comst. *Khim. Sel'skom Khoz.,* **1,** 26–27 (1964) (In Russian)
243. Unterberger, V. K. Use of phosphorous insecticides in the form of oil emulsions against the San Jose scale. *Khim. Sel'skom Khoz.,* **7,** 34–41 (1964) (In Russian)
244. Unterstenhöfer, G. On two new insecticidally active carbamates. *Mededel. Landbouw-Hogeschool Opzoekingssta. Staat Gent,* **28,** 758–66 (1963) (In German)
245. Van Der Merwe, C. P. A note on the introduction and spread of perniciosus scale, *Aspidiotus (Quadraspidiotus) perniciosus* (Comstock), in South Africa. *J. Entomol. Soc. S. Africa,* **25,** 328–31 (1963)
246. Vasev, A. Improving the technique of the warning service and new chemical compounds for the control of the codling moth. *Rastit. Zasht.,* **9,** 26–31 (1961) (In Bulgarian with English summary)
247. Weiden, M. H. J., Moorefield, H. H., Payne, L. K. O-(methylcarbamoyl)

oximes: a new class of carbamate insecticide-acaricides. *J. Econ. Entomol.,* **58,** 154–55 (1965)

248. Westigard, P. H., Gentner, L. G., Berry, D. W. Present status of biological control of the pear psylla in southern Oregon. *J. Econ. Entomol.,* **61,** 740–43 (1968)

249. Westigard, P. H., Madsen, H. F. Pear psylla in abandoned orchards. *Calif. Agr.,* **17** (1), 6–8 (1963)

250. Westigard, P. H., Madsen, H. F. Oviposition and egg dispersion of the apple aphid with observations on related mortality factors. *J. Econ. Entomol.,* **57,** 597–600 (1964)

251. Wildbolz, T. On the influence of horticultural practices on apple aphids. *Proc. Intern. Congr. Entomol., 12th, London, 1964,* 589 (1965)

252. Wilde, W. H. A., Watson, T. K. Bionomics of the pear psylla, *Psylla pyricola* Foerster, in the Okanagan Valley of British Columbia. *Can. J. Zool.,* **41,** 953–61 (1963)

253. Wood, T. G. Field observations on flight and oviposition of codling moth, (*Carpocapsa pomonella* (L.)) and mortality of eggs and first instar larvae in an integrated control orchard. *New Zealand J. Agr. Res.,* **8,** 1043–59 (1965)

254. Zech, E. Beitrag zur Kenntnis einiger in Mitteldeutschland aufgetretener Parasiten des Apfelwicklers (*Carpocapsa pomonella* (L.)). *Z. Angew. Entomol.,* **44,** 203–20 (1959)

255. Zech, E. 5 jahrige Untersuchungen über den Schupfverlauf von *Carpocapsa pomonella* (L.) mit besonder Berücksichtigung der 2 Generation. *Nachrbl. Deut. Pflanzenschutzdienst. (Berlin),* **12,** 143–50 (1961)

ULTRALOW VOLUME APPLICATIONS OF CONCENTRATED INSECTICIDES IN MEDICAL AND VETERINARY ENTOMOLOGY

By C. S. Lofgren

Entomology Research Division, Agricultural Research Service, U. S. Department of Agriculture, Gainesville, Florida

After the discovery of DDT's insecticidal activity in 1942, entomologists and chemists made a concentrated effort to find similar highly effective insecticides. Their efforts resulted in the discovery of hundreds of potentially effective chemicals and, as a result, we have achieved the best insect control ever known, but these advances did not come about without the raising of concurrent problems. For example, insect resistance to chemicals and the potential hazards of pesticides to other organisms are now basic facts of life for economic entomologists. However, these problems have caused us to reevaluate all our methods of insect control and to concentrate some of our effort on the development of more efficient and safer uses of pesticides. The use of undiluted concentrated pesticides in what is commonly referred to as the ultralow volume application technique has been one of the more interesting new developments in insect control in recent years. The very nature of the requirements for this method has caused a renewed interest in the basic mechanics of producing insecticidal sprays and in the means of increasing their efficiency in killing insects. A breakthrough in this area came when Himel & Moore in Montana investigated methods of controlling the spruce budworm, *Choristoneura fumiferana* (23), with aerial sprays. By using a fluorescent tracer in the spray formulation, they determined that 93 per cent of the larvae were contacted directly by droplets 50 μ or less in diameter. Since the major portion of the insecticide in most conventional aerial sprays is applied as droplets larger than 50 μ, the greatest amount of the insecticide in conventional sprays may be wasted. The ramifications of these findings have many applications in the fields of medical and veterinary entomology and have caused an increase in research on droplet size in relation to insect kill.

Terminology

With few exceptions, ultralow volume in entomological literature means the application of small quantities of liquid concentrated insecticides. The term "ultralow volume" was probably first used by workers in Africa for concentrated sprays of dieldrin used to control a locust, *Schistocerca gregaria* (67). Subsequently, the terms low volume and ultralow volume (ULV) were used in the United States to describe concentrated sprays of dieldrin for control of grasshoppers and eventually sprays of undiluted malathion (51, 69). However, the term, ULV, is strictly relative and must be related to some specific higher volume since the standard volumes vary, depending on such things as insect susceptibility, unit area to be treated, and type of application. In the United States, less than one-half gallon of spray per acre is considered ULV by the U. S. Department of Agriculture for labeling insecticides for insect control. This definition provides a convenient reference point. However, naled and malathion applied as ULV aerosols outdoors will kill mosquitoes when they are applied at a rate of 4 ml per acre (58). In contrast, 14 gallons per acre would be less than the conventional amount applied to control ticks, and 24 gallons per acre is less than the conventional amount used to control chiggers. It is, therefore, easy to see how a proliferation of terms could occur to describe the various rates used for low volume application. Actually, the need for a volume reference is questionable because a "concentrated insecticide spray" should imply that an ultralow or minimal volume was used. If not, an overdose of insecticide is being used for the control purpose involved. At any case, it is not within the purview of this paper to propose terminology so I will use the following guidelines: one-half gallon or less per acre is ULV; a concentrated formulation is one containing 20 per cent or more insecticide as per Sayer (67) in his first reference to ULV; and a technical formulation contains no less than 70 per cent insecticide. (These definitions force me to omit work on dilute sprays for some livestock pests that have been or could be classified as ULV (17, 50.)

Advantages and Disadvantages of Ultralow Volume

Most advantages of ULV applications of insecticides revolve, in one way or another, around savings in money and time. The most obvious savings result from the reduction or elimination of the need for diluents. For example, conventional aerial sprays to control adult mosquitoes in Florida are applied at the rate of one to six quarts per acre (32 to 192 fl oz), depending on the vegetative overgrowth. The common mosquito adulticides (malathion, naled, and fenthion) are usually applied at rates of from 0.5 to 3 fl oz per acre as ULV sprays.

Reduced volumes also mean that the size of dispensing equipment can usually be reduced; thus, smaller and less expensive trucks or aircraft are required for transport and application. Considerable time is also conserved in loading and reloading insecticide onto such spray equipment.

Also, the savings in time and cost achieved with the ULV technique have permitted researchers and persons involved in actual control to think in terms of treating much larger acreages. The possibility is especially important for vector control programs set up during epidemics because efficient and rapid control methods are required if the epidemics are to be halted. These requirements can be met, at least partially, by aerial application of ULV insecticides. Thus, vector control now is dependent primarily on the ingenuity used in adapting the techniques to the specific vector. Another possible benefit of ULV applications could be increased effectiveness of the insecticide. Some evidence exists that ground and aerial ULV applications of several insecticides are more effective for tick control (54) and ULV ground aerosol applications of naled (58) are more effective than conventional procedures for mosquito control. The differences, however, are not great (about twofold). Mount et al. (58) concluded that better kill with ULV naled resulted from better drift characteristics because mosquitoes were killed at greater distances from the applicator. In contrast, ULV sprays of fenthion at 0.2 pounds per acre and malathion at 0.2 and 0.4 pounds per acre were reported to be less effective than conventional sprays (3 quarts per acre) against mosquitoes (55); however, in the same tests, naled was equally effective in ULV and conventional sprays. Fenthion also gave slower kill of mosquitoes as a ULV spray (55, 71). The reason is not readily apparent.

Ultralow volume application of insecticides actually has few disadvantages though car spotting or damage to some automobile finishes does occur because of the solvent or the corrosive properties of some of the insecticides. However, such damage is noticeable only when large droplets (> 100 μ) are present in the spray. Also, greater care must be taken in handling concentrated or technical insecticides because of the increased degree of exposure. However, this is primarily a hazard to the applicators. Hazards to nontarget organisms should not be any greater than with conventional sprays, but studies of effects on such organisms are being conducted (43, 73), and more extensive studies are needed. In addition, drift of ULV sprays can be a serious problem if high doses of insecticides are used. However, low doses are normally used for contact sprays in mosquito control, and small droplets and drift are essential to produce the necessary coverage for wide swaths and good penetration of the insecticide into the insect's environment.

Equipment for Application

Initially, standard spray systems were used on aircraft for ULV applications of insecticides. However, when technical or concentrated insecticides were placed in this equipment, they often loosened accumulations of dirt, rust, and other deposits that eventually clogged the lines and nozzles used to deliver the insecticide, and the small orifice tips used to meter it. Many researchers therefore attempted to develop special spray systems. Some used compressed air or carbon dioxide to propel the insecticide (15, 21, 36); others used pumps driven by electric motors (1/10 to 1 hp) operated from the electrical system of the aircraft or from separate storage batteries (1, 44). Stainless steel 2- or 3-gallon tanks were generally used as the reservoir for the insecticide, but the tank size was actually limited only by the space available in the aircraft. The pump, motor, and tank were sometimes mounted in compartments in front of or behind the cockpit with the controls placed in the cockpit. Belly-tanks that strap to the fuselage of aircraft are also available and have been used. Plastic or metal tubing is generally used to convey the insecticide from the pump out along the standard booms on the aircraft because it is more resistant to the chemicals. Nozzles were placed in these lines, as needed, and the lines and nozzles were fastened to the boom with tape or clamps.

More recently, investigators have developed ULV spray equipment for multi-engine aircraft. One of the more useful rigs was designed for the C-47 aircraft (19, 47). This system has a short boom mounted directly under the fuselage and depends on drift to achieve swath width. An electric motor and a gear pump are used to propel the insecticide. The entire spray rig can be mounted within a few hours without modifying the aircraft. Then, when the spray operation is completed, it can be removed in no more than one hour. Also, a similar system was built for the Piper Aztec® (10); to avoid contamination of this aircraft with insecticide it was necessary to position the boom under the tail section.

Several types of nozzles have been used for aerial ULV spraying. Early researchers in the United States (20, 37) followed the lead of investigators working with agricultural insects and used spinning disc or screen cage nozzles such as the Mini-Spin® nozzle. While these nozzles are capable of producing droplets of fairly uniform size (26), they tend to be expensive, hard to install and, because they are more complex, more prone to breakdown. Hydraulic nozzles with flat fan tips are now most commonly used because when they are oriented properly to the wind, they produce droplets of satisfactory sizes, and they are relatively inexpensive and easily replaced. However, they are not trouble-free. The diaphragms degrade quickly from contact with concentrated insecticides, and the edges of the tip orifice erode after prolonged use. Hollow-cone nozzles are also used, but the droplets are not as small as those produced with the flat fan tips (56).

Ground equipment for the application of ULV insecticides has also been developed in the last few years. Treatment of small areas has been made with modified, motorized backpack sprayers (13, 54). The main problem in converting these sprayers was the regulation of the flow rate of the insecticide. This regulation was accomplished by using drilled discs or small metering valves mounted in the insecticide lines. Stainless steel or insecticide-resistant plastic tanks were used for containing the insecticide. Also, in some models, filters were installed to help eliminate nozzle stoppage. During operation, the insecticide was allowed to flow into the airstream near the end of the air nozzle; thus, formation of the droplets depended primarily on the velocity of the air flowing past the nozzle. Aside from these modifications the basic design of conventional backpack sprayers was used. No definite information is available on the size of the droplets produced by these apparatuses; however, droplets of 5 to 50 μ were deposited on glass slides 80 feet from one such machine (13).

High volume cold foggers have also been modified or redesigned for

ULV applications. The first such modification (58) involved the Curtis® cold fogger. A small nozzle head was substituted for the large boom assembly, and three pneumatic nozzles from the cold fogger were mounted in this head. Insecticide was pumped to the nozzles with a carbon dioxide pressurized system, and rate of flow was controlled by a flowmeter that delivered volumes as low as a few milliliters per minute. With these modifications, it was possible to apply less than one gallon per hour instead of the 40 to 120 gallons per hour commonly applied with conventional cold foggers. The size of the droplets produced ranged from 6 to 20 μ mass median diameter (MMD), depending on rate of flow and the volume of air passing through the nozzles.

At least one other similar system has been described (70), and others will undoubtedly be developed or adapted from conventional aerosol machines.

Droplet Size

The most desirable droplet size for any ULV spray or aerosol is related to the purpose of the spray, for example, whether it is a mosquito larvicide or adulticide spray, a residual spray, or a ground aerosol or space spray. Each type of application probably has a particular range of droplet sizes that will give maximum efficiency. In some instances, such as in mosquito control, kill of both larvae and adults is required. Sprays for this purpose may require two different droplet size ranges in the same spray. Also, the environment of the insect must be considered since insects located under vegetation, in tree holes, or in other protected areas may be reached only by droplets of a certain size, if at all. The problem then is to determine the ideal size for each procedure so that minimum doses of insecticide can be used, and unnecessary contamination of the ecosystem can be avoided.

Aerial sprays.—Medical entomologists have long been aware that the relationship between droplet size and insect kill is important in ground aerosols (4), but few have related the droplet size of aerial sprays to insect kill.

However, recent studies of spruce budworm larvae (23) with aerial sprays tagged with fluorescent particles showed vividly the need to control size to obtain maximum kill. By this technique, the size of droplets deposited on larvae was determined by counting the number of fluorescent particles in clusters on the insect cuticle and relating this to the volume of insecticide that would theoretically contain this number of particles. Droplets below 50 μ were found to be deposited most readily on the spruce budworm larvae. (Since the technique does not detect droplets less than 20 μ, it was impossible to determine the minimum size of droplets impinging on the insects.) Similar studies have not been made with medical or veterinary insects.

Much emphasis has been placed by some researchers on the rates of deposition of droplets on dye cards as an indication that an adequate and uniform spray was applied. In fact, it has been specifically suggested that 10 or more droplets per square inch are necessary for adequate kill of mosquitoes

(28). Observations by other workers cast doubt on the validity of such guidelines as hard and fast rules for adult mosquito control (49, 56, 57, 68). For example, one report stated that naled gave a better kill of natural populations of mosquitoes when the deposits on horizontally placed cards averaged only seven droplets per square inch than when they averaged 13 to 35 per square inch (49). In other studies (64), an average of 67 per cent of caged mosquitoes was obtained when they were placed adjacent to dye cards on which no droplets were collected; mortalities above 67 per cent correlated directly with the number of droplets on dye cards and complete kill was always obtained when 50 or more droplets per square inch were collected. This correlation is not surprising since it probably indicated that a large number of droplets of all sizes were present. The obvious conclusion is that droplets too small to detect or deposit on dye cards (probably 20 μ or less) are of major importance for kill of mosquito adults.

To this date, no one has actually determined the ideal droplet size spectrum to use for adult mosquito contact sprays or larval sprays. One set of guidelines recommends that the MMD of aerial sprays for adult and larval control should be 50 to 60 μ with not more than 10 per cent of the droplets less than 25 μ nor more than 125 μ (28); however, specific data were not reported from the tests on which these recommendations were based. Information from other sources suggest that the lower limit could probably not be met if flat fan tips were used since, at speeds of 100 or 150 miles per hour and line pressures of 40 to 60 pounds per square inch, the percentage of total spray volume in droplets less than 25 μ ranges from 15 per cent for flat fan 8006 tips spraying malathion to 56 per cent for flat fan 8001 tips spraying naled (57). Considerable variations can occur in evaluating the droplet spectrums of aerial sprays depending upon the methods used. Thus, any set of guidelines needs to include specific information on the methods to be used for checking the droplet spectrum.

Studies of toxicity made in combination with chemical analyses showed that the LD_{100} of malathion for *Aedes taeniorhynchus* (Wiedemann) is about 10 nanograms (74). This amount of insecticide is contained in a droplet 25 μ in diameter. Thus, for contact sprays against mosquito adults, droplets of malathion that are larger than 25 μ results in waste of insecticide unless a less concentrated spray is used. For more toxic compounds, the minimum size of droplet that kills would be less than 25 μ. However, other factors such as the number of droplets per unit volume of air need to be considered and more efficient applications of some compounds can perhaps be made with concentrated solutions instead of with the technical chemical (63).

Mount (53) critically reviewed the data published during the past 25 years and concluded that in aerial sprays for contact kill of adult mosquitoes, the ideal droplet ranges from 10 to 25 μ. (This range is a compromise between small droplets that would be excessively influenced by turbulent air currents and winds, and large droplets that would contain more insecticide than required for kill.) Droplets larger than 50 μ would not give adequate coverage and are apt to cause other problems, for example, car spotting. Research to determine the validity of these conclusions is vitally needed. However, the delivery of droplets of particular size ranges from aircraft is extremely difficult.

Studies of the size of droplets produced by flat-fan tips when the operating conditions were similar to those that are normal with single engine aircraft, were simulated with ground equipment by Mount et al. (56) by mounting the nozzle in the air blast of the Buffalo Turbine® mist blower. When the droplets were collected on silicone-treated glass slides and measured in the usual manner, line pressures of 20 to 80 pounds per square inch and tips with different spray angles (15° to 110°) were found to have no effect on the MMD of the sprays produced. The smallest droplets were produced when the nozzles and tips were angled 45° forward into the wind, and the largest were produced when they were angled 45° backward. Speed of the air blast also had a definite influence. Thus, the MMD at 95 miles per hour was 36 μ compared with 45 μ at 50 miles per hour. In addition, tip size had a definite influence although not as much as might have been expected. Tips with over a 100-fold difference in flow rate (0.023 gpm versus 2.0 gpm) caused only a twofold difference in the MMD (32 μ to 64 μ). In general, these results and other information obtained with sprays from aircraft (57) indicate that the droplets produced by flat fan tips (Spraying System 730023 to 8020) directed forward on aircraft flying at 80 to 150 mph range in size from about 30 to 60 μ MMD.

Kruse et al., in 1949 (41), also found that variations in nozzle position, air speed, and tip size influenced the size of droplets produced by flat fan tips; however, the MMD's reported for their sprays were considerably higher than those listed by Mount et al. (56, 57) for the same spraying tips. This difference in results might be due to the fact Mount and colleagues either collected droplets in strong winds or flooded the collection area with spray, thus increasing the chances for small droplets to impinge on the collection surfaces (silicone-treated glass slides). Droplet collections made during normal spray operations or with single swaths generally miss some of the finest droplets because they drift out of the collection area or they do not impinge in low wind velocities. This rationale is speculative; however, it is obvious that researchers in this field need to standardize on consistent and effective methods for collecting representative samples of droplets from aerial sprays.

Droplets that measure less than 5 μ MMD and from 30 to 50 μ MMD have been reported when thermal fogging devices were used on aircraft (41, 66). However, no study on the range of sizes that can be produced by this method of application has been noted. These devices might be used to produce sprays with drops of 5 to 30 μ MMD. Then, it would be possible to test the entire range of droplet size.

Not as much information is available on the size of droplets used to control mosquito larvae. The primary problem is finding a size capable of ready penetration of the canopies of vegetation that may cover the breeding areas. Kruse et al. (41), in 1949, concluded that the most favorable size of droplet for DDT sprays (20 per cent in a hydrocarbon solvent) was 70 to 80 μ. Workers in California (12) suggested that droplets too small to impinge easily on plant surfaces (drops under 100 μ) were best for sprays applied over dense stands of rice; however, the majority of spray droplets needed to be over 10 μ so swath and drift could be controlled. Tests have shown that with a spray having an MMD of about 70 μ, only 10 per cent of the chemical penetrated very dense stands of rice (75).

Penetration of the vegetative canopy by droplets to control adults and larvae is a problem that needs much more investigation. Tests in Panama (44) showed that control of anopheline mosquitoes in jungles with malathion or fenthion required about two to three times the dose normally used in open-to-moderate canopy. Certainly, for the most efficient use of insecticide for either larval or adult control, a more accurate determination of optimum size for penetration of vegetation must be made.

Ground aerosols.—The size of droplet has also been studied in relation to ground ULV applications. The findings generally agree with those that have long been established for ground equipment used in mosquito control (4). This type of application can be described best as drift spraying or aerosoling. From theoretical considerations based on Stokes Law, the droplets should be 10 μ or less to achieve adequate drift (300 to 400 feet) when wind velocities are low (<3 mph). Mount (53) concluded that 5 to 10 μ was probably optimum for ground applications of aerosols. It has been shown that malathion droplets of 6 to 10 μ MMD gave almost twice the kill of caged mosquitoes as those of 11 to 20 μ MMD. These results were explained primarily by the fact that the smaller droplets had better drift characteristics and gave better kill of mosquitoes in cages located further from the applicator. However, the drift of droplets in this size range is entirely dependent on the vagaries of the wind or air currents, and insecticide applied from ground equipment has drifted and killed mosquitoes and house flies, *Musca domestica* one to two miles from the applicator (70). However, these kills were obtained in open areas at high rates of application, and it is rather doubtful that the amount of insecticide used under normal operating conditions would drift in an intact cloud for this distance and still kill insects.

Correlation of droplet size with insect kill appears to be one of the most interesting areas of research for the future, and it actually holds the promise of greatly improving and refining our methods of control for insects of medical and veterinary importance. As mentioned previously, minimum doses need to be applied to save money and eliminate unneeded contamination of the environment. Comparisons of the doses required to kill mosquitoes by ground and aerial application illustrate this point. Almost 10 times as much insecticide is required to obtain comparable kill of caged mosquitoes by aerial application: aerial sprays of malathion are normally applied at a rate of 3 fluid ounces or 0.225 pounds per acre; ground ULV applications at 0.018 pounds per acre give comparable kill (58).

Control of Medically Important Arthropods

Adult mosquitoes.—More emphasis has been placed on the control of adult mosquitoes with ULV sprays than on the control of any other medically important insect. Thus, more than 10 years before the initial work on ULV sprays to control grasshoppers, studies were made in the southeastern United States and in Alaska with ULV application of concentrated DDT formulations (7, 41). For these tests, the investigators used 20 per cent solutions of DDT in fuel oil-hydrocarbon solvents. This concentration is close to the maximum amount of this insecticide that can be placed in solution for aerial spraying (therefore, they could not use a more concentrated spray). Application at rates as low as 3.45 to 4 fluid ounces per acre (0.05 pounds of DDT per acre) gave good control of anopheline larvae in the Tennessee River Valley area but only fair control of adult mosquitoes in Alaska. The research received limited attention, and no effort was made to continue the use of the low volume applications.

The first recent aerial ULV applications for control of mosquitoes were made against *Aedes nigromaculis* in California in 1963 (61) and *Aedes taeniorhynchus* and/or *Aedes sollicitans* in Kentucky (39) and Florida (20) in 1964. In all three tests, technical malathion (2 to 9 fl oz per acre) gave satisfactory control. Numerous tests have been made since and many other insecticides have been evaluated as concentrated sprays. Table I contains a listing of the chemicals evaluated, the mosquito species and stage against which they were tested, the dose applied, and the relative effectiveness of the treatments. This tabulation gives a good idea of the wide variety of materials that have been tested and certainly makes it evident that the technique can be used effectively.

Probably the most important potential use of the ULV technique is the control of disease vectors during epidemics and emergencies. Very large-scale vector control programs involving the treatment of hundreds of square miles have always been difficult. For success these programs must be carried out quickly so the infected vectors are eliminated, and control must be maintained long enough to prevent newly emerging vectors from becoming infected. If both aims can be accomplished, transmission of the disease can be halted. However, the expense of the insecticide and the diluent, the high cost of aerial application, and the limited favorable weather conditions during which applications can be made have previously made large-scale programs either difficult or impractical. The ULV technique may not reduce the insecticidal cost appreciably but it is certainly a partial answer to the other problems, and both the U. S. Public Health Service and the World Health Organization have undertaken programs to evaluate its potential for controlling disease vectors during epidemics or natural disasters.

The first large-scale demonstration of the practicality of the technique

TABLE I

Tests of Aerial ULV Application of Insecticides Against Natural Populations of Mosquito Larvae and Adults[1] (Unless Noted[2])

Insecticide and type of formulation	Amount applied		Species	Effectiveness[3]	Test location	Date reported	Reference
	Fl oz/acre	lb/acre					
			MOSQUITO ADULTS				
Bay 39007	1.2–3.2	0.0375–0.075	*A. sollicitans*	F-VG	Kentucky	67, 68	36,40
concentrate	1.6, 2.4	0.05, 0.075	*A. stimulans*	VG	Michigan	67	72
Fenthion	0.8, 1.6	0.05, 0.1	*A. sollicitans*	VG	Kentucky	66	37
technical	1.6–6–4.8	0.1–0.32	*A. albimanus*	F-VG	Canal Zone	58	44
			A. triannulatus	F-VG	Canal Zone	58	44
	0.75, 1.5	0.05, 0.1	*A. stimulans*	VG	Michigan	66	71
	3.2	0.2	*A. taeniorhynchus*	G	Florida	67	55
concentrate	4	0.05	*A. sollicitans*	G-VG	Kentucky	66	37
	1.6, 3.2	0.05, 0.1	*A. taeniorhynchus*	G	Florida	67	55
Malathion	2–6	0.15–0.46	*A. sollicitans*	G-VG	Kentucky	65, 66, 67	36, 37, 39
technical				G-VG	Louisiana	66	5
technical	2–9.6	0.15–0.72	*A. taeniorhynchus*	F-VG	Florida	65, 67	20, 55
	3	0.225	*C. tarsalis*	G-VG	Texas	67	52
	3, 6	0.225, 0.45	*A. aegypti*	G-VG	Thailand	69	32, 46, 48
			C. quinquefasciatus	G-VG	Thailand	69	32, 46, 48
	6, 20	0.45, 1.5	*A. simpsoni*	F-VG	Ethiopia	69	8
	3	0.225	*A. albimanus*	G	Salvador	69	31
	3, 8	0.225, 0.62	*A. albimanus*	F-G	Canal Zone	68	44
			A. triannulatus	F-G	Canal Zone	68	44
	3	0.225	*C. quinquefasciatus*	VG	Texas	67	30
	—	0.314	*Aedes* sp.	VG	Alaska	69	59
	—	0.5	*Anopheles* sp.	VG	S. Vietnam	68	24
technical and concentrate[a]	7	0.27, 0.52	*A. vexans*	VG	Minnesota	67	13
Naled technical	0.5, 1.0	0.05, 0.1	*A. sollicitans*	VG	Kentucky	67, 68	36, 40
					Louisiana	69	49
	1.8	0.2	*A. taeniorhynchus*	VG	Florida	67	55
	0.56	0.055	*A. nigromaculis*	VG	California	67	18
	0.5–0.75	0.05–0.076	*C. nigripalpus*	G-VG	Florida	69	68
			Psorophora confinnis	G-VG	Florida	69	68
	—[b]	0.018, 0.036	*A. taeniorhynchus*	G-VG	Florida	68	58
concentrate	1.6, 3.2	0.05, 0.01	*A. taeniorhynchus*	VG	Florida	67	55
Trichlorfon	9.7, 12	0.3, 0.375	*A. sollicitans*	G-VG	Kentucky	67, 68	35, 40
			MOSQUITO LARVAE				
Abate concentrate	25	0.031	*A. sollicitans*	VG	Kentucky	66	38
Dursban	6.5–9.2	0.0125–0.05	*A. freeborni*	VG	California	68	12, 42
concentrate			*C. tarsalis*	VG	California	68	12, 42
			A. melanimon	VG	California	68	12, 42
			A. nigromaculis	VG	California	68	12, 42
Fenthion							
technical	0.8, 1.6	0.05, 0.1	*A. sollicitans*	P-VG	Kentucky	66	37, 38
concentrate	6.4–11	0.07–0.12	*A. nigromaculis*	G-VG	California	65	61
	4	0.05	*A. sollicitans*	VG	Kentucky	66	37

[1] This summary was prepared to give a representative sample of the insecticides evaluated and the test species involved and was not intended to be all-conclusive; no tests with caged insects are included.

[2] [a] = Backpack sprayer; [b] = cold aerosol.

[3] Very good (VG) 90–100%; good (G) 75–89%; fair (F) 50–74%; poor (P) <50%.

TABLE I—(*Continued*)

Insecticide and type of formulation	Amount applied Fl oz/acre	lb/acre	Species	Effectiveness[3]	Test location	Date reported	Reference
Malathion							
technical	6	0.5	*C. tarsalis*	F	California	65	61
			A. nigromaculis	F	California	65	61
concentrate	10.2	0.1	sp. not specified	P-VG	California	67	12
	—	0.5	*Anopheles* sp.	VG	S. Vietnam	68	24
Trichlorfon	3	0.1	*C. pipiens*	VG	Kentucky	66	38
concentrate			*A. vexans*	VG	Kentucky	65	38

was made in 1966 during an outbreak of St. Louis Encephalitis (SLE) in Dallas, Texas (29, 30). At this time, the U. S. Public Health Service in cooperation with the Special Aerial Spray Flight, 4500 Air Base Wing, of the U. S. Air Force Tactical Command successfully treated 475,000 acres in 8 days. The treatments were highly effective in controlling the vector of the disease, *Culex quinquefasciatus*: the density of the species was reduced 90 to 95 per cent, and the rate of infective mosquitoes dropped from 1 in 160 specimens before treatment to 1 in 29,000 (checked during a post-treatment period of 45 days). The actual applications were made by six C-123 cargo aircraft modified as spray planes, and 3 fluid ounces of malathion were applied per acre. At the operating conditions used, each plane was capable of treating 150 acres per minute and carried a payload of 600 gallons of malathion, which means that each aircraft was capable of operating for 2 hours. Also, on most mornings, weather conditions are favorable for spraying (winds under 10 mph without thermal convection currents) for an average of about 2 to 3 hours. Thus, the time when the applications could be made compared favorably with the operating time of the airplane, indicating that they were being used at maximum efficiency.

In the late summer of 1967, the Public Health Service again demonstrated the great benefits that could be derived from ULV applications of insecticides. At that time, hurricane Betsy struck the lower Gulf Coast area of Texas and deposited over 20 inches of rain. The extreme flooding that followed these rains caused displacement of many residents, and this displacement, in conjunction with lack of adequate sanitation and protection from disease vectors, caused public health officials to fear that an epidemic might occur. An emergency program was initiated, and the Special Aerial Spray Flight of the U. S. Air Force and a private contractor were called on to treat large areas in the flooded regions. The primary concern was to control *Culex tarsalis,* the vector of Western Equine Encephalitis (WEE) and *C. quinequefasciatus* the vector of SLE. The actual application of insecticide was made between about October 9 and October 31, 1967. About 3,200,000 acres were treated with 74,000 gallons of technical malathion (29). It is impossible to prove that the treatments prevented an epidemic, but the high degree of mosquito control reported may have prevented a major one from occurring.

In early 1967, the World Health Organization began to study the possibility of using ULV applications of insecticide for control of *Aedes aegypti,* the vector of dengue haemorrhagic fever in Southeast Asia, at the *Aedes* Research Unit in Bangkok, Thailand. The first tests were made in February, 1967 (32). A single-engine Cessna 180 was used to apply malathion to plots of about 1 square mile or less. Three fluid ounces of malathion were insufficient for adequate control of *A. aegypti;* however, 6 fluid ounces per acre gave very good control of the vector when three applications were made within a single week. The following November, additional tests were made with a C-47 aircraft of the Royal Thai Air Force in an attempt to control the population of *A. aegypti* in Nakhon Sawan, Thailand, a city of about 50,000 people (48). In this test, about 7 square miles were treated with two applications of 6 fluid ounces of malathion applied 4 days apart. The population of *A. aegypti* was reduced by 88 per cent or more during the 10 days of observation. Also ovitraps and dissections of female mosquitoes collected during the post-treatment period indicated that the pretreatment population had been eliminated very quickly since no eggs were deposited in the traps after 4 days, and the majority of the females collected were nulliparous.

Subsequent tests near Bangkok showed that 3 fluid ounces per acre of malathion gave good control of *A. aegypti,* but the pretreatment population was not reduced as dramatically as in the tests with the higher dose (46). Nevertheless, the results of this series of tests indicate that an epidemic of dengue haemorrhagic fever probably could be halted by controlling the vector with ULV applications of malathion. The fact that a relatively high dose of 6 ounces per acre would have to be used for each treatment to achieve adequate control is associated with the indoor resting and biting habits which decrease its chance of exposure to the insecticide.

Another study was made in Hale County, Texas, in 1967 in which an attempt was made to correlate the control of *C. tarsalis* with the incidence of WEE (22, 52). Two towns were treated about once a week from June to August with 3 fluid ounces per acre of malathion; a third received two treatments in July. Because the total areas treated were not sufficiently large (6.25 to 25 square miles), vector mosquitoes migrated into the test sites from untreated areas within about 48 hours. The seasonal rates of infection by the WEE virus in mosquitoes collected from the treated and untreated areas were not significantly different during the test period. However, the rate of infection in nestling house sparrows was higher in the check town than in two of the three treated towns, and there were no confirmed cases of WEE in the test towns during 1967, though only a few human cases were found anywhere in Texas or in the United States that year. The results were therefore suggestive of a reduction in the number of subclinical infections, but the data were inconclusive.

It is very unfortunate that conditions were such that the test did not give more definitive data on the efficiency of the ULV technique in halting the transmission of the virus. Control of a vector should indicate control of the associated disease, but data showing a positive correlation is needed. Unfortunately, experiments of this nature are expensive and require extensive backup facilities and manpower.

The results of the tests in Thailand and Hale County, Texas, demonstrate how important it is to know the biology of the vector and to put this information to use when setting up a control program. Thus, with a vector such as *C. tarsalis* that can migrate for several miles, a sufficiently large barrier must be provided to prevent rapid migration of mosquitoes back onto the treated area. The opposite situation occurs with *A. aegypti*: this species does not disperse over long distances, which undoubtedly accounts for the much greater success in controlling this species even though the areas treated were not as large as those in the Hale County tests.

Ultralow volume applications of malathion have also been used extensively in South Vietnam to control the anopheline vectors of malaria (24). However, the spray operations were difficult to handle, and the assessment of the results was equally difficult. It is therefore hard to determine the efficacy of these applications. A test on Con Son Island produced good control of adults and larvae at the rate of application used (0.5 lb/acre of 57 per cent emulsion concentrate), and tests in Panama (44) showed that anopheline mosquitoes can be controlled in dense jungles by the ULV application of 8 fluid ounces per acre (0.62 lb/acre) of malathion.

Recently, several new approaches to the ULV application of insecticides have been evaluated. If they are successful, they could be extremely useful in large-scale vector control programs. For example, a new fuselage-boom spraying system (19, 47) was shown to have many advantages. As noted previously, it can be assembled and installed on the aircraft (C-47) in less than 4 hours; and disassembly requires even less time. The advantage for large-scale programs during disease epidemics is considerable since any one of these widely available aircraft could be quickly converted to a spray plane; then, when the insecticide has been applied, it could be quickly reconverted to its original use. The principle behind use of this equipment is drift spraying. This type of spraying is not new. A number of years ago, it was called the Porton method and was used extensively in Canada (9). Also, as noted, drift spraying has been used to control locusts in Africa (14). Essentially, this type of spraying leaves it to cross winds to carry and distribute the spray droplets over the desired swath. The primary difference between the older methods of drift spraying and the present method is the smaller size of the droplets which permits much greater drift, wider swaths, and more droplets in the size ranges that can most easily impinge on the mosquitoes.

Drift spraying has also been used in California in what has been referred to as "hi-lo" spraying (2, 60). In these tests, aircraft were flown at altitudes as high as 2000 feet, and the swaths were 2 miles wide or more in crosswinds of 6 to 10 miles per hour. Actually, if the droplets are small enough, so high an altitude is not necessary to obtain this drift as was demonstrated in tests with ground equipment in which swaths of 1 to 2 miles were obtained with aerosol droplets of 5 to 10 μ MMD (70).

In other tests in California, the "hi-lo" principle has been used to provide what is referred to as "stacked swaths." Thus, the town of Colusa, California was treated by flying an airplane over the same swath several times at different altitudes. If the insecticide drifted for different distances from each altitude and if weather conditions were uniform, a uniform application should have been achieved. It proved difficult, however, to estimate the distance of the drifts, but satisfactory mosquito control was obtained (2).

The general idea involved in drift spraying is sound; the limiting factors are the ability to produce small droplets and the dependence on satisfactory meteorological conditions. Also, high altitude applications (over 1000 feet) are difficult to make because weather conditions are more varied over the wider range of altitudes. In addition, there is a limit to the width of swath that can be efficiently produced since wide swaths require that large amounts of insecticide be dispensed in a short time. Actually, it seems most practical to fly swaths that are 1000 to 5000 feet wide. Swaths of these widths can probably be controlled adequately, and the dispensing equipment does not need to be as large or as cumbersome as the equipment required for wider swaths.

Night application of ULV aerial sprays is also being investigated in Louisiana (49) and in Florida (68). During this time, mosquitoes are most active, and contact sprays should be most efficient. Also, meteorological conditions tend to be more favorable at night, i.e., less wind and convection currents, and lights can be used to guide the aircraft.

The fuselage booms, the method of drift spraying, and night flying, if put together, could greatly increase the efficacy of large-scale vector control programs.

Applications of ULV sprays or aerosols from ground equipment have lagged behind applications from aircraft. However, as noted, in Minnesota, a backpack sprayer has been used to treat small areas (13). Experiments in Florida (58) showed that ULV aerosols of naled proved superior to conventional high volume thermal aerosols against *Aedes taeniorhynchus* in orange groves. Thus, these tests indicated that the ULV technique is applicable to ground equipment. The main limiting factor up to now has been the lack of suitable or commercially available equipment for application.

Mosquito larvae.—Table I also lists insecticides evaluated against natural populations of mosquito larvae. The majority of the investigations were made in California; but some studies were made in Kentucky and Michigan, and numerous investigators have exposed mosquito larvae in artificial containers to ULV sprays. Mosquito larvae breed in a variety of situations, so control with aerial spray is dependent on the amount of protection the breeding site affords. For example, *A. aegypti* breed in Thailand in water storage containers, most of which are covered, or in ant traps and other containers in houses. With these conditions, larval control with aerial sprays is almost impossible, even with a highly effective larvicide such as Abate® (O,O-dimethyl phosphorothioate O,O-diester with 4,4′-thiodiphenol) (45). However, aerial sprays are effective in open breeding sites.

Since larval breeding in natural pools, ponds, lakes, and rivers normally is associated with vegetation, the amount of vegetative cover is of prime importance in determining the effectiveness of the sprays, as noted by a number of investigators (12, 38, 72, 75). As a result, though drift spraying works well for control of adult mosquitoes, it is not useful for larval control. Also, since larval breeding is limited to relatively small, distinct areas, the main objective is to deposit the insecticide at these sites, and drift spraying is not an efficient way to accomplish this goal.

Researchers in California investigated the use of both helicopters and fixed-wing aircraft for the control of larvae in rice fields (11, 42). The aircraft were flown at low altitudes (8 to 14 feet) to keep drift at a minimum, but swath widths of 150 to as much as 250 feet were attained with the helicopter and the fixed-wing aircraft, respectively. Rates of application varied from 4 to 12 fluid ounces per acre compared with conventional sprays of one-half gallon or more. Excellent control was obtained in rice fields with both Dursban® (O,O-diethyl O-3,5,6-trichloro-2-pyridyl phosphorothioate) (42) and fenthion (12, 61).

The success of sprays applied by helicopter is especially encouraging. Although the operating expenses of helicopters are high and payload capacity is low, the helicopters can be used both for survey and treatment. Thus, possible breeding sites can be located, surveyed, and treated on the spot. In addition, the lower volumes applied increase the payload capacity.

Ticks.—Control of ticks on an area-wide basis with ULV sprays has proved practical (54). Present recommendations call for the use of very high volumes of liquid (15 to 50 gallons per acre). In comparison, Bay 39007 (*o*-isopropoxyphenyl methylcarbamate) and fenthion applied at a rate of 0.5 pounds per acre (32 and 8 fl oz/acre, respectively) as ULV sprays with ground equipment gave good control of the lone star tick, *Ablyomma americanum* for six weeks. The applications were made with a modified Mighty-Mite backpack sprayer (see Equipment section) in late May and June when the nymphs were searching for a blood meal and were more susceptible to insecticide than just after engorging.

Ultralow volume aerial applications have also been made with technical fenthion. Good control of ticks was obtained at a rate of 2 pounds per acre (32 fl oz/acre); however, only fair control was obtained at 0.5 and 1 pounds per acre. The researchers noted that wind conditions were adverse for aerial application and that the plots were relatively small (40 acres) so some of the insecticide probably drifted out of the plots.

House flies.—The effect of two aerial applications of 6 fluid ounces per acre of malathion on natural populations of adult house flies was tested in Thailand in November, 1968 (48). Before the first treatment, the average number of adults landing on 61-square centimeter grids was 1180 during 1 hour. Within 2 days post-treatment, the population had been reduced by 95 per cent. The second application, made 4 days after the first did not give a further reduction; however, the control remained above 90 per cent for 2 more days and then dropped to 76 per cent 4 days later.

Ultralow volume applications of insecticide have also been used against house fly larvae breeding in chicken manure (6). Undiluted emulsion concentrates of dimethoate and formothion at 2 g per square meter of actual insecticide (5.0 to 8.4 ml per square meter) gave larval control that was comparable to that achieved with higher-volume conventional applications (100 ml per square meter). The applications were made with an applicator designed by the researchers. The tank of insecticide was pressurized with carbon dioxide, and the liquid was sprayed through Teejet® 730023 flat fan tips.

Tsetse flies.—A ULV aerial spraying experiment using technical fenthion was conducted in northern Tanganyika in 1966 (25). Three applications of approximately 0.52 fluid ounces or 0.044 lbs/acre were applied at 3-week intervals with a Cessna 185 aircraft fitted with rotary atomizers. Post-spray catches indicated that over 90 per cent reduction in the numbers of *Glossina swynnertoni* and *G. pallidipes* was obtained during the first 10 weeks after spraying; *G. morsitans* populations were reduced by 87 per cent during the same time period. It was felt that a fourth application would have been justified to obtain over 90 per cent control of *G. morsitans,* which is the more important vector of sleeping sickness in the area.

Stable flies.—Studies of *Stomoxys calcitrans* were made in northwest Florida in an area where large numbers of these flies congregate along the beaches, especially during late summer and fall, and are a serious pest of man. The adaption of ULV application to the control of this pest is not easy because these flies migrate long distances and are most active during the daytime when high winds and temperatures prevail, conditions which make it most difficult to apply aerial sprays. However, tests were attempted (65) with ULV sprays of concentrated naled (actual volume varying from about 9 fluid ounces per acre to 32 fluid ounces per acre) applied from a Stearman aircraft. The applications were made upwind of the beach and the insecticide was allowed to drift over the beaches. In some tests kill was obtained 1000 feet from the flight path of the plane; however, consistent control was not possible because of the adverse weather conditions at time of application (9:00 to 10:00 a.m.).

Midges.—ULV aerial applications of technical malathion (2 and 4 fluid ounces per acre) for control of adult midges, *Glyptotendipes paripes* were made in Florida (62) from a Stearman 450 or a Call-air 190 aircraft from a height of 20 to 30 feet. One to three swaths were applied over the margins of the lake in which the midges were breeding. The higher rate gave good control in 3 hours, and the residue was effective for 4 days. However, it took almost 6 hours for satisfactory kill with the lower dose, and the residual activity lasted only 3 days. Little difference in initial kill was attributable to the number of swaths used; however, residual activity appeared to be less with only one swath.

Also, control of the midge, *Chironomus fulvipilus* was observed when larvae were exposed to malathion sprays in shallow enamel pans in about 1 inch of water. After exposure to 2 or 6 fluid ounces per acre, 90 per cent to 100 per cent of the larvae were dead or moribund in 24 hours (62).

Sand flies.—Aerial ULV sprays of technical fenthion and malathion (4 fl oz/acre) were applied to control *Culicoides guyanensis* and *C. furens* breeding in a swamp on the west bank of the Pacific entrance to the Panama Canal. Temporary control of adults and good control of larvae were achieved in two tests (3).

Control of Veterinary Arthropods

The only use of ULV sprays reported against arthropod pests of livestock is concerned with the control of horn flies, *Haematobia irritans* and face flies, *Musca autumnalis,* on rangeland cattle. The technique has been very successful in the mid-western United States and is being recommended for horn and face fly control in North and South Dakota.

The first published report was by Dobson & Sanders (16). They applied 8 fluid ounces per acre of technical malathion over beef cattle in a 35-acre pasture in Indiana with Stearman aircraft equipped with 80015 flat fan tips and flying at an air speed of 85 miles per hour and an altitude of 20 to 40 feet. Good control was obtained for several days and adequate control for 1 week.

Similar tests were made in 1966 in Kentucky (34) with trichlorfon (4 lb/gal) applied at 0.1 pounds per acre (3 fluid ounces). The chemical gave good control of both species; however, flies were counted for only 24 hours. Additional test made in the fall of 1966 (35) with cattle in a 5-acre pasture gave very good control of horn flies for 10 days but poor control of face flies.

In South Dakota, ULV sprays of 95 per cent malathion at 6 ounces per acre were applied with a Piper Cub P-11 aircraft to range cattle (27). The plane flew at a height of 50 feet and delivered a swath 150 feet wide. Sprays to both the pasture and the herd gave the best control; however, satisfactory control was obtained when only the herd or the herd and the loafing areas were treated. Levels of control below that causing economically important damage (50 horn flies per side and 5 face flies per face) was achieved on the experimental herds when four applications were made during the summer of 1965 and six applications in 1966. Populations of horn flies on untreated herds in 1965 averaged from 150 to 200 per side; herd bulls had extremes of 2000 to 10,000 flies in some instances.

Attempts to use ULV technique for control of horn and face flies in New Mexico have not been successful (33). In these tests, technical fenthion and malathion applied at 2 fluid ounces and 6.4 fluid ounces per acre, respectively, to unrestrained range cattle averaged from 66 per cent to 74 per cent control 2 days after application.

LITERATURE CITED

1. Akesson, N. B., Burgoyne, W. E., Mulhern, T. D., Phillips, K. L. A low volume spray system for small aircraft used in mosquito control. *Calif. Vector Views,* **13,** 63–66 (1966)
2. Akesson, N. B., Burgoyne, W. E., Yates, W. E. Technical volume formulations and aerial spray application, *ASAE Paper No. 68–616, Am. Soc. Agr. Eng.,* Chicago, Ill. (Dec. 1968)
3. Altman, R. M., Keenan, C. M., Pearson, W. G. Control of *Culicoides* sandflies, Fort Kobbe, Canal Zone in 1968. Presented at *25th Ann. Meet. Am. Mosquito Control Assoc., Williamsburg, Virginia, 1969*
4. American Mosquito Control Association. Ground equipment and insecticides for mosquito control. *Bull. No. 2* (Revised ed., 101 pp., 1968)
5. Anderson, C. H. Test of low volume aerial application of malathion for adult mosquito control. *Rept. Florida Anti-Mosquito Assoc., 37th Ann. Meet.,* 20–24 (1966)
6. Bailey, D. L., LaBrecque, G. C., Whitfield, T. L. Low volume and conventional sprays with various insecticides for house fly larvae control in poultry houses. *J. Econ. Entomol.* (Submitted for publication, 1969)
7. Blanton, F. S., Travis, B. V., Smith, N., Husman, C. N. Control of adult mosquitoes in Alaska with aerial sprays. *J. Econ. Entomol.,* **43,** 347–50 (1950)
8. Brooks, G. D. Preliminary studies on the use of ultra-low volume applications of malathion for control of *Aedes simpsoni. Presented at 40th Ann. Meet. Florida Anti-Mosquito Assoc., Winter Park, Florida, 1969*
9. Brown, A. W. A. *Insect Control by Chemicals,* Chap. VI, 447–53. (John Wiley & Sons, New York, 817 pp., 1951)
10. Burgoyne, W. E., Akesson, N. B. A low volume spray system for twin engine aircraft used in mosquito control. *Calif. Vector Views,* **16,** 17–18 (1969)
11. Burgoyne, W. E., Akesson, N. B., Mulhern, T. D. The introduction of helicopters into California mosquito control. *Proc. Ann. Conf. Calif. Mosquito Control Assoc., 33rd.,* 68–71 (1966)
12. Burgoyne, W. E., Akesson, N. B., Mulhern, T. D., Phillips, K. The present status of low volume (LV) air sprays for California mosquito control. *Mosquito News,* **27,** 398–406 (1967)
13. Buzicky, A. W. How we modified a back-pack sprayer for ULV insecticide applications. *Pest Control,* **35,** 42, 44, 46 (1967)
14. Courshee, R. J. Drift spraying for vegetation baiting. *Bull. Entomol. Res.,* **50,** 355–70 (1959)
15. Dearman, A. V., Powell, H. F., Thompson, R. K. Portable sprayer for aerial LVC applications. *J. Econ. Entomol.,* **58,** 1050–52 (1965)
16. Dobson, R. C., Sanders, D. P. Low-volume high concentration spraying for horn fly and face fly control on beef cattle. *J. Econ. Entomol.,* **58,** 379 (1965)
17. Eschle, J. L., Berry, I. L. Low-volume application of insecticide to cattle for control of the horn fly. *J. Econ. Entomol.,* **60,** 293–94 (1967)
18. Geib, A. F., DeWitt, R. H., White, A. C. Ultra-low volume application of naled to control adult mosquitoes in California pastures. *Calif. Vector Views,* **14,** 70–73 (1967)
19. Glancey, B. M., Ford, H. R., Lofgren, C. S. Evaluation of a fuselage-mounted spray boom for ultra-low volume application of insecticides for mosquito control. *Mosquito News* (Submitted for publication, 1969)
20. Glancey, B. M., Lofgren, C. S., Salmela, J., Davis, A. N. Low volume aerial spraying of malathion for control of adult salt-marsh mosquitoes. *Mosquito News,* **25,** 135–37 (1965)
21. Glancey, B. M., White, A. C., Husman, C. N., Salmela, J. Low volume applications of insecticides for control of adult mosquitoes. *Mosquito News,* **26,** 356–59 (1966)
22. Hess, A. D., Hayes, R. O. Recent developments on encephalitis. *Proc. Biennial Public Health Vector Control Conf., 7th, Atlanta, Georgia, 1968,* 34, 35

23. Himel, C. M., Moore, A. D. Spruce budworm mortality as a function of aerial spray droplet size. *Science,* **156,** 1250–51 (1967)
24. Holway, R. T., Morrill, A. W., Santana, F. J. Mosquito control activities of the U. S. Armed Forces in the Republic of Vietnam. *Mosquito News,* **27,** 297–307 (1967)
25. Irving, N. S., Beesley, J. S. S. Airspray experiment with fenthion against *Glossina morsitans,* Westw., *G. Swynnertoni* Aust., and *G. pallidipes* Aust. *Trop. Pest. Res. Inst., Arusha, Misc. Rept.,* 621 (1967)
26. Isler, D. A. Atomization of low-volume malathion spray. *J. Econ. Entomol.,* **59,** 688–90 (1966)
27. Kantack, B. H., Berndt, W. L., Balsbaugh, E. U., Jr. Horn fly and face fly control on range cattle with aerial applications of ultra-low volume malathion sprays. *J. Econ. Entomol.,* **60,** 1766–67 (1967)
28. Kilpatrick, J. W. Performance specifications for ultra-low volume aerial application of insecticides for mosquito control. *Pest Control,* **35,** 80, 82, 84 (1967)
29. Kilpatrick, J. W. The role of ULV aerial applications of insecticides for mosquito control in disaster areas. *Proc. Biennial Public Health Vector Control Conf., 7th, Atlanta, Georgia, 1968,* 84–87
30. Kilpatrick, J. W., Adams, C. T. Emergency measures employed in the control of St. Louis encephalitis epidemics in Dallas and Corpus Christi, Texas. *Proc. Ann. Conf. Calif. Mosquito Control Assoc., 35th,* 53 (1967)
31. Kilpatrick, J. W. Pletsch, D. J., Breland, S. G., Lewellan, L. L. Preliminary evaluation on the role of ULV aerial application of malathion for malaria control in Central America. Presented at *Ann. Meet. Am. Mosquito Control Assoc., 25th, Williamsburg, Virginia, 1969*
32. Kilpatrick, J. W., Tonn, R. J., Jatanasen, S. Evaluation of ultra-low volume insecticide dispensing systems for single-engine aircraft and their effectiveness against *Aedes aegypti* populations in Southeast Asia. *Bull. World Health Organ.* (In press, 1969)
33. Kinzer, H. G. Aerial applications of ultra-low volume insecticides for controlling the horn fly on unrestrained range cattle. *J. Econ. Entomol.* (In press, 1969)
34. Knapp, F. W. Aerial application of trichlorfon for horn fly and face fly control on cattle. *J. Econ. Entomol.,* **59,** 468 (1966)
35. Knapp, F. W. Ultra-low volume aerial application of trichlorfon for control of adult mosquitoes, face flies, and horn flies. *J. Econ. Entomol.,* **60,** 1193 (1967)
36. Knapp, F. W., Gayle, C. H. ULV aerial insecticide application for adult mosquito control in Kentucky. *Mosquito News,* **27,** 478–82 (1967)
37. Knapp, F. W., Pass, B. C. Low volume aerial sprays for mosquito control. *Mosquito News,* **26,** 22–25 (1966)
38. Knapp, F. W., Pass, B. C. Effectiveness of low volume aerial sprays for mosquito control. *Mosquito News,* **26,** 128–32 (1966)
39. Knapp, F. W., Roberts, W. W. Low volume aerial application of technical malathion for mosquito control. *Mosquito News,* **25,** 46–47 (1965)
40. Knapp, F. W., Rogers, C. E. Low volume aerial insecticide applications for control of *Aedes sollicitans* (Walker). *Mosquito News,* **28,** 535–40 (1968)
41. Kruse, C. W., Hess, A. D., Ludvik, G. F. The performance of liquid spray nozzles for aircraft insecticide application. *J. Natl. Malaria Soc.,* **8,** 312–34 (1949)
42. Lembright, H. W. Dosage studies with low volume applications of Dursban insecticide. *Down Earth,* **24,** 16–17 (1968)
43. Linn, J. D. Effects of low volume aerial spraying of Dursban and fenthion on fish. *Down Earth,* **24,** 28–30 (1968)
44. Lofgren, C. S., Altman, R. M. Glancey, B. M. Control of Anopheline species in the Canal Zone with ultra-low volume sprays of malathion and fenthion. *Mosquito News,* **28,** 353–55 (1968)
45. Lofgren, C. S., Ford, H. R., Tonn, R. J., Bang, Y. H. (Unpublished data, 1969)
46. Lofgren, C. S., Ford, H. R., Tonn, R. J., Bang, Y. H., Sirobodhi, P. Control of *Aedes aegypti* (L.), the vector of dengue haemorrhagic fever in Asia, with ULV mala-

thion applied from a C-47 aircraft. 3. Effectiveness of 3 fluid ounces per acre. *Bull. World Health Organ.* (In press, 1969)

47. Lofgren, C. S., Ford, H. R., Tonn, R. J., Jatanasen, S. Control of *Aedes aegypti* (L.), the vector of dengue haemorrhagic fever in Asia, with ULV malathion applied from a C-47 aircraft. 1. Configuration and installation of the fuselage-mounted spray system. *Bull. World Organ.* (In press, 1969)
48. Lofgren, C. S., Ford, H. R., Tonn, R. J., Jatanasen, S. Control of *Aedes aegypti* (L.), the vector of dengue haemorrhagic fever in Asia, with ULV malathion applied from a C-47 aircraft (2), a large-scale test in Nakhon Sawan, Thailand. *Bull. World Health Organ.* (In press, 1969)
49. Machado, W. C., Bordes, E. S., Blake, A. J., Carmichael, G. T. A technique for ULV insecticide application from high altitudes. Presented at *Ann. Meet. Am. Mosquito Control Assoc., 25th, Williamsburg, Virginia, 1969*
50. Matthysse, J. G., Pendleton, R. F., Padula, A., Nielson, G. R. Controlling lice and chorioptic mange mites on dairy cattle. *J. Econ. Entomol.*, **60,** 1615–23 (1967)
51. Messenger, K. Low volume aerial spraying will be boon to applicator. *Agr. Chem.*, **18,** 63–66 (1963)
52. Mitchell, C. J., Hayes, R. O., Holden, P., Hill, H. R., Hughes, T. B. Effects of ultra-low volume applications of malathion in Hale County, Texas. I. Western enchephalitis virus activity in treated and untreated towns. *J. Med. Entomol.* (In press, 1969)
53. Mount, G. A. Optimum droplet size for adult mosquito control with space applications of insecticides. Presented at *Ann. Meet. Florida Anti-Mosq. Assoc., 40th, Winter Park, Florida, 1969*
54. Mount, G. A., Hirst, J. M., McWilliams, J. G., Lofgren, C. S., White, S. A. Insecticides for control of the lone star tick tested in the laboratory and as high- and low-volume sprays in wooded areas. *J. Econ. Entomol.* **61,** 1005–7 (1968)
55. Mount, G. A., Lofgren, C. S. Ultra-low volume and conventional aerial sprays for control of adult salt-marsh mosquitoes, *Aedes sollicitans* (Walker) and *Aedes taeniorhynchus* (Wiedemann) in Florida. *Mosquito News,* **27,** 473–77 (1967)
56. Mount, G. A., Lofgren, C. S., Pierce, N. W., Baldwin, K. F. Effect of air speed, line pressure, nozzle position, orifice type, and orifice size on droplet size of ultra-low volume aerial sprays. Presented at *Ann. Meet. Am. Mosquito Control Assoc., 25th, Williamsburg, Virginia,* 1969
57. Mount, G. A., Lofgren, C. S., Pierce, N. W., Baldwin, K. F., Ford, H. R., Adams, C. T. Droplet size, density, distribution and effectiveness of ultra-low volume aerial sprays dispersed with TeeJet® nozzles. Presented at *Ann. Meet. Florida Anti-Mosquito Assoc., 40th, Winter Park, Florida, 1969*
58. Mount, G. A., Lofgren, C. S., Pierce, N. W., Husman, C. N. Ultra-low volume nonthermal aerosols of malathion and naled for adult mosquito control. *Mosquito News,* **28,** 99–103 (1968)
59. Mount, G. A., McWilliams, J. G., Adams, C. T. Control of adult mosquitoes in Alaska with malathion. *Mosquito News,* **29,** 84–86 (1969)
60. Mulhern, T. D. Recent experiences with low volume spraying for the control of mosquitoes in California. Presented at *Intern. Congr. Trop. Med. and Malaria, 8th, Teheran, Iran, 1968*
61. Mulhern, T. D., Gjullin, C. M., Lopp, O., Ramke, D., Frolli, R., Reed, D. E., Murray, W. D. Low volume airplane sprays for control of mosquito larvae. *Mosquito News,* **25,** 442–47 (1965)
62. Patterson, R. S., von Windeguth, D. L., Glancey, B. M., Wilson, F. L. Control of the midge, *Glyptotendipes paripes* with low volume aerial sprays of malathion. *J. Econ. Entomol.,* **59,** 864–66 (1966)
63. Rathburn, C. B., Boike, A. H., Jr., Rogers, A. J. Progress report of low-volume aerial spray tests against adults of *Aedes taeniorhynchus* (Wied.) and *Culex nigripalpus* Theob. *Rept. Ann. Meet. Florida Anti-Mosquito Assoc., 39th, Naples, Florida, 1968,* 26–32
64. Rathburn, C. B., Rogers, A. J., Boike,

A. H., Lee, R. M. Evaluation of the low-volume aerial spray technique by use of caged adult mosquitoes. Presented at *Ann. Meet. Am. Mosquito Control Assoc., 25th, Williamsburg, Virginia, 1969*
65. Roger, A. G., Clements, B. W. (Unpublished data, 1969)
66. Salmela, J., Sidlow, A. J., Davis, A. N. A new thermal-aerosal generator for dispensing insecticidal fogs from a Stearman airplane. *Mosquito News,* **20,** 275–80 (1960)
67. Sayer, H. J. An ultra-low volume spraying technique for the control of the desert locust, *Schistocerca gregaria* (Forsk.). *Bull. Entomol. Res.,* **50,** 371–86 (1959)
68. Shepard, M. The effects of ultra-low volume Dibrom® from a C-47 aircraft on adult mosquitoes. Presented at *Ann. Meet. Am. Mosquito Control Assoc., 25th, Williamsburg, Virginia, 1969*
69. Skoog, F. E., Cowan, F. T., Messenger, K. Ultra-low volume aerial spraying of dieldrin and malathion for rangeland grasshopper control. *J. Econ. Entomol.,* **58,** 559–65 (1965)
70. Stains, G. S., Fussell, E. M., Keathley, J. P., Murray, J. A., Vaughan, L. M. Swath kills of one to two miles utilizing a new low volume aerosol ground dispersal unit. Presented at *Ann. Meet. Am. Mosquito Control. Assoc., 25th, Williamsburg, Virginia, 1969*
71. Stevens, L. F., Stroud, R. F. Control of mosquito adults and larvae with ultra-low volume aerial applications of Baytex® and Baytex-Baygon® mixture. *Mosquito News,* **26,** 124–28 (1966)
72. Stevens, L. F., Stroud, R. F. Control of mosquito adults and larvae with ultra-low volume aerial applications of Baygon® and Baygon-Baytex® mixture. *Mosquito News,* **27,** 482–85 (1967)
73. Washino, R. K., Whitesell, K. G., Womeldorf, D. J. The effect of low volume application of Dursban on non-target organisms. *Down Earth,* **24,** 21–22 (1968)
74. Weidhaas, D. E., Bowman, M. C., Mount, G. A., Lofgren, C. S., Ford, H. R. Relationship of droplet size and dosage to adult mosquito control with space sprays. Presented at *Ann. Meet. Am. Mosquito Control Assoc. 25th, Williamsburg, Virginia, 1969*
75. Womeldorf, D. J., Gillies, P. A. Bioassay determination of aircraft swaths and rice-canopy penetration by low volume insecticidal sprays at Colusa, California. *Down Earth,* **24,** 23–27 (1968)

MITE TRANSMISSION OF PLANT VIRUSES

By G. N. Oldfield[1]

Entomology Research Division, Agricultural Research Service,
United States Department of Agriculture
Riverside, California

The subject of mite transmission of plant viruses has been reviewed several times by Slykhuis (122, 124, 126, 128). A considerable number of papers on this subject have been published since Slykhuis's last review; several constitute significant additions to the knowledge of the relationships between mites and the plant viruses they transmit.

Except, perhaps, for the apparent case of transmission of potato virus Y by *Tetranychus telarius* (107), the only proven mite vectors of plant viruses are certain members of the family Eriophyidae. In 1927, Amos et al. (2) reported a positive correlation between the infestation of black currants with the black currant gall mite (*Cecidophyopsis ribis*) and the development of reversion disease. Each of the reports by Massee (69), Smith (135), and Thresh (145), which appeared much later, contributed to the incrimination of *C. ribis* as the vector of currant reversion virus. Shortly after Massee's report appeared in 1952 several other eriophyid mites were incriminated as vectors of plant viruses. Slykhuis (118–120) reported that *Aceria tulipae* transmitted both wheat streak mosaic virus and wheat spot mosaic virus. Flock & Wallace (34) reported that *Aceria ficus* transmitted fig mosaic virus. Wilson et al. (166) reported an undescribed species of *Eriophyes* [later described by Keifer & Wilson and named *E. insidiosus* (60)] as a vector of the peach mosaic virus. Then Mulligan (79, 80) showed that *Abacarus hystrix* transmitted ryegrass mosaic virus.

Since Mulligan's report in 1958, a few additional cases of apparent eriophyid transmission of plant viruses have been reported; however, several of these need confirmation.

CHARACTERISTICS OF ERIOPHYID MITES OF SIGNIFICANCE TO THEIR ROLE AS VECTORS OF PLANT VIRUSES

Obviously, eriophyids are extremely small arthropods. They range from about 90 μ long in the case of *Heterotergum wilsoni* to well over 300 μ long in the case of *Novophytoptus stipae*. Most species range between 150 and 225 μ long. Because of their diminutive size, they have often been overlooked as vector candidates. The small size allows them to occupy areas on

[1] The author wishes to acknowledge the help extended by Prof. I. M. Newell and Mr. N. S. Wilson during the course of this study.

the plant that are inaccessible to larger organisms and allows the production of huge populations in comparatively small areas.

Eriophyids are well known for their ability to cause a wide array of galls, each type of which is usually characteristic of a certain mite species. However, many species do not cause galls and are often considered as causing no damage to their hosts. Many species cause rather subtle changes in their hosts such as the slight curling or twisting of leaves. Certain of these changes are often overlooked. Under certain circumstances, as in the case of *Aculus* mites on *Prunus* spp. (40, 158, 164), symptoms of feeding injury on leaves may closely resemble symptoms of virus infection. In cases such as these, precautions must be taken to distinguish between feeding injury and symptoms of virus infection.

Eriophyids exhibit several types of life cycles (44, 53, 89); however, all of the types of life cycles are completed on a single host. While in certain cases (deutogynes) eriophyids do not feed for several months, they nevertheless do not leave the host plant except to establish themselves on another plant.

Eriophyids may be disseminated by wind (119, 132) or may be transported by other insects (39, 67, 132, 162). They often are observed in an upright position resting on the caudum and apparently become airborne from this position. Eriophyids exhibit varying degrees of host specificity. There appears to be some basis for generalizing that species that have a more intimate relationship with their hosts (i.e., gall formers and bud inhabitants) are usually more host-specific than species that function as leaf vagrants. Often, gall-forming species are limited to a single species or group of species within a genus of plants. Leaf-vagrant species may be equally host-specific or, as in the case of *Calacarus citrifolii* in South Africa (155), they may have a host range including plants in several families. As will be discussed in this paper, *Aceria tulipae* also has a wide host range, but all of its hosts are monocots.

All eriophyids are essentially parasites of perennial plants. However, some annuals may become infested from closely related perennials. In the case of *A. tulipae,* wheat, an annual, may become infested by mites from perennial plants or, in areas where wheat grows in overlapping sequence, *A. tulipae* may be perpetuated on this annual.

The mouthparts of eriophyids are specifically adapted for piercing plant cells. The length of the chelicerae suggests that they are able to penetrate only a very few cell layers of plant tissue. *Phyllocoptruta oleivora,* the citrus rust mite (100), usually penetrates only the epidermal cell layer of orange rinds. Orlob (92) suggested that because of the structure and attachment of the stylets and rostrum, *A. tulipae* probably only penetrates five μ into plant tissue; i.e., only into the epidermal cells.

Because of their diminutive size, eriophyids pose problems not often encountered in working with other arthropod taxa that transmit plant viruses. In most other taxa, individuals are large enough to be transferred and subsequently recovered from test plants after a prescribed transmission period. In this case, the appearance—and more importantly, the persistence and development of disease symptoms—is usually considered evidence of transmission of an etiological agent by the arthropod. On the other hand, the failure of symptoms to persist and spread would suggest that they were caused by a phytotoxic substance injected by and inherent to the arthropod.

Because of their propensity for finding their way into buds and other protected places on their hosts (and their small size), it is extremely difficult to recover even the majority of individual eriophyids after transferring them from one plant to another. For this reason, special techniques must be used to show that abnormalities which may subsequently appear in a test plant after introducing eriophyids are due to transmission of a virus by the mites and not due to injection of a phytotoxic substance.

Slykhuis (126) suggested the following three conditions that must be met to prove that an eriophyid is a vector of a plant virus: (*a*). The presence of mites must be correlated with the appearance of the disease in nature. (*b*). The development of disease symptoms must not depend on the continued presence of the mites. It is preferable if the causal virus can also be transmitted by artificial means without mites. (*c*). The mites must not be able to induce the disease symptoms on healthy plants until after they have fed on diseased plants or have acquired virus in some other way.

In regard to the second condition, certain mite-transmitted viruses are sap-transmissible. In this case, the test plant can be tested for the presence of virus with minimum precautions against transfer of mites in the sap to another plant. Other mite-transmitted viruses are not sap-transmissible but can be mechanically transmitted only by grafting. In this case, far greater measures must be taken to insure that eriophyids are not transferred on the grafted tissue since what may in reality be continued transfer of a toxocogenic eriophyid from plant to plant, may appear to be evidence of virus transmission if all eriophyids are not destroyed. Flock & Wallace (34) used sulfur dust to kill *Aceria ficus* after allowing them to feed on test plants for a prescribed time. There are arguments against complete reliance on this technique to prove virus transmission since it is almost impossible to certify that the population of mites is completely destroyed. Nevertheless, the extension of this technique or other similar techniques is useful to preclude the transfer of live mites from a test plant with disease symptoms to a healthy mite-free plant during grafting.

WHEAT STREAK MOSAIC

The Vector

Geographical distribution.—*Aceria tulipae* was originally described from tulips in California by Keifer (52). The tulips were said to have originated in Holland. Reports from several areas in the United States (4, 24, 27, 92, 140) and from Canada (95, 123) of transmission of wheat streak mosaic virus (WSMV) by *A. tulipae* indicate that this species is widespread in North America. In South America, it has been reported from Venezuela (29) and Argentina (105). Puttarudriah & Channabasavanna (102) reported an eriophyid mite as a pest of garlic in Mysore, India. The general description of damage, mite anatomy, and distribution of mites on the plant suggested that it may be *A. tulipae*. Shtein-Margolina et al. (111) and Oliinyk (91) reported studies using *A. tulipae* from Russia. It is probably widespread throughout the Holarctic Region.

Feeding injury.—In 1961, *A. tulipae* was implicated as the cause of "silver top" disease of grasses in southern Alberta, Canada (45); however, a later report (3) indicated that silver top is caused by certain insects and not by *A. tulipae*. Recently, Nault et al. (84) reported that a salivary phytotoxin injected by *A. tulipae* caused red streak of corn. It also damages garlic in California (62) and Venezuela (29). According to Smalley (131), it can cause virus-like symptoms on garlic.

Host range.—*A. tulipae* has a wide range of hosts in the Gramineae. Slykhuis (119) reared it on *Poa compressa* and *Oryzopsis hymenoides* as well as on wheat. Connin (23) found *A. tulipae* naturally infesting western wheat grass (*Agropyron smithii*), Canada wild rye (*Elymus canadensis*), green foxtail (*Setaria viridis*), smooth crabgrass (*Digitaria ischaemum*), and oversummering volunteer wheat (*Triticum aestivum*). In greenhouse studies (24), *A. tulipae* reproduced on all 27 varieties of wheat tested, all 6 barley varieties tested, all 10 corn varieties tested, all 5 sudan grass varieties tested, and 12 of 24 wild grass species tested.

According to Slykhuis (128), as yet no perennial has been shown to function as an important source of spread of the virus because apparently none are good hosts of both *A. tulipae* and wheat streak mosaic virus. Gibson (37) observed that sorghum, sprouted under wheat that was infested with *A. tulipae*, was apparently not a good host. Slykhuis (119) collected *A. tulipae* from the field on foxtail barley (*Hordeum jubatum*), Canada wild rye, and western wheat grass and found that they did not survive when placed on wheat, and *A. tulipae* from wheat did not survive when placed on any of the other three grasses. According to Del Rosario & Sill (27), physiological strains of *A. tulipae* that are adapted to wheat, western wheat grass, and onion do not readily colonize the other two hosts in each case. Such a phenomenon might account for the apparent unimportance of wild grasses as sources of spread of the virus.

Biology.—According to Staples & Allington (140), the life cycle of *A. tulipae* includes the egg, two nymphal stages and the adult. The nymphs are incapable of gross movements for several hours before molting. On wheat, reproduction is apparently entirely parthenogenetic. The life cycle described by Staples & Allington is typical of many eriophyid species. According to Gibson [cited by Somsen (139)], no specialized overwintering female (deutogyne) is produced, but Somsen (139) reported the existence of a "migratory form" which was larger and less prone to injury during transfer. Somsen stated that differences between the so-called "migratory form" and the usual adult were difficult to quantify, but he suggested that the "migratory form" might account for the appearance of sudden epiphytotics of wheat streak mosaic. Somsen's report is indeed interesting; inasmuch as it is, as far as I know, the first report of such a form in the Eriophyidae. I have observed extremely large adults in certain *Eriophyes* species, but I have not attempted to show that they represent a significantly different morphological type or relate their appearance to any annual event or change in the host plant.

Slykhuis (119) showed that eggs of *A. tulipae* from southern Alberta, Canada, could survive when exposed to −30.7°C for 2.5 minutes. In general, *A. tulipae* survived lower temperatures than host wheat plants. Slykhuis also showed that hatchability of *A. tulipae* eggs was highest at 100 per cent relative humidity. Del Rosario & Sill (25) found that successful rearing required rather high humidity. They were able to herd the negatively phototactic mites with a beam of light. Del Rosario & Sill (26) studied reproductive potential and other aspects of the biology of *A. tulipae.* They were able to keep adult mites alive on a wheat-decoction-dextrose agar culture media for periods up to 80 days but none reproduced.

The Virus

Geographical distribution.—The disease now known as wheat streak mosaic was first reported in 1929 (70) from Kansas. It caused heavy losses of wheat yield in western Kansas during 1949 and 1954 (31, 116). Also, it has caused serious losses to wheat in other great plains areas including Nebraska (140), South Dakota (117), Wyoming (161), Colorado (5), Montana (77), and southern Alberta, Canada (118). In 1965, Slykhuis (128) reported finding it in southwestern Saskatchewan, Canada. It has also been reported from wheat in North Dakota (153), Iowa (35, 160), California (46), Oklahoma (159), Ohio (72), Washington (78), and Texas (4). In Europe, wheat streak mosaic has been reported from Yugoslavia (142), U.S.S.R. (104), and Rumania (98). Slykhuis (125) found wheat streak mosaic virus in Jordan but not in Australia, New Zealand, West Pakistan, India, Iran, or Egypt. In 1957, Finley (33) found corn infected with wheat streak mosaic in Idaho. More recently, wheat streak mosaic virus was found naturally infecting corn in Nebraska (47), Iowa (35), and Ohio (72, 163).

Host range.—Several workers have studied the host range of wheat streak mosaic virus. McKinney (71) found that all 39 varieties of wheat tested were susceptible as were certain varieties of barley, oats, and corn. McKinney & Sando (74) found each of 18 varieties of wheat to be very susceptible but found none of 16 species of *Agropyron* grasses to be systemically infected (only six showed local lesions). Hybrids of *Agropyron* and wheat were generally less susceptible than wheat. McKinney & Fellows (73) found certain annual and perennial representatives of 13 additional genera of grasses to be susceptible. Later, Fellows & Schmidt (32) reported a wide diversity of reactions among several *Agropyron*-wheat hybrids, but even the symptomless hybrids carried the virus.

Apparently, while wheat streak mosaic virus has a wide host range in the Graminae, it does not infect species in other families. Sill & Connin (113) listed 41 species in 20 dicot families that were shown to be immune. Sill & Agusiobo (112) reported certain varieties of oats, barley, rye, and corn as susceptible, but none of 27 species in other 13 monocot families were susceptible. Notably, these included onion and tulip—two species long recognized as hosts of *A. tulipae*. These investigators also tested several dicot species; all were immune.

Morphology and other characteristics.—Much more is known about wheat streak mosaic virus than any other eriophyid mite-borne virus. Gold et al. (42) observed elongated particles about 15 mμ $\times$ 670 mμ in electron micrographs of juice extracts from WSMV-infected wheat leaves but none in juice extracts of healthy wheat leaves. Brakke (11) succeeded in purifying the virus from infected wheat. Brakke & Staples (12) showed that particles shorter than 650 mμ were not infective and Brandes & Wetter (13) found that WSMV particles were consistently much longer than those of barley stripe mosaic virus and soil-borne wheat mosaic virus. Shepard & Carroll (110) recently observed typical rod-shaped particles of WSMV in transverse and tangential section in the cytoplasm but not in the nucleus, chloroplasts, or mitochondria of infected cells of wheat and barley. No such particles were found in cells of healthy plants. Particle morphology agreed with that reported earlier by Brakke & Staples (12).

Recently, Paliwal & Slykhuis (95) found particles corresponding in size to WSMV particles in infected wheat plants and infective *A. tulipae* but not in healthy wheat plants or noninfective *A. tulipae*. Also, Oliinyk (91) found similar particles in macerates of infective *A. tulipae*.

Mite-Virus Relationships

Several characteristics of the vector, host plant, and virus have enabled workers to learn much about the transmission of WSMV that remains unanswered for other mite-transmitted viruses. The vector, *A. tulipae,* is relatively easy to colonize on wheat plants. The relatively small size and the fact that it is an annual enables easy manipulation, culturing, and isolation of the test plant.

Recently, Slykhuis (129) reviewed methods and apparatus for experimenting with eriophyid mite-transmitted viruses. Many of these are applicable to studies of eriophyid mite transmission of grass viruses but are of little value for studying those eriophyid mite-transmitted viruses of woody perennials. Transmission studies of WSMV are also facilitated by the fact that it is readily sap-transmissible, and the latent period in wheat is very short. Slykhuis (119) reported that symptoms appeared about a week after inoculation. Sill & Fellows (115) reported that, following sap inoculation, the time required for sympton expression was five days at 82° F and nine days at 68° F. Del Rosario & Sill (27) indicated that symptoms appeared about a week after inoculation, but they were able to reisolate the virus both by sap inoculation and using *A. tulipae* after two full days following inoculation.

Wheat streak mosaic virus is the only eriophyid mite-transmitted virus in which several investigators have effected transmission using the vector. After Slykhuis's (118) original report, several others in North America (4, 24, 27, 92, 95, 140) and U.S.S.R. (91, 111) successfully transmitted this virus using *A. tulipae*.

Connin (24) showed that *A. tulipae* was able to transmit WSMV to all 27 wheat varieties tested plus several varieties of oats, barley, and several species of wild grass. Sill & Del Rosario (114) showed that *A. tulipae* could transmit WSMV from wheat to corn and back to wheat. These authors (27) found wide differences in vector efficiency among populations of *A. tulipae* collected in the field from various host species. *A. tulipae* from *A. smithii* was an extremely inefficient vector to wheat (1 per cent) but improved after the mites "adapted" to wheat (32 per cent); however, this strain was still a far less efficient vector from wheat to wheat than a strain that occurred naturally on wheat. The latter strain showed an 84 to 92 per cent vector efficiency from wheat to wheat. *A. tulipae* that occurred naturally on wheat was an efficient vector to virus-susceptible corn varieties and the mites from wheat adapted to corn easily.

In Slykhuis's (119) report of transmission of WSMV using *A. tulipae,* he transferred eggs to separate groups of healthy wheat seedlings and manually inoculated one group with WSMV. Then he demonstrated that *A. tulipae* from the infected plants transmitted the virus, but those from healthy plants did not. Slykhuis (119), Del Rosario & Sill (27), and Orlob (92) all showed that WSMV was not transmitted transovarially but by the adult and both nymphal stages. They agreed that both nymphal stages acquired the virus. Also, by successfully inoculating wheat plants with WSMV using macerates of nymphs, Orlob (92) demonstrated that they acquired the virus. Slykhuis (119), Del Rosario & Sill (27), and Orlob (92) found that the adult was unable to transmit WSMV unless it had access to the virus before reaching the adult stage. Slykhuis and Del Rosario & Sill considered this evidence that the adult could not acquire the virus. Yet, Orlob succeeded in inoculating plants manually with WSMV using macerated adults that had access to the virus only after becoming adults and thus showed that they do acquire the virus.

Orlob showed that young adults are efficient vectors but soon become poor vectors. Young adults and second-stage nymphs were quite efficient. In transfers of one individual per plant, efficiencies ranged from about 40 per cent to 67 per cent. First-stage nymphs were less efficient, but this was probably partly due to an observed higher mortality among first instars than among older stages following transfer.

Orlob also studied acquisition of the virus by *A. tulipae*. When mites were given a 10-minute virus-acquisition feeding period and then transferred one per plant to 116 susceptible plants, none of the plants became infected. With a 15-minute virus-acquisition period, two of 173 plants became infected. There was a linear relationship between length of acquisition-feeding period and percentage of plants that became infected. With a 16-hour acquisition-feeding period (the longest period tested), about half of the plants to which mites were transferred became infected.

Orlob obtained very similar results in studies of the inoculation-feeding period. When mites were given a 10-minute inoculation-feeding period, no infection of the plants resulted. A very low percentage of plants became infected when mites were given a 15-minute inoculation feeding period. The percentage of infections increased linearly as the inoculation feeding period was increased until at 16 hours almost half of the plants became infected. Orlob suggested that the increased transmission with an increase in time was due to an increased probability that the virus was deposited in the proper site.

It is generally agreed that WSMV persists in *A. tulipae* for at least a few days. By transferring infective mites to immune hosts and then transferring groups of mites from the immune host to susceptible wheat plants at daily intervals, Slykhuis (119) showed that WSMV was retained in *A. tulipae* for at least six days. Del Rosario & Sill (27) used a similar technique and found that there was no loss in infectivity of *A. tulipae* for at least four days after being transferred to immune hosts. Also, they transferred *A. tulipae* to a wheat-dextrose decoction agar that was found to sustain adults (but upon which no reproduction occurred) and, upon removing mites to susceptible wheat plants at various intervals, they found that the virus had persisted for 18 days. Infectivity of the mites remained quite high through the 11th day on agar, then it rapidly decreased through the next week. Using Slykhuis's technique, Orlob (92) demonstrated transmission by *A. tulipae* seven days after transfer to immune hosts when held at 23° to 28° C and 61 days after transfer to immune hosts when held at 3° C.

Orlob offered additional evidence of the persistence of WSMV. He demonstrated persistence through the molt by transferring immobile, molting nymphs from infected plants to healthy plants and effecting transmission. Also, he immersed infective *A. tulipae* in a 1 per cent formaldehyde bath for 2 minutes (a treatment that inactivates WSMV in leaf extracts) and found that no loss in ability to transmit the virus had occurred. Orlob considered these results as evidence of "the persistent or circulatory type of virus-vector relationship."

Orlob successfully inoculated plants with WSMV using macerates of *A. tulipae* nymphs and adults that were reared on infected wheat plants, but he was unable to inoculate plants using macerates of *Aculus mckenziei* or *Abacarus hystrix* treated similarly. (These two eriophyid species occur on wheat in nature but do not transmit WSMV.) Orlob found particles analogous to WSMV particles in homogenates of both *A. tulipae* and the other two species when they fed on infected wheat. (Oliinyk (91) reported finding similar particles in macerates of infective *A. tulipae* and infected wheat.) Orlob failed to find virus-like particles in the two nonvectors after they had fed on healthy wheat. He suggested that perhaps the virus was inactivated by the nonvectors. Paliwal & Slykhuis (95) reported a positive reaction between antiserum for WSMV and extracts of *A. tulipae* from infected plants but reported that WSMV antiserum did not react serologically with extracts of *A. tulipae* from healthy plants or extracts of the same two nonvector species used by Orlob that had been reared on infected plants. Paliwal & Slykhuis corroborated Orlob's results in which he successfully inoculated wheat plants with macerates of infective *A. tulipae* but not with noninfective *A. tulipae* or the two nonvector species that were reared on infected wheat. As Orlob had, they observed particles analogous to WSMV particles in *A. tulipae* from infected plants but failed to find them in *A. tulipae* from healthy plants. Unfortunately, they did not study homogenates of the two nonvector species; consequently, that part of Orlob's work has not yet been duplicated.

Paliwal & Slykhuis also observed WSMV-like particles in ultrathin sections of infective *A. tulipae.* They observed large numbers of WSMV-like particles in 7 of 13 individuals from infected wheat but failed to find similar particles in 10 individuals from healthy wheat. The greatest concentration of particles was found in the lumens of the hindgut and posterior part of the midgut. They did not find similar particles within any of the mite tissues. They also prepared whole mounts of *A. tulipae* and examined the internal organs with the light microscope. They found that the alimentary canal is essentially a simple tube in which the midgut and hindgut are connected by a narrow tube that becomes indistinct, perhaps because of degeneration, in older adults. In the adults, the anterior part of the midgut almost closes due to the pressure caused by the maturation of eggs and the development of nutritive tissue. They suggested that the inability of the adult to become infective may be related to these differences between nymphs and mature adults. These investigators pointed out that the concentration of the virus in the lumen of the gut and the absence of virus in any tissue of the vector indicated that while the virus is persistent, it is not circulative in the vector. They suggested that backflow to the mouthparts or elimination of infective virus from the anus or both, might be involved in its transmission. In the case of defecation, they suggested that feeding punctures or the action of the anal setae or anal sucker could cause abrasion adequate to introduce freshly eliminated virus into epidermal cells.

Shtein-Margolina et al. (111) found polygonal particles in ultrathin sections of infected plant tissue and tissues of infective *A. tulipae.* They found rods analogous to WSMV in suspensions of infected wheat tissue but failed to explain the relationship between the polygonal particles and the rods. The diameter of the polygonal particles and the electron transparency of the central area suggested that these were transverse sections of WSMV rods; however, the authors did not discuss this possibility. They considered their findings evidence that WSMV reproduces in tissues of *A. tulipae*. This, of course, is contrary to Paliwal & Slykhuis's report.

The investigation of latency in the vector (as well as additional electron micrographic studies) might help to elucidate the relationship between *A. tulipae* and WSMV. Orlob (92) showed that *A. tulipae* can acquire enough virus within 10 minutes to cause infection of a healthy plant; however, we do not know if there is a period after acquisition in which the mite is unable to transmit the virus. The demonstration of a considerably long latent period after acquisition would suggest that the virus passes through tissues of the mite—and perhaps multiplies—before being transmitted. No latent period or a short latent period would appear to favor Paliwal & Slykhuis's hypothesis. It is well to keep in mind that these relationships are unknown for other eriophyid mite-transmitted viruses. Perhaps the case of WSMV and its vector will prove to be uniquely different than any of the relationships reported for insects and the viruses they transmit.

EPIDEMIOLOGY

Although both WSMV and *A. tulipae* have many hosts among the grasses, and *A. tulipae* has other monocot hosts, none of the other monocot hosts of *A. tulipae* are hosts of WSMV; consequently, their role in its spread is, at most, only one of sustaining populations of the vector. Furthermore, Del Rosario & Sill (27) showed that *A. tulipae* from onion did not readily colonize wheat, but it finally became "adapted" to wheat. They showed that even after adaptation it was not an efficient vector of WSMV compared to *A. tulipae* which occurred naturally on wheat. As mentioned earlier, Slykhuis (119) was unable to rear *A. tulipae* from wheat on several grass species upon which *A. tulipae* had been found reproducing. Likewise, *A. tulipae* occurring naturally on several wild grass species would not colonize wheat. Also, Orlob (93) reported that *A. tulipae* from several grasses did not really adapt to wheat.

The identity of the various populations that are considered to be *A. tulipae* should be investigated in greater detail inasmuch as the reports cited above suggest that some of the taxa which, by present criteria, are considered to be *A. tulipae* may be closely related but distinct from it. A further elucidation of the relationships among the various populations currently referred to as *A. tulipae* might contribute considerably to an understanding of the epidemiology of WSMV.

Several investigators have sought to explain the factors that accounted for epiphytotics in certain areas. According to Slykhuis (119), the important phases in the annual cycle of wheat streak mosaic appeared to be associated with living wheat plants. Since both the vector and the virus overwinter in winter wheat and move to spring wheat or volunteer wheat the following spring, summer fallow must be kept free of volunteer wheat for a sufficient period to prevent a subsequent serious outbreak of WSMV. Also, Connin (23) recognized the epidemiological importance of oversummering volunteer wheat. King & Sill (61) stressed the importance of the unusually long warm period during fall that allowed the development of huge populations of *A. tulipae* and resulted in an epiphytotic in Kansas during 1959. Slykhuis et al. (130) emphasized the importance of diseased immature wheat as foci from which infective *A. tulipae* spread to newly planted winter wheat. In southern Alberta, Canada, losses due to wheat streak mosaic were greatest in winter wheat seeded before nearby winter or spring wheat had matured.

In Washington (15), the lack of summer rain and the resulting small reservoir of oversummering host plants apparently precludes appreciable spread of *A. tulipae* and serious losses from wheat streak mosaic. Atkinson & Slykhuis (6) related a severe outbreak of wheat streak mosaic in southern Alberta to a spring drought which delayed development of spring grains to the extent that *A. tulipae* and WSMV spread from the immature wheat to the newly planted winter wheat crop.

According to Gibson & Painter (38), Kantack and Knutson found large numbers of *A. tulipae* on ripening kernels of winter wheat and observed that volunteer wheat resulting from hail immediately became infested with *A. tulipae.* Gibson & Painter (38) showed that *A. tulipae* infesting wheat kernels could move directly from the kernels to the resulting wheat seedlings. Perhaps this phenomenon and the demonstrated persistence of WSMV in *A. tulipae* might result in immediate infection of new wheat seed lings with the virus.

Most of the evidence suggests a minor epidemiological role for wild perennial and annual grasses (93, 119), but corn may be epidemiologically important where both corn and wheat are grown, inasmuch as many varieties are hosts of WSMV (47). On the other hand, Sill & Del Rosario (114) found that all field corn varieties that they tested were resistant to WSMV and only a few sweet corn varieties were damaged, yet they found that *A. tulipae* could transmit WSMV from wheat to corn and from corn to wheat. They suggested that corn might be a possible oversummering reservoir for *A. tulipae* and WSMV. Other investigators (72, 163) recently reported the isolation of WSMV from diseased corn and wheat in Ohio.

WHEAT SPOT MOSAIC

This virus disease was discovered by Slykhuis (118, 119) in the course of studies that proved that *A. tulipae* transmitted WSMV. Unlike wheat streak mosaic, wheat spot mosaic has not been found outside southern Alberta, Canada. Slykhuis (127) reported that he reisolated it in 1958, but its severe pathogenicity complicated lengthy culturing in wheat.

While studying WSMV transmission, Slykhuis (119) transferred *A. tulipae* from certain wheat plants that showed severe chlorotic mottling in the field to healthy wheat plants in the greenhouse. Over half of the greenhouse plants developed symptoms similar to the field plants from which *A. tulipae* had been collected; however, WSMV could be sap-transmitted to healthy wheat plants from only about half of the plants that showed symptoms after transferring *A. tulipae* to them. In another experiment, he transmitted WSMV both manually and with *A. tulipae* from five of ten severely chlorotic, stunted wheat plants in the field. *A. tulipae* from the five other plants induced severe chlorotic symptoms, but the presence of WSMV could not be demonstrated by manual inoculation of healthy plants. Slykhuis suggested that a nonsap-transmissible virus was probably involved.

The inability to sap-inoculate this entity necessitated that other measures be taken to show conclusively that symptoms were due to a virus transmitted by *A. tulipae* and not to feeding injury by the mite. Slykhuis (120) accomplished this by destroying the populations of *A. tulipae* on the plants that showed wheat spot mosaic symptoms and observing a subsequent reappearance of symptoms in the absence of mites.

Also, he established colonies of *A. tulipae* by transferring eggs from plants showing wheat spot mosaic symptoms to healthy plants. The absence of symptoms indicated that the virus was not transovarially transmitted. Then, he transferred mites from the healthy plants and mites from plants showing wheat spot mosaic symptoms to separate groups of healthy plants and found that symptoms of wheat spot mosaic appeared only on plants receiving *A. tulipae* from diseased plants. He further showed that *A. tulipae* from healthy plants could transmit wheat spot mosaic virus when given a seven-day acquisition-feeding period on infected wheat. Also, he found that the adult and both nymphal stages were able to transmit this virus and it was retained through the molts.

In one series of comparative tests with individual mites, 65 per cent transmitted wheat spot mosaic virus and 34 per cent transmitted wheat streak mosaic virus (128). Mites remained infective for 13 days on *Lolium perenne,* a species that is immune to the virus. Single *A. tulipae* could simultaneously carry both WSMV and wheat spot mosaic virus.

In greenhouse tests, Slykhuis (120) showed that wheat spot mosaic virus infected several graminaceous hosts including cultivated varieties of wheat, corn, barley, and Hungarian millet as well as the wild grass species, *Setaria verticillata, S. viridis,* and *Eragrostis cilianensis.* Oats and several species of wild grasses showed no symptoms of wheat spot mosaic.

While its severe pathogenicity and the inability to sap-transmit it are complicating factors in studying wheat spot mosaic virus, the ease with which both the host plant and vector can be reared and the basic attractiveness of studying the interrelationships between a mite vector and two plant viruses that can be carried simultaneously, would appear to be ample reason to conduct further studies of this virus.

RYEGRASS MOSAIC

Ryegrass mosaic virus (RMV) causes pale-green streaks on ryegrass leaves in England, Wales, Scotland, and several countries of northern Europe (121). Slykhuis (121) and Mulligan (80) credited each other with transmitting ryegrass mosaic virus with a mixed colony of eriophyid mites. After obtaining identifications of the various species in the mixed colony, Mulligan (79, 80) obtained pure colonies of *Abacarus hystrix* by single transfers from infected plants and succeeded in demonstrating its ability to transmit the virus. He showed that *A. hystrix* could transmit the virus after as little as a 2-hour acquisition-feeding period. He also stated that the various instars transmitted ryegrass mosaic virus equally often but showed no supporting data. Also, Mulligan studied persistence of the virus in *A. hystrix* by transferring infective mites to virus-immune wheat plants that supported *A. hystrix*. On three occasions the virus was transmitted after 6 hours, but not after 12 hours on the immune wheat.

Mulligan studied electron micrographs of clarified sap from healthy and infected plants and found that only the infected plants contained flexuous rods. He presumed these were ryegrass mosaic virus particles. He was able to sap-inoculate many British grasses with ryegrass mosaic virus. *A. hystrix* transmitted RMV to Blenda oats but not to Proctor barley or Capelle wheat, although both were suitable hosts for infective *A. hystrix*. Also, timothy was immune to the virus. Mulligan tested several dicotyledonous species; all were immune. He did not list the dicotyledons that he tested.

According to Keifer (54) *A. hystrix* lives on the upper surface furrows of the leaves and has a wide distribution on perennial grasses throughout the Northern Hemisphere. Keifer suggested the common name "cereal rust mite" for this species. According to Keifer (56), *Abacarus oryzae*, a species that infects rice plants in the Philippines, is very similar to *A. hystrix*. In the description of the closely related *A. oryzae*, Keifer stated "The mites . . . are said to have come from plants (rice) affected with dwarf disease called 'tungro.' "

Although there are no other reports of diseases of rye being transmitted by eriophyids, and no account of the role of *A. hystrix* as vector of ryegrass mosaic virus has appeared since Mulligan's paper, a mosaic disease of Italian ryegrass was reported from western Washington in 1957 (16). A year later, *A. hystrix* was found in Washington (15). The relationship of this disease to the eriophyid-borne mosaic of ryegrass apparently has not been investigated.

CURRANT REVERSION

Reversion of black currant is so named because plants with this disease resemble wild uncultivated *Ribes nigrum;* i.e., they revert to the wild type. Reversion has long been recognized as a serious disorder of black currants in the British Isles. According to Smith (138), it is widespread in the British Isles and probably occurs throughout northern Europe. Slykhuis (128) cited reports of currant reversion from several northern European countries. McLarty (75) reported "reversion" on currants in British Columbia, but his evidence was fragmentary at best.

The study of the epidemiology of reversion has been mainly a study of the relationship between reversion and the black currant gall mite (or big bud mite), *Cecidophyopsis ribis,* inasmuch as the relationship between the incidence of big buds caused by *C. ribis* and reversion of black currants was recognized quite early.

The Vector

Geographical distribution and host range.—This species apparently occurs in most areas where black currants are grown. According to Mumford (81), it occurs in the British Isles, Denmark, Germany, the Netherlands, Russia, Sweden, Norway, Finland, and British Columbia. Mumford listed the following *Ribes* spp. as the only known hosts of *C. ribis: R. nigrum, R. rubrum, R. alpinum, R. grossularia,* and *R. sanguineum.* Both Warburton & Embleton (162) and Massee (67) reported finding *C. ribis* in buds of *R. rubrum* and *R. nigrum,* but they agreed that on *R. rubrum* the buds were not swollen as in *R. nigrum.*

Recently, van Eyndhoven (156) described *Cecidophyopsis selachodon* as a new species of gall mite from *R. rubrum* in the Netherlands. Later, Boczek (10) reported this species from Poland. Morphologically, *C. selachodon* and *C. ribis* are quite similar. These reports naturally raise the question of the identity of eriophyid mites reported earlier from other *Ribes* spp. and considered to be *C. ribis.*

Biology.—*C. ribis* spends most of the year in the buds, but during the spring it leaves the buds and subsequently enters new buds. There has been considerable disagreement regarding the activities of mites during emigration from old buds to new ones. Earlier, workers gave seemingly authoritative but often conflicting accounts of the behavior of *C. ribis* during this period. Warburton & Embleton (162) found individuals in leaf axils during the migration period and noted that those that left the older buds were mostly adults. Massee (68) reviewed earlier accounts of the big bud mite's life history and reported his own observations. Taylor (cited by Massee) said that the mites distributed themselves on the outer surface of the big buds and dispersed by leaping. When one alighted on a leaf, it proceeded to the petiole and disappeared between the upper surface of the petiole and the twig where a new bud eventually emerged. Massee reported that during migration to new buds in the spring, eggs are laid on the young shoots and flowers. He also reported that eggs were laid on the leaves and shoots during summer and stated "It has been noted that the mites (*C. ribis*) copulate on the leaves prior to entering buds." He said that both immatures and adults migrated. Amos et al. (2) reported that the *C. ribis* adult population consisted of about 98 per cent females.

Recently, Collingwood & Brock (21) and Smith (132, 134) studied various aspects of the biology of *C. ribis.* During December, no eggs were laid; egg production began in January and reached a peak in March (21). Mites were ready to leave the buds by mid-February but were unable to escape until the buds opened in March (132). At that time populations averaged about 30,000 per bud (21). Populations in the buds decreased rapidly once emergence began. Using a suction apparatus connected to a timer, Smith (132) found that at 80°F, the rate of emergence from the buds was about 800 per hour.

Upon emerging from the buds, the mites often stood erect on their anal suckers and leaped from the bud by contracting the muscles on one side and springing. In still air, mites were capable of leaping about two inches. Under experimental conditions, mites left the bud in increasing numbers at wind velocities up to 24 mph; but as velocities increased above that air speed, the mites showed a decreasing tendency to become erect and fewer left the bud (132).

Warburton & Embleton (162), Massee (68), and Smith (132) mentioned finding *C. ribis* attached to several arthropods during the period of migration. Smith (132) found that mites remained attached to a tethered aphid for 5 to 10 minutes at a wind velocity of 20 mph. At 3 mph, mites remained attached to the aphid for 6 hours. The mites immediately released themselves upon contacting an object.

Mites that did not leap from the buds or attach to animals crawled in all directions on the bud; but upon reaching the stem, they exhibited a directional response and moved upward (66). Smith (132) stated that the upward movement was a positive phototactic response and no geotaxis was involved.

Mites were detected moving along the new shoot growth for a three-month period during the spring. The protracted migration period resulted from the gradual opening and drying of infested buds which exposed an increasing portion of the mite population. During this period the buds swelled, the blossoms appeared, and the fruit set and attained most of its size (21).

Individual mites were unable to exist outside the buds for long periods. Smith (132) removed buds during March and found that individuals that had reached the stem could survive for only a few days even under the shelter offered by normally developing leaf tissue. He found mites in appreciable numbers only on stems and leaf axils, i.e., the shortest routes between old infested buds and new axillary buds. Unlike Massee (68), he found mites on exposed leaves and blossoms only occasionally. Neither Collingwood & Brock (21) nor Smith (132, 134) mentioned finding *C. ribis* eggs outside bud galls. Thresh (151) never found eggs until the dispersal period was over and then he found them only on leaf primordia and meristems of new buds. Smith (132) stated that *C. ribis* moved toward higher humidity and suggested that was the reason they moved toward leaf axils. When mites arrived at the leaf axils, they proceeded to penetrate the new bud tissue by crawling inside the outer scale leaves and continuing in a circular direction until they reached the center of the bud. As a result they were usually concentrated near the apical meristem. Penetration took an average of 32 hours during postblossom time.

According to Smith (134), desiccation is probably the most important mortality factor during the migration period, but starvation is probably also important. Smith considered predation by polyphagous arthropods of little importance as a mortality factor during this period. Many mites perished during penetration of the buds.

Smith (132) found that mites which emerged from infested buds later in the spring were less likely to produce galls because they had to move farther up the shoot to reach suitable buds and were, therefore, more likely to fall victim to any of several mortality factors. Even so, when placed on suitable buds, later migrants formed big buds as readily as do earlier migrants.

Apparently a considerable time elapses between the first entry into the new bud and resumption of egg laying. Smith (134) first observed eggs approximately six weeks after the first entry into the new buds. He suggested (133) that this delay might be due to unfavorable nutritional conditions existing during the period of flower bud initiation.

During the summer, the mites reproduce in the new buds and these buds grow considerably during July and August. By October, a peak population averaging 4000 mites per gall is attained. Oviposition declines rapidly until, from late November through December, no eggs are laid (21). Smith (134) estimated that at least six generations per year are produced.

Mite-Virus-Host Relationships

A correlation between the incidence of big bud and reversion—or "going wild" as it was sometimes called—was noted by Lees in 1917 (63); however, he thought that some reversion was caused by a factor or factors unrelated to big bud since he observed that not all reverted bushes had big buds. In 1925, Lees (65) reportedly transmitted reversion to one of eight plants by grafting and concluded that since he found no microorganisms associated with the disease a virus must be the cause.

Amos et al. (2) initiated studies on transmission of reversion in 1921. They transferred *C. ribis* to one branch of a two-branched black currant bush and attempted to isolate the branches from each other. Reversion symptoms and mite-infested big buds developed on both branches. They also performed rather extensive grafting experiments and showed a positive correlation between the presence of big bud and reversion on plants, and a lack of positive correlation between the number of big buds and the degree of reversion. These workers were unable to demonstrate seed transmission or sap transmission of reversion.

Several years later, Massee (69) transferred large numbers of *C. ribis* from reverted bushes to each of 24 healthy black currant bushes over a two-year period. Six healthy plants received no mites. Colonies were established on the 24 test plants during the first year, and each of these plants developed reversion symptoms in three years or less. The six check plants showed no symptoms of reversion. Massee concluded that *C. ribis* "can be regarded as a vector of reversion."

Unfortunately, Massee's work was hardly more indicative of virus transmission than of injection of a phytotoxin by *C. ribis*. Also, the work reported earlier by Amos et al. (2) did not conclusively show that reversion is graft-transmissible since they apparently made no attempt to control *C. ribis*. The apparent graft transmission could have been nothing more than phytotoxic effects of *C. ribis* that had moved from scion to stock or vice versa.

Proof that *C. ribis* transmits a virus causing reversion was difficult to obtain owing to several factors. First, *C. ribis* feeds in terminal and axillary buds. As a result of feeding, developing of flowers is often prevented and leaves issuing from terminal buds exhibit an abnormality that resembles reversion and probably has been confused with reversion in the past (146). Also, as indicated by Massee (69), two to three years may elapse before symptoms of reversion appear. Recently, Thresh (151) pointed out that there have been few attempts to find a suitable indicator plant for reversion, yet an indicator is essential for diagnosis and to distinguish the effects of reversion virus from those of the vector and other viruses.

Several recent studies have contributed to a substantiation of earlier reports that *C. ribis* transmitted a virus that caused reversion. Smith (135) transferred 1, 5, or 20 *C. ribis* to each of several healthy black currant plants then fumigated them to kill the mites. Only a few plants developed populations, and these apparently were limited to buds close to the point where mites were introduced. Some populations were initiated with a single mite. After two years, 46 per cent of the plants exhibited typical symptoms of reversion.

Thresh (145) reported what must be considered the most conclusive evidence that *C. ribis* transmits a virus which causes reversion. Thresh transferred *C. ribis* from plants that exhibited symptoms of reversion or vein pattern (an early symptom of reversion) to healthy black currant seedlings and then dipped the seedlings in 0.05 per cent endrin after four days to destroy the mites. No live mites were found in subsequent observations. Within a month, vein pattern appeared on the seedlings. At this point, Thresh grafted patches of bark of the seedlings to healthy black currants. These plants subsequently showed symptoms of reversion but check plants did not.

Both Smith (136) and Thresh (150) conducted field experiments which showed natural spread of *C. ribis* and reversion. Smith showed that both mites and reversion spread along rows much more readily than across rows. Apterous and alate currant aphids (*Hyperomyzus lactucae*) were important in spreading the mites. Thresh demonstrated that *C. ribis* and reversion spread predominantly in the direction of winds prevailing during the mite's dispersal season. Thresh stated that more bushes developed galls than later produced symptoms of reversion. In both healthy and virus-infected bushes, the incidence of galls decreased with increasing distance from the source.

In the course of studies of *C. ribis* and its relation to reversion, certain workers suggested that reverted bushes were more susceptible than healthy bushes to attack by *C. ribis* since big buds occurred almost exclusively on reverted bushes, and big buds were seldom observed prior to symptoms of reversion (64, 65, 143). Thresh's recent studies (147, 148, 151) of the spread of *C. ribis* in the field plots showed rather conclusively that reverted bushes were many times more susceptible to infestation by *C. ribis* than were healthy bushes. Also, Thresh reported differences in susceptibility among varieties of healthy black currants and greater susceptibility in plants infested with a virulent strain of reversion than in plants infected with a mild strain of reversion. He largely attributed the degrees of susceptibility of the plants to the relative densities of epidermal hairs that impeded the movement of dispersing mites. Numerous hairs developed early in the growth of leaves and stems of healthy bushes; however, hairs were quite sparse on reverted bushes. Differences in susceptibility among varieties and between plants infected with different strains could be accounted for similarly. Thresh considered that infection with reversion caused an increase in the proportion of susceptible buds by decreasing the density of hairs on leaves developing around shoot apices. Thresh further pointed out that the increased susceptibility to infestation by *C. ribis* was due to the specific anatomical changes resulting from virus infection and not just to the presence of virus, since bushes with reversion symptoms on one part of the plant only developed much higher infestations on those branches than on those without symptoms. This increased susceptibility is indeed interesting, inasmuch as it represents a mutually advantageous relationship between the vector-mite and the virus it transmits.

Control of Vector and Virus

Thresh's work indicated that control of *C. ribis* depended upon control of reversion and vice versa. Slykhuis (128) reviewed the subject of acaricidal control of *C. ribis.* Roguing of infected black currant bushes in conjuction with chemical control of *C. ribis* is the standard practice in England. Chemical applications are intended to kill the mites during the spring migration period.

Recently, Smith & Corke (137) reported control of *C. ribis* using (2-chloroethyl) trimethyl ammonium chloride, a plant growth retardant. This gave control comparable to that resulting from accepted applications of endosulfan. According to the authors, the growth retardant had no direct toxic effect on the mites but altered the habit of the plant and made it more resistant to successful colonization by the mite.

Certain investigators have used extreme temperatures to control *C. ribis* or reversion virus, or both. Taksda (144) studied cold hardiness in populations of *C. ribis* from eastern Norway, western Norway, and England and found marked differences in the ability to produce eggs at 6° C and the ability to survive at −18.5° C. The degree of cold hardiness was greatest in the populations from the coldest area and least in the population from the warmest area, where this characteristic would be of least survival value.

Cold treatment of infested cuttings was not recommended for the Norwegian populations since temperatures that killed mites also damaged the cuttings. In the English population, there appeared to be a reasonable safety margin between the lethal temperature for mites and that causing damage to the cuttings.

Thresh (149) used warm water treatments to eliminate *C. ribis* infestations of dormant black currant cuttings. Effective treatments did not affect subsequent growth of the cuttings. Apparently, both mites and eggs were destroyed. Campbell (19) succeeded in obtaining reversion-free clones of black currants by exposing infected bushes to hot air (34° C) and grafting the soft apex of shoots (1 cm long) to currant seedlings. The heat treatment masked symptoms of reversion on infected bushes only temporarily, but apparently the virus was destroyed in the apices since bushes resulting from the grafts remained healthy.

In still another approach to control of reversion, Tiits (152) attempted to graft-inoculate various varieties of black currant, hybrids, and other *Ribes* spp. and concluded that reversion was limited to black currant. He suggested breeding reversion-resistant varieties by crossing black currants with other *Ribes* spp.

FIG MOSAIC

Fig mosaic disease was studied first in California by Condit & Horne (22). Apparently, these investigators immediately suspected that *Aceria ficus,* an eriophyid that was widespread on figs in California, might cause the disease itself or transmit a virus that caused the disease. Although they recognized that the disease might be the direct result of feeding by *A. ficus* on young, tender leaves, they also found figs in Oregon that were heavily infested with *A. ficus* but which had deep green foliage.

In greenhouse tests, they rooted 100 cuttings from plants showing mosaic and observed that 74 developed mosaic symptoms. In contrast, when they grew trees from seeds of trees that showed mosaic symptoms, the seedlings showed no symptoms of mosaic. They performed graft-transmission tests, but since the trees were infested with mites the resulting appearance of symptoms of mosaic could hardly be considered proof of graft-transmisision of a causative virus. Also, they infested two healthy fig seedlings with *A. ficus,* and both seedlings developed mosaic symptoms. Unfortunately, they did not attempt to confirm transmission by destroying the mites and graft-inoculating healthy trees from the seedlings.

Although *A. ficus* was considered the probable vector of a virus causing fig mosaic (43), no proof was provided until 1955. According to Flock & Wallace (34), in 1944 Wallace performed tests similar to those of Condit & Horne (22) in which eriophyid mite-infested bud scales from field trees were placed on small fig seedlings. Several of the seedlings developed mosaic symptoms, but no attempt was made to destroy the mites; consequently, the possibility of direct feeding injury remained. Then, in 1955, Flock &

Wallace (34) demonstrated that mosaic symptoms persisted on figs in the absence of *A. ficus* by treating infested cuttings with sulphur to kill the mites, rooting the seedlings, and observing the development of mosaic symptoms on new growth. Also, they successfully graft-inoculated healthy fig seedlings by implanting diseased, mite-free plant tissue, and thus showed that fig mosaic was, indeed, caused by a transmissible virus. Having established that fig mosaic virus was transmissible by grafting, they transferred various numbers of *A. ficus* from fig trees infected with mosaic to healthy seedlings that were kept in mite-free cages. After three to five days the plants were dusted with sulphur to eliminate the mites. *A. ficus* proved to be an efficient vector. Seven of ten plants that received one mite developed fig mosaic. Higher percentages of infection resulted when greater numbers of mites were used. Flock & Wallace recognized that feeding injury by *A. ficus* might cause early symptoms that could be confused with symptoms of infection by the virus. To differentiate between the two, they established a virus-free colony of *A. ficus* by transferring eggs to healthy seedlings. Then they transferred virus-free mites to one group of healthy seedlings and infective mites to another group and compared the symptoms that appeared on the two groups. The virus-free mites caused leaf distortion, chlorosis, and russeting, but these symptoms were distinguishable from symptoms of mosaic that appeared on the group that received infective mites. This experiment also showed that fig mosaic virus was not transmitted through the egg of *A. ficus*. Also, according to Blodgett & Gomec (9) the virus is not transmitted through the seed or by sap inoculation.

Fig mosaic virus has been reported only from the family Moraceae. Condit & Horne (22) listed four *Ficus* spp. as hosts. Burnett (17, 18) added 13 more *Ficus* spp. and *Cudranea tricuspidata* to the host list. Vashisth & Nagaich (157) showed that it also infected mulberry, *Morus indica*. These investigators also cited unpublished experiments in which they transmitted fig mosaic virus using *A. ficus*. This apparently is the only confirmation of Flock & Wallace's incrimination of *A. ficus* as the vector.

Fig mosaic is quite likely present in all countries where figs are grown. It has been reported from countries on every continent except South America (9). According to Condit & Horne (22), the vector is widespread in California and also occurs in Oregon. *A. ficus* has also been reported from Italy (43) and India (157). It seems likely that this mite will be found in most areas where figs are grown.

The life history of *A. ficus* was studied by Baker in California (7). He reported that all stages and both sexes were found throughout the year. Mites spent the dormant season in buds and were exposed as the buds burst in the spring. Following bud burst, eggs were laid on the stems and on both surfaces of the leaves, although as the leaves matured a greater proportion of eggs were laid on the lower leaf surface. During July, many mites left the leaves and entered the fruits. In addition to transmitting fig mosaic virus, *A. ficus* feeds on leaves and kills epidermal cells (7). Also, it causes russeting and scarring of the eye scales and seeds of the fruit and it occasionally causes stunting of twigs and immature-leaf drop (30).

PEACH MOSAIC

This disease was first recognized in Texas and Colorado. It is now known to occur also in southern California, southern Utah, Arizona, New Mexico, southern Oklahoma, western Arkansas, and Mexico (48).

THE VIRUS

Although peach mosaic virus is readily transmitted by grafting, it is not sap-transmissible (48) and attempts to purify it have been unsuccessful (96). Its host range is limited to certain species of the genus *Prunus* (48). All of the 209 peach varieties tested by Cochran & Pine were susceptible (20). Of these , most clingstone varieties showed only slight symptoms, but most freestone varieties showed definite symptoms. Forty-two of 43 horticultural varieties of plum tested by Pine & Cochran were susceptible (97). Also, nectarines, almonds, and apricots are susceptible. Several other species of *Prunus* have been experimentally infected (20, 48). On the other hand, cherries (*P. avium, P. cerasus,* and *P. mahaleb*) are immune (20, 48).

In Texas, wild *P. angustifolia* is an important reservoir of the virus (48). In New Mexico, Arizona, and Utah, *P. munsoniana,* planted along irrigation canals, is often infected (48). In addition to these two species, three other species native to areas east of the Rocky Mountains *P. americana, P. mexicana,* and *P. hortulana,* are also susceptible (20). In contrast, Cochran & Pine (20) tested six *Prunus* species native to western North America and found that only *P. subcordata* (Sierra plum) was susceptible.

THE VECTOR

In the case of currant reversion and fig mosaic, certain eriophyids were suspected of being vectors many years before their role as vector was proven. While this was not the case with the eriophyid species that transmit grass viruses, nevertheless, these species were described before their vector capabilities were demonstrated.

In the case of peach mosaic, the discovery of *Eriophyes insidiosus* was the result of the search for a vector of the virus. This species was first found in retarded buds of mosaic-infected peach trees by Wilson in 1955, and within a few months of its discovery, Wilson et al. (166) demonstrated its ability to transmit the virus. Keifer & Wilson (60) described it shortly after it was shown to be a vector.

On most commercial peach varieties, *E. insidiosus* is usually limited to retarded adventitious buds found near the base of large scaffold branches. They cause considerable cell hypertrophy in these buds as well as in buds of all their other known hosts (165). On wild plums (*P. hortulana, P. mexicana,* and *P. angustifolia*), and on some flowering peaches, they occasionally are found unprotected on petioles and green stem tissue near leaf axils as well as in buds; however, on commercial peach varieties they have been found only inside retarded buds. On wild plums and flowering peaches, they infest axillary buds and are thus more generally dispersed on the host plant (165).

In southern California, reproduction occurs on peach throughout the year, but it is quite low during the winter. From March to May, populations increase rapidly and then remain high in retarded buds throughout the summer. Usually, by October, the buds die and the mites either leave or die (165). Little direct information on migration is available; however, Jones & Wilson (49) showed that when groups of healthy potted peach trees were exposed in a peach-mosaic infected orchard for 2-month periods from March to October, natural spread of the virus occurred as early as April and continued at least September. Presumably, *E. insidiosus* left the buds throughout this period.

Prior to the discovery of *E. insidiosus,* another eriophyid mite, *Aculus cornutus* (the peach silver mite), had been shown to be incapable of transmitting peach mosaic virus. After *E. insidiosus* was discovered and incriminated as a vector, *Eriophyes prunandersoni,* a species that closely resembles *E. insidiosus,* came under close scrutiny. Although *E. prunandersoni* causes erinea on leaves of *P. andersoni, P. fasciculata,* and *P. fremontii*—three xerophytic species native to western North America—attempts to rear it on peach and *P. hortulana* failed. Also, attempts to rear *E. insidiosis* on *P. andersoni* and *P. fremontii* failed. Perhaps more significantly, in mixed plots, *E. insidiosus* developed heavy populations on *P. hortulana, P. mexicana,* and peach in one growing season, but failed to develop detectable populations on *P. andersoni* or *P. fremontii* over a six-year period (165).

Several other *Prunus* species are hosts of *Eriophyes* spp. that are closely related to *E. insidiosus.* Keifer (55, 58) described two species from *P. subcordata,* the only native western North American species that is known to be susceptible to the peach mosaic virus. At present, it is unknown whether these species are capable of transmitting the virus.

The discovery of *E. insidiosus* led to a survey of peach orchard environs in southern California where none of the early recognized *Prunus* hosts of *E. insidiosus* occurred naturally. Several new species of Eriophyidae have been found as a result of the survey; however, *E. insidiosus* has not been found on any additional plant species (165).

The known host range of *E. insidiosus* includes many varieties of commercial and flowering peaches as well as *P. cerasifera, P. simonii, P. hortulana, P. munsoniana, P. mexicana,* and *P. angustifolia.* The latter four species are native to southeastern United States and, as mentioned earlier, two of them are reservoirs of peach mosaic virus as well as *E. insidiosus* in certain south central states west of the Mississippi River (165).

Since 1955, *E. insidiosus* has been found in several other states. To date, it has been found in western Colorado, Arizona, New Mexico, Texas, Arkansas, and Utah. Also, it has been collected from *P. angustifolia* outside the range of peach mosaic virus in Mississippi and Georgia. Efforts to find it in Illinois, Maryland, Missouri, Louisiana, and Wisconsin have failed (165).

Until recently, in California, *E. insidiosus* was known to occur only south of the Tehachapi Mountains; however, it is now found on flowering peaches in a few locations in the San Joaquin Valley. In spite of concerted efforts by state and federal agencies to find *E. insidiosus* in commercial varieties of peaches in the San Joaquin Valley, it is still unknown except on flowering peaches. Also, the virus is not known to occur in that area (165).

Mite-Virus Relationships

In the studies in which Wilson et al. (166) showed that *E. insidiosus* transmitted peach mosaic virus, they reported transferring several mites from infected peach to each of several small Rio Oso Gem peach seedlings. Several of the seedlings developed typical symptoms of peach mosaic. They also transferred *E. insidiosus* from healthy peach trees to seedlings, but no symptoms of mosaic appeared on any of the latter group. To further check that mites from the infected trees had actually transmitted a virus, they grafted pieces of bark from the several seedlings showing symptoms to other healthy Rio Oso Gem seedlings. The appearance of symptoms on all of the latter plants showed conclusively that *E. insidiosus* had, in fact, transmitted a virus causing peach mosaic.

Subsequent tests in which single adults were transferred to each of 80 plants resulted in two cases of transmission. When two to ten mites were transferred to each healthy plant, 11 of 56 plants became infected. When 50 or more mites were transferred to each plant, 18 of 25 plants became infected. In all, several hundred plants have been experimentally inoculated with peach mosaic virus using *E. insidiosus* from sources in southern California and New Mexico (165).

Wilson & Jones (165) transferred approximately 5400 eggs to 28 healthy peach plants and allowed the mites to hatch and feed. The lack of any cases of transmission indicated that the virus was not transmitted through the egg of the vector. *E. insidiosus* was shown to be able to transmit the peach mosaic virus from peach to peach, apricot and *P. mexicana,* from *P. mexicana* to peach, and from *P. hortulana* to peach. Although *E. insidiosus* can transmit the virus to apricot, it does not reproduce on apricot (165).

Several of the details of transmission have, as yet, eluded discovery owing to exceptional technical difficulties in rearing and manipulating the vector. Nevertheless, by holding infective mites on glass slides and then transferring them to healthy plants and effecting transmission, Wilson & Jones showed that the virus persisted in the vector for at least 48 hours (165).

In order to perform critical vector tests on eriophyid-transmitted viruses, it is necessary to have adequate laboratory or greenhouse cultures of the vector. This has been difficult to attain with *E. insidiosus;* however, a recent technique used by Oldfield & Wilson (90) may facilitate the rearing of large numbers of *E. insidiosus* in the greenhouse. They established greenhouse cultures of *E. insidiosus* by inducing root formation on infested flowering peach cuttings and then planting them.

Control of Vector and Virus

In southern California, control of the spread of peach mosaic has involved systematic surveying of peach orchards and removal of infected trees. Jones et al. (51) recently concluded an experiment in which they attempted to evaluate the effectiveness of chemical control of *E. insidiosus* as a means of controlling the spread of the virus. Each spring, for five successive years, a single treatment of diazinon was applied at petal fall to each of several peach orchards in which peach mosaic was spreading rapidly.

Over 100 new cases of peach mosaic appeared in one year after the first treatment. The number of new cases decreased to 46 the second year. For the next five years, there were never more than two new cases per year. Based on known cases of spread in other southern California areas, the authors concluded that spread of the virus had been significantly arrested by controlling the vector.

CHERRY MOTTLE LEAF

This disease was first reported from Oregon in 1920 and its virus nature was established in 1935. It occurs in sweet cherry-growing districts of Washington, Oregon, Idaho, California, and British Columbia. In Washington, it is most often found in foothill or canyon orchards, often in close association with wild bitter cherry, *Prunus emarginata* (76).

In 1958, L. S. Jones found an unidentified species of eriophyid in abnormally enlarged buds of *P. emarginata* bordering commercial cherry orchards near Wenatchee, Washington. This species was later described and named *Eriophyes inaequalis* by Wilson & Oldfield (167).

Later, *E. inaequalis* was found in the vicinity of wild *P. emarginata* in a few buds of commercial cherries that were infected with mottle leaf virus (50). Still later, Jones et al. (50) showed that *E. inaequalis* transmitted the cherry mottle leaf virus. *P. emarginata* was found to be the principle host of the mite and a common reservoir of the virus.

In the initial series of tests, Jones et al. transferred 50 *E. inaequalis* from mottle leaf-infected *P. emarginata* to each of 20 newly germinated peach seedlings. Since peach is not a host of *E. inaequalis* and is a symptomless carrier of cherry mottle leaf virus, buds from each of the 20 peach trees were grafted to healthy potted Bing cherry trees in the following spring. Typical cherry mottle leaf symptoms developed on 12 of the Bing cherry trees. Subsequent tests corroborated these findings. Also, in 1968, Wilson & Oldfield (168) transferred *E. inaequalis* from mottle leaf-infected *P. emarginata* to each of 13 Bing cherry trees. Apparent symptoms of mottle leaf developed on eight of these plants but not on any of the check plants that received no mites. Verification of these results by graft transmission to other healthy Bing cherries is in progress at the time of this writing.

ROSE ROSETTE

Recently, Allington et al. (1) reported that they had transmitted rose rosette virus in Nebraska with *Phyllocoptes fructiphilus* Koch. (The correct name is *Phyllocoptes fructiphilus* Keifer 1940.) In one series of tests, ten *P. fructiphilus* from infected wild rose were transferred to each of ten healthy *Rosa eglanteria*. According to the authors, five plants became infected with rose rosette virus. The authors state that several species of *Rosa* were proved to be infected with rose rosette virus "either by grafting or by mite transmission"; however, they did not specifically state that the virus was graft-transmitted from those test plants to which *P. fructiphilus* had been transferred. This may only be an error of omission. Unless they graft-transmitted a virus from the plants that received mites, the appearance of symptoms on plants to which *P. fructiphilus* was transferred could be attributed to a mite-induced toxemia that resembled rose rosette. A definite statement that both types of transmission were accomplished in sequence would greatly substantiate their claim of transmission of a virus by *P. fructiphilus*. Also, a comparison of the effects on healthy rose plants of populations from infected and healthy roses would further substantiate their case.

Recently, Keifer (59) described *Phyllocoptes slinkardensis* from a wild species of rose in Mono County, California. He stated that *slinkardensis* "was extremely close to *fructiphilus*" except for the shape of the microtubercles. Also, he said that the mites were collected from roses showing witch's broom and "grafting tests have shown that this broom is virus induced and the *Phyllocoptes*, which was found on the native rose, could be the vector." The relationship between *P. fructiphilus* and *P. slinkardensis* probably should be investigated.

A LATENT VIRUS OF PLUM

In 1966, Proeseler & Kegler (101) reported that *Aculus fockeui* transmitted a latent virus of plum trees from plum to *Chenopodium foetidum*. Generally, the mites perished within 24 hours after being transferred to *Chenopodium*. Local lesions developed on the leaves after seven to ten days. Juice inoculations from these plants to other *C. foetidum* and to *C. quinoa* resulted in the appearance of local lesions on the inoculated plants. Proeseler & Kegler found particles about 750 mμ long in preparations from *Chenopodium* leaves to which mites had been transferred and which subsequently showed lesions, and from *Chenopodium* leaves that had been juice-inoculated. They concluded that these particles were the virus that the mites transmitted.

PIGEON PEA STERILITY

In 1963, Seth (108) conducted transmission tests of pigeon pea sterility virus with aphids, leafhoppers, whiteflies, and two unidentified mite species but failed to get positive results. However, a third mite, an eriophyid, reportedly transmitted the virus. In one test he transferred healthy pigeon pea leaves with eriophyids to one group of healthy plants, and diseased pigeon pea leaves also with eriophyids to another group of healthy plants. In a second test, he transferred 5 to 20 eriophyids from healthy plants to one group of healthy plants and 5 to 20 eriophyids from diseased plants to a second group of healthy plants. After five days, the plants were sprayed regularly with a pesticide to kill the mites. None of the 53 plants in the two groups that received mites from healthy pigeon pea became diseased. Seven of 30 plants that received diseased pigeon pea leaves with eriophyids showed sterility symptoms. Four of 24 plants that received eriophyids from diseased plants showed sterility symptoms. Seth concluded that the eriophyids had transmitted sterility virus.

Although Seth included a photomicrograph of one eriophyid and stated that he found them buried between the hairs on the undersurface of the leaves, there is no evidence that he used just one species of eriophyid and he apparently did not identify the species. The demonstration of graft transmission of the virus from plants inoculated by the mites would greatly strengthen Seth's claim of transmission by eriophyids.

No studies corroborating Seth's work have yet been published. In 1965, Narayanasamy & Ramakrishnan (82) reported negative results in attempts to transmit sterility virus using certain aphids, leafhoppers, and "an eriophyid mite, *Tetranychus* sp." Obviously, from this statement we cannot be sure that they used an eriophyid mite. Also, these workers offered evidence which suggested that nematodes or other soil-borne organisms transmitted the sterility virus.

MANGO MALFORMATION

This disease is characterized by the transformation of the inflorescence into a compact mass of sterile flowers in adult trees and production of numerous vegetative shoots at the growing point or in the axil of the leaf in the case of seedlings (83).

No adequately controlled vector tests using eriophyids have been reported for this disease; however, Nariani & Seth (83) pinned eriophyid-infested bud scales taken from diseased and apparently healthy plants on mango seedlings and found that seedlings that received bud scales from either source later developed symptoms of malformation. In a later paper, these authors (109) reported on methyl bromide fumigation of mango seedlings affected with malformation and commented that feeding by *Aceria mangiferae* caused malformation.

Other recent reports reviewed the status of this disease and cast some doubt on the role of *Aceria mangiferae* as the direct cause of the disease. Prasad et al. (99) were unable to find any correlation between populations of *A. mangiferae* and the degree of malformation, and they concluded that the disease was not the result of direct feeding injury by the mite. They were also unable to transmit the disease by budding or grafting and were thus unable to suggest a viral nature of the disease. Then, Ginai (41) suggested that malformation was caused by a virus that is spread by "mites and other insects as vectors," but he offered no experimental evidence to support his contention. More recently, Rai & Singh (103) reported recovery from malformation in mango saplings treated with 0.1 per cent Diazinon® to kill *A. mangiferae*. This evidence favors a hypothesis that *A. mangiferae* causes malformation; however, critical tests are obviously necessary to clarify the role of *A. mangiferae* in malformation disease.

AGROPYRON MOSAIC

Agropyron mosaic virus has been reported from *A. repens* in the United States and Canada (128). Staples & Brakke (141) reported that this entity was indistinguishable from WSMV on the basis of particle size, sedimentation rate, and stability toward selective denaturation procedures, but had a slightly different host range than WSMV. Later, Slykhuis (128) cited unpublished tests which showed other important differences between these entities. Also, Slykhuis observed that wheat seedlings became infected with Agropyron mosaic virus when they were grown in pots covered with cages made of 72-mesh per inch screen and exposed in the field near naturally infected *A. repens*. This suggested that a very small vector was involved, but Slykhuis's tests with eriophyids were inconclusive.

GRAPEVINE PANASCHURE

Ochs (86–88) reported that *Eriophyes vitis* and several insects transmitted a virus that reportedly caused panaschure of vines in Germany. Ochs claimed that she was able to sap-transmit the virus to several herbaceous plants; however, she offered practically no experimental evidence for mite transmission. Niemeyer & Bode (85) refuted the above claims after extensive attempts failed to duplicate her sap inoculations of herbaceous plants. Until more convincing evidence is reported, this must remain a doubtful case of transmission by an eriophyid mite.

CADANG CADANG

Cadang cadang is a degenerative disease of coconut palms which is widespread in the Philippines. A virus is suspected to be the causative agent (128). The pattern of spread of cadang cadang led Briones & Sill (14) to consider eriophyids as vector suspects; however, vector tests with four species from coconut palms were negative. Nevertheless, according to Bigornia (8), eriophyids are continuing to receive attention as possible vectors of a causative virus of this disease.

CONCENTRIC RING BLOTCH OF CITRUS

In 1958, Dippenaar (28) reviewed the history of this disease in South Africa and gave experimental evidence to indicate that it was caused by the feeding of *Calacarus citrifolii.* However, in 1963, Rossouw & Smith (106) performed tests with what reportedly was *C. citrifolii* and maintained that *C. citrifolii* transmitted a virus causing the disease. Their conclusion was largely based on tests in which they transferred mites from diseased and healthy plants to different groups of healthy rough lemon seedlings. Those that received mites from healthy plants did not develop symptoms of concentric ring blotch, but those that received mites from diseased plants did develop symptoms. Rossouw & Smith did not state what measures were taken to identify mites from the various sources and they did not state whether they had obtained transmission by grafting or sap inoculation. Also, they stated that the symptoms appeared only on the localized spots where the mites were confined. No subsequent corroborative reports have appeared, but a later report by Van der Merwe & Coates (155) on the biology of *C. citrifolii* failed to mention Rossouw and Smith's work. Instead, they noted that Dippenaar had demonstrated that *C. citrifolii* was the cause of concentric ring blotch.

PLANT VIRUS TRANSMISSION BY TETRANYCHID MITES

Slykhuis (128) cited several reports of suspected virus transmission by tetranychids; however, in most cases little or no experimental evidence is offered.

In 1963, Schulz (107) reported successful transmission of potato virus Y (PVY) by *Tetranychus telarius* (L.) [= *T. urticae* Koch (154)]. The highest rate of transmission was obtained when mites were given a 5-minute acquisition-feeding period followed by a 5-minute transmission-feeding period. In this case, 12 of 32 test plants developed symptoms of infection with potato virus Y. Mechanical inoculations from the test plants confirmed the presence of the virus.

Recent reports by Fritzsche et al. (36) and Orlob (94) failed to confirm Schulz's work, but elucidated some interesting relationships between *T. urticae* and several plant viruses including potato virus Y. Beside being unable to demonstrate transmission of PVY by the method employed by Schulz, Fritzsche et al. were unable to transmit PVY by rubbing leaves with homogenates of mites that had fed on PVY-infected plants. They also attempted to inoculate healthy plants with PVY by rubbing the leaves with feces of mites that had fed on infected plants. This also failed. Neither Fritzsche et al. nor

Orlob were able to detect PVY particles in electron micrographs of homogenates of mites that had fed on PVY-infected plants.

Orlob was also unable to demonstrate transmission of eight other plant viruses by *T. urticae* although he demonstrated by bioassay, electron microscopy, or serology, or all three methods, that tobacco mosaic virus, potato virus X, onion yellow dwarf virus, and tomato bushy stunt virus were in the mites that fed on plants infected with these viruses. Orlob showed that TMV could be acquired in 10 seconds and most mites acquired it within a 16-hour acquisition-feeding period. Feces were infectious, but feeding and moving by mites did not result in infection of the plant. On the other hand, mites were able to inoculate either TMV or PVX sprayed onto the leaf surface. The results of Fritzsche et al. generally agreed with or complemented Orlob's results using TMV. Although mites did not transmit TMV, Fritzsche and colleagues found that the virus was recovered from healthy plants upon which mites from TMV-infected plants had fed. This was accomplished by successfully inoculating healthy plants with macerates of the plants to which the mites were transferred. Fritzsche et al. found TMV particles in electron micrographs of macerates and feces of mites that had fed on TMV-infected plants. As Orlob had, they, too, found that mites were able to inoculate plants with TMV that was sprayed onto the leaves. Orlob summarized the results as providing an example of the failure of a mite to transmit plant viruses in the absence of immediate obvious reasons why it should fail to do so.

In spite of two rather extensive studies of the relationships between *T. urticae* and several plant viruses including potato virus Y, Schulz's report remains uncorroborated at this time.

CONCLUDING REMARKS

Other than wheat streak mosaic virus and *A. tulipae,* few specific relationships between eriophyid mites and the viruses they trasmit have been elucidated. I have mentioned some reasons for the paucity of information in certain cases. Certainly, the size of eriophyids and their ability to cause virus-like symptoms by their feeding activities have, in many cases, slowed progress toward discovering many of these relationships. Nevertheless, we can make certain generalizations at this time,, although perhaps some will be invalidated as more information regarding eriophyid transmission becomes available.

The available evidence points to a high degree of specificity between vector eriophyids and the viruses they transmit. As yet, no virus is known to be transmitted by more than one eriophyid species, and there is no substantiated case of transmission of any of the eriophyid-borne viruses by any other taxonomic group. As far as is known, only one eriophyid species, *A. tulpiae,* transmits more than one virus and there appears to be reason to suspect that the two viruses (WSMV and wheat spot mosaic virus) might be related. As yet, only these two viruses have been shown to be transmitted by *A. tulipae,* although mites currently considered to be *A. tulipae* occur on many monocotyledenous species, the total number of which are affected by many viruses.

In the two cases studied (WSMV and peach mosaic virus), the virus persists in the vector for at least a few days. Also, no case of transovarial transmission in the mite has been reported. In fact, in several instances virus-free colonies are commonly established by transferring eggs of infective mites to healthy, mite-free plants. This technique has been used successfully in vector studies of WSMV, wheat spot mosaic virus, ryegrass mosaic virus, peach mosaic virus, and fig mosaic virus.

Keifer (57) recently designated the long-recognized family Eriophyidae as superfamily Eriophyoidea and included three families: Rhyncaphytoptidae, Phytoptidae, and a more restricted Eriophyidae. The proven vector species of eriophyids all belong to Eriophyidae in the restricted sense. One of the other two families (Phytotidae) is largely restricted to conifers. The other family (Rhyncaphytoptidae) includes only species that are free-living on leaves. This may be a reason for the absence of any reported vectors in this family since most of the well-studied cases of transmission by eriophyids involve species that are relatively intimately associated with their hosts. Transmission of ryegrass mosaic virus by *A. hystrix* is an exception; *A. hystrix* is a rust mite. *A. fockeui,* reportedly a vector of a latent virus of plum, is also a rust mite.

The correct identification of eriophyids used in vector studies is of paramount importance. Since often more than one species live on the same plant, the establishment of pure colonies is a prerequisite to any critical study. Periodic sampling of colonies for identification is also important.

At present, *A. fockeui,* the species recently reported as a vector of a latent virus of plum, is the only reported vector in which the diapausing deuterogynous generation is known to occur in its life cycle. An investigation of retention of the virus through the deutogyne stage might be a valuable addition to the knowledge of mite-virus relationships.

Wheat streak mosaic virus particles and ryegrass mosaic virus particles are about 700 mμ long. The oral opening of eriophyids appears to be somewhat less than one μ (1000 mμ) in diameter (perhaps ½ μ). The relative sizes of virus particle and oral opening are such that particles may not be able to enter if they are oriented with the long axis across the oral opening.

In most of the substantiated cases of transmission by eriophyids, virus infection results in the appearance of irregular chlorotic areas on the leaves. This condition is usually called "mosaic." In this group are included wheat streak mosaic, wheat spot mosaic, ryegrass mosaic, fig mosaic, peach mosaic, and cherry mottle leaf. The best known symptoms of currant reversion constitute an exception; however, according to Thresh (145), even currant reversion includes a "vein pattern" on leaves as an early symptom. Apparently, some of the diseases that have been recently reported as caused by eriophyid-borne viruses exhibit quite different symptoms.

From the reports by various investigators, it is obvious that there are several reports of virus transmission by eriophyids that need much substantiative work. Similarly, the case of transmission of potato virus Y by *T. urticae* needs corroboration.

Eriophyids have been shown to be capable of transmitting viruses of monocots and viruses of dicots. Undoubtedly, they will continue to command attention as vector candidates.

LITERATURE CITED

1. Allington, W. B., Staples, R., Viehmeyer, G. Transmission of rose rosette virus by the eriophyid mite *Phyllocoptes fructiphilus. J. Econ. Entomol.,* **61,** 1137–40 (1968)
2. Amos, J., Hatton, R. G., Knight, R. C., Massee, A. M. Experiments in the transmission of reversion of black currants. *Ann. Rept. East Malling Res. Sta., Kent, 1925, II Suppl.,* 126–50 (1927)
3. Arnott, D. A., Berges, I. Causal agents of silver top and other types of damage to grass seed crops. *Can. Entomologist,* **99,** 660–70 (1967)
4. Ashworth, L. J., Futrell, M. C. Sources, transmission, symptomatology and distribution of wheat streak mosaic virus in Texas. *Plant Disease Reptr.,* **45,** 220–24 (1961)
5. Atkinson, D. E. Western wheat mosaic in Colorado and its transmission by the grain aphid *Toxoptera gramineum. Phytopathology,* **39,** 2 (1949)
6. Atkinson, T. G., Slykhuis, J. T. Relation of spring drought, summer rains and high fall temperatures to the wheat mosaic epiphytotic in southern Alberta. *Can. Plant Disease Surv.,* **43,** 154–59 (1963)
7. Baker, E. W. The fig mite, *Eriophyes ficus* Cotte and other mites of the fig tree (*Ficus carica* L.). *Bull. Calif. Dept. Agr.,* **28,** 266–75 (1939)
8. Bigornia, A. E. (Personal communication, 1969)
9. Blodgett, E. C., Gomec, B. Fig mosaic. *Plant Disease Reptr.,* **51,** 893–96 (1967)
10. Boczek, J. Studies on mites (Acarina) living in Poland. VIII. *Bull. Acad. Polon. Sci., Sér. Sci. Biol.,* **16,** 631–36 (1968)
11. Brakke, M. K. Properties, assay and purification of wheat streak mosaic virus. *Phytopathology,* **48,** 439–45 (1958)
12. Brakke, M. K., Staples, R. Correlation of rod length with infectivity of wheat streak mosaic virus. *Virology,* **6,** 14–26 (1958)
13. Brandes, J., Wetter, C. Classification of elongated plant viruses on the basis of particle morphology. *Virology,* **8,** 99–115 (1959)
14. Briones, M. L., Sill, W. H. Habitat, gross morphology and geographical distribution of four new species of eriophyid mites from coconuts in the Philippines. *Food Agr. Organ., U.N., FAO Plant Protect. Bull.,* **11,** 25–30 (1963)
15. Bruehl, G. W., Keifer, H. H. Observations on the wheat streak mosaic in Washington, 1955–57. *Plant Disease Reptr.,* **42,** 32–5 (1958)
16. Bruehl, G. W., Toko, H., McKinney, H. H. Mosaics of Italian ryegrass and orchard grass in western Washington. *Phytopathology,* **47,** 517 (1957)
17. Burnett, H. C. Species of *Ficus* susceptible to the fig mosaic virus. *Proc. Florida State Hort. Soc., 1960,* **73,** 316–20 (1961)
18. Burnett, H. C. Additional hosts of the fig mosaic virus. *Plant Disease Reptr.,* **46,** 693 (1962)
19. Campbell, A. I. The inactivation of black currant reversion virus by heat therapy. *Ann. Rept. Long Ashton Res. Sta., 1964,* 89–92 (1965)
20. Cochran, L. C., Pine, T. S. Present status of information on host range and host reactions to peach mosaic virus. *Plant Disease Reptr.,* **42,** 1225–28 (1958)
21. Collingwood, C. A., Brock, A. M. Ecology of the black currant gall mite (*Phytoptus ribis* Nal.). *J. Hort. Sci.,* **34,** 176–82 (1959)
22. Condit, I. J., Horne, W. T. A mosaic of fig in California. *Phytopathology,* **23,** 887–96 (1933)
23. Connin, R. V. Oversummering volunteer wheat in the epidemiology of wheat streak mosaic. *J. Econ. Entomol.,* **49,** 405–6 (1956)
24. Connin, R. V. The host range of the wheat curl mite, vector of wheat streak mosaic. *J. Econ. Entomol.,* **49,** 1–4 (1956)
25. Del Rosario, M. S., Sill, W. H., Jr. A method of rearing large colonies of an Eriophyid mite, *Aceria tulipae* (Keifer), in pure culture from single eggs or adults. *J. Econ. Entomol.,* **51,** 303–6 (1958)
26. Del Rosario, M. S., Sill, W. H., Jr. Additional biological and ecological characteristics of *Aceria tulipae. J. Econ. Entomol.,* **57,** 893–96 (1964)

27. Del Rosario, M. S., Sill, W. H., Jr. Physiological strains of *Aceria tulipae* and their relationships to the transmission of wheat streak mosaic virus. *Phytopathology*, **55**, 1168–75 (1965)
28. Dippenaar, B. J. Concentric ring blotch of citrus; its cause and control. *S. African J. Agr. Sci.*, **1**, 83–106 (1958)
29. Doreste, S. E. Advances gained in the control of the garlic mite, *Aceria tulipae*, a pest new to this crop in Venezuela. *Ing. Agron.*, **11**, 13–18 (1963)
30. Ebeling, W., Pence, R. A severe case of an uncommon type of injury by the fig mite. *Bull. Calif. Dept. Agr.*, **39**, 47–48 (1950)
31. Fellows, H. A survey of the wheat mosaic disease in western Kansas. *Plant Disease Reptr.*, **33**, 356–58 (1949)
32. Fellows, H., Schmidt, J. Reactions of *Agrotricum* hybrids to the virus of yellow streak mosaic of wheat. *Plant Disease Reptr.*, **37**, 349–51 (1953)
33. Finley, A. M. Wheat streak mosaic, a disease of sweet corn in Idaho. *Plant Disease Reptr.*, **41**, 589–91 (1957)
34. Flock, R. A., Wallace, J. M. Transmission of fig mosaic by the eriophyid mite *Aceria ficus*. *Phytopathology*, **45**, 52–54 (1955)
35. Ford, R. E., Lambe, R. C. Wheat streak mosaic virus incidence in Iowa. *Plant Disease Reptr.*, **51**, 389 (1967)
36. Fritzsche, R., Schmelzer, K., Schmidt, H. Evaluation of the ability of *Tetranychus urticae* Koch as a vector of plant viruses. *Arch. Pflanzenschutz*, **3**, 89–100 (1967)
37. Gibson, W. W. Observations on the wheat curl mite, *Aceria tulipae* (K.) on wheat and sorghum sprouted under ripening wheat. *J. Kansas Entomol. Soc.*, **30**, 25–28 (1957)
38. Gibson, W. W., Painter, R. H. The occurrence of wheat curl mites, *Aceria tulipae* (K.), a vector of wheat streak mosaic on wheat seedlings grown from infested kernels. *Trans. Kansas Acad. Sci.*, **59**, 492–94 (1956)
39. Gibson, W. W., Painter, R. H. Transportation by aphids of the wheat curl mite, *Aceria tulipae* (K.) a vector of wheat streak mosaic virus. *J. Kansas Entomol. Soc.*, **30**, 147–53 (1957)
40. Gilmer, R. M., McEwen, F. L. Chlorotic fleck, an eriophyid mite injury of myrobalan plum. *J. Econ. Entomol.*, **51**, 335–37 (1958)
41. Ginai, M. A. Malformation of mango inflorescence. *West Pakistan J. Agr. Res.*, **3**, 248–51 (1965)
42. Gold, A. H., Houston, B. R., Oswald, J. W. Electron microscopy of elongated particles associated with wheat streak mosaic. *Phytopathology*, **43**, 458–59 (1953)
43. Graniti, A. Fig mosaic in Italy and its possible vector. *Riv. Frutticolt. Ortic.*, **16**, 23–25 (1954)
44. Hall, C. C., Jr. A look at eriophyid life cycles. *Ann. Entomol. Soc. Am.*, **60**, 91–94 (1967)
45. Holmes, N. D., Swailes, G. E., Hobbs, G. A. The eriophyid mite, *Aceria tulipae* (K.) and silver top of grass. *Can. Entomologist*, **93**, 644–47 (1961)
46. Houston, B. R., Oswald, J. W. A mosaic disease of wheat, barley and oats new to California. *Phytopathology*, **42**, 12 (1952)
47. How, S. C. Wheat streak mosaic virus on corn in Nebraska. *Phytopathology*, **53**, 279–80 (1963)
48. Hutchins, L. M., Bodine, E. W., Cochran, L. C., Stout, G. L. Peach mosaic. *In* Virus diseases and other disorders with viruslike symptoms of stone fruits in North America. *U. S. Dept. Agr., Handbook, 10*, 26–36 (1951)
49. Jones, L. S., Wilson, N. S. Peach mosaic spreads throughout the growing season. *Bull. Calif. Dept. Agr.*, **40**, 117–18 (1951)
50. Jones, L. S., Wilson, N. S., Anthon, E. W. Transmission of cherry mottle leaf virus by an eriophyid mite. (In preparation)
51. Jones, L. S., Wilson, N. S., Burr, W., Barnes, M. M. Restriction of peach mosaic virus spread through control of vector mite. (In preparation)
52. Keifer, H. H. Eriophyid studies. *Bull. Calif. Dept. Agr.*, **27**, 181–206 (1938)
53. Keifer, H. H. Eriophyid studies XII. *Bull. Calif. Dept. Agr.*, **31**, 117–29 (1942)
54. Keifer, H. H. Eriophyid studies XIV. *Bull. Calif. Dept. Agr.*, **34**, 18–38 (1945)

55. Keifer, H. H. Eriophyid studies. B–8. *Calif. Dept. Agr., Spec. Publ.,* 20 pp. (1962)
56. Keifer, H. H. Eriophyid studies. B–9. *Calif. Dept. Agr., Spec. Publ.,* 20 pp. (1963)
57. Keifer, H. H. Eriophyid studies. B–11. *Calif. Dept. Agr., Spec. Publ.,* 20 pp. (1964)
58. Keifer, H. H. Eriophyid studies. B–14. *Calif. Dept. Agr., Spec. Publ.,* 20 pp. (1965)
59. Keifer, H. H. Eriophyid studies. B–21. *Calif. Dept. Agr., Spec. Publ.,* 24 pp. (1966)
60. Keifer, H. H., Wilson, N. S. A new species of eriophyid mite responsible for the vection of peach mosaic virus. *Bull. Calif. Dept. Agr.,* **44,** 145–46 (1955)
61. King, C. L., Sill, W. H., Jr. 1959 wheat streak mosaic epiphytotic in Kansas. *Plant Disease Reptr.,* **43,** 1256–57 (1959)
62. Lange, W. H., Jr. *Aceria tulipae* (K.) damaging garlic in California. *J. Econ. Entomol.,* **48,** 612–13 (1955)
63. Lees, A. H. Reversion of black currants. *Ann. Rept. Long Ashton Res. Sta., 1916,* 31–34 (1917)
64. Lees, A. H. Statistical studies on the propagation of big bud and reversion disease of black currants. *Ann. Rept. Long Ashton Res. Sta., 1922,* 53–57 (1923)
65. Lees, A. H. Reversion disease of black currants: Means of infection. *Ann. Rept. Long Ashton Res. Sta., 1924,* 66 (1925)
66. Lloyd-Jones, C. P., Smith, B. D. The use of radioactive phosphorous to follow the movement of the black currant gall mite. *Ann. Rept. Long Ashton Res. Sta., 1960,* 133–34 (1961)
67. Massee, A. M. The black currant gall mite on red currants. *Ann. Rept. E. Malling Res. Sta., Kent, 1926,* 151–52 (1928)
68. Massee, A. M. The life history of the black currant gall mite, *Eriophyes ribis* (Westw.) Nal. *Bull. Entomol. Res.,* **18,** 297–309 (1928)
69. Massee, A. M. Transmission of reversion of black currants. *Ann. Rept. E. Malling Res. Sta., Kent, 1951,* 162–65 (1952)
70. McKinney, H. H. Mosaic diseases of wheat and related cereals. *U.S. Dept. Agr. Circ. 442,* 23 pp. (1937)
71. McKinney, H. H. Tests of varieties of wheat, barley, oats and corn for reaction to wheat streak mosaic virus. *Plant Disease Reptr.,* **33,** 359–69 (1949)
72. McKinney, H. H., Brakke, M. K., Ball, E. M., Staples, R. Wheat streak mosaic virus in the Ohio Valley. *Plant Disease Reptr.,* **50,** 951–53 (1966)
73. McKinney, H. H., and Fellows, H. Wild and forage grasses found to be susceptible to the wheat streak mosaic virus. *Plant Disease Reptr.,* **35,** 441–42 (1951)
74. McKinney, H. H., Sando, W. J. Susceptibility and resistance to the wheat streak mosaic virus in the genera, *Triticum, Agropyron, Secale,* and certain hybrids. *Plant Disease Reptr.,* **35,** 476–79 (1951)
75. McLarty, H. R. Currant. In *22nd Ann. Rept. Can. Plant Disease Surv.,* 82 (1942)
76. McLarty, H. R., Lott, T. B., Milbrath, J. A., Reeves, E. L., Zeller, S. M. Mottle leaf. In Virus diseases and other disorders with viruslike symptoms of stone fruits in North America. *U.S. Dept. Agr., Agr. Handbook* 10, 106–11 (1951)
77. McNeal, F. H., Dubbs, A. L. Influence of wheat streak mosaic on winter wheat in an area of Montana in 1955. *Plant Disease Reptr.,* **40,** 517–19 (1956)
78. Meiners, J. P. Wheat streak mosaic found in Washington. *Plant Disease Reptr.,* **38,** 714–15 (1954)
79. Mulligan, T. E. Transmission of ryegrass mosaic virus. *Ann. Rept. Rothamstead Expt. Sta., 1957,* 110–11 (1958)
80. Mulligan, T. E. The transmission by mites, host range and properties of ryegrass mosaic virus. *Ann. Appl. Biol.,* **48,** 575–79 (1960)
81. Mumford, E. P. On the fauna of the diseased big bud of the black currant *Ribes nigrum* L. with a note on some fungus parasites of the gall mite *Eriophyes ribis* (Westw.) Nal. *Marcellia,* **27,** 29–62 (1931)
82. Narayanasamy, P., Ramakrishnan, K. Studies on the sterility mosaic disease of pigeon pea. 1. Transmission of the disease. *Proc. Indian Acad. Sci., Sect. B,* **62,** 73–86 (1965)
83. Nariani, T. K., Seth, M. C. Role of eriophyid mites in causing malfor-

mation disease of mango. *Indian Phytopathol.*, **15**, 231–34 (1963)
84. Nault, L. R., Briones, M. L., Williams, L. E., Barry, B. D. Relations of the wheat curl mite to kernel red streak of corn. *Phytopathology*, **57**, 986–89 (1967)
85. Niemeyer, L., Bode, O. Über den Virusnachweis bei Reben (Bemerkungen zu einer Veröffentlichung von G. Ochs) *Z. Pflanzenkrankh. Pflanzenschutz*, **66**, 640–44 (1959)
86. Ochs, G. Über drei Viren als Erreger von Rebkrankheiten. *Z. Pflanzenkrankh., Pflanzenschutz*, **65**, 11–17 (1958)
87. Ochs, G. Untersuchungen über die Verbreitung der Rebenviren durch Vektoren. *Naturwissenschaften*, **45**, 193 (1958)
88. Ochs, G. Ubertragungsversuche von drei Rebviren durch Milben und Insekten *Z. Angew. Zool.*, **47**, 485–91 (1960)
89. Oldfield, G. N. The biology and morphology of *Eriophyes emarginatae*, a *Prunus* finger gall mite, and notes on *E. prunidemissae*. *Ann. Entomol. Soc. Am.*, **62**, 269–77 (1969)
90. Oldfield, G. N., Wilson, N. S. A method of establishing colonies of *Eriophyes insidiosus*, the vector of the peach mosaic virus. (In preparation)
91. Oliinyk, A. M. Transmission of wheat streak mosaic virus by mites: *Aceria tulipae*. *Mikrobiol. Zh., Acad. Nauk Ukr RSR*, **29**, 338–41 (1967)
92. Orlob, G. B. Feeding and transmission characteristics of *Aceria tulipae* Keifer as vector of wheat streak mosaic virus. *Phytopathol. Z.*, **55**, 218–38 (1966)
93. Orlob, G. B. Epidemiology of wheat streak mosaic in South Dakota 1962–66. Host range studies. *Plant Disease Reptr.*, **50**, 819–21 (1966)
94. Orlob, G. B. Relationships between *Tetranychus urticae* Koch and some plant viruses. *Virology*, **35**, 121–33 (1968)
95. Paliwal, Y. C., Slykhuis, J. T. Localization of wheat streak mosaic virus in the alimentary canal of its vector, *Aceria tulipae* K. *Virology*, **32**, 344–53 (1967)
96. Pine, T. S. Host range and strains of peach mosaic virus. *Phytopathology*, **55**, 1151–53 (1965)
97. Pine, T. S., Cochran, L. C. Peach mosaic virus in horticultural plum varieties. *Plant Disease Reptr.*, **46**, 495–97 (1962)
98. Pop, I. V. The streak mosaic of wheat in the Rumanian People's Republic. *Phytopathol. Z.*, **43**, 325–36 (1962)
99. Prasad, A., Singh, H., Shukla, T. N. Present status of mango malformation disease. *Indian J. Hort.*, **22**, 254–64 (1965)
100. Pratt, R. M. (Personal communication, 1967)
101. Proeseler, G., Kegler, H. Transmission of a latent virus from plum by gall mites (Eriophyidae). *Deut. Akad. Wiss. Berlin, Monatsber.*, **8**, 472–76 (1966)
102. Puttarudriah, M., Channabasavanna, G. P. An eriophyid mite as a new pest of garlic in Mysore. *Food Agr. Organ, U.N., FAO Plant Protect Bull.*, **6**, 123–24 (1958)
103. Rai, B., Singh, N. An observation on recovery from malformation in mango sapling. *Current Sci. (India)* **36**, 525–26 (1967)
104. Razvyazkina, G. M., Kopkova, E. A., Belyanchikova, Y. V. Wheat streak mosaic. *Zashchita Rastot Vreditelei Boleznei*, **8**, 54–55 (1963)
105. Rosi-DeSimons, N. E. Four new mites reported in Argentina. *Idia (Buenos Aires)*, 204 (1964)
106. Rossouw, D. J., Smith, A. J. The relation of *Calacarus citrifolii* Keifer to concentric ring blotch of citrus. *S. African Citrus J.*, **354**, 7–9 (1963)
107. Schulz, J. T. *Tetranychus telarius* (L.) new vector of Virus Y. *Plant Disease Reptr.*, **47**, 594–96 (1963)
108. Seth, M. L. Transmission of pigeon pea (*Cajanus cajan*) sterility (virus) by an eriophyid mite. *Indian Phtytopathol.*, **15**, 225–27 (1963)
109. Seth, M. L., Nariani, T. K. A note on methyl bromide fumigation of mango seedlings affected with the malformation disease. *Indian Phytopathol.*, **19**, 390 (1966)
110. Shepard, J. F., Carroll, T. W. Electron microscopy of wheat streak mosaic virus particles in infected plant cells. *J. Ultrastruct. Res.*, **21**, 145–52 (1967)
111. Shtein-Margolina, V. A., Cherni, N. E., Razvyazkina, G. M. Wheat streak mosaic virus in cells of the plant and mite vector. *Dokl. Akad.*

Nauk. SSSR., **169**, 1446–48 (1966)
112. Sill, W. H., Jr., Agusiobo, P. C. Host range studies of the wheat streak mosaic virus. *Plant Disease Reptr.*, **39**, 633–42 (1955)
113. Sill, W. H., Jr., Connin, R. V. Summary of the known host range of the wheat streak mosaic virus. *Trans. Kansas Acad. Sci.*, **56**, 411–17 (1953)
114. Sill, W. H., Jr., Del Rosario, M. S. Transmission of wheat streak mosaic virus to corn by the eriophyid mite *Aceria tulipae*. *Phytopathology*, **49**, 396 (1959)
115. Sill, W. H., Jr., Fellows, H. Symptom expression of the wheat streak mosaic virus disease as affected by temperature. *Plant Disease Reptr.*, **37**, 30–33 (1953)
116. Sill, W. H., Jr., Fellows, H., King, C. L. Kansas wheat mosaic situation (1953–4). *Plant Disease Reptr.*, **39**, 29–30 (1955)
117. Slykhuis, J. T. Virus diseases of cereal crops in South Dakota. *S. Dakota State Coll. Agr. Expt. Sta. Tech. Bull. 11*, 29 pp. (1952)
118. Slykhuis, J. T. The relation of *Aceria tulipae* (K.) to streak mosaic and other chlorotic symptoms of wheat. *Phytopathology*, **43**, 484–85 (1953)
119. Slykhuis, J. T. *Aceria tulipae* Keifer in relation to the spread of wheat streak mosaic. *Phytopathology*, **45**, 116–28 (1955)
120. Slykhuis, J. T. Wheat spot mosaic caused by a mite-transmitted virus associated with wheat streak mosaic. *Phytopathology*, **46**, 682–87 (1956)
121. Slykhuis, J. T. A survey of virus diseases of grasses in northern Europe. *Food Agr. Organ., U.N., FAO Plant Protect. Bull.*, **6**, 129–34 (1958)
122. Slykhuis, J. T. Current status of mite-transmitted plant viruses. *Proc. Entomol. Soc. Ontario*, **90**, 22–30 (1960)
123. Slykhuis, J. T. Eriophyid mites in relation to the spread of grass viruses in Ontario. *Can. J. Plant Sci.*, **41**, 304–8 (1961)
124. Slykhuis, J. T. Mite transmission of plant viruses. In *Biological transmission of disease agents*, 41–61. (Maramorosch, K., Ed., Academic Press, New York, 192 pp., 1962)
125. Slykhuis, J. T. An international survey for virus diseases of grasses. *Food Agr. Organ., U.N., FAO Plant Protect. Bull.*, **10**, 1–16 (1962)
126. Slykhuis, J. T. Mite transmission of plant viruses. In *Advances in Acarology*, **1**, 326–40. (Naegele, J. A., Ed., Comstock Press, Ithaca, N.Y., 479 pp., 1963)
127. Slykhuis, J. T. Current research on mites in relation to plant virus transmission. *Phytoprotection*, **45**, 101–7 (1964)
128. Slykhuis, J. T. Mite transmission of plant viruses. In *Advances in Virus Research*, **11**, 97–137. (Smith, K. M., Lauffer, M. A., Eds., Academic Press, New York and London, 425 pp., 1965)
129. Slykhuis, J. T. Methods for experimenting with mite transmission of plant viruses. In *Methods in Virology*, **I.**, Chap. 10, 347–68. (Maramorosch, K., Kobrowski, H., Eds., Academic Press, New York, 640 pp., 1967)
130. Slykhuis, J. T., Andrews, J. E., Pittman, U. J. Relation of date of seeding winter wheat in southern Alberta to losses from wheat streak mosaic, root rot and rust. *Can. J. Plant Sci.*, **37**, 113–27 (1957)
131. Smalley, E. B. The production on garlic by an eriophyid mite of symptoms like those produced by viruses. *Phytopathology*, **46**, 346–47 (1956)
132. Smith, B. D. The behavior of the black currant gall mite (*Phytoptus* i.e. *Cecidophyes ribis* Nal.) during the free living phase of its life cycle. *Ann. Rept. Long Ashton Res. Sta., 1959*, 130–36 (1960)
133. Smith, B. D. Effect of temperature and photoperiod on black currants and on the behavior of the gall mite (*Phytoptus* i.e. *Cecidophyes ribis* Nal.). *Ann. Rept. Long Ashton Res. Sta., 1959*, 137–38 (1960)
134. Smith, B. D. Population studies of the black currant gall mite (*Phytoptus ribis* Nal.). *Ann. Rept. Long Ashton Res. Sta., 1960*, 120–24 (1961)
135. Smith, B. D. Experiments in the transfer of the black currant gall

mite (*Phytoptus ribis* Nal.) and of reversion. *Ann. Rept. Long Ashton Res. Sta., 1961,* 170–72 (1962)

136. Smith, B. D. A field study of the spread of the black currant gall mite (*Phytoptus ribis* Nal.) and of the virus disease reversion. *Ann. Rept. Long Ashton Res. Sta., 1962,* 124–29 (1963)
137. Smith, B. D., Corke, A. T. K. Effect of (2-chloroethyl) trimethylammonium chloride on the eriophyid gall mite *Cecidophyopsis ribis* Nal., and three fungus diseases of black currant. *Nature,* **212,** 643–44 (1966)
138. Smith, K. M. *A textbook of plant virus diseases.* (Little, Brown & Co., Boston, 652 pp., 1957)
139. Somsen, H. W. Development of migratory form of wheat curl mite. *J. Econ. Entomol.,* **59,** 1283–84 (1966)
140. Staples, R., Allington, W. B. Streak mosaic of wheat in Nebraska and its control. *Nebraska Agr. Expt. Sta. Bull.,* **178,** 1–41 (1956)
141. Staples, R., Brakke, M. Relation of *Agropyron repens* mosaic and wheat streak mosaic viruses. *Phytopathology,* **53,** 969–72 (1963)
142. Sutic, D., Tosic, M. The virus of wheat streak mosaic in Yugoslavia. *Zastita Bilja,* **79,** 307–14 (1964)
143. Swarbrick, T. Berry, W. E. Further observations on the incidence and spread of reversion and big bud on black currants. *Ann. Rept. Long Ashton Res. Sta., 1936,* 124–32 (1937)
144. Taksda, G. The ecology of cold hardiness in different populations of the black currant gall mite, *Cecidophyopis ribis. Entomol. Exptl. Appl.,* **10,** 377–86 (1967)
145. Thresh, J. M. A vein pattern of black currant leaves associated with reversion disease. *Ann. Rept. E. Malling Res. Sta., Kent, 1962,* 97–98 (1963)
146. Thresh, J. M. Abnormal black currant foliage caused by the gall mite, *Phytoptus ribis* Nal. *Ann. Rept. E. Malling Res. Sta., Kent, 1962,* 99–100 (1963)
147. Thresh, J. M. Increased susceptibility of the mite vector (*Phytoptus ribis* Nal.) caused by infection with black currant reversion virus. *Nature,* **202,** 1028 (1964)
148. Thresh, J. M. Association between black currant reversion and its gall mite vector (*Phytoptus ribis* Nal.). *Nature,* **202,** 1085–87 (1964)
149. Thresh, J. M. Warm water treatments to eliminate the gall mite *Phytoptus ribis* Nal. from black currant cuttings. *Ann. Rept. E. Malling Res. Sta., Kent, 1963,* 131–32 (1964)
150. Thresh, J. M. Field experiments on the spread of black currant reversion virus and its gall mite vector (*Phytoptus ribis* Nal.). *Ann. Appl. Biol.,* **58,** 219–30 (1966)
151. Thresh, J. M. Increased susceptibility of black currant bushes to *Phytoptus ribis* following infection with virus. *Ann. Appl. Biol.,* **60,** 455–67 (1967)
152. Tiits, A. Some observations on black currant reversion when the virus is transmitted by grafting. *Izvest. Akad. Nauk, Eston. SSR., Ser. Biol.,* **4,** 267–71 (1964)
153. Timian, R. G., Bissonette, H. L. Wheat streak mosaic virus in North Dakota. *Plant Disease Reptr.,* **48,** 703 (1964)
154. Tuttle, D. M., Baker, E. W. *Spider mites of Southwestern United States and a revision of the family Tetranychidae.* (U. Arizona Press, Tucson, 143 pp., 1968)
155. Van der Merwe, G. G., Coates, T. J. Biological study of the grey mite *Calacarus citrifolii* Keifer. *S. African J. Agr. Sci.,* 817–24 (1965)
156. Van Eyndhoven, G. L. The red currant gall mite, *Cecidophyopsis selachodon* n. sp. *Entomol. Ber.,* **27,** 149–51 (1967)
157. Vashisth, K. S., Nagaich, B. B. *Morus indica,* an additional host of fig mosaic. *Indian Phytopathol.,* **18,** 315 (1965)
158. Vukovits, G. Beobachtungen and Untersuchungen über die an Prunus Arten vorkommende Sternflecken-(Kräusel) Krankheit. *Pflanzenschutz Ber.,* **26,** 1–17 (1961)
159. Wadsworth, D. F., Barley and wheat mosaic in Oklahoma. *Plant Disease Reptr.,* **33,** 482–83 (1949)
160. Wallin, J. R. Field observations of

wheat mosaic in Kansas, Nebraska and Iowa. *Plant Disease Reptr.*, **34,** 211–12 (1950)

161. Walters, H. J. Virus diseases of small grains in Wyoming. *Plant Disease Reptr.*, **38,** 836–37 (1954)

162. Warburton, C., Embleton, A. L. The life history of thc black currant gall mite *Eriophyes (Phytoptus) ribis* Westw. *Linnean Soc. J. Zool.*, **28,** 366–78 (1901–03)

163. Williams, L. E., Gordon, D. T., Nault, L. R., Alexander, L. J., Bradfute, O. E., Findley, W. R. A virus of corn and small grains in Ohio and its relation to wheat streak mosaic virus. *Plant Disease Reptr.*, **51,** 207–11 (1967)

164. Wilson, N. S., Cochran, L. C. Yellow spot, an eriophyid mite injury on peach. *Phytopathology*, **42,** 443–47 (1952)

165. Wilson, N. S., Jones, L. S. Unpublished data Fruit Disease Vector Investigations, Fruit Insects Research Branch, *U. S. Dept. Agr., A.R.S., Entomol. Res. Div.* (1955–65)

166. Wilson, N. S., Jones, L. S., Cochran, L. C. An eriophyid mite vector of the peach mosaic virus. *Plant. Disease Reptr.*, **39,** 889–92 (1955)

167. Wilson, N. S., Oldfield, G. N. New species of eriophyid mites from western North America with a discussion of eriophyid mites on *Populus. Ann. Entomol. Soc. Am.*, **59,** 585–99 (1966)

168. Wilson, N. S., Oldfield, G. N. Unpublished data Fruit Disease Vector Investigations, Fruit Insects Research Branch, *U. S. Dept. Agr., A.R.S., Entomol. Res. Div.* (1969)

RESISTANCE OF TICKS TO CHEMICALS[1,2]

By R. H. Wharton and W. J. Roulston
Division of Entomology, CSIRO, Long Pocket Laboratories, Indooroopilly, Brisbane, Australia

The increasing problem of arthropod resistance to chemicals jeopardises advances that have been achieved in the fields of public health and agricultural and livestock production (26, 27, 123). Although there are only a few species of ticks which have become resistant, as they are chiefly of veterinary importance, their existence poses a threat to livestock production, either because they transmit diseases or affect productivity, or both. Resistance shows its highest incidence in one-host ticks of the genus *Boophilus,* probably because a much larger fraction of the total population of such species is under chemical challenge at any one time than two- or three-host ticks. The latter also have much longer life cycles [a single generation may extend over one, two or three years compared with two to three months in *Boophilus* (37, 53)] and they also tend to be less host-specific to domestic animals. Resistance has nevertheless developed in two- and three-host ticks and may be expected to develop in other species controlled primarily by chemicals.

The resistant tick species include few that attack man, and this probably reflects the fact that chemicals are used less against them than against livestock pests. Man is invariably a casual host and, to go beyond the personal protection of individuals, chemical measures involve the costly procedure of extensive area control. Thus, the selective pressure is generally lower than that with livestock pests.

There are special difficulties in defining and studying problems of chemical resistance in ticks, the more important of which are the collection of adequate samples from the field, the long life cycles of most ticks, expense and labour involved in maintaining colonies, and the difficulty in devising satisfactory standard laboratory test methods.

This review supplements recent contributions to our understanding of

[1] The term, acaricide, is frequently applied to chemicals used for the control of phytophagous mites while insecticide, tickicide, ixodicide and acaricide are used for chemicals developed for the control of ticks. In this review, the term, acaricide, has been used since it is logical to apply this term for all *Acarina* in the same way as insecticides for Insecta and nematocides for Nematoda.

[2] The following abbreviations are used in this chapter: BHC for 1,2,3,4,5,6-hexachlorocyclohexane; DDE for 1,1-dichloro-2,2-*bis*(*p*-chlorophenyl)ethylene; DDT for 1,1,1-trichloro-2,2-*bis*(*p*-chlorophenyl)ethane; AChE for acetylcholinesterase; OP for organophosphorus.

the resistance problem published in the *Annual Review of Entomology* (14, 60, 63), and the review by Whitehead (117) on resistance of ticks to chemicals.

Present Status of Resistance

Argasidae.—There are no confirmed records of resistance to chemicals in argasid ticks. It is generally recognised that DDT is not effective against *Ornithodoros* species (9, 13, 17); γBHC has been used for over 20 years in East Africa against *O. moubata* with no evidence of resistance (54), and is used against *O. tholozani* and *O. lahorensis* (9). The report of resistance to DDT in *O. rudis,* and to DDT, dieldrin and γBHC in *Argas columbarum* in Colombia (13, 67), lacks adequate documentation. Organophosphorus chemicals are recommended for the control of the fowl tick *A. persicus* (104) but this reflects a desire to minimise problems of undesirable residues rather than change enforced by resistance to chlorinated hydrocarbons. The spinose ear tick, *Otobius megnini,* has been controlled for many years in North America with γBHC which is still recommended for its control (104).

Ixodidae.—Resistance has been confirmed in *Rhipicephalus sanguineus* in the United States, in *Boophilus decoloratus* in Africa, in *B. microplus* in Australia, South American countries, Malagasy and India; and in *Rhipicephalus appendiculatus* and *R. evertsi* in South Africa. Resistance has not been recorded in the genera *Ixodes, Hyalomma* or *Haemaphysalis.*

Amblyomma americanum, the lone star tick of the United States, has been reported as resistant to dieldrin (16, 27), and *Dermacentor variabilis,* the American dog tick, as resistant to DDT and dieldrin (16, 27, 107, 123). In the absence of adequate documentation these reports have been discounted. Despite the widespread use of DDT, toxaphene and lindane on livestock in the United States against *A. maculatum, D. andersoni, D. occidentalis* and *D. albipictus,* no resistance has been reported in these species (31, 39) and γBHC is still recommended for the control of *D. andersoni* in Canada (56). BHC has also been used for many years against *Ixodes ricinus* on sheep in Europe without evidence of resistance (122). *Anocentor nitens,* the tropical horse tick, is one species in which resistance could occur—it has a life history similar to *Boophilus* and will be subject to heavy chemical challenge because of its importance as a vector of equine piroplasmosis in southeastern United States (23).

The area control treatments of forests and pastures which have been undertaken for the control of ixodid vectors of human disease in Europe, Soviet Union and the United States have apparently not led to the development of resistance, but laboratory tests to determine the response of these ticks to chemicals are exceedingly sparse (see below).

The common dog tick, *Rhipicephalus sanguineus,* which is widely distributed in practically all countries between 50°N and 35°S (46) was shown to be resistant to chlordane in the United States (34) but, whereas 3 per cent chlordane sprayed in kennels failed to kill ticks, 0.5 per cent lindane was effective. Subsequently, ticks resistant to chlordane were found to be widespread in Texas and resistance was shown to extend to lindane in a few of the strains (40). The only other record of resistance is from Panama (2). Although there seems to be no doubt that resistance to chlordane and to lindane has developed, laboratory tests so far reported have been inadequate for the determination of resistance levels or for the investigation of the apparent lack of the usual cross-resistance to lindane in some strains. United States Department of Agriculture recommendations (104) indicate that *R. sanguineus* may also be resistant to DDT, but early tests indicated that DDT was not satisfactory for its control (34). No published evidence of induced DDT-resistance has been found.

Resistance to toxaphene in the two-host red tick, *Rhipicephalus evertsi,* was recognised on a property in East London Province, South Africa, in 1959 after four years of continuous treatment of cattle with toxaphene (118). Dosage-mortality tests on engorged females and larvae established that resistance extended to γBHC and dieldrin, but not to sodium arsenite, carbaryl, dioxathion or to DDT (no evidence of "vigour tolerance" to DDT, which has been stated to occur (117), is apparent in the tests).

Toxaphene resistance was also recognised in the three-host brown ear tick, *R. appendiculatus,* on a number of cattle properties (including the one where toxaphene resistance to *R. evertsi* was found) in East London Province, and in the Transvaal in South Africa in 1965 (7). Dosage-mortality tests on larvae, engorged nymphs and engorged females established that resistance extended to γBHC, but not to the OP chemicals dioxathion and chlorfenvinphos (though it may be significant that a slight increase in LC_{50} of dioxathion was recorded on the property where toxaphene resistance to *R. evertsi* was recorded earlier, and where a composite dioxathion/toxaphene dip-wash had been used for five years). Resistance was recognised after some 10 to 15 years' use of toxaphene or γBHC or both. Although there are no published records of resistance from other parts of Africa, Baker (5) states that resistance to chlorinated hydrocarbons in *R. appendiculatus* and *R. evertsi* is widespread throughout South Africa and is present also in Rhodesia and East African countries. There has been no indication of resistance in the *Amblyomma* and *Hyalomma* species that infest cattle in Africa.

The cattle tick of Australia and South America, *Boophilus microplus,* and the blue tick of Africa, *B. decoloratus,* have developed resistance to arsenic,[3] DDT, cyclodiene compounds and to OP and carbamate chemicals. The history of the development of resistance in these ticks has been reviewed frequently (9, 57, 117).

Five species of *Boophilus* are recognised: *B. annulatus* (= *calcaratus*)

[3] Arsenic in this context refers to an alkaline solution of arsenious oxide.

from North America, the southern part of Europe, Russia, west and central Africa; *B. microplus* [regarded as a subspecies of *B. annulatus* by Graham & Price (32)] from the tropics and subtropics of Asia, Australia, central and southern America and from east and southern Africa; *B. decoloratus* from Africa; *B. kohlsi* from the Near East (Jordan); and *B. geigyi* from west Africa (1, 3, 46). *B. kohlsi* is found mainly on sheep and goats, but all the other species occur predominantly on cattle. Their importance lies in the loss of production which results from heavy infestations (52, 58) and in their ability to transmit *Babesia* and *Anaplasma* infections (71).

Boophilus annulatus was eradicated from the United States in a campaign extending from 1906–1943, based mainly on dipping cattle in arsenic at intervals of 14 days for periods varying from 6 to 12 months (25, 103). In similar campaigns, *B. microplus* has been eradicated from Florida (1923–1961), Puerto Rico (1936–1952) (103), and from small areas toward the southern limit of its distribution in Australia (53, 82, 103). The most recent campaign in 1956–1957, based on dipping cattle in DDT every 14 days for 15 months, failed to eradicate *B. microplus* from an area in New South Wales which was more favourable to the tick (53). Resistance to DDT was not the reason for the failure (92).

The development of resistance in *B. microplus* and *B. decoloratus* has followed a remarkably similar pattern. Wherever cattle have been subjected to frequent treatments with chemicals, resistance has occurred and, except for resistance to OP-chemicals, has been recognised at about the same time. The first record of chemical tick control was the use of arsenic in Queensland, Australia, in 1895, and its use rapidly spread to other countries (82). Arsenic apparently gave satisfactory control for about forty years when complaints were heard that, in Australia, South Africa and Argentina, the standard of tick control had declined (9, 82, 117). The presence of arsenic-resistant strains of *Boophilus* ticks was confirmed by field and laboratory tests in South Africa (24, 119) and in Australia (43, 51). Failure of arsenic to control ticks has also been reported from Argentina, Brazil, Colombia, Jamaica (4) and Venezuela (65).

DDT became available commercially for tick control in 1946 and the other chlorinated hydrocarbons, BHC, toxaphene and dieldrin were marketed within a few years. Resistance to DDT was reported in Australia in 1955 (50) and confirmed by Stone (91) who showed a 20-fold resistance in larvae and engorged females. Resistance developed slowly, however, and satisfactory control was maintained on many properties, despite resistance to DDT, until its withdrawal in 1962 (57). In South Africa, DDT preparations were introduced as a substitute for BHC to control ticks resistant to both arsenic and BHC, but about five years later resistance to DDT was reported in *B. decoloratus* (113). Ticks resistant to DDT were found to be resistant also to Dilan® and to pyrethrins (115). In both Australia and South Africa, resistance to BHC was found some eighteen months after it was introduced for tick control (42, 120). Resistance to BHC in *Boophilus* species has always extended to toxaphene and the cyclodienes (59, 117).[4] Ticks resistant to all chlorinated hydrocarbons were reported in Argentina (11), Brazil (29) and in Venezuela (65). Resistance to BHC and other chlorinated hydrocarbons was found in Malagasy (102) and in India (19). Resistance to BHC has also been found in ticks from Kluang, Malaya and Sabah, Borneo (84). Resistance to DDT is independent of resistance to the other types of chlorinated hydrocarbons, but strains of *B. microplus* and *B. decoloratus* which are resistant to both groups have been described (100, 114).

There is comparatively little information regarding OP-resistance in *Boophilus* except from Australia where, in recent years, the emergence of strains exhibiting resistance to a wide range of OP and carbamate chemicals has created a serious problem for cattle owners and control authorities. Diazinon was the first OP compound to be approved[5] for use against cattle tick in Australia (1956), followed by dioxathion (1958), coumaphos (1959), carbophenothion (1961), ethion (1962) and the carbamate, carbaryl (1963). Thus, when all chlorinated hydrocarbons were banned for use on cattle in 1962 an array of excellent acaricides was available (78). However, in the latter part of 1963, following the failure of dioxathion to control ticks in the field, a strain of ticks from Ridgelands in central Queensland was shown to have some resistance to all OP and carbamate acaricides in current use (83, 86). Another strain from the same area was also investigated (78). The strains were tested by different methods [Shaw (83) by dipping larvae in the formulated acaricide, Roulston et al. (78) by exposing larvae in filter papers impregnated with olive solutions of the test chemical] but exhibited similar resistance patterns. The resistance levels established by the two methods were respectively as follows: dioxathion 25, 7.2; carbophenothion 62, 17; ethion 3, 2.7; carbaryl 38, 13; diazinon 15, 10; and coumaphos 1.8, 1.6. The name Ridgelands has been adopted in Australia to describe resistant ticks of this type.

To determine what the resistance levels meant in terms of field control, trials were undertaken in which calves infested with either susceptible (Yeerongpilly) or resistant (Ridgelands) ticks were sprayed with the recommended concentration of commercially formulated acaricides and held in stalls for 22 days after spraying (78). The efficiency of treatment was determined from a comparison of the numbers of engorged ticks falling daily from sprayed and unsprayed cattle. The results of spraying trials with dioxathion (0.075 per cent) and coumaphos (0.025 per cent) are illustrated in Figure 1. Resistance was expressed mainly in the survival of many more nymphs, which dropped as engorged females 10 to 15 days after the calves

[4] Some field reports that BHC-resistant ticks were controlled by toxaphene are believed to have originated when control failures were attributed to resistant ticks rather than an ineffective dipping bath from which the BHC had been differentially lost (74).

[5] As in many countries, before a pesticide may be marketed in Australia, it must be approved by a registration board.

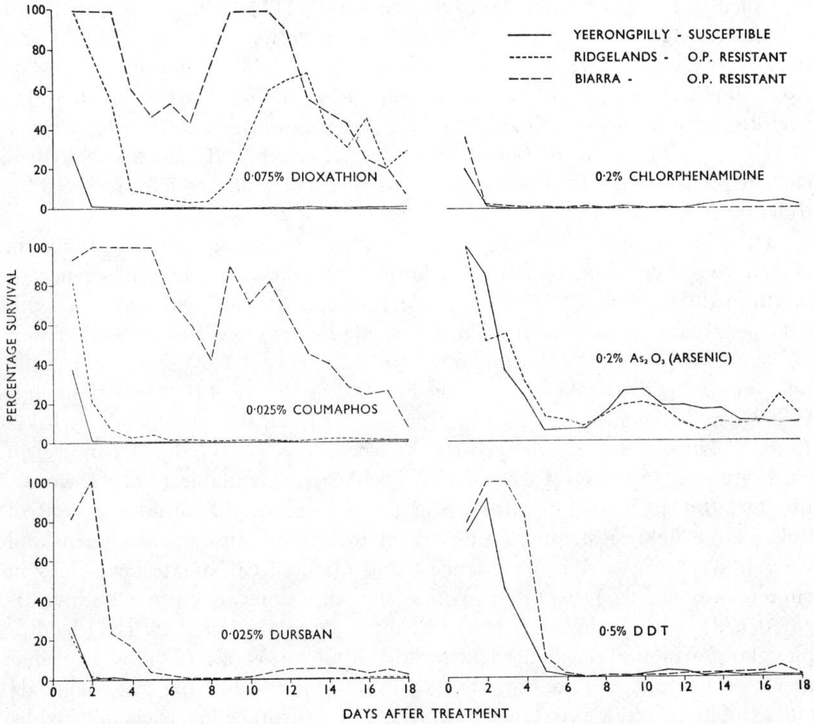

FIG. 1. Percentage survival of susceptible (Yeerongpilly) and OP-resistant (Ridgelands and Biarra) ticks on calves sprayed with dioxathion, coumaphos, Dursban®, chlorphenamidine, arsenic and DDT.

were sprayed with dioxathion. With coumaphos, little difference between the two strains was apparent and this also applied to ethion, whereas the other chemicals in current use gave unsatisfactory control. Potential acaricides were also tested in the laboratory and those with resistance factors lower than six-fold were further evaluated in spraying trials. The following promised effective control of the Ridgelands strain: bromophos, 0.15 per cent; bromophos-ethyl, 0.1 per cent; chlorfenvinphos, 0.06 per cent; chlorxylam, 0.5 per cent; Ciodrin®, 0.1 per cent; Dursban®, 0.025 per cent; fenchlorphos, 0.2 per cent; Imidan®, 0.075 per cent and S.D. 8448, 0.025 per cent. These results were in broad agreement with the assessment of these chemicals as potential acaricides against *B. annulatus* and *B. microplus* by Drummond et al. (22).

A new strain of OP-resistant *B. microplus* was recognised from Biarra in the Brisbane Valley in southeastern Queensland in 1966 (79, 85). This strain exhibited enhanced resistance to the acaricides in current use including ethion and coumaphos (Fig. 1), and laboratory tests showed that Biarra larvae were resistant to all OP and carbamate chemicals tested. Of the chemicals previously considered promising for the control of Ridgelands ticks, those with the lowest resistance factors were tested in spraying trials, and of these, bromophos-ethyl, Dursban and Imidan appeared most promising (79). A feature of the spraying trials against Biarra ticks was that many partly engorged females survived treatment and dropped as engorged females from cattle for several days after spraying (see results with coumaphos, Fig. 1). Dursban, bromophos-ethyl and, to a limited extent, Imidan have been used successfully in the field to control the Biarra strain—Dursban and bromophos-ethyl at twice the registered concentrations of 0.025 and 0.05 per cent, respectively.

Spraying trials were also carried out with arsenic, DDT and chlorphenamidine against Biarra and susceptible ticks (Fig. 1). There was little or no difference between the effect of these chemicals on OP-resistant and on susceptible ticks. However, chlorphenamidine (CIBA 8514/Schering 36268) is unstable in aqueous suspension and produces undesirable toxic effects in cattle; DDT cannot be used because of residues and recent experiments have shown clearly that arsenic, even against arsenic-susceptible strains, is a less effective acaricide than modern chemicals.

A third OP-resistant strain has been recognised from Mackay, some 600 miles north of the Biarra area. The ticks were provisionally diagnosed as Biarra type when found to be resistant to ethion and coumaphos (64). Further examination showed that both the spectrum and mechanism of resistance in this strain differ from those of Ridgelands and Biarra ticks (77). Dosage-mortality tests and spraying trials indicate that the Mackay strain exhibits resistance to all registered OP-chemicals and is potentially as serious a threat to cattle tick control as the Biarra strain.

Ticks similar to the Ridgelands strain have been found at widely separate localities along the coastal region of Queensland (Fig. 5).[6] On the other hand, ticks of the Biarra and Mackay strains have, to date, been found only in restricted areas, in the southeast and north of Queensland, respectively.

The dosage-mortality responses of the three strains are quite distinct. Ridgelands-type larvae can be recognised because of high resistance to dioxathion but low resistance to coumaphos; Biarra-type larvae because of high level resistance to coumaphos, ethion and very high level resistance to other OP chemicals including Cyanox® (× 220), and Mackay-type larvae because of high level resistance to coumaphos and ethion, but only low level resistance to Cyanox (× 4). The pattern of resistance to a range of OP and

[6] Ticks resistant to dioxathion and provisionally classified as of the Ridgelands type were detected in the Bonalbo region of northern New South Wales in June, 1969. A resistant strain of the Biarra type has also been selected from a sample of ticks submitted for testing in March, 1968. The nature of the resistance was not clear initially and was revealed only after five generations of high selection in the laboratory.

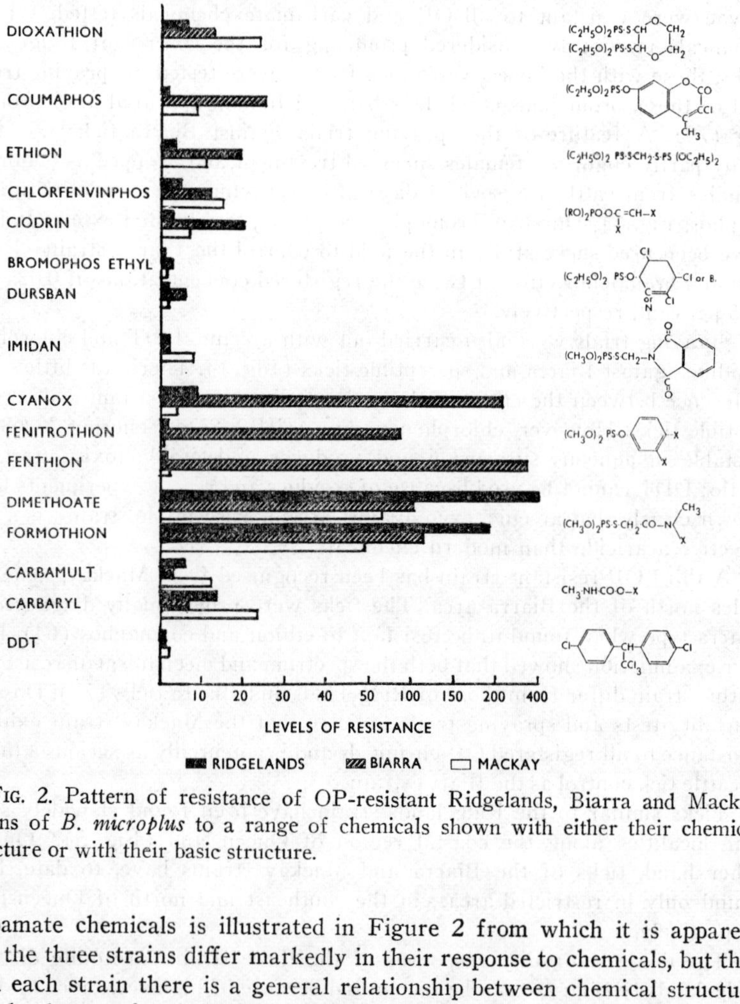

FIG. 2. Pattern of resistance of OP-resistant Ridgelands, Biarra and Mackay strains of *B. microplus* to a range of chemicals shown with either their chemical structure or with their basic structure.

carbamate chemicals is illustrated in Figure 2 from which it is apparent that the three strains differ markedly in their response to chemicals, but that with each strain there is a general relationship between chemical structure and the degree of resistance which can be expected.

Information on OP resistance in *Boophilus* from other countries is restricted to the description by Shaw, Thompson & Baker (87) of a strain of *B. decoloratus* from South Africa which is obviously of the Ridgelands type, and a report of similar (but not Biarra or Mackay) strains of *B. microplus* from Venezuela, and Brazil (85). In view of the closely parallel development of resistance to arsenic and chlorinated hydrocarbons in *B. microplus* and *B. decoloratus* in different countries, extension and intensification of OP resistance in these species must be expected in African and South American countries.

Detection of Resistance

The World Health Organization Expert Committee on Insecticides (123) issued tentative instructions for determining the resistance of adult ticks to insecticides, based on the topical application method using a microcapillary apparatus. There has been little response and the only records of tests are those reported in mimeographed WHO documents by Busvine & Srivastava (17) on *O. moubata* and *Hyalomma dromedarii,* Uspenskij (105) on *I. persulcatus,* and Privora et al. (68) on *D. marginatus.* These indicated that the method was satisfactory for unfed adult *I. persulcatus* and for unfed adult *Hyalomma,* but that recently fed adult *O. moubata* were remarkably tolerant to insecticides (LC_{50} of DDT, 25 per cent; dieldrin, 4.3 per cent; malathion, 25 percent). Difficulties were also encountered with unfed adult *D. marginatus,* mainly because ticks failed to die in a relatively short period after treatment.

Resistance in ticks of veterinary importance has usually been recognised because of the failure to obtain a satisfactory kill of the parasitic stages on treated animals. This poses a problem to those who are concerned with the early recognition of resistance, since failure of control in the field may be due to inadequate treatment, and many reports of resistance are unfounded (83). The quantitative assessment of the efficiency of an acaricide in the field is virtually impossible to achieve and although it is possible to demonstrate resistance by comparing the results of spraying cattle infested with either suspected resistant or known susceptible ticks (7, 118), the confirmation of resistance must be based on dosage-response tests on the nonparasitic stages of the tick. Methods of testing have been reviewed briefly (30) and include topical application (48), immersion (42), injection (100) or exposure to impregnated papers (99) of engorged females and immersion (83) or exposure of unfed larvae to impregnated papers (99). For routine testing, the use of unfed larvae is preferable since these are more readily standardised, the criterion of response is death and is more easily established and an adequate sample can usually be obtained. It may be possible to treat larvae topically with microcapillary pipettes as is done with spider mites (8) but the method would appear to be too laborious under most circumstances. The resistance factors recorded from larval immersion tests on the Biarra OP-resistant strain of *B. microplus* (85) were higher than those recorded from larval exposure to impregnated papers (79) but the cross-resistance pattern was similar. The latter technique has been used successfully to detect OP resistance in Queensland in recent years (64).

It is unlikely that the same test method would be applicable to ticks of medical and veterinary importance. The former are normally collected as questing unfed adult females, whereas the latter are normally collected as engorged females from infested animals and either engorged females or unfed larvae are tested.

BIOCHEMISTRY OF TICK RESISTANCE

Studies on tick biochemistry and metabolism of chemicals have been carried out on susceptible strains of ticks (20, 44, 45) on strains resistant to arsenic (35, 36, 75, 101, 116), to BHC and dieldrin (73) and to OP compounds (49, 72, 77, 81).

The spectacular success of BAL (British antilewisite) as an antidote for arsenical poisoning in vertebrates demonstrated that at least one mode of action of arsenicals was reaction with sulphydryl compounds (66, 90). It was logical, therefore, to deduce that arsenic-resistant ticks might contain greater amounts of sulphydryl compounds than susceptible ticks, and South African workers claimed this was so in *B. decoloratus* (117). However, in similar studies Roulston & Schuntner (75) found no difference in the amounts of sulphydryl compounds in susceptible and resistant *B. microplus.*

The mode of action of DDT is still uncertain (61). However, it is generally assumed that in insects the major resistance mechanism is detoxification of the DDT due to increased levels of the enzyme DDT dehydrochlorinase which degrades DDT to the relatively nontoxic DDE (13, 61, 63, 89). Early studies on the metabolism of DDT in susceptible and DDT-resistant *B. decoloratus* failed to demonstrate DDE (117). Later studies with [^{14}C] DDT showed that in similar strains of *B. microplus,* DDT was metabolised to DDE, but there was no demonstrable causative relationship between DDT metabolism and resistance (80). WARF antiresistant (N-di-*n*-butyl-*p*-chlorobenzenesulphonamide) is believed to inhibit DDT dehydrochlorinase and to restore the effectiveness of DDT against some DDT-resistant strains of arthropods (55), but it was ineffective when used with DDT against DDT-resistant ticks (73). This result is consistent with the finding that metabolism of DDT to DDE in cattle ticks is of little significance as a resistance mechanism.

The mode of action of γBHC and the cyclodienes is still unknown but both are regarded as neurotoxicants (61). There is no clear evidence to associate resistance to insects with reduced penetration or increased metabolism (61). Studies on γBHC and dieldrin-resistant strains of *B. microplus* have also shown that these factors are not related to resistance (73).

Studies on the mode of action of OP and carbamate chemicals and on the mechanism of resistance in *B. microplus* have been more rewarding. It is generally agreed that the mode of action of OP compounds involves inhibition of AChE. Phosphorothionates per se are weak inhibitors of AChE but *in vivo* they are oxidized to the phosphates which are potent AChE inhibitors (61). This was confirmed for the phosphorothionate, coumaphos, in cattle tick larvae (76). Further studies showed that, while there was close correlation between AChE inhibition and death of ticks, AChE inhibition had to be of the order of 90 per cent before high mortality occurred in larvae treated with coumaphos—an indication that ticks can survive on approximately 10 per cent of the normal amount of AChE. Lee & Batham (49) and Schuntner et al.

(81) found that larval homogenates of the Ridgelands (M) strain of *B. microplus* exhibited only 20 to 25 per cent of the AChE activity shown by larvae of susceptible strains. The former authors showed that, of the total activity of the enzyme in the resistant larvae, there were two portions, one of which was inhibited at a slower rate than the other, and that the relative insensitivity of this portion to inhibition by several phosphates correlated with the order of resistance of the strain to the parent OP compound. The latter authors showed that the penetration and metabolism of coumaphos, diazinon and the main individual components of dioxathion were similar in susceptible and resistant larvae. However, the percentages of original AChE activities in the two strains 6 hours after treatment were different, and the ratios of these activities were directly proportional to the order of resistance (81). On the basis of these investigations OP-resistance in the Ridgelands strain is considered to be due to the presence of AChE which is relatively insensitive to inhibition. The enzyme is still sufficiently sensitive to many OP chemicals, e.g., coumaphos, ethion, Dursban, and bromophos-ethyl to allow them to reduce the AChE level below that critical for tick survival.

The finding that only a proportion of the AChE present in the Ridgelands (M) strain is relatively insensitive to inhibition (49, 81) conflicts with genetic observations indicating that the level of OP resistance and of low brain AChE activity are controlled by one or more closely linked genes (see below). The most likely explanation is that both research groups studied larvae which included heterozygotes—OP resistance is incompletely dominant (95) and the presence of heterozygotes would have been difficult to demonstrate in dosage-mortality tests at that time. Smissaert (88) found that homogenates of resistant spider mites *Tetranychus urticae* contained only one type of cholinesterase which was less sensitive to diazoxon or paraoxon inhibition than the enzyme from susceptible mites, but the F_1-cross mites possessed two cholinesterases, one sensitive and one insensitive.

Studies by Roulston et al. (72) have shown that there is no difference in the penetration or metabolism of diazinon in susceptible (Yeerongpilly) or OP-resistant Biarra ticks. Their *in vitro* and *in vivo* results indicate that resistance is again due to the presence of relatively insensitive AChE and that the wider spectrum of resistance in this strain is due to the presence of a larger amount of a more insensitive enzyme than in the Ridgelands strain. Shaw, Cook & Carson (85) claim that there is less AChE than in the Ridgelands strain, but agree that the resistance mechanism is insensitive AChE and report that 70 per cent of the AChE in Biarra ticks is insensitive, compared with 40 per cent in Ridgelands ticks.

A comparison of the AChE activity in the Yeerongpilly, Ridgelands, Biarra and Mackay strains, together with the bimolecular rate constants for the inhibition of AChE from these strains with the oxygen analogues of coumaphos (coroxon) and diazinon (diazoxon) are shown in Table I. This shows that the level of AChE activity in the Mackay strain is about the same as in the Biarra strain but the sensitivity of AChE in the Mackay

TABLE I

THE ACTIVITY OF AChE IN LARVAL HOMOGENATES OF YEERONGPILLY (SUSCEPTIBLE), RIDGELANDS, BIARRA AND MACKAY OP-RESISTANT STRAINS OF *B. microplus* AND THE BIMOLECULAR RATE CONSTANT (*k*) FOR THE REACTION OF AChE IN THESE HOMOGENATES WITH THE OXYGEN ANALOGUES OF COUMAPHOS (COROXON) AND DIAZINON (DIAZOXON) (77).

Strain of *B. microplus*	AChE activity as percentage of Yeerongpilly strain	Bimolecular rate constant (*k*) for reaction between AChE and inhibitor (litre mole^{-1} min^{-1})	
		Coroxon as inhibitor	Diazoxon as inhibitor
Yeerongpilly	100[a]	2.6×10^5	9.2×10^4
Ridgelands	14	8.7×10^4	4.3×10^3
Biarra	30	6.6×10^2	8.3×10^2
Mackay	27	2.2×10^5	6.9×10^4

[a] 100% = 440 μmoles acetylthiocholine hydrolysed/g/hr.

strain is approximately equal to that in the susceptible Yeerongpilly strain (77). When Yeerongpilly, Biarra and Mackay larvae were examined 6 hours after treatment with a low concentration of coumaphos, approximately 30 times more coroxon was found in the Yeerongpilly and Biarra larvae than in the Mackay larvae. This indicated that either coroxon was being produced at a lower rate or else the coroxon was being detoxified at a faster rate in Mackay larvae. Higher levels of diethyl phosphate, a degradation product of coroxon, were found in the Mackay strain indicating that coroxon was metabolised at a faster rate and that this was the resistance mechanism (77).

In *B. microplus* and *T. urticae,* both an insensitive enzyme system and detoxification have been shown to be resistance mechanisms. Resistance in the Ridgelands and Biarra strains is similar to that in the OP-resistant Leverkusen strain of *T. urticae* in which resistance is associated with a low level of AChE in insensitive form (88, 106). On the other hand, resistance in the Mackay strain resembles that in the OP-resistant Blauvelt strain of *T. urticae* in that resistance is associated with a detoxifying mechanism (106). An insensitive enzyme system has not been identified as the cause of OP resistance in insects in which detoxification has been shown to be the usual cause (13).

GENETICS OF RESISTANCE

Investigation of the genetics of resistance in ticks poses many problems. The long life history delays selection procedures, and, combined with the relatively large number of chromosomes in most genera [*Argas* 2n = 26, *Ornithodoros* 2n = 12 – 32, *Ixodes* 2n = 23 – 28, *Boophilus, Hyalomma, Dermacentor, Amblyomma* and *Rhipicephalus* 2n = 21, 22 (62)] and the

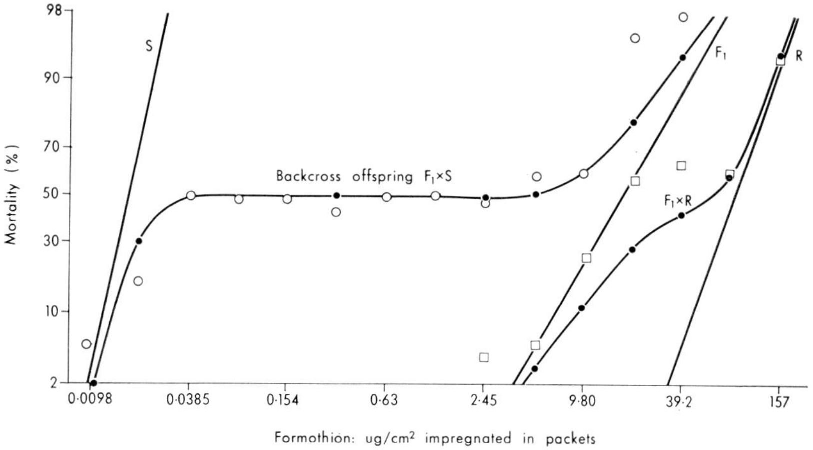

FIG. 3. Observed dosage-mortality relationships to Formothion (iso) for larvae of *B. microplus* obtained from the back-crosses F_1 X S (circles) and F_1 X R (squares), as compared to those expected for monofactorial inheritance (points joined by lines). From *Genetics of Insect Vectors of Disease* (95, 124) by permission of the editor.

necessity to maintain species such as *B. microplus* on stalled cattle, means that it would be a long, arduous and expensive procedure to develop standardised strains (124). Nevertheless, the mode of inheritance of resistance to DDT, dieldrin and to OP chemicals in *B. microplus* has been reasonably well established. Arsenic resistance has not lent itself to genetic studies because the low order of resistance involved (43) makes it difficult to characterise phenotypes.

Studies by Stone (93–95) which have been reviewed recently (15) showed that resistance in all cases is due to single or closely linked autosomal genes, with DDT resistance incompletely recessive, dieldrin resistance fully dominant and OP resistance in the Ridgelands strain incompletely dominant. Dosage-mortality tests to determine the mode of inheritance in the Ridgelands strain were conducted on engorged females and on unfed larvae. The results were in broad agreement, but the larval tests on the progeny of parental-susceptible and -resistant ticks, on F_1 and back-crosses to susceptible and resistant ticks using Formothion (iso) (>2000-fold resistance) provided the best evidence of monofactorial inheritance of OP resistance (Fig. 3). The degree of dominance estimated on the basis of Stone's formula (97) was 0.775. OP resistance in the Biarra strain in Australia is also incompletely dominant (98). The genetic relationship between the Ridgelands and Biarra OP-resistant strains is unknown. It was originally thought that Biarra resistance developed on a property where, several years earlier, Ridgelands resistance had been discovered (79, 85) but the initial tests would not have distinguished between the two strains and it is possible

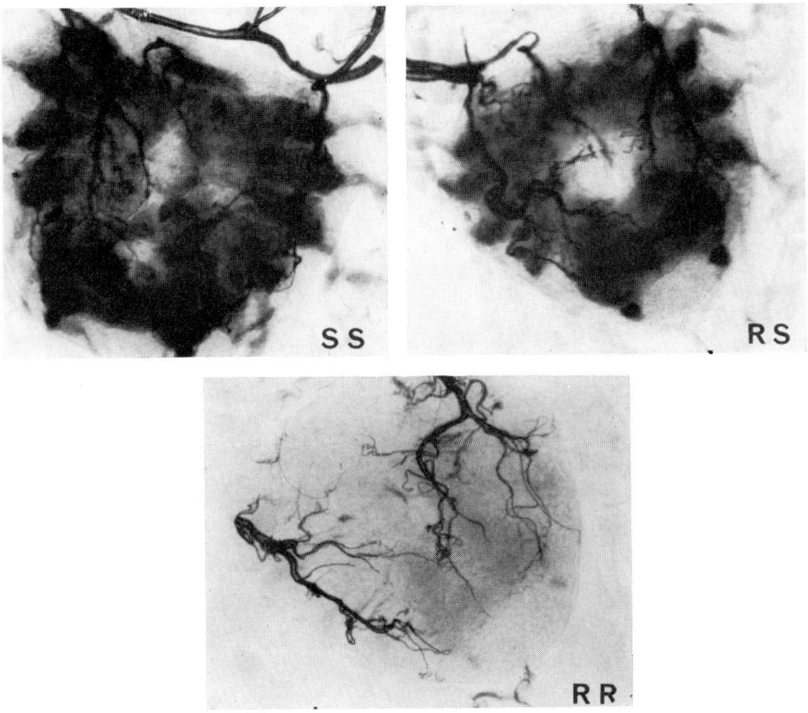

FIG. 4. Brains of homozygous OP-resistant Ridgelands, hybrid and homozygous-susceptible *B. microplus* after treatment by a histochemical method to indicate AChE activity (98).

that Biarra resistance arose directly from a susceptible population. No information is available regarding the inheritance of resistance in the OP-resistant Mackay strain.

Recognition of the fact that larval homogenates of ticks of the OP-resistant Ridgelands strain exhibited reduced cholinesterase activity (49, 81) led to an examination of brain AChE activities in adult ticks by histochemical and biochemical methods (96). Brains of homozygous-resistant ticks were found to have about 12 per cent of the AChE activity of the brains of susceptible ticks while brains of hybrid ticks had about 78 per cent of the activity of their susceptible parents. Thus, hybrids resembled the resistant parent in their response to chemicals, but resembled the susceptible parent in brain AChE activity (Fig. 4). This provided an opportunity for selection of a homozygous-resistant strain on the basis of low brain AChE activities in parent ticks. Subsequently, the inheritance of low brain AChE activity was shown to be controlled by one or more closely linked autosomal genes, and low brain AChE levels and resistance to OP chemicals to be controlled by the same or closely linked genes (96).

The relationships between genetic control of enzyme activity and chemical resistance in the Biarra and Mackay strains is being investigated by histochemical and biochemical methods by Stone (98). Homogenates of larval ticks of the Biarra strain exhibit higher AChE activity to acetylthiocholine than ticks of the Ridgelands strain (72), and brains from Biarra and susceptible ticks are less readily distinguished in their AChE activity toward acetylthiocholine. However, there is a differential brain esterase activity toward another substrate, indoxyl acetate, and the brains are further distinguished by the presence of inhibitor-insensitive AChE in Biarra ticks. Brains from Mackay ticks exhibit low AChE activity like those of Ridgelands ticks but differ in that their AChE is inhibitor-sensitive (98).

Apart from the indication that chlordane-resistant *R. sanguineus* were not cross-resistant to lindane (34, 40), resistance in ticks conforms to the usual spectrum of insecticide resistance. Resistance to DDT is distinct from resistance to the cyclodiene derivatives and γBHC, and resistance to OP chemicals is not related genetically to either DDT or cyclodiene resistance (93–95). Multiresistant strains are not uncommon. Thus, Whitehead (115) has described an arsenic/DDT/γBHC-resistant strain of *B. decoloratus* (larvae of this strain also exhibited an 18-fold resistance to pyrethrum); a strain of *B. microplus* resistant to DDT, γBHC and dieldrin was described by Stone & Webber (100); a significant proportion of the Biarra OP-resistant strain of *B. microplus* was found to be resistant to toxaphene (79), and ticks with OP resistance of the Ridgelands type have been shown to include a proportion resistant to DDT (108).

Biological comparisons between the resistant and susceptible strains have been meagre. The parasitic life cycles of DDT-resistant and of Biarra OP-resistant *B. microplus* are slightly longer than that of a susceptible strain (10, 98). The persistence of DDT- and dieldrin-resistant phenotypes in field populations of *B. microplus* many years after the use of these chemicals had been discontinued indicates that any biotic disadvantage related to resistance to these chemicals must be minor. Laboratory culture of DDT- and dieldrin-resistant strains in the absence of chemicals resulted in the reversion from nearly 100 per cent homozygous DDT-resistant to about 55 per cent in 13 generations (thereafter remaining stable for at least another 18 generations), and from about 55 per cent dieldrin-resistant to an undetectable percentage in 24 generations (93, 94).

Significance of Resistance in Relation to Control

The major impact of resistance in ticks has been in the cattle industries of southern Africa, South America and in Australia, where tick control is still mainly dependent on chemicals. The situation is at present most critical in Australia but a similar situation would appear to be inevitable in Africa and South America, and may be even more critical in those localities in Africa where resistance to toxaphene in *R. evertsi* and *R. appendiculatus* has developed. It is expected that OP resistance in *B. decoloratus* will spread rapidly because of the necessity, through drought, to move cattle from the original focus to other parts of the country (5). Toxaphene resistance in *R.*

evertsi and *R. appendiculatus* is also known to be widespread in cattle-raising districts of South Africa and to be an increasingly important problem in Rhodesia, Zambia, Kenya and Uganda (5). A wide variety of acaricides has been used on many properties in South Africa since DDT replaced arsenic for the control of *B. decoloratus* (7, 9). DDT was not effective against two- and three-host ticks and mixtures were introduced, e.g., DDT/arsenic, DDT/toxaphene and, more recently, dioxathion/toxaphene and dioxathion/γBHC mixtures. It has now become necessary to use OP chemicals to control resistant *Rhipicephalus* ticks along much of the eastern coastal belt of South Africa and a mixture of dioxathion/chlorfenvinphos has proved effective (5). A side effect of toxaphene resistance in *R. evertsi* has been an increased incidence of strikes by the screw-worm, *Chrysomya bezziana*, in some areas (6).

In Australia the early recognition of OP resistance in *B. microplus* provided an opportunity for the assessment and development of alternative acaricides. However, the spectrum of resistance in the Biarra and Mackay strains is so wide that it appears unlikely that any chemicals from the organophosphorus or carbamate groups will provide the standard of control achieved against OP-susceptible ticks. There is also the distinct possibility that strains combining insensitive AChE and a detoxification mechanism, and thus exhibiting enhanced resistance, will evolve. Chemicals with a different mode of action are needed. Chlorphenamidine offered prospects of a practical alternative (79), but has proved too unstable for use in dips and too toxic for use on cattle at the concentrations required to produce satisfactory control of *B. microplus*. However, when added at very low concentrations to existing OP-acaricides, a marked improvement in tick control is achieved (111). This chemical could prove to be the forerunner of a new group of acaricides. It has also been reported that some thiadiazolone compounds are effective against normal and resistant *Boophilus* (125).

Efforts were made to restrict the spread of the Ridgelands strain, mainly by attempting to ensure that cattle from infested properties were tick-free prior to movement. When the Biarra strain was recognised, rigid quarantine measures were imposed on infested and adjoining properties (57). This unique effort to contain a resistance problem has met with limited success —the Biarra strain is still confined to southeastern Queensland (Fig. 5) and the resistant tick appears to have been eliminated from many infested properties by dipping cattle for 6 to 9 months at 10- to 17-day intervals with Dursban or bromophos-ethyl. However, the number of known infested properties increased from 24 in September, 1966, to 437 in February, 1969 (33). The presence of free-roaming red deer *Cervus elaphus* which are known to be good hosts for *B. microplus* and from which resistant ticks have been collected (64), illustrates one of the many problems associated with the containment of the resistant strain. The reduction of tick numbers to very low levels also means that cattle must be inoculated against tick fever. Both *Babesia argentina* and *B. bigemina* are endemic in Queensland but calves

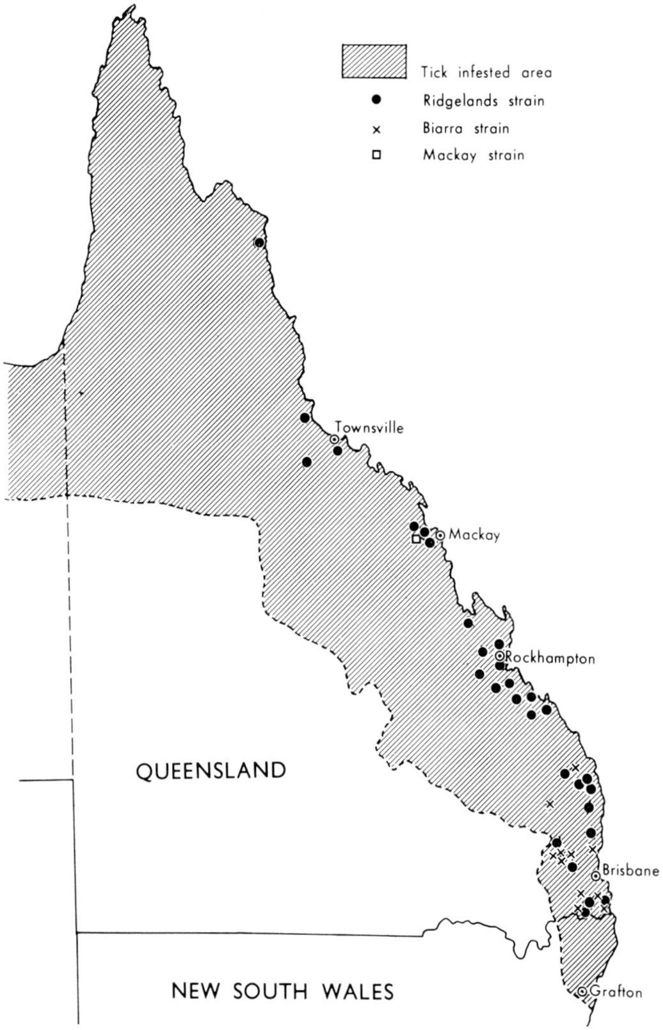

FIG. 5. Distribution of OP resistance in eastern Australia related to the distribution of *B. microplus.*

are normally born to immune mothers and are thus protected during their early life so that they rarely exhibit clinical symptoms when they become infected (71). A standardised *B. argentina* vaccine is provided for cattle owners who are subject to quarantine (18).

Although the measures to restrict the spread of OP resistance have been at least partially successful, it is possible that the high degree of selection pressure applied in infested areas could lead to the more rapid emergence of a strain with even greater resistance. On the other hand, experience in northern New South Wales (Fig. 5) where control of *B. microplus* is the responsibility of the government and where tick fever transmission does not occur (21), suggests that very efficient control measures (and thus high selection pressure) may have at least delayed the development of resistance. Arsenic resistance developed in New South Wales (43, 51) but resistance to DDT or to OP chemicals has not been recognised,[7] although resistance to both developed in ecologically similar areas in Queensland where control was less efficient and cattle owners normally prefer some ticks to be present to maintain tick fever immunity in their cattle.

Resistance is not universal, even in areas where OP resistance has been known to occur for several years (Fig. 5). Surveys in 1968 for OP resistance in the Rockhampton region, where Ridgelands resistance was found in 1963, and in northern Queensland, revealed that resistance was present on 13/23 and 3/20 properties, respectively (111). There is little doubt also that ticks can still be controlled by the more efficient use of existing OP acaricides (109). This raises a problem for the chemical industry in that the market for new chemicals which may be required to control a highly resistant strain will be limited, leaving little incentive to production (85).

The impact of resistance in South American countries has been similar to that in Australia but Biarra OP resistance has not yet been recognised (85), so that a number of organophosphorus compounds should still be useful. In Africa, the presence of two- and three-host ticks requires that chemical treatments be applied at more frequent intervals, usually weekly, at certain times of the year (9, 69) and it has been suggested that this may have delayed the development of resistance to OP chemicals in *B. decoloratus* (83). However, resistance to arsenic, DDT and γBHC/toxaphene was closely paralleled in South Africa and Australia (117), and, although *B. decoloratus* may receive more frequent chemical challenge than *B. microplus,* chemicals are usually employed at a lower concentration in South Africa than in Australia. The pressure for selection could therefore be similar under the two regimes.

The introduction of efficient chemicals for the control of ticks on livestock has undoubtedly reduced losses in productivity but there is an urgent need for the development and application of alternative methods of control which are less dependent on chemicals. Barnett (9) has summarised approaches based on alteration of the microhabitat, rotational grazing (pasture-spelling), grass burning and the use of resistant hosts. All require a more detailed knowledge of the ecology of ticks than is available for most species but this knowledge is essential if the advances which have been

[7] OP-resistance was detected in New South Wales in June, 1969 (see Footnote 6 above) Resistance has thus developed in New South Wales some six years later than in Queensland, in an area where tick populations are normally low and where acaricides are used less frequently than in areas more favourable for the tick.

achieved are to be maintained. It is also necessary to integrate these approaches with the efficient use of pastures and cattle management (38, 110). Cattle which are resistant to ticks offer the most attractive solution to the *B. microplus* problem. Although resistance is usually associated with Zebu (*Bos indicus*) cattle (28, 70, 110), Wilkinson (121) demonstrated that improved control followed selection of cattle for resistance in a dairy herd of European (*Bos taurus*) cattle. There is strong evidence that resistance to ticks is highly heritable (41, 112) and, although breeding for tick resistance within cattle of European origin would be a slow process, selection within cross-bred herds should lead to the relatively rapid development of highly resistant cattle. Earlier studies by Bonsma (12) drew attention to the greater resistance of Africander cattle to enzootic diseases such as heartwater. There are suggestions also that Zebu-type cattle are less susceptible to *Babesia* infection (28) and the obvious advantages that these cattle offer, apart from resistance to ticks, suggests that the trend toward their use will increase in tropical and subtropical areas where tick problems and tick resistance to chemicals are of greater significance.

The attention which has been focused in this review on ticks of veterinary importance and particularly on *B. microplus* reflects the dearth of information on resistance in ticks, especially in ticks of public health importance. It is unlikely, in fact, that resistance has developed in species other than those reviewed but it would be unwise to be complacent. There is an obvious need for the development and acceptance of standardised methods for testing ticks and for the documentation of the dosage-mortality responses of species such as *Argas persicus, Amblyomma americanum, Anocentor nitens, Dermacentor andersoni, D. albipictus, Ixodes ricinus, Rhipicephalus sanguineus* and *Haemaphysalis longicornis* [formerly *H. bispinosa* (47)].

ACKNOWLEDGEMENTS

We are grateful to Dr. B. F. Stone for permission to publish the photographs of *B. microplus* brains, and to him and other colleagues for unpublished information.

LITERATURE CITED

1. Aeschlimann, A., Morel, P. C. *Boophilus geigyi* n. sp. une nouvelle tique du bétail de l'Ouest Africain. *Acta Trop.*, **22,** 162–68 (1965)
2. Altman, R. M. *Rept. U.S. Armed Forces Pest Control Board, Wash.*, **31,** 7–8 (1958)
3. Anastos, G. The ticks, or *Ixodides,* of the U.S.S.R.—a review of the literature. *U.S. Public Health Serv. Publ.. 548,* 397 pp. (1957)
4. Ault, C. N. Investigaciones sobre las dificultades de combatir la garrapata *Boophilus microplus. Rev. Med. Vet., Buenos Aires,* **30,** 174–211, 254–97 (1948)
5. Baker, J. A. F. (Personal communication)
6. Baker, J. A. F., McHardy, W. M., Thorburn, J. A., Thompson, G. E. *Chrysomyia bezziana*—some observations on its occurrence and activity in the Eastern Cape Province. *J. S. African Vet. Med. Assoc.,* **39,** 3–11 (1968)
7. Baker, J. A. F., Shaw, R. D. Toxa-

phene and lindane resistance in *Rhipicephalus appendiculatus,* the brown ear tick of equatorial and southern Africa. *J. S. African Vet. Med. Assoc.,* **36,** 321–30 (1965)
8. Ballantyne, G. H., Harrison, R. A. Genetic and biochemical comparisons of organophosphate resistance between strains of spider mites (*Tetranychus* species: Acari). *Entomol. Exptl. Appl.,* **10,** 231–39 (1967)
9. Barnett, S. F. The control of ticks on livestock. *Food Agr. Organ. U.N. FAO Agr. Studies,* **54,** 115 pp. (1961)
10. Bennett, G. F. (Unpublished data, 1968)
11. Boero, J. J. La resistencia de la garrapata a los clorados. *Rev. Med. Vet., Buenos Aires,* **35,** 169–74 (1953)
12. Bonsma, J. C. Hereditary heartwater-resistant characters in cattle. *Farming S. Africa,* **19,** 71–96 (1944)
13. Brown, A. W. A. Insecticide resistance in arthropods. *World Health Organ. Monograph Serv.,* **38,** 240 pp. (1958)
14. Brown, A. W. A. Mechanisms of resistance against insecticides. *Ann. Rev. Entomol.,* **5,** 301–26 (1960)
15. Brown, A. W. A. Genetics of insecticide resistance in insect vectors. In *Genetics of Insect Vectors of Disease,* Chap. 17, 505–52. (Wright, J. W., Pal, R., Eds., Elsevier, Amsterdam, 794 pp., 1967)
16. Brown, A. W. A. Insecticide resistance comes of age. *Bull. Entomol. Soc. Am.,* **14,** 3–9 (1968)
17. Busvine, J. R., Srivastava, S. C. (Unpublished data, 1968)
18. Callow, L. L., Mellors, L. T. A new vaccine for *Babesia argentina* infection prepared in splenectomised calves. *Australian Vet. J.,* **42,** 464–65 (1966)
19. Chaudhuri, R. P., Naithani, R. C. Resistance to BHC in the cattle tick *Boophilus microplus* (Can.) in India. *Bull. Entomol. Res.,* **55,** 405–10 (1964)
20. Clark, A. G., Hitchcock, M., Smith, J. N. Metabolism of Gammexane in flies, ticks and locusts. *Nature,* **209,** 103 (1966)
21. Curnow, J. (Unpublished data, 1968)
22. Drummond, R. O., Ernst, S. E., Trevino, J. L., Graham, O. H. Insecticides for control of the cattle tick and the southern cattle tick on cattle. *J. Econ. Entomol.,* **61,** 467–70 (1968)
23. Drummond, R. O., Graham, O. H. Insecticide tests against the tropical horse tick. *Dermacentor nitens,* on horses. *J. Econ. Entomol.,* **57,** 549–53 (1964)
24. Du Toit, R., Graf, H., Bekker, P. M. Resistance to arsenic as displayed by the single host blue tick, *Boophilus decoloratus* (Koch) in a localised area of the Union of South Africa. *J. S. African Vet. Med. Assoc.,* **12,** 50–58 (1941)
25. Ellenberger, W. P., Chapin, R. M. Cattle-fever ticks and methods of eradication. *U.S. Dept. Agr., Farmers' Bull.* 1057 (1919, revised 1932)
26. Report of the First Session of the F.A.O. Working Party of Experts on Resistance of Pests to Pesticides, Rome, Italy, *Food Agr. Organ. U.N., FAO PL/1965/18.* (106 pp., mimeo., 1967)
27. Third meeting of the F.A.O./O.I.E. Expert Panel on Tick-Borne Diseases of Livestock, Hamilton, Montana, U.S.A. *Food Agr. Organ. U.N., FAO Rept. 1966/10.* (70 pp., mimeo., 1967)
28. Francis, J. Resistance of Zebu and other cattle to tick infestation and babesiosis with special reference to Australia; an historical review. *Brit. Vet. J.,* **122,** 301–7 (1966)
29. Freire, J. J. Arseno e cloro resistencia emprego do tiofosfato dietilparanitrofenila (parathion) na luta antigarrapato *Boophilus microplus* (Canestrini, 1888). *Bol. Dir. Prod. Anim. Rio de Janeiro,* **9,** 3–31 (1953)
30. Graham, O. H., Drummond, R. O. Laboratory screening of insecticides for the prevention of reproduction of *Boophilus* ticks. *J. Econ. Entomol.,* **57,** 335–39 (1964)
31. Graham, O. H., Harris, R. L. Recent developments in the control of some arthropods of public health and veterinary importance. Livestock insects. *Bull. Entomol. Soc. Am.,* **12,** 319–25 (1966)
32. Graham, O. H., Price, M. A. Some morphological variations in *Boophilus annulatus microplus* from

northern Mexico. *Ann. Entomol. Soc. Am.*, **59**, 450–52 (1966)

33. Grant, K. D. (Personal communication)
34. Hansens, E. J. Chlordane-resistant brown dog ticks and their control. *J. Econ. Entomol.*, **49**, 281–83 (1956)
35. Harington, J. S. Contents of cystine-cysteine, glutathione and total free sulphydryl in arsenic-resistant and sensitive strains of the blue tick *Boophilus decoloratus*. *Nature*, **184**, 1739–40 (1959)
36. Harington, J. S. A suggested role for copper in the arsenic resistance of the blue tick *Boophilus decoloratus* Koch. *J. S. African Vet. Med. Assoc.*, **32**, 373–79 (1961)
37. Harley, K. L. S. Studies on the survival of the nonparasitic stages of the cattle tick *Boophilus microplus* in three climatically dissimilar districts of north Queensland. *Australian J. Agr. Res.*, **17**, 387–410 (1966)
38. Harley, K. L. S., Wilkinson, P. R. A comparison of cattle tick control by "conventional" acaricidal treatment, planned dipping and pasture spelling. *Australian J. Agr. Res.*, **15**, 841–53 (1964)
39. Harris, R. L., Graham, O. H., McDuffie, W. C. Resistance of livestock insects to insecticides in the United States. *Agr., Vet. Chem.*, **6**, 78–81 (1965)
40. Hazeltine, W. Chemical resistance of the brown dog tick. *J. Econ. Entomol.*, **52**, 332–33 (1959)
41. Hewetson, R., Resistance of cattle to cattle tick, *Boophilus microplus*. II. The inheritance of resistance to experimental infestations. *Australian J. Agr. Res.*, **19**, 497–505 (1968)
42. Hitchcock, L. F. Resistance of cattle tick (*Boophilus microplus* (Canestrini)) to benezene hexachloride. *Australian J. Agr. Res.*, **4**, 360–64 (1953)
43. Hitchcock, L. F., Roulston, W. J. Arsenic resistance in a strain of cattle tick (*Boophilus microplus* (Canestrini)) from northern New South Wales. *Australian J. Agr. Res.*, **6**, 666–71 (1955)
44. Hitchcock, M., Smith, J. N. Detoxication mechanism in the tick *Boophilus decoloratus*. *Biochem. J.*, **87**, 34–35P (1963)
45. Hitchcock, M., Smith, J. N. Comparative detoxication. 13. Detoxication of aromatic acids in arachnids: arginine, glutamic acid and glutamine conjugations. *Biochem. J.*, **93**, 392–400 (1964)
46. Hoogstraal, H. African Ixodoidea I. Ticks of the Sudan. *U.S. Naval Med. Res. Unit No. 3, Cairo*. 1101 pp. (1956)
47. Hoogstraal, H., Roberts, F. H. S., Kohls, G. M., Tipton, V. J. Review of *Haemaphysalis (Kaiseriana) longicornis* Neumann (resurrected) of Australia, New Zealand, Japan, China, Korea and U.S.S.R. misidentified as *H. bispinosa* and *H. neumanni* Donitz and its parthenogentic and bisexual populations. *J. Parasitol.*, **54**, 1197–1213 (1968)
48. Kitaoka, S., Yajima, A. Comparison of effectiveness between pesticides against *Boophilus microplus* by topical application and spraying. *Natl. Inst. Animal Health. Quart., Tokyo*, **1**, 41–52 (1961)
49. Lee, R. M., Batham, P. The activity and organophosphate inhibition of cholinesterases from susceptible and resistant ticks. *Entomol. Exptl. Appl.*, **9**, 13–24 (1966)
50. Legg, J., Brooks, O. H., Joyner, C. N. A note on the appearance of a DDT-resistant cattle tick *Boophilus microplus* (Canes.) in Queensland. *Australian Vet., J.*, **31**, 148 (1955)
51. Legg, J., Shanahan, G. J. The appearance of an arsenic-resistant cattle tick (*Boophilus microplus*) in a small area of New South Wales. Its reaction to some of the newer insecticides. *Australian Vet. J.*, **30**, 95–99 (1954)
52. Little, D. A. The effect of cattle tick infestation on the growth rate of cattle. *Australian Vet. J.*, **39**, 6–10 (1963)
53. Mackerras, I. M., Waterhouse, D. F., Maiden, A. C. B., Edgar, G. The cattle tick problem in N.S.W. *Dept. Agr. N.S. Wales, Sci. Bull.*, **78**, 100 pp. (1961)
54. Msangi, A. S. Seminar on the ecology and control of ticks and mites of public health importance. Problems in the control of *Ornithodoros* vectors of human disease in Africa. *World Health Organ., U.N., WHO/VBC/68.57*, 175–81 (1968)
55. Neeman, M. WARF Antiresistant

for DDT. *Wisconsin Alum. Res. Found. Tech. Rept.*, 15 pp. (1960)
56. Neilson, C. L., Rich, G. B., Proceter, P. J. Live stock pests. *Bull. Dept. Agr. Brit. Columbia,* **66,** 4 (1966)
57. Newton, L. G. Acaricide resistance and cattle tick control. *Australian Vet. J.,* **43,** 389–94 (1967)
58. Norman, M. J. T. Grazing and feeding trials with beef cattle at Katherine, N. T. *Commonwealth Sci. Ind. Res. Organ., Australia, Div. Land Res. Reg. Surv. Tech. Paper,* **12,** 15 pp. (1960)
59. Norris, K. R., Stone, B. F. Toxaphene-resistant cattle ticks (*Boophilus microplus* (Canestrini)) occurring in Queensland. *Australian J. Agr. Res.,* **7,** 211–26 (1956)
60. O'Brien, R. D. Mode of action of insecticides. *Ann. Rev. Entomol.,* **11,** 369–402 (1966)
61. O'Brien, R. D. Resistance. In *Insecticides: Action and Metabolism.* Chap. 15, 231–52. (Academic Press, New York, London, 332 pp., 1967)
62. Oliver, J. A. Cytogenetics of acarines. In *Genetics of insect vectors of disease,* Chap. 13, 417–39. (Wright, J. W., Pal, R., Eds., Elsevier, Amsterdam, 794 pp., 1967)
63. Oppenoorth, F. J. Biochemical genetics of insecticide resistance. *Ann. Rev. Entomol.,* **10,** 185–206 (1965)
64. O'Sullivan, P. J. (Personal communication)
65. Pacheco Torres, M. F. Lucha contra ectoparasitos que afectan la ganadeira en Venezuela—comportamiento de los insecticidas utiligados. *Rev. Vet. Venezuela,* **21,** 202–23 (1966)
66. Peters, R. A. The biochemistry of some toxic agents. Part I. The present state of knowledge of biochemical lesions induced by trivalent arsenical poisoning. *Bull. Johns Hopkins Hosp.,* **97,** 1–20 (1955)
67. Pinotti, M. Recent findings of insect resistance and behaviouristic modifications of certain insects in South American areas. *Intern. Symp. Control Insect Vectors of Disease,* **1,** 184–200 (1954)
68. Privora, M., Rupes, V., Cerny, V. Toxicity of certain insecticides for the tick *Dermacentor marginatus. World Health Organ., WHO/VBC/67.48* (1967)
69. Purchase, H. S. Some thoughts on ticks and their practical control. Part I. *Bull. Epizoot. Diseases Africa,* **3,** 226–30 (1955)
70. Riek, R. F. Studies on the reactions of animals to infestation with ticks. VI. Resistance of cattle to infestation with the tick *Boophilus microplus* (Canestrini). *Australian J. Agr. Res.,* **13,** 532–50 (1962)
71. Riek, R. F. The cattle tick and tick fever. *Australian Vet. J.,* **41,** 211–15 (1965)
72. Roulston, W. J., Schnitzerling, H. J., Schuntner, C. A. Acetylcholinesterase insensitivity in the Biarra strain of the cattle tick *Boophilus microplus* as a cause of resistance to organophosphorus and carbamate acaricides. *Australian J. Biol. Sci.,* **21,** 759–67 (1968)
73. Roulston, W. J., Schnitzerling, H. J., Schuntner, C. A. (Unpublished data, 1968)
74. Roulston, W. J., Schuntner, C. A. Depletion of gamma isomer from benzene hexachloride cattle dipping fluids. *Australian J. Agr. Res.,* **9,** 402–20 (1958)
75. Roulston, W. J., Schuntner, C. A. Sulphydryl content of the embryos of the Australian cattle tick. *Nature,* **186,** 1069–70 (1960)
76. Roulston, W. J., Schuntner, C. A., Schnitzerling, H. J. Metabolism of coumaphos in larvae of the cattle tick (*Boophilus microplus*). *Australian J. Biol. Sci.,* **19,** 619–33 (1966)
77. Roulston, W. J., Schuntner, C. A., Schnitzerling, H. J., Wilson, J. T. Detoxification as a resistance mechanism in a strain of *Boophilus microplus* resistant to organophosphorus and carbamate compounds. *Australian J. Biol. Sci.* (In press, 1969)
78. Roulston, W. J., Stone, B. F., Wilson, J. T., White, L. I. Chemical control of an organophosphorus- and carbamate-resistant strain of *Boophilus microplus* (Can.) from Queensland. *Bull. Entomol. Res.,* **58,** 379–92 (1968)
79. Roulston, W. J., Wharton, R. H. Acaricide tests on the Biarra strain of organophosphorus-resistant cattle ticks *Boophilus microplus* from southern Queensland. *Australian Vet. J.,* **43,** 129–34 (1967)
80. Schnitzerling, H. J., Roulston, W. J.,

Schuntner, C. A. The absorption and metabolism of ^{14}C DDT in DDT-resistant and susceptible strains of the cattle tick *Boophilus microplus. Australian J. Biol. Sci.* (In press, 1969)

81. Schuntner, C. A., Roulston, W. J., Schnitzerling, H. J. A mechanism of resistance to organophosphorus acaricides in a strain of the cattle tick *Boophilus microplus. Australian J. Biol. Sci.,* **21,** 97–109 (1968)
82. Seddon, H. R. Diseases of domestic animals in Australia. Part 3. Tick and mite infestations. *Serv. Publs Dept. Health Australian Vet. Hyg.,* **7,** 200 pp. (1951)
83. Shaw, R. D. Culture of an organophosphorus-resistant strain of *Boophilus microplus* (Can.) and an assessment of its resistance spectrum. *Bull. Entomol. Res.,* **56,** 389-405 (1966)
84. Shaw, R. D. (Personal communication)
85. Shaw, R. D., Cook, M., Carson, R. E. Developments in the resistance status of the southern cattle tick to organophosphorus and carbamate insecticides. *J. Econ. Entomol.,* **61,** 1590–94 (1968)
86. Shaw, R. D., Malcolm, H. A. Resistance of *Boophilus microplus* to organophosphorus insecticides. *Vet. Record,* **76,** 210–11 (1964)
87. Shaw, R. D., Thompson, G. E., Baker, J. A. F. Resistance to cholinesterase-inhibitors in the blue tick, *Boophilus decoloratus,* in South Africa. *Vet. Record,* **81,** 548–49 (1967)
88. Smissaert, H. R. Cholinesterase inhibition in spider mites susceptible and resistant to organophosphate. *Science,* **143,** 129–31 (1964)
89. Smith, J. N. Detoxication mechanisms. *Ann. Rev. Entomol.,* **7,** 465–80 (1962)
90. Stocken, L. A., Thompson, R. H. S. British Anti-Lewisite 2. Dithiol compounds as antidotes for arsenic. *Biochem. J.,* **40,** 535–48 (1946)
91. Stone, B. F. Resistance to DDT in the cattle tick, *Boophilus microplus* (Canestrini). *Australian J. Agr. Res.,* **8,** 424–31 (1957)
92. Stone, B. F. The DDT-susceptibility of fourteen strains of the cattle tick, *Boophilus microplus,* from northern New South Wales. *Australian Vet. J.,* **36,** 343–47 (1960)
93. Stone, B. F. The inheritance of DDT-resistance in the cattle tick, *Boophilus microplus. Australian J. Agr. Res.,* **13,** 984–1007 (1962)
94. Stone, B. F. The inheritance of dieldrin-resistance in the cattle tick, *Boophilus microplus. Australian J. Agr. Res.,* **13,** 1008–22 (1962)
95. Stone, B. F. Inheritance of resistance to organophosphorus acaricides in the cattle tick, *Boophilus microplus. Australian J. Biol. Sci.,* **21,** 309–19 (1968)
96. Stone, B. F. Brain cholinesterase activity and its inheritance in cattle tick (*Boophilus microplus*) strains resistant and susceptible to organophosphorus acaricides. *Australian J. Biol. Sci.,* **21,** 321–30 (1968)
97. Stone, B. F. A formula for determining degree of dominance in cases of monofactorial inheritance of resistance to chemicals. *Bull. World Health Organ.,* **38,** 325–26 (1968)
98. Stone, B. F. (Unpublished data, 1969)
99. Stone, B. F., Haydock, K. P. A method for measuring the acaricide susceptibility of the cattle tick *Boophilus microplus* (Can.). *Bull. Entomol. Res.,* **53,** 563–78 (1962)
100. Stone, B. F., Webber, L. G. Cattle ticks, *Boophilus microplus,* resistant to DDT, BHC and Dieldrin. *Australian J. Agr. Res.,* **11,** 105–19 (1960)
101. Thompson, M. E., Johnston, A. M. Total sulphydryl content of embryos of arsenic-resistant and sensitive strains of the blue tick *Boophilus decoloratus. Nature,* **181,** 647–48 (1958)
102. Uilenberg, G. Résistance à l'hexachlorocyclohexane d'une souche de la tique *Boophilus microplus* (Canestrini) à Madagascar—Essais préliminaires sur sa sensibilité à quelques autres ixodicides. *Rev. Élev. Méd. Vét. Pays Trop.,* **16,** 137–50 (1963)
103. Report of cooperative cattle fever tick eradication activities. *U.S. Dept. Agr., ARS.* (Fiscal year, 1962)
104. Suggested guide for the use of insecticides to control insects affecting crops, livestock, households,

stored products, forests and forest products. *U.S. Dept. Agr., ARS. Agr. Handbook,* 331, (U.S. Govt. Printing Office, 273 pp., 1968)

105. Uspenskij, I. V. On the susceptibility of *Ixodes persulcatus* P. Sch. ticks to DDT and other pesticides in west Sajan, U.S.S.R. *World Health Organ., WHO/VBC/67.74* (1967)
106. Voss, G., Matsumura, F. Resistance to organophosphorus compounds in the two-spotted spider mite: two different mechanisms of resistance. *Nature,* **202,** 319–20 (1964)
107. Ward, N. R. PCOs and Veterinarians must work together to control pet ectoparasites. *Pest Control,* **28,** 14, 32 (1960)
108. Webber, L. G. (Personal communication)
109. Wharton, R. H. Acaricide resistance and cattle tick control. *Australian Vet. J.,* **43,** 394–99 (1967)
110. Wharton, R. H., Harley, K. L. S., Wilkinson, P. R., Utech, K. B. W., Kelley, B. M. A comparison of cattle tick control by pasture spelling, planned dipping and tick-resistant cattle. *Australian J. Agr., Res.,* **20,** 783–97 (1969)
111. Wharton, R. H., Roulston, W. J. (Unpublished data, 1968)
112. Wharton, R. H., Utech, K. B. W., Turner, H. G. Resistance to the cattle tick *Boophilus microplus* in a herd of Australian Illawarra shorthorn cattle—its assessment and heritability. *Australian J. Agr. Res.* (In press, 1969)
113. Whitehead, G. B. DDT resistance in the blue tick, *Boophilus decoloratus,* Koch. *J. S. Africa Vet. Med. Assoc.,* **27,** 117–20 (1956)
114. Whitehead, G. B. The development and mechanism of insecticide resistance in the blue tick, *Boophilus decoloratus,* Koch. *J. S. Africa Vet. Med. Assoc.,* **30,** 221–34 (1959)
115. Whitehead, G. B. Pyrethrum resistance conferred by resistance to DDT in the blue tick. *Nature,* **184,** 378–79 (1959)
116. Whitehead, G. B. Investigation of the mechanism of resistance to sodium arsenite in the blue tick *Boophilus decoloratus* Koch. *J. Insect Physiol.,* **7,** 177–85 (1961)
117. Whitehead, G. B. Resistance in the Acarina: Ticks. *Advan. Acarol.,* **2,** 53–70 (1965)
118. Whitehead, G. B., Baker, J. F. A. Acaricide resistance in the red tick *Rhipicephalus evertsi* Neumann. *Bull. Entomol. Res.,* **51,** 755–64 (1960)
119. Whitnall, A. B. M., Bradford, B. An arsenic-resistant tick and its control with gammexane dips. *Bull. Entomol. Res.,* **38,** 353–72 (1947)
120. Whitnall, A. B. M., Thorburn, J. A., McHardy, W. M., Whitehead, G. B., Meerholz, F. A BHC-resistant tick. *Bull. Entomol. Res.,* **43,** 51–65 (1952)
121. Wilkinson, P. R. Selection of cattle for tick resistance, and the effect of herds of different susceptibility on *Boophilus* populations. *Australian J. Agr. Res.,* **13,** 974–83 (1962)
122. Wood, J. C. Recent developments in the control of external parasites. *Vet. Record,* **78,** (22) 1–4 Clin. Suppl. (1966)
123. World Health Organization. Insecticide Resistance and Vector Control. Thirteenth Report of the W.H.O. Expert Committee on Insecticides. *World Health Organ., Tech. Rept. Ser.,* **265,** 227 pp. (1963)
124. Wright, J. W., Pal, R. Standardized strains of insects of public health importance. *Genetics of Insect Vectors of Disease.* Appendix B, 735–58. (Elsevier, Amsterdam, 794 pp., 1967)
125. Anon. Veterinary pesticides. *Nature,* **222,** 418 (1969)

MYCOPLASMA AND PHYTARBOVIRUSES AS PLANT PATHOGENS PERSISTENTLY TRANSMITTED BY INSECTS[1]

BY R. F. WHITCOMB[2] AND R. E. DAVIS[3]

[2]*Entomology Research Division and* [3]*Plant Virology Laboratory, Crops Research Division, Agricultural Research Service, U. S. Department of Agriculture, Beltsville, Maryland*

INTRODUCTION

The taxonomic position of vectors has long been a major criterion for the classification of plant viruses transmitted by insects. Thus, many reviews have dealt separately with transmission by aphids or with transmission by leafhoppers, even though the nature of the agents and the mechanisms of their transmission overlapped. Transmission mechanisms have also often been used to categorize agents transmitted by insects. Persistent transmission, however, is the most common mechanism, and its variations are correspondingly both so numerous and so wide that it is not generally suitable as a major category of disease agent classification. The need for a classification based on the nature of the agents themselves is clear, and the feasibility of such a scheme has become increasingly apparent as our knowledge of the pathogens has become more refined.

Only a few years ago, a small handful of plant disease agents transmitted persistently by their vectors had been purified, isolated, and characterized. Now, however, a sufficiently large number of plant viruses (177) have been characterized to make two outstanding features evident. There is a wide range of diversity in structure of these viruses and an obvious analogy between many of them and those viruses not pathogenic to plants. Viruses with totally different ecologies may be virtually indistinguishable in shape and size, symmetry of capsid and number of capsomeres, and type, strandedness, and molecular weight of nucleic acid. Although it is not immediately apparent whether such similarities indicate evolutionary relationship or simply reflect laws governing virus assembly, the potential value of classifying plant disease agents in a scheme recognizing their similarities with viruses of other hosts is clear. Classifications of animal viruses (7, 284, 315, 316), tentative as they may have been, have greatly facilitated the progress of research in animal virology and have enabled workers to predict the possible utility of various experimental approaches. The recently discovered

[1] Mention of specific equipment, trade products, or a commercial company does not constitute its endorsement by the U. S. Department of Agriculture over similar products or companies not named.

novel etiologies of potato spindle tuber disease (117, 118, 362) and scrapie disease (141, 227) demonstrate that not all pathogens necessarily belong to classes already described. Yet, although classification of viruses is still, as taxonomic sciences go, in its infancy, we have already learned that a given unknown biological entity is likely to be related in some way to others already characterized. The experience of the entire field of virology thus must be relied upon in the establishment of taxonomic criteria. For this reason, grouping of the insect-borne plant disease agents according to their nature, taking into account broad analogies with other pathogens, seems long overdue.

In this review, we begin the task of grouping the insect-borne plant viruses by their structures. At present, virion size, symmetry of the capsid, number of capsomeres, the presence or absence of an envelope around the nucleocapsid, and the nature and the strandedness of nucleic acid (269) constitute the most useful characters for major groupings of viruses, while the ultimate molecular taxonomy, based on descriptions of the viral genome, requires elaborate technology and must be considered as a goal of the future. Initial examination of unknown insect-borne plant pathogens, however, should be broader in scope than the assumption that they must be viruses. The leafhopper-borne yellows "viruses" were belatedly recognized (124) as probable mycoplasmalike organisms, illustrating that the criteria previously applied to these disease agents were not taxonomic. "Viruslike" properties such as transmission by leafhoppers, graft transmission, and "viruslike" symptoms induced in plants obviously are not necessarily indicative of viral nature of a pathogen. The demonstration by Yarwood et al. (508) that the beet latent virus (430, 431) was actually a bacterium provides a fairly recent example in plant pathology illustrating that "viruslike" biological properties are not good taxonomic criteria. In the absence of direct information on the nature of a pathogen, tentative placement of an agent in broad groupings may be achieved by determining stability to heat, phenol, formaldehyde, or lipid solvents, intracellular site of subunit synthesis and virion assembly, the length of the single-step intracellular replication cycle, filterability through membranes with controlled pore sizes, sedimentation rate of infectious entities, and the effect of antibiotic substances upon the replicative process.

Phytarboviruses.—Plant pathogenic viruses borne persistently by insects constitute a major category of disease agents for which we propose the term "phytarboviruses." The term phytarbovirus is descriptive of biological relationships and is meant in a broad sense to include all plant pathogenic viruses circulative in their arthropod vectors, whether or not their multiplication in arthropods has been demonstrated. It is hoped the term will draw attention to possible morphological counterparts (Table I) of these plant pathogenic viruses among vertebrate pathogens. Yet, we stress that phytarbovirus is proposed as a descriptive designation and not as a concept in a natural system of virus classification.

TABLE I.

Some Analogies Between Phytarboviruses and Certain Viruses Affecting Other Hosts

Viral Type	Phytarbovirus	Arbovirus	Insect virus	Plant virus	Animal virus	Bacterial virus
Diplornavirus	wound tumor rice dwarf maize rough dwarf	Colorado tick fever bluetongue EDIM virus	cytoplasmic polyhedrosis	—	reovirus	—
Rhabdovirus	potato yellow dwarf lettuce necrotic yellows wheat striate mosaic sowthistle yellow vein rice transitory yellowing northern cereal mosaic	vesicular stomatitis Flanders virus Kern Canyon virus	sigma virus	—	rabies	—
Picornavirus	barley yellow dwarf potato leafroll pea enation oat blue dwarf	—	bee sacbrood cricket paralysis silkworm flaccidity European foul brood	turnip yellow mosaic cucumber mosaic brome mosaic many others	poliovirus echovirus coxsackievirus	f 2 R 17 MS 2 others
Small DNA viruses	—	—	*Galleria* densonucleosis	—	—	ϕx fd
Group A, Group B virus	—	Eastern equine encephalomyelitis Western equine encephalomyelitis yellow fever	—	—	bat salivary gland virus	—
Poxvirus	—	myxoma is transmitted by a stylet-borne mechanism	vagoiavirus	—	vaccinia	—
None of above	tomato spotted wilt hoja blanca carrot mottle	many	many	many	many	many

—: No known representative.

Possible relationships with animal viruses.—The possibility of relationships between the viruses of hosts of different kingdoms has been raised often in recent years. The most frequent foci of comparison have been the reoviruses (379, 396, 436) and the rhabdoviruses (316, 413). Some of the details of these two groups are discussed in separate sections.

Other animal viruses such as poxviruses (240) may have counterparts in the vagoiaviruses (489), e.g., the virus of *Amsacta* described by Roberts & Granados (369). Best (34) called attention to a superficial similarity between the tomato spotted wilt virus (TSWV) and a myxovirus infecting terns (18), although the necessary evidence for a helical nucleocapsid of TSWV is apparently not available. Myxoviruses and paramyxoviruses may be highly pleomorphic and, in fact, might even be mistaken for organisms like *Mycoplasma gallisepticum,* which has surface projections similar to those of Newcastle disease virus (94). The internal detail of such agents is required to distinguish them.

No viruses of the plant kingdom have been found which correspond to the large number of group A or B arboviruses. Since these animal viruses are usually persistently transmitted by their arthropod vectors, it may be worthwhile to note a few of their properties. Ultrathin sections of the group A viruses reveal particles 400 to 480 Å across, whereas group B particles are that size or smaller. The virions consist of a dense nucleoid [which may differ (414) somewhat with each virus], that is surrounded by a membrane. A high lipid content is accounted for by the mode of elaboration of the virion at the cell membrane, in which the triple-layered unit membrane envelopes the nucleoid (2) and projections are added somehow on the exterior. The 4 to 6 per cent RNA of the particles is single-stranded and can be extracted in infectious form by conventional methods.

Other animal viruses transmitted by mosquitoes or ticks may have widely different properties. Vesicular stomatitis virus, for example, the prototype rhabdovirus (316), multiplies in and is apparently vectored by mosquitoes (32, 331), as is the similar Flanders virus (82). Other mosquito-borne animal viruses such as Colorado tick fever virus (328) may resemble the reoviruses. These authors (328) also discuss the resemblance of the viruses of bluetongue, African horse sickness, rift valley fever, and epizootic diarrhea of infant mice to reoviruses. Plant pathogens, as well as vertebrate pathogens, may be found in this group of double-stranded RNA viruses for which the name diplornavirus has recently been proposed (480). Finally, there are a number of ungrouped arboviruses which may have morphologies unlike any of the other arboviruses (381).

The picornaviruses form a large group (poliovirus, echovirus, coxsackievirus, rhinovirus) of animal viruses. They are 270 to 300 Å across, are usually stable to heat or lipid solvents, have 32 capsomeres, and contain single-stranded RNA which often has a molecular weight of about 2×10^6 daltons. Many plant viruses, such as turnip yellow mosaic, and possibly certain of the viruses discussed in this review might satisfy this description (242) and would certainly be termed picornaviruses if they infected animal hosts. Certain insect viruses also may be rather typical small RNA viruses (469).

In no case has structural analogy led to the demonstration of serological relationship. For example, careful efforts (167) have failed to detect a relationship between any of the serotypes of reovirus and the plant virus causing clover wound tumor, whose structure is nearly identical. Nucleic acid homology experiments may some day answer the obvious evolutionary question of relationship between these and other diplornaviruses. Until such time, although analogies between newly purified viruses and known viruses will no doubt continue to be inferred, it would be advisable to be somewhat cautious. Such analogies have seemed, at times, to have relied heavily on the imagination, such as a report (197) suggesting that the fine structure of a nuclear polyhedrosis virus resembled that of lettuce necrotic yellows virus. This suggested resemblance overlooks the likelihood that the two particles in question would be expected to contain DNA and RNA, respectively.

NATURE OF THE VIRAL PATHOGENS

Large Polyhedral Viruses (Diplornaviruses)

General.—A number of leafhopper- or planthopper-borne viruses have polyhedral or spherical morphologies with diameters of 60–70 mμ. Two of these viruses, wound tumor virus and rice dwarf virus, contain double-stranded RNA. Several other plant viruses, whose nucleic acid has not been characterized, probably also belong to this group; the viruses of Fiji disease of sugarcane (464), maize rough dwarf mosaic (481–483), and rice black-streaked dwarf (252, 253).

These large polyhedral plant viruses share many characteristics with a number of other viruses including the reoviruses (379, 396, 436) which infect a wide array of mammalian species, the cytoplasmic polyhedrosis viruses of insects (10, 435), and several arboviruses (64, 136, 328, 450, 480). The name diplornavirus has recently been proposed for these viruses (480).

Reoviruses.—By far the best studied members of the diplornavirus group are the reoviruses (379, 396, 436). Reovirus virions are polyhedra measuring 55 to 77 mμ, depending on the suspending media and method of examination (395, 436). The particles are resistant to lipid solvents. They have an inner core of about 40 mμ and an inner protein layer about 5 mμ wide (477) which may be the locus of the RNA polymerase activity in purified virions (63, 396). The outer layer of 92 capsomeres is assembled according to rules of icosahedral symmetry (477). The particles contain 14.6 per cent RNA, with a calculated molecular weight of 10.2×10^6 daltons (182). Much of the RNA is double-stranded, but single-stranded RNA, comprising 25 per cent of total viral RNA, is also present (395). Electron microscopy of the reovirus RNA showed three modal classes of double-stranded lengths, which were interpreted as stable subunits (132). Similar fragments (14–15S, 12S, and 10.5S) isolated from infected cells did not hybridize, indicating that they are independent and perhaps functional segments of the genome (20, 21, 359, 486, 487).

The eclipse period of reoviruses in tissue culture cells is 6 hr (183). Degradation of particles, after their penetration by viropexis, may be intermediated by cellular enzymes (108), since cores devoid of capsomere protein were envisioned. This process can presumably be simulated *in vitro* by such enzymes as chymotrypsin, with resulting enhancement of infectivity (434). Reovirus-infected cells contain cytoplasmic matrices with microtubules (410, 433). So far as is known, they do not multiply in any invertebrates, or in cultured invertebrate cells (447).

Cytoplasmic polyhedroses.—The cytoplasmic polyhedroses of insects (10, 435) are a large group of viruses with icosahedral symmetry (222) about 50 to 70 mμ across, with an inner core 35 mμ in diameter. The outer diameter does not take into account projections, 17 to 30 mμ in length, located at the vertices of the polyhedral particles. The particles are stable to deoxycholate. The particles contain 25 to 28 per cent RNA (203) which is double-stranded (204, 322). The RNA has been extracted in infectious form (245), in contrast to failures with reovirus. The most notable feature of infected cells is the polyhedra, which vary in size and shape, but whose crystalline bodies may enclose as many as 10,000 virions (10). Virions from such polyhedra, and free virions as well, can be isolated by conventional techniques such as density gradient centrifugation (202). Abnormal cytoplasmic features of the infected cell include microtubules, virogenic stroma containing largely incomplete particles, and crystalline matrices containing complete virions (10). This evidence suggests that assembly of capsomeres into capsids precedes the final events in assembly of complete particles. Latency and transovarial passage are known for some of the cytoplasmic polyhedrosis viruses.

Wound tumor virus (WTV).—This virus (51) is about 60 mμ in diameter and is transmitted by several leafhopper species of the subfamily Agalliinae. The virions have 92 capsomeres, each 7.5 mμ in diameter, with smaller units measuring 2.5 mμ (38). As with reoviruses, there is an inner layer of protein between the core and the capsomeres (448). The virions contain about 20 per cent double-stranded RNA. Double-strandedness of the RNA was demonstrated by electron microscopy (256), by base pairing studies (56), and by X-ray diffraction (466). RNA isolated by the phenol method was fragmented (256). Whether the fragments have biological significance, as in the reoviruses (20, 21, 486, 487), is not known. The virions require divalent cations and amino acids for maximum stability (51).

In plants, WTV induces enations, veinal hyperplasia, and tumors at wound sites on stems and roots. Mechanical transmission of the virus to plants may be achieved by pinpricking infective suspensions into the crown of crimson clover (68). In insect vectors, the minimum latent period before transmission is eight days (293). The virus is passed transovarially (47). In the insect vector, the virus induces certain pathological changes, especially in nerve cells (213). WTV was the first plant virus to be grown in invertebrate tissue culture (92), in which it can be assayed by a fluorescent antibody cell-counting method (88, 89, 92). Multiplication of the virus in both plant and insect hosts is accompanied by the appearance of tubular structures (213) and viroplasmic loci (404).

Rice dwarf virus. (RDV).—Rice dwarf virus (155–158) is borne by two deltocephaline leafhopper species. It is often passed transovarially to progeny (155, 335, 336), and continued passage of this virus to succeeding generations provided the first evidence that plant viruses could multiply in their insect vectors (159). The particles, 70 mμ in diameter (163, 164), are surrounded by a membranous coat (399) which can be removed by snake venom phospholipase without destroying the infectivity of the particle (467). Purification of RDV has been reviewed by Suzuki (453). Base pairing studies (323) and X-ray diffraction (387) indicate that the RNA (11 per cent of the virion by weight) is double-stranded. Recently, Kimura & Shikata (248) showed the icosahedral particles to have 32 capsomeres, each composed of 5 or 6 structural subunits. The presence of 32 capsomeres contrasts with the 92 reported for WTV and reoviruses, but is the same as that of bluetongue virus (480). RDV has a lower temperature optimum (18° C) in insects than does WTV (27° C), but the latent periods in the vectors are similar. A detailed comparison of the two viruses has recently been presented (406). Rice dwarf virus systemically invades the vector (162, 399, 406) and induces pathological effects in the mycetome and fat body (334) but is not known to be oncogenic.

Maize rough dwarf virus (MRDV).—The size of MRDV (169, 482) is similar to that of RDV (60 to 70 mμ), but the nucleic acid composition has not been determined. Like WTV, MRDV induces veinal hyperplasia in plant hosts. The virus is transmitted by several delphacid planthoppers (196). Like WTV and RDV, MRDV can be detected by electron microscopy in various organs of its vectors (170, 481–483) and its multiplication is accompanied by the appearance of viroplasmic matrices and microtubules. Vidano (483) has recently outlined a series of steps, beginning with stripping of the capsid from virions in lysosomelike structures (phagocytic vesicles) that is accompanied by production of granular microtubules as a necessary intermediate step before assembly of virions in fibrillar viroplasmic matrices. As with some other viruses (10, 242), the suggestion was made that "empty" virions residing largely in the fibrillar viroplasm represent immature particles awaiting completion by addition or maturation of core material.

Bacilliform Viruses (Rhabdoviruses)

A number of plant viruses—as well as certain viruses of *Drosophila*, fish, and other animals—resemble, in their bacilliform morphology, rabies or vesicular stomatitis (413). The sigma virus of *Drosophila* (389) is of special interest because it was first considered a cytoplasmically transmitted genetic factor. The virus has deleterious effects on its host (388), and many aspects of the infection have been studied (357, 388, 389). In addition to those discussed, plant pathogenic rhabdoviruses include a *Plantago* virus (221), *Gomphrena* virus (255), transitory yellowing virus of rice (90, 91, 400), and Japanese northern cereal mosaic virus (283, 401). Other animal viruses include Flanders virus (329) and Kern Canyon virus (330). The name rhabdovirus has been proposed for this interesting group of viruses (316).

Vesicular stomatitis virus (VSV).—VSV is a bacilliform virus about 180 mμ long and 60 mμ in diameter. Frequently, one end of the particle appears rounded while the other end is flat, giving it a "bullet shape" (223). Shorter particles have been observed, particularly under conditions of high multiplicity (193, 225). These shorter particles can produce a homologous interference (101) mediated by the RNA (226). VSV virions are sensitive to treatments directed against lipids such as ether, deoxycholate, and phospholipase (413). The virus particles may mature at the membranes of cytoplasmic vacuoles. Host cell components may be incorporated into the membrane (79). The outer membrane of the virion characteristically possesses projections about 100 Å in length. Simpson & Hauser (413) concluded that the internal structure of the particles consisted of a single-stranded helical array, 50 mμ in diameter, with a smaller helix 15 to 18 mμ within. Nakai & Howatson (333) reported the dimensions of the subunits of the helix to be 90 $\times$ 30 $\times$ 30 Å and concluded they were held together by a single-stranded thread of RNA. The helix consisted of 30 coils 49 mμ in width, with 4 coils of diminishing width forming a hemispherical cap. The molecular weight of the RNA was calculated to be 3.5 $\times$ 10^6 daltons for the complete particles and 1.1 $\times$ 10^6 daltons for the truncated particles. Wagner et al. (485) found an array of proteins which were produced by degradation of virions, the proportions of which were the same for the truncated and complete particles. Infectious RNA has not been obtained from VSV or any other rhabdovirus. In the mosquito (331) the virus invades many tissues and multiplies to high titer (32). The transmission of VSV to infant mice was independent of total titer in the mosquitoes over a wide range of virus concentrations (32).

Potato yellow dwarf (PYDV).—There are two major strains of PYDV in eastern North America, one adapted to *Agallia constricta* as a principal vector, and another adapted to *Aceratagallia sanguinolenta* (43). PYDV is bacilliform, with dimensions of 380 $\times$ 75 mμ (288). Upon extraction in conventional aqueous buffers, this characteristic morphology is apparently lost (287). Prominent surface projections are visible in negatively stained preparations (57). The virions have at least 20 per cent lipid (3) and are stabilized by divalent cations and amino acids (67). The virus can be mechanically transmitted to leaves of *Nicotiana rustica,* in which the induction of local lesions is followed by systemic invasion of the vascular tissues and subsequent invasion of parenchymal tissue (39). PYDV is transovarially passed to progeny of several of its vectors (47). Virus and antigen are in relatively low concentration in the vector, and no adverse effect of PYDV infection has been demonstrated in the vector despite apparent systemic invasion (418). The minimum latent period in the vector is about four or five days (50).

Wheat striate mosaic virus (WSMV).—WSMV (427) has a minimum latent period of three to five days in its deltocephaline leafhopper vector (274, 428). Partial purification (274) was achieved in a stabilizing buffer containing mannitol, EDTA, $MgCl_2$, and bovine serum albumin. The particles, 260 mμ × 80 mμ, had a tendency to fragment into two pieces, 170 mμ × 80 mμ and 90 mμ × 80 mμ. Whether these fragments are of biological significance is not known. WSMV was visualized in cytoplasm, between nuclear membranes, and in nuclei of both parenchyma and phloem companion cells in the plant (271, 273). It has not been mechanically transmitted to plant hosts.

Maize mosaic virus (MMV).—MMV is transmitted by a delphacid planthopper. Its latent period in the insect may be as short as four days (77) but often is longer (312). The purified particles are 242 mμ ± 10 × 48 mμ ± 10 (209). Similar particles have been envisioned in ultrathin sections of salivary glands and intestinal epithelia of insect hosts (210). Particles were found in tubules and cisterns of the endoplasmic reticulum, where virion assembly may occur (210).

Sowthistle yellow vein virus (SYVV).—SYVV is transmitted by an aphid (127) after a minimum latent period of eight days. This latent period may be shortened to a mean of about six days by injection of infectious hemolymph (461). Such hemolymph passages were used to demonstrate multiplication in the vector (461). The particles were estimated to be 220 mμ × 80 mμ, with conspicuous cross-striations in negatively stained preparations (367). In the plant, cross sections of particles showed a double band of electron-dense material surrounding an electron-lucent core, but in the insect only a single electron-dense ring could be detected. Both nucleus and cytoplasm were invaded by the virus. Hackett et al. (194) compared SYVV and VSV under identical conditions and found that SYVV particles had a greater width. There was no serological reaction between the two viruses.

Lettuce necrotic yellows virus (LNYV).—LNYV, an aphid-borne virus of lettuce in Australia, is mechanically transmissible to *Nicotiana glutinosa* (449). Little is known of the dynamics of its aphid transmission, but factors in its mechanical transmission have been carefully studied (104). The virus is localized in the cytoplasm of mesophyll, epidermis, and glandular hair cells (83). Virions have also been observed in immature tracheids and vessels of the xylem tissue, and infectious virus has been recovered from the transpiration stream (84). The particles in leaf cells (83) were 52 mμ × 340–380 mμ and frequently occurred in bundles enclosed in a double membrane. It was suggested that the inner membrane played a role in viral assembly. In the aphid vector, O'Loughlin & Chambers (340) found two kinds of particle; one similar to those seen in plant cells, and another which seemed to lack the outer coat. Particles were observed in muscle, fat body, brain, mycetome, trachea, epidermis, salivary glands, and cells of the alimentary tract. LNYV virions are most stable at pH 8.6 in a glycine buffer, and infectivity is further stabilized by bovine serum albumin (104). The virions are unstable to clarification treatments with most organic solvents, but are stable to limited treatment with Freon 113 (105). Chloroform, butanol, or phenol destroy LNYV activity (199), as do certain chelating agents (11). Particles purified by Freon clarification and density gradient centrifugation had a buoyant density in sucrose of 1.20 g/ml and were about 66 mμ × 227 mμ, with cross-striations at intervals of 4.5 mμ (199). Purification of virus was enhanced by chromatography on calcium phosphate gels (313). In negatively stained preparations, uniformly arranged projections could be detected (504), and an internal helix seemed to underlie the cross-striations (504). The appearance of particles after negative staining was a function of the pH; rounded forms appearing at higher or lower pH were assumed to result from changes in permeability of the membrane (503).

Small Viruses (23 to 33 mμ)

Many small viruses of animals and plants (242) have in common an icosahedral morphology and sizes ranging from 22 to 33 mμ, and possess single-stranded RNA, usually with a molecular weight between 1 and 2 million daltons. Well-characterized plant viruses which fit this description may be transmitted by such diverse vectors as fungi, nematodes, beetles, or aphids. Vectorship may provide certain clues for generic groupings, as with the nematode-borne nepoviruses, or with the beetle-borne viruses like turnip yellow mosaic, wild cucumber mosaic, squash mosaic, cowpea mosaic, and others (177, 201). It has long been known that some small viruses, e.g., cucumber mosaic virus are transmitted nonpersistently by aphids. Ling (279) recently suggested that leafhoppers also may transmit small viruses nonpersistently. In experimental tests, *Circulifer tenellus* transmitted sowbane mosaic virus, a seed-borne virus transmitted experimentally by other insects including leaf miner flies (30). The presence of RNA has been demonstrated in only one of the small viruses discussed in this review article. It is likely, however, that many of the viruses in the 23 to 33 mμ size class will prove, when they can be isolated in amounts large enough to permit nucleic acid analysis, to contain single-stranded RNA.

Beet western yellows virus (BWYV).—BWYV is not mechanically transmissible and has a latent period of 12 to 24 hr in its aphid vector (128). Development of a membrane feeding technique enabled Duffus & Gold (129) to assay BWYV fractions after density gradient centrifugation. Virus infectivity was associated with a slowly sedimenting zone, indicating that BWYV is a small virus. Neutralizing antibodies could be prepared from such zones, even after prior clarification by a butanol-chloroform mixture (181). Such antibodies neutralized infectivity of several isolates suspected to be BWYV but did not affect infectivity of PLRV (131). Ruppel (380) recently reported the presence, in intestines of viruliferous aphids, of particles 25 to 30 mμ in diameter believed to be BWYV virions.

Barley yellow dwarf virus (BYDV).—BYDV is a serious disease of grains in the United States, and its biology has been extensively studied (72, 371). The existence of a number of strains has made BYDV a model for the intensive study of host specificity of persistent aphid transmission (375). The virions are polyhedra about 30 mμ in diameter, which respond poorly to negative staining techniques (377). Similar particles were obtained from several isolates of BYDV. Two serological techniques, neutralization (376) and agar double diffusion tests (1), however, demonstrated that two of the BYDV isolates were serologically related distantly, if at all. The virions are stable for four months at 3° C, and are stable to chloroform, chymotypsin, trypsin, ribonuclease, and deoxyribonuclease (377). Although only low concentrations can be obtained from diseased plant material, Jensen (239) had no difficulty observing probable BYDV particles in phloem tissue. Large numbers of presumed virions, 24 mμ in diameter, were present and were occasionally in crystalline array. Vertical, but not lateral, spread of virus in the vascular tissue seemed possible. Cell organelles were disrupted, and a fine fibrous network filled the background cytoplasm.

Pea enation mosaic virus (PEMV).—PEMV can be purified from infected plant tissue by sucrose density gradient centrifugation (66, 73, 178, 232). Two components were obtained, one sedimenting at 94S, another at 113S (66, 178). Izadpanah & Shepherd (232) also obtained both components and found that even after three cycles of density gradient centrifugation, each was capable of producing infections yielding the typical ratio of bottom to top component particles. Bozarth & Chow (66), though using a similar 3-cycle procedure, were able to obtain a top component preparation which had a typical nucleoprotein ultraviolet absorbance spectrum, but was not infectious. The RNA content of the bottom component was estimated to be 27 per cent (397); that of the top component was 18 per cent (66).

In plants, hyperplastic growths, interpretable as true neoplasms, occur on pods and leaves. Ultrathin sections of these growths revealed large accumulations of virions in nuclei whose nucleoli were destroyed (403). It was concluded that PEMV probably multiplies in the nucleus and is released into the cytoplasm, where lesser accumulations and, rarely, microcrystals are observed in later stages of infection. Similar bodies were observed in the gut lumens and fat body cells of aphid vectors (407). In these thin sections, the virions were estimated to be about 28 mμ in diameter, with hexagonal contours suggestive of cubic symmetry. Size estimates from the various purified virus preparations ranged from 22 to 37 mμ.

Pea enation mosaic virus is unusual for a persistently borne virus in its mechanical transmissibility to plants. It is not known whether the virus is propagative in any of its several aphid vectors. Osborn (344) showed that inoculativity is retained through molting. The excreta ("honeydew") of viruliferous aphids is highly infectious (366). Although the minimum latent period may be as short as 4 to 6 hr, the average latent period (LP_{50}) is about 24 hr (456). Aphids retain the ability to transmit PEMV for long periods, but the inoculativity decreases rapidly, in association with a similar decline in feeding rate and reproductive activity (457). The decay of inoculativity can be reversed temporarily by a second exposure to infected plants (459). Sylvester (458), weighing evidence for and against multiplication, concluded that limited multiplication may occur.

Potato leafroll virus (PLRV).—PLRV, isolated by Peters (349, 350) from infected aphids, is a small polyhedron, 23 mμ in diameter, stable to chloroform and to partitioning in butoxyethanol-ethoxyethanol phase systems. Kojima et al. (258) were able to purify similar particles (24 to 25 mμ polyhedra) from infected plants after chloroform emulsification and density gradient centrifugation. In ultrathin sections of plant material, particles 23 mμ in diameter were observed in degenerated phloem cells of petioles and veins (8, 258). Earlier claims that PLRV was a DNA virus have been refuted by several workers, and claims that infectious RNA could be extracted from PLRV infected potato have been questioned (349).

Potato leafroll virus is not mechanically transmissible, and lack of a suitable test plant hindered earlier vector work before *Physalis floridana* was shown to be a better indicator plant than potato (249). The virus has been assayed largely by injection of extracts into nonviruliferous aphids. Harrison (198) employed this method to bioassay virus content of aphids after they had fed on diseased plants. Because the titer of extracts from insects increased during periods of continuous feeding up to six days, and because virus concentration decreased when the aphids were held on an immune host plant, Harrison concluded that PLRV probably did not multiply in its vector. Stegwee & Ponsen (439), however, showed that PLRV could be maintained in 15 serial hemolymph transfers, with retention of inoculativity in the final transfers. Calculations demonstrated that the virus multiplied in its vector. Peters & Van Loon (351) were able to transmit PLRV by membrane feeding. Transmission of PLRV after acquisition feeding (286) occurs after a latent period reported to be 1.5 hr to 4.5 hr (249) or 9.5 to 120 hr (285). Transovarial passage of the virus has been claimed (324).

Viruslike particles.—Whether all isolates studied and reported to be PLRV by various workers were the same virus may be questioned. Peters (348, 350) showed that at least one, and perhaps two viruses other than PLRV were present in certain of his virus stocks. These presumed viruses were transmissible by aphids and had sizes and properties similar to those of PLRV. These entities, termed "viruslike particles" pending further scrutiny of their nature, could be separated from PLRV by passage through differentially susceptible hosts. Not all viruslike particles visualized in insects are small (211, 272). The various viruslike particles observed possibly represent latent infections in insects, plants, or both.

Beet curlytop virus (CTV) and related viruses.—Of all the agents discussed in this review, that of beet curlytop is perhaps the most perplexing. Almost a decade before the isolation of any virus, CTV was known to be filterable (394). In 1933, Severin & Freitag (393) reported that the agent was stable to temperatures of 80° C, was stable under aerobic conditions at room temperature for 8 days, and remained infectious for at least 100 days under anaerobic conditions. Fresh extracts from viruliferous beet leafhoppers were still infectious at dilutions of 1:24,000. Despite these early indications that the virus was stable and present in reasonably high concentration in the insect vector, the agent has not yet been isolated. Further evidence of its stability is the infectivity of resuspended ethanol precipitates. This property also characterizes an Argentinian and a Brazilian isolate (28, 29). CTV and its deltocephaline leafhopper vector probably originated in the Mediterranean region, where they occur together (143). Both were apparently introduced into the western United States, where CTV is a serious disease, especially in beets. The Brazilian and Argentinian curly top viruses are specifically transmitted by agalliian leafhoppers (28, 29). Another possible member of the group is transmitted by a membracid (412). The possible evolution of CTV-vector relationships has recently been discussed by Oman (341). None of the curlytop viruses are transovarially passed. Estimates of latent period in the vector vary from 20 to 30 minutes to 22 hr, but are confounded by the possibility of occasional stylet transmissions. Bennett & Wallace (31) estimated a minimum latent period of 4 hours. All of the small phytarboviruses have proved difficult to isolate, and the best estimate to date of the nature of CTV is that it is probably an exceptionally stable small virus with particular features which make monomers difficult to isolate (J. E. Duffus, A. H. Gold, and S. H. Smith, personal communication).

An anomaly as perplexing as the difficulties in purification of CTV is the question of its multiplication in the vector. In spite of thorough studies of its vector relations, careful experiments (22–24, 31) have failed to provide evidence for multiplication of CTV in its vector. Lack of multiplication is suggested by the rapid decay of inoculativity once the vector has left an infected plant, by an increase in CTV titer of vectors feeding continuously on diseased plants, by the failure of strains to cross-protect in the vector (24), by the presence of CTV in a number of nonvector species after feeding on infected plants (31), and by the apparent difficulty in achieving serial passage by insect injection (297). Yet, the hypothesis of multiplication seems favored by lengthy retention of inoculativity by the vector (Brazilian curlytop may be retained up to 85 days in its vector), by increase in inoculativity during the first 48 hr after exposure of vectors to low doses of CTV by membrane feeding (26), by the possible dependence of the latent period on injected dose of CTV (297), and by the relatively high titer of CTV in insects compared with that in plants (393).

Curlytop virus has a shorter incubation period in the plant (under some conditions the average can be as little as 2.7 days) (23) than do other agents discussed in this review, and its replication cycle, in plants at least, must be short. Thus, the relevant period during which multiplication would be most easily detectable may occur during the first 12 hours after the first acquisition of virus. This period, however, has not yet been closely examined. In a later section (Persistent Transmission of Small Viruses), we discuss possible multiplication in vectors of CTV and other agents whose biological relationships with their vectors seem to be similar.

Oat blue dwarf virus (OBDV).—Oat blue dwarf virus (14) is transmitted by the leafhopper *Macrosteles fascifrons,* but the latent period and other details of transmission have yet to be determined. The particles could be purified (15) from plant hosts by a combination of polyethylene glycol chromatography and density gradient centrifugation. The purified virus measured 28 to 30 mμ in diameter and was stable for at least three weeks in 0.01 *M* phosphate buffer pH 7.0.

Rice tungro virus (RTV).—This small (30 to 33 mμ) virus is stable to such treatments as heating for 1 hr at 40° C or standing for 1 day at room temperature (165). RTV is stable at temperatures below 63° C for 10 min, and at room temperature for 24 hr. The virus persists in its deltocephaline leafhopper vector for less than a week and thus may be stylet-borne (278, 279). If so, it would represent the only known case of natural transmission by this mechanism by a species of the auchenorrhynchous Homoptera. RTV is not mechanically transmissible to plants or insects and must be bioassayed by allowing insects to feed on infectious solutions (165).

Maize streak virus (MSV).—On the basis of its short latent period in its deltocephaline leafhopper vector and the sensitivity of its transmission efficiency to dosage (445), it is reasonable to predict that MSV will prove to be a small, relatively stable, RNA virus.

Miscellaneous Phytarboviruses

Tomato spotted wilt virus (TSWV).—TSWV (34) is a virus whose morphology, while as yet incompletely defined, appears pleomorphic by all methods used to date. Particles, purified by a sequence of methods including density gradient centrifugation, range in diameter from 55 to 125 mμ (36, 54, 476). In ultrathin sections of plant tissues (229, 254, 310), particles 50 to 140 mμ in diameter have been reported. Best (34) pointed out a similarity in morphology of TSWV with that of tern virus (18), although a helical nucleocapsid has not been noted for TSWV. Chemical analysis of purified TSWV preparations demonstrated the presence of about 19 per cent lipid (35) and an RNA content of about 5 per cent (34). On these and other grounds, Best considered TSWV to be a "pleomorphic myxovirus." TSWV has a low thermal inactivation point (33) and, although its stability *in vitro* is poor, longevity is improved by the addition of such reducing substances as cysteine and Na_2SO_3 (37). The virus is transmitted persistently by thrips after a latent period of 4 to 12 days (382–384), but must be acquired by immature stages of the vector (386).

Carrot mottle virus (CMotV).—CMotV is apparently dependent upon a second virus, carrot redleaf virus, for its transmissibility by an aphid vector. Infectivity is destroyed by ether, chloroform, and other organic solvents. Recently, the morphology of CMotV was discussed by Murant et al. (327). Spherical particles about 50 mμ across were observed in partially purified preparations and in ultrathin sections. The particles appeared to be bounded by a unit membrane and were occasionally seen budding from the tonoplast. Details of the core were not presented. The density of the particles was estimated to be 1.154.

Hoja blanca virus.—Hoja blanca is a destructive disease of rice transmitted by two species of planthoppers (Fulgoroidea: Delphacidae) in the New World (140). It is not mechanically transmissible and is transovarially passed. Recently, Herold et al. published electron micrographs of particles 42 mμ in diameter which may represent the virions (212).

NATURE OF THE "YELLOWS" PATHOGENS

HISTORICAL

The yellows diseases of plants comprise a group of 40 or more known diseases which affect many important food, forage, and horticultural plants. Abnormalities in floral parts, including virescence and phyllody, are characteristic for many yellows diseases; but yellowing of leaves, proliferation of axillary buds, reduction of leaf lamina, and general stunting are common symptoms. The hyperplasia and other fundamental pathological alterations (61, 138, 139, 385, 465) induced by the agents in plants pose fundamental biological questions about regulation of plant growth. Prior to the reports of Doi et al. (124) that mulberry dwarf disease might be caused by a mycoplasmalike agent susceptible (231) to tetracycline antibiotics, causal agents of these diseases had long been considered to be viruses, based on such circumstantial evidence as their filterability and transmission by grafting.

Early work by Kunkel demonstrated that aster yellows (AY) disease was incited by an infectious agent (259) that was transmitted in nature by leafhoppers. This agent could be transmitted by grafting but was not mechanically transmissible (260). Kunkel also demonstrated that the AY agent in plants (263) and in insects (262) was sensitive to heat treatment. After such treatments, completion of a latent period was necessary before vectors were able to resume transmission. Kunkel offered his data in support of the hypothesis that the AY agent multiplied in insects as well as in plants (266). Black (40) gave additional evidence for the multiplication of aster yellows agent in insects; and in 1952, Maramorosch (294) definitively demonstrated that the AY agent multiplied in its leafhopper vector. In the meantime, a search for possible organisms associated with AY and cranberry false blossom disease had given negative results (121–123). Hartzell (200) envisioned intracellular bodies in vascular tissue of peach infected with the peach yellows agent and in the gut of the vector, but the bodies were also observed in the epidermal cells. It is possible that certain of the bodies Hartzell observed may have been large forms of the peach yellows disease agent.

Demonstration that AY agent multiplied in its vector cast no further light on its nature. On the basis of symptomatology, mode of transmission, and the lack of success in attempts to culture a yellows agent, a viral nature was still favored (45). Other disease agents persistently transmitted by their insect vectors, those of rice dwarf (159, 160, 161), wound tumor (53), and clover club leaf (44) were shown almost 20 years ago to multiply in their insect vectors. Although the symptoms induced in plants by these agents were different, they were all considered to be viruses, and eventually the viral morphology of RDV (399, 467), WTV (38, 68), MRDV (170, 481), PYDV (48, 288), and other persistently transmitted agents was described. The purification of these agents proved difficult, so it was no surprise that the AY agent also resisted purification. There seemed no good reason to suspect a possible fundamental difference in the nature of the yellows agents. Sensitivity *in vivo* to heat treatments at relatively low temperature was not only demonstrated for AY and other yellows diseases (236, 264, 265, 442, 471) including peach yellows, potato witches' broom, cranberry false blossom, X-disease of peach, western X-disease, and others, but for certain nonyellows diseases also (19, 228, 243, 415). Thus, heat sensitivity of the yellows agents *in vivo* seemed only to indicate that their infectivity was more fragile than that of most other viruses. Now, however, this heat sensitivity of the yellows agents recalls the sensitivity of mycoplasma to heat treatments (205), used to rid them as contaminants (16) from tissue cultures.

Based on his observations that the AY agent passed through bacterial filters with difficulty and that infectious units sedimented rapidly at low centrifugal speeds, Black (42) concluded that the agent must be large. Later, Lee & Chiykowski (275) attributed the rapid sedimentation of AY in their studies to aggregation of the presumed causal virus. AY and other yellows disease agents thus seemed to be similar to other viruses borne by leafhoppers, with the exception that their lability *in vitro* and low apparent concentration *in vivo* (48) consequently made them more difficult to isolate and purify than some of the other leafhopper-borne agents. The search for the causal agents in yellows disease by purification or electron microscopy, or both, resulted in reports of viruslike particles in aster yellows disease (360), stolbur disease (59, 484), oat sterile dwarf (70), and clover phyllody disease (354), but pathogenicity of the particles was never established. Other efforts at purification, in which fractions were tested for their infectivity (500), failed to identify any promising virus fraction.

None of the approaches in yellows disease research, based on the assumption that these diseases were induced by viruses, succeeded in isolation of the pathogens or in definitively proving their nature. These failures may now be explained by a new hypothesis that the presumed causal virus was an illusory target. In August 1967, Doi et al. (124) proposed that yellows diseases were caused not by viruses but by microorganisms resembling *Mycoplasma* (133, 206) (Pleuropneumonialike organisms, PPLO), or PLT (psittacosis-lymphogranulona-trachoma) group (*Chlamydia* or *Bedsonia*)-like agents (233). The hypothesis was based on electron microscopic observation of pleomorphic bodies in phloem of diseased plants. We now know that these pleomorphic mycoplasmalike bodies were observed in the phloem of yellows-diseased plants by a number of workers as early as 1964 (305), but their significance in yellows disease etiology evidently was not recognized until the publication of findings by Doi et al. (124) and Ishiie et al. (231). It now seems likely that mycoplasmalike microorganisms are involved in disease production of most, if not all, the yellows diseases. The partial list (Table II) of diseases probably belonging to the yellows group gives some indication of the wide variety of host plants affected and the worldwide distribution of the yellows diseases.

Evidence on Etiology of Yellows Diseases

Electron microscopy.—The discovery of pleomorphic bodies in the phloem of infected plants and the recognition of the bodies as possible mycoplasmalike or chlamydialike organisms provided the first substantive indications that nonviral agents might be responsible for yellows disease production. The bodies described (124) were present in plants with mulberry dwarf, aster yellows, paulownia witches' broom, or potato witches' broom disease. Similar pleomorphic bodies were subsequently described by other workers in plants with any of several other yellows diseases (61, 65, 171–173, 175, 176, 191, 277, 289, 305, 307, 355, 356, 405, 408, 446, 505, 506), and in yellows-infected insect vectors (69, 171, 176, 191, 220, 305, 307, 405). Examination of these bodies reveals morphological similarity with *Mycoplasma* and bacterial L forms (4, 5, 120, 154, 276). Like the *Mycoplasma* and L forms, the bodies in yellows-infected plants and insects are bounded by a single unit membrane, are devoid of a cell wall, and are highly pleomorphic. The various forms of the presumed organisms resemble stages in the reproductive cycles of *Mycoplasma* and L forms (120, 276). In some respects, the larger round bodies in infected hosts resemble the reticulate bodies of the Chlamydeae (463), but these organisms possess, in at least some stages of the life cycle, a cell wall that contains a mucopeptide layer similar to that of bacterial cell walls (326). Since the presumed organisms found in yellows-infected hosts are not known to be derived from bacterial parents, apparently lack a cell wall, and evidently can be cultured in cell-free media (86), they are referred to here as mycoplasmalike organisms.

In plants, chains of bodies, budding forms, and elongated and filamentous forms have been described. Round bodies, 75 to about 110 mμ in diameter, which resemble the elementary bodies of *Mycoplasma* or L forms, are also seen, although some may represent thin filamentous forms in cross section.

TABLE II. Some Plant Diseases Whose Causative Agents May Be Mycoplasma-like Organisms

Aster yellows	U. S., Far East, Europe
Mulberry dwarf	Far East
Potato witches' broom	Far East
Paulownia witches' broom	Far East
Clover phyllody	Europe, Canada
Tomato big bud	U. S., Australia, Europe
Potato stolbur	Europe
Parastolbur	Europe
Corn stunt	U. S.
Sugarcane whiteleaf	Taiwan
Rice yellow dwarf	Philippines, Far East
Legume little leaf	Australia
Clover dwarf	Europe
Apple proliferation	Europe
Oat sterile dwarf	Europe
Western X-disease	U. S.
Peach X-disease	U. S.
Alfalfa witches' broom	U. S., Europe
Cranberry false blossom	U. S.
Sandalwood spike	India, Far East
Flavescence dorée	Europe
Strawberry green petal	Europe
Peach yellows	U. S.
Little peach	U. S.
Cherry buckskin	U. S.
Cotton virescence	Europe
Crimean yellows	Soviet Union
Eggplant little leaf	Europe
Papaya bunchy top	Puerto Rico
Witches' broom of groundnut	Java, Far East
Clover stolbur	Europe
Blueberry stunt	U. S.
Tobacco yellow dwarf	Australia
Yellow wilt of sugar beet	Argentina
Little cherry	U. S.
Rubus stunt	Netherlands
Mal azul	Portugal
Sweet potato witches' broom	Japan and Taiwan
Witches' broom of legumes	Japan and Taiwan
Cryptotaenia japonica witches' broom	Japan

Particle diameters of rounded bodies in various diseases range from 75 or 80 mμ to more than 800 mμ. The internal structure of many of the larger pleomorphic bodies reveals netlike strands of material, possibly nucleic acid.

and numerous ribosomelike granules. Particles resembling elementary bodies are generally filled completely with dense material, and the strands of material like nucleic acid cannot be distinguished. The internal organization of bodies, whether found in insects or plants, is similar, and is typical of the *Mycoplasma* (6, 125).

The mycoplasmalike organisms in plants are apparently restricted to the phloem and are generally found only in sieve elements devoid of recognizable organelles, but they have been seen in sieve cells containing mitochondria, dictyosomes, endoplasmic reticulum, plastids, or degenerate nuclei (506). Intracellular residence in phloem parenchyma was first noted by Doi et al. (124), and pleomorphic bodies were seen in phloem companion cells by Lin & Lee (277). Later, Worley (506) found that the cytoplasm of occasional parenchyma cells with the full complement of intact subcellular organelles contained many mycoplasmalike bodies, especially the filamentous forms. The occurrence of long filamentous forms in cytoplasm of phloem parenchyma cells, and their relative scarcity in sieve cells, may indicate an influence of differences in nutritional factors or physical properties of the environment on morphology—a reportedly significant effect with both mycoplasmas and L forms (119, 120, 311).

Mycoplasmalike bodies seen in insects also were found within the cytoplasm of cells with a full complement of organelles. Although *Mycoplasma* commonly reside extracellularly in tissue culture, both extracellular and intracellular residence *in vitro* in tissue culture can occur in the same culture (208), and both extracellular and intracellular residence in whole animals has been demonstrated (314, 452). While the presence of mycoplasmalike bodies in the cytoplasm of yellows-affected host cells is consistent with findings on vertebrate *Mycoplasma,* the frequency of its occurrence in both insect and plant hosts may prove useful in the general study of intracellular residence.

Prevalence of the mycoplasmalike bodies in phloem of yellows-diseased plants is correlated with stage of plant infection and severity of symptoms (277, 506). At early stages of plant infection, only a few mycoplasmalike bodies are present; but, at later stages, the cells become densely packed. Although several cells in a vascular bundle may contain many bodies, other sieve cells in the same bundle may contain none. Proportions of various forms in cells and tissues vary with stage of infection and severity of symptoms. In AY (506), tissues with severe symptoms contained relatively few round bodies but did contain many forms which were electron-dense, elongated, and apparently degenerate. In tissues with less advanced symptoms, however, the round forms predominate. Worley (506) tentatively interpreted the electron-dense elongated forms as degenerate round bodies. Large round bodies that appear to be disintegrating are also seen in published micrographs (176, 289, 506). Some of these forms may be similar to autolysing large bodies observed by Dienes & Bullivant (120). Some of the material observed in infected cells may represent sieve cell slime fibers or a rudimentary extracellular matrix (184) produced by certain of the mycoplasmalike organisms.

Presumably, the mycoplasmalike bodies usually pass from cell to cell through the pores in sieve plates of sieve cells, but in some cases their passage presumably may be blocked by callose (61). The small round bodies may be carried passively through sieve pores large enough to accommodate them, and some elongated forms and filaments may pass from cell to cell in the same way. Shikata & Maramorosch (405) have reported finding elongated forms of mycoplasmalike bodies in the pores connecting adjacent sieve elements in infected plants. Passage of the larger forms of mycoplasmalike bodies through sieve pores also seems possible. The results of van Boven et al. (474) with streptococcal L forms indicate that particles like L forms and *Mycoplasma* can undergo considerable deformation and pass through pores smaller than their diameters. This observation testifies to the extreme pliability of organisms that lack a rigid cell wall. It is also possible that development of fine filaments from large bodies may permit passage, even through plasmodesmata, of material originating in these bodies. Streptococcal L forms reportedly (475) can pass through small pore size filters by development of filaments and subsequent formation of microcolonies on the opposite side of the filters placed on appropriate growth media.

In yellows-infected insect vectors, mycoplasmalike bodies have been found in the salivary glands (171, 176, 220), intestine (69, 171, 176, 191, 305), fat body (405), and nervous system (191, 305), in insects carrying the aster yellows (220), corn stunt (191, 305), clover phyllody (171, 176), or oat sterile dwarf (69) agents. In the case of AY disease, Hirumi & Maramorosch (220) noted four general types of pleomorphic bodies in the cytoplasm of salivary glands of infected leafhoppers. One type of body, 500 to 800 mμ in diameter, contained inclusions resembling the pleomorphic bodies of smaller size also seen in the tissue. The remaining three types of bodies were large bodies without inclusions, intermediate sized bodies (around 120 to 400 mμ), and small (80 to 100 mμ) bodies. The authors proposed a hypothetical reproductive cycle based on similarities with stages in the life cycle of the free-living organism, *Mycoplasma laidlawii* (325). Pleomorphic bodies 110 to 600 mμ in diameter that resembled *Mycoplasma* were observed in salivary glands and gut epithelium of vectors carrying the clover phyllody agent (171, 176). The appearance of strongly Feulgen-positive material in the salivary glands coincided with the beginning of transmission by the vector and can probably be accounted for by the presence of the mycoplasmalike bodies. Distribution of the various morphological types varied with the degree of parasitism of infected cells (176).

Maniloff et al. (290) reported an unusual morphology and fine structure of *Mycoplasma gallisepticum*, by which it may be distinguished from other species of *Mycoplasma*. Differences in mycoplasmalike bodies in different yellows diseases so far envisioned are small and difficult to assess, especially since different hosts have been used for study of most of the various diseases. Further comparison of mycoplasmalike bodies associated with different diseases in a given host, such as the study by Ploaie et al. (355) on crimean yellows, European clover dwarf, European aster yellows, stolbur, and parastolbur in periwinkle, seem desirable. These authors noted differences in the size range of bodies associated with these diseases. Differences in the appearance of limiting membranes and pleomorphism of bodies were also noted, but some of them may have been fixation artifacts.

Even in the same host, mycoplasmalike bodies associated with a given disease may vary somewhat, depending upon the nature of the infected host cell. Worley (506) noted compact masses of long filamentous forms in cytoplasm of phloem parenchyma in AY-infected plants. In sieve elements, however, filamentous forms were less prevalent, shorter, and more disperse. Also, inclusions reported (220) in mycoplasmalike bodies in salivary glands of AY-infected vectors evidently are either absent in plants or occur with lower frequency. Moreover, these inclusions have been reported only in salivary glands (220). Maramorosch et al. (305) noted differences in the internal structure of bodies in cells of the ventral ganglia and in the filter chamber of the intestine of leafhoppers carrying the corn stunt agent. These observations may be related to striking differences in morphology of a given species of *Mycoplasma* that may occur when the organism grows in different milieu (311, 363). The great variety of forms in a given yellows-infected host, of course, are not necessarily of a single species of mycoplasmalike organism.

Mycoplasmalike bodies have also been reported (289) in insects free of any known plant pathogen. Occurrence of these bodies in healthy insects or plants might be anticipated, paralleling the occurrence of both nonpathogenic and pathogenic species of bacteria in plants or insects. Mycoplasmalike bodies have been found associated with viruslike particles in insect vectors carrying the oat sterile dwarf agent (69). Not all oat sterile dwarf-infected insects contain viruslike particles, but the mycoplasmalike particles are apparently consistently associated with the disease. In this disease (361), the presumed microorganism may be primarily responsible for the symptom syndrome; but, in other cases, an association between mycoplasma and virus may be responsible for a given disease syndrome. Pleomorphic bodies interpreted as rickettsiae have also been observed in healthy and viruliferous leafhoppers (305).

So far, the pleomorphic mycoplasmalike bodies have not been observed in extracellular residence either in insects or in plants. Presumably, however, there must be at least a temporary extracellular residence in the alimentary tract, hemolymph, and saliva of the insect host. The presence of mycoplasmalike bodies in phloem sieve elements, and their apparent limitation to phloem tissue, where the leafhopper vectors feed (115), correlates well with transmission of the yellows disease agents.

Antibiotic chemotherapy.—The yellows diseases are markedly suppressed in plants (109–112, 187, 231, 437) and in insects (109, 111, 112, 187, 493) by antibiotics that are effective against diseases caused by *Mycoplasma* in mammals and avian species (207, 337). This suppression is accompanied in plants by disappearance of the pleomorphic mycoplasmalike bodies from the phloem (231), by a reduction in the titer of the disease agent in plants as judged by its availability to leafhopper vectors (110, 112, 493), and by bioassay of extracts from treated plants (110, 112).

Prior to 1967, attempts to achieve antibiotic therapy of diseases caused by leafhopper-borne agents failed. Now it is apparent that these failures were due to unfortunate choices of disease or antibiotic. In studies by Beale & Jones (17), neither penicillin, streptomycin, chloramphenicol, chlortetracycline, nor tetracycline inhibited the expression of symptoms induced in *Nicotiana rustica* by potato yellow dwarf virus. Jensen et al. (238) added antibiotics to extracts containing the WX agent in an effort to control contaminating bacteria that caused mortality of injected insect vectors. Streptomycin, penicillin, or chloramphenicol did not interfere with transmission of the WX agent. These results agree with those reported by Whitcomb & Davis (493) on the AY agent. In tests by Maramorosch (302), streptomycin, penicillin, chlortetracycline, 2,6-diaminopurine, 6-mercaptopurine, and a benzimidazole riboside were tested for an effect on the inoculativity of insect vectors carrying any one of four plant disease agents. The benzimidazole riboside was ineffective against curly top, corn stunt, wound tumor, and aster yellows, and 2,6-diaminopurine and 6-mercaptopurine were both ineffective against aster yellows and curly top. It is unclear, however, just which disease agents were subjected to streptomycin, penicillin, or chlortetracycline in the preliminary tests noted (302). Evidently, in view of later work, the CS and AY agents were not subjected to chlortetracycline, for this antibiotic is effective against yellows diseases in both insects and plants (110, 112, 187, 437, 493). In more recent work, penicillin and streptomycin were not among the antibiotics found to be effective against the AY agent in insects (493).

In retrospect, a re-examination of the very early attempts at chemotherapy of peach disease (247, 440–442) might prove interesting. In one type of test, Stoddard (440, 442) soaked X-infected peach buds in various solutions for one hour and then budded immediately into healthy peach seedlings. Quinhydrone, 8-hydroxyquinoline sulfate, and hydroquinone were among the compounds that reduced transmission of X-disease agent from treated diseased bud to healthy stock seedling. Interestingly, 8-hydroxyquinoline sulfate produced a remission of disease, with symptoms appearing in new growth after plants had undergone a period of winter dormancy, similar to that observed in treatments of Dutch elm disease and *Verticillium* wilt of eggplant with the same material (510). Tetracycline antibiotics, but not cycloheximide, endomycin, neomycin, or streptomycin, induced a temporary recovery from symptoms in young peach rosette-infected trees (246). This demonstration that peach rosette (246) responds to antibiotic chemotherapy may have depended, at least in part, upon a nonviral nature of the causal agent. Possible nonspecific effects, however, due to tissue damage and delayed or poor graft unions with X-disease are difficult to assess.

The work of Ishiie et al. (231) provided the first good evidence that yellows diseases are susceptible to antibiotic chemotherapy. These workers demonstrated a therapeutic effect of tetracycline and chlortetracycline on dwarf-diseased mulberry plants. Kanamycin, on the other hand, was ineffective against the disease. Remission of yellows disease in plants by antibiotics was confirmed by Davis et al. (112) who reported the effectiveness of tetracycline, chlortetracycline, and chloramphenicol against AY disease. The lack of effect of penicillin against yellows disease, first demonstrated with AY (112), is consistent with the absolute penicillin resistance characteristic of the *Mycoplasma* (207), although sensitivity of *Mycoplasma neurolyticum* to penicillin has been reported (507). CS disease has also been demonstrated (187) to be susceptible to tetracycline antibiotics, but not to kanamycin or penicillin. In parallel tests with AY, Davis et al. (112) failed to demonstrate an effect by chlortetracycline on transmission of, or on symptoms induced in plants by, WTV or PYDV.

Blockage of infection by the AY agent (112) or by the CS agent (187) has been achieved in some plants of a group by treatment with chlortetracycline or tetracycline prior to inoculation. Staron et al. (437) confirmed the efficacy of tetracycline against AY in plants and demonstrated an effect of antibiotic against stolbur. The indication by these authors (437) of an effect of tylosin or erythromycin against yellow disease in plants, and the apparent low activity or lack of effect of these antibiotics in tests by Davis & Whitcomb (110), may reflect poor translocation of antibiotics or other differences due to host plant, yellows agent strain, antibiotic concentrations, or method of testing.

Both suppression of symptom development and remission of existing symptoms of AY disease in plants by tetracycline antibiotics were accompanied by a reduction in the titer of extractable AY agent (110, 112), as well as by a marked reduction in the availability of the AY agent to acquisition by insects (493). This contrasts with the lack of an apparent effect on AY agent titer during reversal of AY-induced stunting in plants by gibberellin (299).

Sensitivity of yellows agents to tetracycline antibiotics has also been demonstrated in insect vectors. Davis et al. (112) showed that leafhoppers injected with a mixture of chlortetracycline and AY agent subsequently failed to transmit the agent to plants. In a more extensive study, Whitcomb & Davis (493) demonstrated a dramatic effect of injection of a single dose of antibiotic into insect vectors already transmitting the disease agent to plants. Injection of chlortetracycline, methacycline, tetracycline, oxytetracycline, or tylosin tartrate produced a marked reduction in the ability of infected insects to transmit AY. Inoculativity decreased during the first 5 to 10 days after injection, but after 10 days the insects gradually regained their ability to transmit the AY agent. Resumption of inoculativity probably involves an induced AY incubation period and is reminiscent of the effects of heat treatments of AY-transmitting insects (262). Transmission of the WX disease (D. D. Jensen, personal communication) and the CS (187) agents by their vectors is also inhibited by tetracycline antibiotics. Penicillin, kanamycin, cycloserine, spectinomycin, Vancomycin®, and streptomycin are among the antibiotics without apparent effect on yellows disease either in insects (493) or in plants (110). Gold sodium thiomalate, effective against *Mycoplasma pneumoniae* (309), had no effect on the AY agent.

Antibiotics have become a valuable taxonomic tool in studying unknown disease agents. Their use has been recently extended to studies of "cytoplasmic factors" long thought to be genetic in nature (134, 135). The sensitivity of yellows diseases to certain antibiotics suggests that the causal agents have a complexity greater than that of plant viruses, and more like that of microorganisms. The spectrum of their antibiotic sensitivity fits well with observations by electron microscopy of presumed mycoplasmalike organisms in the phloem of infected plants and in the tissues, including salivary glands, of leafhoppers transmitting yellows disease agents. Possible alternative explanations of the effects of antibiotics on yellows disease in insects and plants have been considered (110, 493). Nevertheless, the efficacious antibiotics most likely exert their suppressive effects on yellows diseases by acting on the common denominators of disease in insect and plant hosts—the yellows disease agents themselves.

Preliminary purification and properties in vitro—Attempts to isolate and purify a presumed causal virus from yellows-diseased hosts have been carried on for many years. Black (42) determined that AY agent passed with difficulty through Berkefeld N and V bacterial filters that retained *Serratia marcescens* but which allowed passage of a contaminant bacterium present in the preparations. Infectivity sedimented rapidly at low centrifugal speeds. These data led Black to suggest that the AY agent was large relative to plant viruses that had been studied up to that time. Later work by Lee & Chiykowski (275) attributed the rapid sedimentation of infectious AY agent to possible aggregation of virus particles. They (275) suggested that AY infectivity was sufficiently stable at 5° C to allow purification procedures that could be completed in less than 4 hr. Purification, however, would remove an agent from the protective factors present in the crude preparations. In tests by Whitcomb et al. (494), AY agent was passed through 7 per cent agar gel filtration columns into various buffers. Infectivity declined rapidly at 4° C and was undetectable after 6 hr. In other work, AY infectivity in crude preparations could be detected after incubation for 48 hr at 5° C (275) and after 24 hr, but not 48 hr, at 0° C in crude preparations diluted $10^{1.5}$ with 0.85 per cent NaCl and adjusted to pH 7.0 with K_2HPO_4 (42). Infectivity lasted less than 4 hr at 25° C (42, 275). Recently, AY infectivity has been reported (219) to be retained for 6 hr at room temperature when a cell-free leafhopper culture medium containing fetal bovine serum was added to extracts from infected tissues.

Some attention has been given to attempts at visualization of the labile yellows agents. Vovk & Nikiforova (484) found spherical particles thought to be the causal virus in extracts from stolbur-infected plants. Blattný (59) observed viruslike particles in crude homogenates from stolbur-infected tobacco and tomato plants, and in 1959, Protsenko (360) claimed visualization by electron microscopy of the aster yellows virus, 38 to 50 mμ in diameter, in homogenates of infected plants. Ploaie (354) described viruslike particles 50 to 100 mμ in diameter in clover phyllody-diseased plants and insect vectors. Following the proposal of mycoplasmalike etiology by Doi et al. (124), Giannotti et al. (174) reported isolation, by differential centrifugation, of mycoplasmalike bodies associated with Flavescence dorée of grape. In all these cases, however, infectivity of the particles visualized was not demonstrated.

Infectivity tests were an integral part of other work designed to develop a method for purification of the presumed AY virus. Steere (438) found that passage of AY infectivity through 7 per cent agar gel filtration columns was reproducibly obtained when the agent was extracted from source insects with 0.3 *M* glycine – 0.03 *M* $MgCl_2$ at pH 8, and when the same buffer was used for eluting the columns. Following the reports of Doi et al. (124), and Ishiie et al. (231), properties of the infectious AY agent in eluates from the 7 per cent agar column (438) were determined by Whitcomb & Davis (492). The demonstrated sensitivity to organic solvents, surfactants, and sonification are consistent with a possible mycoplasmalike etiology of AY disease. In ultrafiltration studies of the AY agent (95), infectivity passed through 300 mμ but not 220 mμ pore size Millipore filters when infected plants were used as the AY agent source. Infectivity could not be detected in 300 mμ filtrates, however, when infected insects were used as source for AY. The general pattern of ultrafiltration resembled that noted previously for *Mycoplasma* from mammalian sources (468) and for bacterial L forms (474). Such a pattern might also be obtained from filtration of the large and pleomorphic paramyxoviruses. Poxviruses or Chlamydeae would also give similar patterns. Smaller viruses, however, would pass 220 mμ filters (80, 224, 479), and it is clear that the valuable technique of ultrafiltration, used to some extent by earlier workers (42, 55, 394), was abandoned by later workers to their own detriment.

Properties of the western X-disease (WX) agent *in vitro* (500) also could be attributed to possible mycoplasmalike nature (432), although a fragile, pleomorphic virus might also explain the findings. The WX agent was sensitive to treatment with butanol or chloroform, and to brief emulsification with Genetron 113®. Although infectivity of AY agent (494) was destroyed by freezing, that of WX agent was not (500). Infectivity of WX was sedimented after 10 min at 25,000 *g*, and in rate zonal density gradient centrifugation, infectivity was found throughout the gradient column after 25 min at 25,000 rpm, with most infectivity in the bottom one-third of the column. WX infectivity was best recovered from gel filtration columns when a buffer containing glycine and Mg^{++} was used.

Cultivation of yellows pathogens in cell-free media.—The hypothesis that the yellows diseases of plants are caused by mycoplasmalike organisms (124, 231) has given rise to attempts to cultivate these agents in cell-free media, in partial fulfillment of Koch's postulates (257). Culture of a *Mycoplasma* in cellfree media was first achieved with the filterable bovine pleuropneumonia organism in 1898 (339), and its morphology was described in 1910 (60, 62). Yet, it was not until 1962 that a *Mycoplasma* pathogenic to humans—the Eaton agent which causes primary atypical pneumonia—was cultured (85). Since that time, *Mycoplasma* have become the subject of considerable attention (206, 207). Now, a little more than two years after the first proposal of the *Mycoplasma* hypothesis of yellows disease etiology, Chen & Granados (86) have probably achieved the first culture of a plant pathogenic *Mycoplasma,* strongly supporting a *Mycoplasma* etiology of the corn stunt (CS) disease. In a liquid medium containing inorganic salts, amino acids, vitamins, organic acids, purines, pyrimidines, coenzymes, steroids, ATP, sugars, and undefined supplemental nutrients, CS infectivity could be detected in cultures and subcultures after incubation for periods up to 50 days at 25° C. This major accomplishment in yellows disease research provides substantial evidence for the etiology of yellows plant disease and opens the way for the development of an entirely new area of plant pathology.

In the course of attempts to culture a possible plant pathogenic *Mycoplasma* from AY-infected insects, we isolated a small bacteriumlike organism (494) from AY-infected *Macrosteles fascifrons.* The organism grew on a bean pod agar medium as a nurse-dependent satellite of a species of *Pseudomonas,* which was isolated from the vector and was similar to an egg-transmitted symbiont reportedly associated with *M. fascifrons* (342). Leafhoppers injected with suspensions of the nurse-dependent organism, however, failed to transmit AY to healthy plants. The authors (494) interpreted the organism as a fastidious microorganism associated with the vector, although a possible relationship with the production of AY disease was not completely excluded. Our continuing efforts to culture the AY agent have been monitored by infectivity assays. Several media containing inorganic salts, amino acids, vitamins, horse serum, sucrose, and cholesterol, as well as various undefined supplements, were developed in which AY agent infectivity was retained during incubation for 48 hr but not 72 hr at 22° C under N_2. Although these results suggested a possible basal medium that might be enriched for eventual cultivation of the agent, AY agent apparently does not multiply in any of the media we have tested.

A possibility that *Mycoplasma* or mycoplasmalike organisms may be involved in the production of plant diseases other than those of the yellows type was suggested by Hampton et al. (195). These workers report isolation and culture of a mechanically transmissible *Mycoplasma* that was observed by thin section electron microscopy between the cell wall and plasmalemma and in the cytoplasm of cells of alfalfa mosaic virus-infected plants. The organism, however, was reportedly 15 to 250 mμ in diameter, which would place almost half its entire size range well below that possible for *Mycoplasma* (325). Serological tests (195) indicated a curious circumstance—a close antigenic relationship with *Mycoplasma salivarium,* from humans, as well as with *M. gallisepticum* and *M. meleagridis,* from avian species. Although the authors claimed to have fulfilled Koch's postulates for claiming pathogenicity of the mycoplasmalike agent in plants, full details are not provided in the text of the published report.

Control

The success of antibiotics in laboratory trials and the resulting suppression of yellows diseases in plants suggests the possibility of their application in practical control. Unfortunately, antibiotics have not only failed to eradicate existing infections (110, 231), but are not wholly effective in blocking infection when applied to plants before inoculation. Perhaps antibiotics or other substances will eventually be found which can eradicate yellows agents in their plant hosts, but the basis for practical control of leafhopper-borne diseases will probably continue to depend for some time, at least, on our understanding and manipulation of their epidemiologies (27, 71).

Yellows diseases are worldwide in distribution and attack many species of crop plants (Table II) and wild hosts (261). In the United States, aster yellows is transmitted mainly by *Macrosteles fascifrons,* a vector which feeds indiscriminately on sedges, grasses, and herbaceous plants, and which exhibits migratory behavior (93, 126). The disease agent overwinters in certain perennial plants and is probably universally present in some of the perennials of the weedy field association. Crop protection under these circumstances may be a matter best suited to management of the agricultural ecology. Replacement of weed acreage with a more desirable cover as advocated by Piemeisel (352), for example, may be desirable. In that case, it was pointed out that the weedy hosts of the CTV vector proliferated in the wake of poor range management, and renovation of grasslands was advocated. In other cases, recourse to indexing, crop rotation, insertion of barriers, roguing, or rational use of insecticides may be necessary (78). The excellent review of Broadbent (71) makes it unnecessary for us to discuss these measures in detail.

BIOLOGICAL CYCLES OF VIRAL AND NONVIRAL PATHOGENS

Transmission Mechanisms

General.—The mechanism by which an insect species transmits a disease agent is intimately dependent upon the biology of the vector. Although aphids (375, 458) transmit certain viruses persistently, for example, the majority of viruses transmitted by aphids are stylet-borne (353).

Most insects, however, are biologically suited to the persistent mode of transmission; leafhoppers and other auchenorrhynchous Homoptera (230, 338), whiteflies (102), beetles (149), and vectors of animal viruses (82, 113) usually transmit in a persistent fashion. The transmission of certain arboviruses may have both a stylet-borne, and circulatory phase (82). Although the infrequent early transmissions of such viruses as PLRV (249) may result from stylet transmissions, this has not been demonstrated. Transmission of plant viruses by the fecal route is unknown for insects, but may occur with certain mite-borne plant viruses (346, 429).

The term, persistent transmission, covers a variety of biological relationships. Common to most of them is the existence of a latent period (the time between acquisition of disease agent and the ability of the vector to transmit). Latent periods may be found to vary from a few hours to 70 days or longer. Transmission usually continues throughout the life of a vector, but the rate of decay of inoculativity (the frequency of transmission per unit time) varies greatly with each vector-pathogen combination (145).

Persistent transmission of small viruses.—In the following discussion of the persistent transmission of small viruses, no attempt will be made to consider "semipersistent transmission" (458) of viruses (81, 116) like cauliflower mosaic (the only DNA-containing plant virus) (398) or rice tungro (165), or the transmission of small viruses (177) by beetles (149). In these cases, there is no evidence that the viruses are propagative in their vectors.

Persistent transmission of small viruses by homopterous vectors may begin after a very short latent period (in some cases less than 6 hr, but often on the order of 24 hr). Transovarial passage of such viruses has been claimed in only a single case (324).

Considerable effort has been expended to determine whether or not certain of these viruses multiply in their vectors. The presence of barley yellow dwarf virus in aphids in quantities which permit purification (377), and certain aspects of BYDV vector specificity (374), suggest that multiplication may occur. Careful and intensive investigations (23, 31) have failed to demonstrate multiplication of CTV in the vector *Circulifer tenellus*. Bennett has hypothesized (23) that CTV is passed through the gut wall into the hemolymph and then into saliva by normal transport mechanisms of the insect in a circulatory, nonpropagative mode of transmission. Although present evidence may appear to favor such a hypothesis, the possibility of limited multiplication cannot be excluded. Even the most careful workers (198) experienced difficulty in demonstrating multiplication of such viruses as PLRV, which eventually was shown to multiply by a serial passage experiment (439).

Frequently, dosage of ingested or injected virus markedly affects subsequent transmission efficiency (114, 147, 198, 445). Passage of virus by injection techniques is often difficult (297, 458) and may appear to contraindicate multiplication.

Certain further observations are of interest. A phenomenon of "recharge" has been shown for CTV and PEMV. The transmission of each of these viruses declines rapidly after a peak of inoculativity is reached. A second exposure to diseased plants, however, restores to some extent the waning inoculativity. With tomato yellow leafcurl virus (TYLCV), a virus with a short (21 hr) latent period in its whitefly vector (99), recharge is possible only at the end of a cycle of increasing and decreasing inoculativity. A factor appeared to be present in whitefly homogenates (96) which reduced TYLCV transmission by whiteflies if taken up through membranes. The factor first became detectable 24 hr following acquisition of TYLCV by whiteflies (97), suggesting that it could be, among other possibilities, an antiviral factor analogous to interferon, which was synthesized in conjunction with TYLCV synthesis.

The weight of evidence favors multiplication of a number of small viruses in their vectors. With arboviruses, there may be a threshold effect, i.e., a dosage sensitivity for infection of the intestinal epithelial cells (82). "Recharge" experiments in homopterous vectors may succeed in infecting cells which may be insensitive to the residual concentrations of virus present after the initial infection has subsided. This is especially true of gut epithelial cells, which may be very short-lived, and which are continuously regenerated either from individual nidi which may be virus-free, or by a stripping and replacement of the entire epithelium. The concept of a mechanism of cellular response (96–98) to infection by small viruses provides further explanation for certain phenomena otherwise difficult to explain. Infection of tissue such as gut epithelium, with its notable regenerative capacity surely represents a special case without analogy in any of the other cells in the insect's body. The maximum titer of a picornavirus or an arbovirus in tissue culture cells may be reached in 4 to 6 hr, with virus first detectable in 2 to 3 hr. The hypothetical events of virus multiplication in gut epithelia, therefore, may take place more rapidly than had been imagined, and may well have been undetectable by any of the conventional means of vector research.

Transmission of diplornaviruses and rhabdoviruses.—Larger viruses are transmitted only after a longer latent period in the vector—often eight or more days but as short as four days in the case of WSMV (274) and PYDV (50). There is no evidence that any agent with such a long latent period is nonpropagative, and multiplication of viruses with such long latent periods can be tentatively assumed. It is interesting to note that the latent periods of many rhabdoviruses may be as short as three to four days, whereas the diplornaviruses require, so far as is known, eight days. This corresponds to the general relationship between the single-step growth curves for prototype viruses of the two groups and reminds us that latent periods are, in a sense, summations of single step viral growth curves for single cells.

Transmission of yellow diseases.—In view of recent research indicating nonviral nature of the yellows agents, it is obvious that plant pathogenic agents with prolonged latent periods in their vectors may have little more in common than multiplication in a similar environment. Yet, it is a remarkable fact that agents, even when they are as diverse as viruses and mycoplasmalike organisms, may have similar dynamics of infection and transmission. Although only a few yellows agents have been shown to be transovarially passed (144, 358), for example, this fact was considered to be an interesting discrepancy between viruses, rather than a signal of profound dissimilarity, and perception of the wide diversity between yellows agents and viruses depended upon the discovery of the morphology of the agents. It is of considerable interest, then, to compare the basic store of biological data of the yellows agents and viruses to determine what differences there might be in their biologies. The format of the following sections will permit such a comparison.

Availability

When noninfective insects are transferred from a healthy to a diseased plant, the probability of acquisition of disease agent in successive time intervals should ideally be equal (270). This is not always so, as in the case of the AY agent (188, 267). Although this anomaly has not yet been explained, a period of adjustment on a new host may be required before normal feeding is resumed. With certain arboviruses (82) there is a threshold effect in acquisition; this effect may occur but has not been demonstrated for yellows agents or plant viruses in insects. Yellows agents may be acquired readily, as with AY (267), or with great difficulty, as with WX agent (235). With viruses (68) or yellows agents (110), the availability usually parallels the titer of extractable agent in the plant. Gill (180) demonstrated that the availability of barley yellow dwarf virus from oats fluctuated cyclically. The chief characteristic of limited attempts to control or measure availability has so far been variability. Some of the variability can be attributed to (*a*) incomplete environmental control, (*b*) incomplete attention to the exact timing of infection, (*c*) small sample sizes. The availability of WTV to *Agallia constricta* during three-day acquisition feeds on infected crimson clover (426) was consistent and reproducible when conditions were carefully controlled.

Multiplication in Insect Vectors

Much effort has been expended in the search for evidence that certain plant pathogens multiply in their vectors. Demonstration of continuous transovarial passage through a number of generations (44, 160) gave the first demonstration of the multiplication of a plant disease agent in an insect vector. A second method, one that may be more useful for agents (the yellows agents, for example) which are only rarely transovarially passed, is serial passage by means of insect injection through a number of generations (294, 295, 439). Both of these methods employ calculation of a dilution of agent theoretically obtained in the absence of multiplication. Careful titrations during the growth curve of an agent in its vector provide similar information (364, 497). By quantitatively estimating yields of extractable agents, increases of 3 or 4 log-units of disease agents can be demonstrated (497). Remissions of disease transmission or symptoms which can be imposed by agents such as heat (236, 262) or certain antibiotics (493), but which, upon removal of pressure, are reversed after expiration of a latent period, are also indicative of multiplication. Significant interactions between strains of disease agents (152) and cytopathic effects of disease agents in their vectors (237, 501) are also considered good indications of multiplication.

Dynamics of Multiplication and Transmission

Multiplication.—The dynamics of infection of plant and insect hosts by yellows agents or viruses is similar. After ingestion or injection of the disease agent, the titer of the agent increases according to a growth curve usually characterized by a steep logarithmic phase in which numbers of mature particles rapidly increase. Transmission of the agent by the insect (491, 496) usually begins after attainment of the plateau level. With wound tumor virus, however, maximum inoculativity and maximum titer of virions occurs some time after the plateau level of soluble antigen has been reached, and after transmission by some insects has begun (166). With wheat striate mosaic virus also, there is a lack of correspondence with virus concentration and transmission (345). Infection of insect tissues after acquisition by the oral route occurs sequentially, as shown by the use of fluorescent antibody (419), infectivity assays (425), or histological methods (499). After attainment of maximum concentration, multiplication apparently continues, but at a reduced rate insufficient to maintain the plateau level (364).

Transmission.—When transmission patterns are summarized by plotting inoculativity as a function of the number of days after the insects were first placed on diseased plants, a curve is obtained which contains most, and in many cases all, the available information. Since individual insects may transmit with different efficiencies, the inoculativity curve is a function related only to a given population of insects. Nevertheless, the overall course of the curve usually varies very little for a given vector-disease agent system. The inoculativity curve presumably follows a course kinetically related to and following the growth curve of the disease agent. Transmission rises to a maximum level within a relatively short time following the latent period. Inoculativity has been shown to decrease, in all cases so far examined, after the maximum level was reached. The slope of this decrease may be precipitous, as with WSMV (428), CTV (147), PEMV (411), and WX (497), or gradual, as with WTV (364, 491), or potato leafroll virus (285). When the maximum level of transmission approaches 100 per cent (267), the inoculativity curve is truncated. Thus, a poorly susceptible test plant, such as celery for the WX agent, may be better than a highly susceptible plant for measuring changes in inoculativity (490). Certain phenomena seem to result from an interaction between vector and plant. For example, with certain yellows agents, a particular species of test plant may be susceptible when inoculated by one species of vector, but not when inoculated by another species (144).

Environment, particularly temperature (291, 293, 460) can have an important influence on the dynamics of infection. Growth curves of disease agents in insects have decreased slopes at lower temperatures. At higher temperatures, multiplication may be totally blocked, or may peak and subside rapidly. Events associated with the attainment of maximum titer in a host such as inoculativity, lethality to the vector, or symptom expression in plants, are delayed correspondingly at suboptimal temperatures. The probability of transmission of AY agent changes throughout the diurnal cycle (304).

The age of plants and vectors (417, 420) has an important effect on their susceptibility to the various plant pathogens. Plants are usually most susceptible to infection when young (454, 455), and in some cases, as with WTV in crimson clover (293), older plants become very resistant. The effect of host plant age is perhaps less noticeable with yellows disease agents than with viruses. Vectors also tend to be maximally susceptible when young, in certain cases transmitting only after nymphal acquisition (386, 451, 509). In insects, as well as plants, the effect of host age is more noticeable with viruses than with yellows agents. Permeability of the gut (444) or possible restriction of the multiplication of disease agents to epithelial cells of the intestine (420) have been proposed as mechanisms for the decrease in susceptibility with aging.

Dosage Relationships and Bioassay

Dosage.—For all vector-disease agent combinations, the length of the latent period in the vector or plant is a function of dosage (292, 296, 490). We have already noted that transmission of a disease agent to a plant usually follows a rapid logarithmic phase of increase in the vector. However, during this logarithmic phase of multiplication, the concentration of agent transmitted to the plant quickly passes through the range in which concentration affects the incubation period in the plant, and the dosage effect is less often noted in the plant. The incubation periods of the first early transmissions by insects may be longer than those of subsequent transmissions (262). In systems in which the test plant is very poorly susceptible, a dosage phenomenon is most easily demonstrated and can be detected throughout the life of the vector (490).

Dosage of the agent injected into insects directly influences the latent periods (292, 496), but by this method "saturation doses" are harder to achieve. It is not known whether saturation or oversaturation produces such effects as abortive or deficient infections (142), or genetic interaction between genomes in the infected cells.

Bioassay.—The examination of all properties relating to the numbers of infective virions or infectious units and the corresponding success of endeavors such as purification depend on the development of an adequate bioassay (490).

Many plant disease agents propagative in their vectors (including those of the yellows diseases) cannot be mechanically transmitted to plants. In rare cases, as with WTV and potato yellow dwarf virus, plants can be inoculated with virus by puncturing the crown with a pin. (68). In certain cases, local lesions can be obtained by rubbing leaves with virus inoculum (34, 39).

Artificial feeding techniques, in which vectors are fed on infectious solution, have proved useful with a number or propagative viruses. Stable viruses, such as CTV (75, 76), BYDV (370), and BWYV (129) have been profitably handled in this way, but the yellows agents and other unstable disease agents have not. Since all the viruses purified to date are relatively stable under optimum ionic conditions, it seems possible that the acquisition of such viruses as WTV, PYDV, and RDV from feeding solutions could now be accomplished. Standardization of the ingested dose, however, is another matter. The best approach to date has consisted of measuring, rather than controlling, uptake volumes (130). Although some leafhoppers have been fed through membranes, most work has been done with aphids. Extensive work on aphid nutrition in recent years has elucidated some of the variables in artificial feeding techniques (12, 13, 106). Among these variables, in addition to medium composition, are microbial contamination of extracts, sucrose concentration, color of the medium, and attractive substances.

The technique of insect injection has been described often (53, 444). Bioassay of infectivity by this technique has proved equally useful for yellows agents and viruses, although certain small viruses are not easily transmitted in this way. Results can be analyzed as quantal responses or graded responses (490). In either case, the analysis depends upon experimental determination of a standard curve from dilutions of a virus preparation.

Theoretically, the standard curve obtained by plotting $\log_{10} \log_{10} 1/1\text{-}p$ versus $\log_{10} c$ (where p is the observed fraction of infected insects, and c is the relative virus concentration) should give a linear measure of quantal response (357, 490). In analyses of dilution curves of leafhopper-borne viruses to date, however, this linearity has not been observed (490). Possible factors responsible for nonlinearity include aggregation of the disease agent, variation in susceptibility among test insects, or more likely, a combination of these and other factors.

Estimates of virus concentration by graded response assays can be made with smaller samples than are required in quantal response assays. This type of analysis requires selection of a parameter of the transmission curve such as the T-50 (Time after injection when 50 per cent of insects have completed their latent period (497). Values of the chosen parameter are plotted as a function of concentration on a standard curve, and observed values can be converted to relative concentration by means of the standard curve. Reddy & Black (364) used the serological titer of insect groups 20 days after injection as the chosen parameter. In our work with the AY agent, we have used the total latent period (the time between injection of the insect and the final appearance of symptoms in plants inoculated by the insects) as the chosen parameter (490). The most meticulous analysis of latent period data has been done with PEMV (456). The computerized program used in those studies should be adaptable to other viruses.

Maintenance and Growth of Pathogens in Tissue Culture

Although the methods of bioassay outlined in the previous section have undergone continual refinement, there are severe inherent limitations involved in assays utilizing whole plants and animals. Among them are the length of time required for experiments, the space and facilities required for the tests, and especially, the inability to study dynamics of infection at the cellular level. The rapid advances of the past decade in animal virology made possible by the development of a tissue culture system attest to the importance of *in vitro* studies. Many arboviruses have been studied in vertebrate cell lines, and as a result much is now known about their intracellular replication. Yet, even with the arboviruses, it has proved useful to develop methods which would permit growth of virus in cultured vector cells (100, 185, 186, 347, 365). Workers with the phytarboviruses have been less fortunate in the failure, to date, of plant tissue cultures to provide a comparable experimental system. The development of a suitable assay for a plant virus in cultured cells of the vector (52, 88, 89, 92) therefore represents an advance of the first magnitude.

The major advance made by Chiu et al. (92) was preceded by many efforts, previously described in detail in several reviews (52, 319, 421, 478). Maramorosch (298) first attempted to culture a plant pathogenic agent in insect vector tissues *in vitro*. Tissues of *Macrosteles fascifrons* that had recently acquired the aster yellows agent from infected plants were suspended in hanging drop culture for 10 days in a defined medium. The titer of AY agent in the tissue pieces increased, but AY was not recovered from the surrounding medium. In later work, AY agent could be recovered (214) only from Malpighian tubules after 14 days incubation following the dissection of various organs from vectors that had fed on AY-infected plants. The data from the single experiment reported, however, were considered (214) to be too meager to permit conclusions regarding possible multiplication of the AY agent *in vitro*. Sinha (419), by use of fluorescent antibody, demonstrated the presence of WTV antigens in various organs maintained in culture for 14 days following their removal from viruliferous leafhoppers. Whole-animal culture, which Seecof (390) found useful for the study of sigma virus in *Drosophila*, has not been accomplished for plant pathogens.

Culture of cells from leafhoppers was achieved by several groups (215, 217, 319, 470), but the maintenance of subcultures proved difficult. A major advance in the culture of vector cells came with the recognition by Hirumi & Maramorosch (216) of a critical stage in the embryonated leafhopper egg at which explants are best removed for culture *in vitro*. Subsequently, by improving several facets of the culture system, monolayers of leafhopper vector cells were cultured and successfully inoculated with wound tumor virus (88) and rice dwarf virus (321). The AY disease agent, however, could not be recovered from cultured epithelial cells of the vector 12 days after their inoculation (306). The rice yellow dwarf agent (RYDA) could not be recovered after inoculation of cultured cells known to be susceptible to rice dwarf virus, even though the cells were derived from the RYDA vector (S. Nasu, personal communication). None of the yellows agents has yet been successfully maintained in tissue culture.

Continuous cell cultures from several leafhopper vectors of plant viruses have been obtained (88), and their susceptibility to infection by the agents for which they serve as natural hosts and vectors has been demonstrated (92, 168).

These developments offer the promise of rapid, precise, and sensitive assay methods for propagative insect-transmitted plant viruses. Also, virus-cell interactions can be studied electron microscopically in a controlled system (218). The breakdown of host specificity frequently encountered in tissue culture offers the hope that the method will not be restricted in its usefulness to viruses propagative in insects whose tissues can be cultured. The multiplication of WTV in nonvector leafhopper cells *in vitro* (89, 218) gives some indication that breakdown in host specificity may be put to use with plant viruses. Also, leafhopper tissues have been infected with the Chilo iridescent virus from a lepidopterous host (318).

The possibility that plant tissue cultures may be useful for the study of plant viruses has received some attention, but the cell wall presents a formidable obstacle to successful inoculation. Wounding of the cell wall, and perhaps of the plasmalemma as well, as generally required for successful inoculation of plant tissues with viruses (49). This severely limits possibilities for quantitative assay and prohibits the synchronous infection of cells which is necessary for study of viral replication. Mitsuhashi & Maramorosch (320) reported the inoculation of callus cells in culture with the aster yellows agent and subsequent recovery of the agent by leafhoppers allowed to feed on the tissues 50 days after inoculation with the AY agent. The corn stunt agent reportedly (302) survives also for a time in cultures of single-cell suspensions derived from infected plants.

Distribution of Pathogens in Tissues

Distribution.—Most of the plant viruses propagative in their insect vectors are restricted to the cytoplasm in both plant and insect hosts; but pea enation mosaic virus has been recorded from nuclei in its plant host (403). In plants, potato yellow dwarf virus was found in cytoplasmic invaginations into the nucleus (288). Maize mosaic virus is thought to be associated with the endoplasmic reticulum (210). Actual sites of virus synthesis in plant and vector cells have been tentatively identified (404, 482). These viroplasms are characterized by the production of ribosomes, other bodies not immediately identifiable, and mature virions. These sites have been identified in fat body, hemocytes, brain, mycetome, and alimentary tract.

In the vector (421, 422), RDV (406), WTV (308, 402), and MRDV (482) occur in many tissues. Measurements of infectivity recovered from various organs also indicate widespread distribution in the vector (25, 418, 419, 424). Inefficient transmission of wound tumor virus is apparently paralleled by low concentrations of particles in salivary glands (190, 402). Hemocytes are active centers of WTV multiplication (192).

The presumed mycoplasmalike yellows agents are primarily restricted to the mature sieve elements of the phloem in their plant hosts (505, 506). It is not clear whether this should be considered an intracellular habitat. Although certain organelles of sieve cells may be degenerate, it is known that paramyxoviruses (87) and Chlamydiae (103) can grow in anucleate cells, so the ability of a cell to support multiplication of a virus or simple organism is not necessarily definable in terms of a complete set of organelles. At present, we do not know whether multiplication of yellows agents in insect vectors is primarily intracellular. The cytopathic effects of WX agent in many tissues were interpreted as indicating widespread intracellular invasion by the agent (501). Mycoplasmalike bodies have been reported from fat body cells (405) and from salivary gland cells (220), but infectivity measurements make it clear that hemolymph is the best source of AY agent (425) and WX agent (497).

Pathology.—Certain viruses, such as rice dwarf virus (334) may produce cytopathological changes in vectors easily visible by light microscopy. Other viruses, such as wound tumor virus, may produce changes most readily studied by electron microscopy (213). The most extensive studies of cytological and histological changes in insects carrying plant pathogens, however, have been made on vectors of yellows agents. The first report of cytological effects of a yellows agent on its vector was that by Littau & Maramorosch (280, 281), who reported nuclear alterations in fat body cells under certain circumstances of infection. Such changes have also been observed with the agent of rice yellow dwarf in *Nephotettix cincticeps* (462), winter wheat mosaic in *Psammotettix striatus* (282), and with WX agent in *Colladonus montanus* (501). The significance of such nuclear alterations is unclear. In the case of WX, an agent which shortens the life of one of its vectors, *C. montanus* (235), the fat body cells in normal insects undergo a complex sequence of depletion (501, 502). Only by studying the cellular changes throughout the life of the insect was it posssible to interpret changes induced by WX infection. The rate of depletion of fat was accelerated by allowing the insects to feed on poor plants (including WX-diseased celery), but was decelerated by infection. Cytopathology induced by WX in fat body tissue included accumulation of unidentified basophilic material, which tended to form inclusions (498, 501), and the formation of swollen or multinucleate cells (495). The dense basophilic material resembled similar material previously envisioned in diseased plants (137). Cytopathology also occurred in salivary glands, neural tissue, alimentary tract (499), mycetome, colleterial gland, corpus cardiacum, pericardial cells, and in connective tissue (501). A strain of the corn stunt agent has recently been shown by R. R. Granados (personal communication) to shorten the life of a vector *Dalbulus elimatus,* and to induce pathological changes in cells of the insect host.

A suggestion that European wheat striate mosaic virus, an agent passed transovarially with high frequency to an inbred line, might be pathogenic to the embryo (416, 488) has been made complex by the demonstration of an effect of inbreeding itself (251).

Certain *Mycoplasma* of vertebrates may produce cytopathology (9, 74). Pathological effects of disease agents on their vectors, however, are rare. In the cases reported here, as in the case of pathogenicity of Semliki forest virus to a mosquito vector (317), it is doubtful whether the affected insect is a primary vector in nature.

Host Ranges

The plant host ranges of viruses and yellows agents alike may be broad, especially when tested experimentally. The aster yellows (261) agent infects plants of many families. Likewise, wound tumor virus has a wide host range (51), as does beet curlytop virus (153). The nature of another agent with a wide host range, that of Pierce's disease of grape (148), is unknown. In the vector, it is common for yellows agents to have broad specificities (391).

Strain Interactions

Previous reviews and research reports (151, 152, 189, 244, 268, 301, 472, 473) provide extensive information on interactions between strains of yellows agents. On the other hand, it is singular that very little work has been done with interactions between strains of viruses which are propagative in their insect vectors. This may reflect a greater proclivity of yellows agents to develop strains, or a peculiarity of yellows symptomology in plants which may increase the ease of recognizing existing strains. Studies with known or probable viruses include: those with CTV (24, 179) whose strains do not interact in plant or vector; rice tungro virus strains which cross-protect in the plant (368); BYDV for which cross-protection in plants (234) can be demonstrated in some cases, but where simultaneous synthesis of strains occurs more frequently (374). Only in the latter case has any attention been directed to virus strain interaction in the vector. Rochow (375) concluded that: "apparent loss of vector specificity has been consistent only following transmissions from plants doubly infected by both virus strains". This result raises the possibility that interactions within infected cells, such as phenotypic mixing or genetic recombination (34) may occur.

Unrelated viruses have been shown to be transmitted independently (332). Interference may eventually be demonstrated, even with unrelated viruses, however, since vaccinia virus has been shown to suppress reovirus multiplication (107).

Infection of insect vectors by strains of yellows agents probably will not be found to interfere with virus infection or with transmission of viruses (146). Even though yellows agent strains exert strong influences on each other, double infections of plantain can easily be obtained by feeding vectors carrying each of two strains of aster yellows (150). More invasive strains of AY (150) and corn stunt (301) are able to replace certain established strains in the vector. It should be pointed out, however, that the status of "strains" of yellows agents is at present unknown, and the degrees of taxonomic diversity between strains of yellows agents transmitted by a single vector such as *Macrosteles fascifrons* may vary widely.

Genetics of Insect Vector and Disease Agent

Clones of vectors, derived from different geographical localities (241, 378, 409) or by selection from laboratory stock (250, 443), can differ widely in the ability to transmit viruses to healthy plants or to their progeny. The inheritance of such characters usually (41, 332), but not always (443), proves to be complex. Curiously, clonal differences in ability to transmit have not been noted, so far as we are aware, in vectors of any of the yellows agents.

Mechanisms of vector incompetence have been carefully investigated. Puncture of the guts of individuals of a nontransmitting race of *Cicadulina mbila* rendered them inoculative after they fed on infected plants, suggesting that a simple "gut barrier" was involved (444). Fluorescent antibody studies indicated that the incompetence of older insects could be explained by a failure of WTV infection to proceed beyond the initial sites in intestinal epithelia (420). Failure of a vector to transmit the WX agent was associated (498) with a lack of pathology, and possible lack of WX multiplication, in the anterior lobe of the salivary glands. AY agent apparently multiplies to a limited extent in the gut but not other tissues of a nonvector agalliian species (423).

Genetics of the pathogens.—Variation in disease agents also has been uncovered by geographical sampling (372, 373) and by selective pressure applied in the laboratory. Generally, continued passage of a disease agent in one host alters its virulence to other hosts. Continual passage of virus through plants may render it totally incompetent in former insect vectors, as has been observed with WTV (58) and PYDV (46). In the case of WTV, insects could not be rendered inoculative by injection with vectorless variants of the virus, nor could viral antigen be detected in the insect after infection (58). Vectorless PYDV could not be acquired by insects feeding on diseased crimson clover. If concentrated suspensions of purified vectorless virus were injected into leafhoppers, some transmission resulted, but the PYDV strain established by such transmission to crimson clover was acquired with lower efficiency than were fully vectorial strains from comparable infected plants (50).

Modification of Host Specificity by Yellows Agent Infection

Certain virus-infected plants are more acceptable to insects (247). The AY agent, however has been shown to modify profoundly the host specificity of certain insects which may acquire it. Severin (392) first noted that AY-infected plants were more favorable for maintaining a number of leafhopper species. Maramorosch (300) made the surprising observation that *Dalbulus maidis* individuals, after feeding on AY-infected asters, were subsequently able to survive on healthy aster, carrot, or rye plants, which are normally alien and unacceptable to them. Because heat treatments, which inactivate AY in vectors destroyed this induced tolerance to new plant hosts, Maramorosch (300) concluded that AY infection was necessary for the effect, and that the "virus" was therefore actually beneficial to the insect (303). Orenski et al. (343), by using radioactive tracers, found that acceptability of a new plant host to *Dalbulus maidis* could not be explained in terms of the amount of food taken up and suggested that a change in the digestibility of food might account for the effect.

The revelation that AY agent may be an organism, rather than a virus, as had been supposed, calls for a re-examination of this phenomenon. It is possible, for example, that metabolites, vital to the insect, but normally obtained only from certain host plants, may be synthesized by the AY agent. It seems doubtful, however, that AY infection should be considered beneficial to the insect. A modification of vector host specificity might increase the incidence of transmission of the agent in the field, but *D. maidis* does not transmit AY. Since AY is not passed transovarially, such modifications would be effective for one generation of leafhoppers only and would constitute a biological dead end. We have noted (unpublished results) in our work that *Macrosteles fascifrons* adults are attracted to those portions of AY-infected asters which show fresh symptoms. Many homopterous insects are attracted to yellow objects of any sort, and one obvious possibility is that the induction of yellowing has been selected by the yellows disease agents as a consequence of this color preference. It is also possible that the AY agent induces the synthesis of an attractant that is most prevalent in yellowed tissues. Whatever the explanation of "beneficial effects," biological significance of the biochemistry of AY infection is more apt to be found in terms of processes beneficial to the AY agent, rather than to nonvector insects.

SUMMARY

Viruses of higher animals transmitted persistently by arthropod vectors, and plant viruses transmitted in the same way, seem to comprise large families whose component subgroups are unrelated. The term "arbovirus" in its broadest usage encompasses the family of mosquito and tick-borne viruses of higher animals. In this review we have proposed the term "phytarbovirus" to describe the corresponding family of plant viruses transmitted persistently by leafhoppers, aphids, and other arthropods. There are viruses in these two groups which have sufficient resemblance to each other to justify placement in certain natural groupings (diplornavirus, rhabdovirus, and picornavirus, for example). Therefore, the concept of phytarbovirus is to be regarded as a provisional term, which possibly may have no place in an ultimate natural classification.

A large class of nonviral plant disease agents (yellows agents) are also transmitted persistently by homopterous insects: these have been tentatively identified as mycoplasmalike organisms, with the exception of the agent of corn stunt disease, which has recently been cultured (86) in cell-free media. Although the basic nature of the yellows agents is profoundly different from that of the phytarboviruses, the dynamics of infection of plant and insect, and the consequent details of transmission by the insects, are similar.

So far as is known, the yellows organisms are restricted to the plant-insect biosystem. The largest group of arboviruses (Group A or B) have no obvious counterparts in the plant viruses. Recent research, however, makes it clear that the disease agents transmitted by arthropods comprise a rich array, and in that array we may expect to find new types of pathogens as well as members of known groups in unexpected hosts.

ACKNOWLEDGMENT

We gratefully acknowledge the helpful suggestions offered by Dr. R. Granados and Dr. R. Purcell in our preparation of this manuscript.

LITERATURE CITED

1. Aapola, A. I. E., Rochow, W. F. Immunodiffusion tests with 3 isolates of barley yellow dwarf virus. *Phytopathology,* **58,** 398 (1968)
2. Acheson, N. H., Tamm, I. Replication of semliki forest virus: an electron microscopic study. *Virology,* **32,** 128–43 (1967)
3. Ahmed, M. E., Black, L. M., Perkins, E. G., Walker, B. L., Kummerow, F. A. Lipid in potato yellow dwarf virus. *Biochem. Biophys. Res. Commun.,* **17,** 103–7 (1964)
4. Anderson, D. R., Barile, M. F. Ultrastructure of *Mycoplasma hominis. J. Bacteriol.,* **90,** 180–92 (1965)
5. Anderson, D. R., Hopps, H. E., Barile, M. F., Bernheim, B. C. Comparison of the ultrastructure of several rickettsiae, ornithosis virus, and *Mycoplasma* in tissue culture. *J. Bacteriol.,* **90,** 1387–1404 (1965)
6. Anderson, D. R., Manaker, R. A. Electron microscopic studies of *Mycoplasma* (PPLO strain 880) in artificial medium and in tissue culture. *J. Natl. Cancer Inst.,* **36,** 139–54 (1966)
7. Andrews, C. H. Viruses and Noah's Ark. *Bacteriol. Rev.,* **29,** 1–8 (1965)
8. Arai, K., Doi, Y., Yora, K., Asuyama, H. Electron microscopy of the potato leafroll virus in leaves of three kinds of host plants and the partial purification of the virus. *Ann. Phytopathol. Soc. Japan,* **35,** 10–15 (1969)
9. Armstrong, D., Henle, G., Somerson, N. L., Hayflick, L. Cytopathogenic mycoplasmas associated with two human tumors. I. Isolation and biological aspects. *J. Bacteriol.,* **90,** 418–24 (1965)
10. Arnott, H. J., Smith, K. M., Fullilove, S. L. Ultrastructure of a cytoplasmic polyhedrosis virus

affecting the monarch butterfly *Danaus plexippus* I. Development of virus and normal polyhedra in the larva. *J. Ultrastruct. Res.*, **24**, 479–507 (1968)

11. Atchison, B. A., Francki, R. I. B., Crowley, N. C. Inactivation of lettuce necrotic yellows virus by chelating agents. *Virology*, **37**, 396–403 (1969)
12. Auclair, J. L. Aphid feeding and nutrition. *Ann. Rev. Entomol.*, **8**, 439–90 (1963)
13. Auclair, J. L., Cartier, J. J. Pea aphid: rearing on a chemically defined diet. *Science*, **142**, 1068–69 (1963)
14. Banttari, E. E., Moore, M. B. Virus cause of blue dwarf of oats and its transmission to barley and flax. *Phytopathology*, **52**, 897–902 (1962)
15. Banttari, E. E., Zeyen, R. J. Chromatographic purification of the oat blue dwarf virus. *Phytopathology*, **59**, 183–86 (1969)
16. Barile, M. F., Malizia, W. F., Riggs, D. B. Incidence and detection of pleuropneumonia-like organisms in cell cultures by fluorescent antibody and cultural procedures. *J. Bacteriol.*, **84**, 130–36 (1962)
17. Beale, H. P., Jones, C. R. Virus diseases of tobacco mosaic and potato yellow dwarf not controlled by certain purified antibiotics. *Contrib. Boyce Thompson Inst.*, **16**, 395–407 (1951)
18. Becker, W. B. The morphology of tern virus. *Virology*, **20**, 318–27 (1963)
19. Bell, A. F. A new disease of cane in North Queensland. *Queensland Agr. J.*, **40**, 460–64 (1933)
20. Bellamy, A. R., Joklik, W. K. Studies on reovirus RNA. II. Characterization of reovirus messenger RNA and of the genome RNA segments from which it is transcribed. *J. Mol. Biol.*, **29**, 19–26 (1967)
21. Bellamy, A. R., Shapiro, L., August, J. T., Joklik, W. K. Studies on reovirus RNA I. Characterization of reovirus genome RNA. *J. Mol. Biol.*, **29**, 1–17 (1967)
22. Bennett, C. W. Interactions of sugarbeet curly top virus and an unusual mutant. *Virology*, **3**, 322–42 (1957)
23. Bennett, C. W. Curly top virus content of the beet leafhopper influenced by virus concentration in diseased plants. *Phytopathology*, **52**, 538–41 (1962)
24. Bennett, C. W. Apparent absence of cross protection between strains of the curly top virus in the beet leafhopper *Circulifer tenellus*. *Phytopathology*, **57**, 207–9 (1967)
25. Bennett, C. W. Studies on properties of the curlytop virus. *J. Agr. Res.*, **50**, 211–41 (1935)
26. Bennett, C. W. Acquisition and transmission of curly top virus by artificially fed beet leafhoppers. *J. Am. Soc. Sugar Beet Technologists*, **11**, 637–48 (1962)
27. Bennett, C. W. Epidemiology of leafhopper-transmitted viruses. *Ann. Rev. Phytopathol.*, **5**, 87–108 (1967)
28. Bennett, C. W., Carsner, E., Coons, G. H., Brandes, E. W. The Argentine curly-top of sugar beet. *J. Agr. Res.*, **72**, 19–47 (1948)
29. Bennett, C. W., Costa, A. S. The Brazilian curly-top of tomato and tobacco resembling North American and Argentine curly-top of sugar beet. *J. Agr. Res.*, **78**, 675–93 (1949)
30. Bennett, C. W., Costa, A. S. Sowbane mosaic caused by a seed-transmitted virus. *Phytopathology*, **51**, 546–50 (1961)
31. Bennett, C. W., Wallace, H. E. Relation of the curly-top virus to the vector, *Eutettix tenellus*. *J. Agr. Res.*, **56**, 31–51 (1938)
32. Bergold, G. H., Suarez, O. M., Munz, K. Multiplication in and transmission by *Aedes aegypti* of vesicular stomatitis virus. *J. Invertebrate Pathol.*, **11**, 406–28 (1968)
33. Best, R. J. Thermal inactivation of tomato spotted wilt virus. *Australian J. Exptl. Biol. Med. Sci.*, **24**, 21–25 (1946)
34. Best, R. J. Tomato spotted wilt virus. *Advan. Virus Res.*, **13**, 65–146 (1968)
35. Best, R. J., Katekar, G. F. Lipid in a purified preparation of tomato spotted wilt virus. *Nature*, **203**, 671–72 (1964)
36. Best, R. J., Palk, B. A. Electron microscopy of strain E of tomato spotted wilt virus and comments on its probable biosynthesis. *Virology*, **23**, 445–60 (1964)
37. Best, R. J., Samuel, G. The effect

of various chemical treatments on the activity of the viruses of spotted wilt and tobacco mosaic. *Ann. Appl. Biol.*, **23**, 759–80 (1936)
38. Bils, R. F., Hall, C. E. Electron microscopy of wound-tumor virus. *Virology*, **17**, 123–30 (1962)
39. Black, L. M. Properties of the potato yellow-dwarf virus. *Phytopathology*, **28**, 863–74 (1938)
40. Black, L. M. Further evidence for multiplication of the aster-yellows virus in the aster leafhopper. *Phytopathology*, **31**, 120–35 (1941)
41. Black, L. M. Genetic variation in the clover leafhopper's ability to transmit potato yellow-dwarf virus. *Genetics*, **28**, 200–9 (1943)
42. Black, L. M. Some properties of aster-yellows virus. *Phytopathology*, **33**, 2 (1943) (Abstr.)
43. Black, L. M. Some viruses transmitted by agallian leafhoppers. *Proc. Am. Phil. Soc.*, **88**, 132–44 (1944)
44. Black, L. M. A plant virus that multiplies in its insect vector. *Nature*, **166**, 852–53 (1950)
45. Black, L. M. Viruses that reproduce in plants and insects. *Ann. N. Y. Acad. Sci.*, **56**, 398–413 (1953)
46. Black, L. M. Loss of vector transmissibility by viruses normally insect transmitted. *Phytopathology*, **43**, 466 (Abstr.) (1953)
47. Black, L. M. Occasional transmission of some plant viruses through the eggs of their insect vectors. *Phytopathology*, **43**, 9–10 (1953)
48. Black, L. M. Concepts and problems concerning purification of labile insect-transmitted plant viruses. *Phytopathology*, **45**, 208–16 (1955)
49. Black, L. M. Viruses and other pathogenic agents in plant tissue cultures. *J. Natl. Cancer Inst.*, **19**, 663–85 (1957)
50. Black, L. M. Biological cycles of plant viruses in insect vectors. In *The Viruses*, **2**, 157–85. (Burnet, F. M., Stanley, W. M., Eds., Academic Press, New York, 1959)
51. Black, L. M. Physiology of virus-induced tumors in plants. In *Handbuch der Pflanzenphysiologie*, **XV**/2, 236–66. (Lang, A., ed., Springer-Verlag, New York, 1965)
52. Black, L. M. Insect tissue cultures as tools in plant virus research. *Ann. Rev. Phytopath.*, **7**, 73–100 (1969)
53. Black, L. M., Brakke, M. K. Multiplication of wound-tumor virus in an insect vector. *Phytopathology*, **42**, 269–73 (1952)
54. Black, L. M., Brakke, M. K., Vatter, A. E. Purification and electron microscopy of tomato spotted-wilt virus. *Virology*, **20**, 120–30 (1963)
55. Black, L. M., Maramorosch, K., Brakke, M. K. Filtration and sedimentation of wound-tumor virus. *Phytopathology*, **40**, 2 (1950)
56. Black, L. M. Markham, R. Abstracts of communications presented at the joint meeting of the "Nederlandse Planteziektenkundige Vereniging" and the "Nederlandse Kring Voor Plantevirologie" on 24 May 1962, Wageningen. *Neth. J. Plant Pathol.*, **69**, 215–17 (1963)
57. Black, L. M., Smith, K. M., Hills, G. J., Markham, R. Ultrastructure of potato yellow-dwarf virus. *Virology*, **27**, 446–49 (1965)
58. Black, L. M., Wolcyrz, S., Whitcomb, R. F. A vectorless strain of wound-tumor virus. *Intern, Congr. Microbiol., 7th, Stockholm, Abstr. Communications*, 255 (1958) (Abstr.)
59. Blattný, C. Bemerkungen zur Epidemiologie des Stolburs und der verwandten Krankheiten. *Proc. Conf. on Potato Virus Diseases (Lisse-Wageningen), 3rd*, 255–63 (1958)
60. Bordet, J. La morphologie du microbe de la péripneumonie des bovidés. *Ann. Inst. Pasteur*, **24**, 161–67 (1910)
61. Borges, M. de L. V., David-Ferreira, J. F. Presence of *Mycoplasma* in *Lycopersicon esculentum* Mill. with *Mal azul. Bol. Soc. Brotheriana, Ser. 2*, **42**, 321–33 (1968)
62. Borrel, Dujardin-Beaumetz, Jeantet, Jouan. Le microbe de la péripneumonie. *Ann. Inst. Pasteur*, **24**, 168–79 (1910)
63. Borsa, J., Graham, A. F. RNA polymerase activity in purified virions. *Biochem. Biophys. Res. Commun.*, **33**, 895–901 (1968)
64. Bowne, J. G., Jones, R. H. Observations on bluetongue virus in the salivary glands of an insect vector, *Culicoides variipennis*. *Virology*, **30**, 127–33 (1966)
65. Bowyer, J. W., Atherton, J. G., Teakle, D. S., Ahern, G. A.

Mycoplasma-like bodies in plants affected by legume little leaf, tomato big bud, and lucerne witches' broom diseases. *Australian J. Biol. Sci.*, **22**, 271–74 (1969)
66. Bozarth, R. F., Chow, C. C. Pea enation mosaic virus: purification and properties. *Contrib. Boyce Thompson Inst.*, **23**, 301–9 (1966)
67. Brakke, M. K. Stability of potato yellow-dwarf virus. *Virology*, **2**, 463–76 (1956)
68. Brakke, M. K., Vatter, A. E., Black, L. M. Size and shape of wound-tumor virus. *Abnormal and Pathological Plant Growth Brookhaven Symp. Biol.*, **6**, 137–56 (1954)
69. Brčák, J., Králík, O. Mycoplasma-like microorganism and virus particles in the leafhopper *Javesella pellucida* (F.) transmitting the oat sterile dwaft disease. *Biol. Plantarum*, **11**, 95–96 (1969)
70. Brčák, J., Králík, O., Vacke, J. Virions of the oat sterile-dwarf virus in the midgut cells of its vector *Javesella pellucida. Intern. Symp. Plant Pathol., New Delhi, Dec., 1966–Jan. 1967*, 93–94 (1966) (Abstr.)
71. Broadbent, L. Disease control through vector control. In *Viruses, Vectors, and Vegetation*, 593–630. (Maramorosch, K., Ed., Interscience, New York, 1969)
72. Bruehl, G. W. Barley yellow dwarf, a virus disease of cereals and grasses. *Am. Phytopathol. Soc.* Monogr. No. 1, 52 pp. (1961)
73. Bustrillos, A. D. *Purification, Serology, and Electron Microscopy of Pea Enation Mosaic Virus.* (Doctoral thesis, Michigan State University, 1964)
74. Butler, M., Leach, R. H. A Mycoplasma which induces acidity and cytopathic effect in tissue culture. *J. Gen. Microbiol.*, **34**, 285–94 (1964)
75. Carter, W. A technic for use with homopterous vectors of plant disease, with special reference to the sugarbeet leaf hopper, *Eutettix tenellus* (Baker). *J. Agr. Res.*, **34**, 449–51 (1927)
76. Carter, W. An improvement in the technique for feeding homopterous insects. *Phytopathology*, **18**, 246–47 (1928)
77. Carter, W. *Peregrinus maidis* (Ashm) and the transmission of corn mosaic. I. Incubation period and longevity of the virus in the insect. *Ann. Entomol. Soc. Am.*, **34**, 551–56 (1943)
78. Carter, W. *Insects in Relation to Plant Disease.* (Interscience Publ., New York and London, 705 pp., 1962)
79. Cartwright, B., Pearce, C. A. Evidence for a host cell component in vesicular stomatitis virus. *J. Gen. Virol.*, **2**, 207–14 (1968)
80. Casals, J. Filtration of arboviruses through "Millipore" membranes. *Nature*, **217**, 648–49 (1968)
81. Chalfant, R. B., Chapman, R. K. Transmission of cabbage viruses A and B by the cabbage aphid and the green peach aphid. *J. Econ. Entomol.*, **55**, 584–90 (1962)
82. Chamberlain, R. W. Arboviruses, the arthropod-borne animal viruses. *Current Topics Microbiol. Immunol.*, **42**, 38–58 (1968)
83. Chambers, T. C., Crowley, N. C., Francki, R. I. B. Localization of lettuce necrotic yellows virus in host leaf tissue. *Virology*, **27**, 320–28 (1965)
84. Chambers, T. C., Francki, R. I. B. Localization and recovery of lettuce necrotic yellows virus from xylem tissues of *Nicotiana glutinosa. Virology*, **29**, 673–76 (1966)
85. Chanock, R. M., Hayflick, L., Barile, M. F. Growth on artificial medium of an agent associated with atypical pneumonia and its identification as a PPLO. *Proc. Natl. Acad. Sci., U. S.*, **48**, 41–49 (1962)
86. Chen, T. A., Granados, R. R. Plant-pathogenic mycoplasma: *in vitro* maintenance and transmission to *Zea mays* L. plants. *Science* (In press)
87. Cheyne, I. M., White, D. O. Growth of paramyxoviruses in anucleate cells. *Australian J. Exptl. Biol. Med. Sci.*, **47**, 145–48 (1969)
88. Chiu, R. J., Black, L. M. Monolayer cultures of insect cell lines and their inoculation with a plant virus. *Nature*, **215**, 1076–78 (1967)
89. Chiu, R. J., Black, L. M. Assay of wound tumor virus by fluorescent cell counting technique. *Virology*, **37**, 667–77 (1969)
90. Chiu, R. J., Jean, J. H., Chen, M. H., Lo, T. C. Transmission of transitory yellowing virus of rice by

two leafhoppers. *Phytopathology,* **58,** 740–45 (1968)

91. Chiu, R. J., Lo, T. C., Pi, C. L., Chen, M. H. Transitory yellowing of rice and its transmission by the leafhopper *Nephotettix apicalis apicalis* (Motsch.). *Botan. Bull. Acad. Sinica,* **6,** 1–18 (1965)
92. Chiu, R. J., Reddy, D. V. R., Black, L. M. Inoculation and infection of leafhopper tissue cultures with a plant virus. *Virology,* **30,** 562–66 (1966)
93. Chiykowski, L. N., Chapman, R. K. Migration of the six-spotted leafhopper in central North America. *Univ. Wisconsin Res. Bull. 261:* Part 2, 21–45 (1965)
94. Chu, H. P., Horne, R. W. Electron microscopy of *Mycoplasma gallisepticum* and *Mycoplasma mycoides* using the negative staining technique and their comparison with *Myxovirus. Ann. N. Y. Acad. Sci.,* **143,** 190–203 (1967)
95. Cohen, R., Purcell, R., Steere, R. L. Ultrafiltration of the aster yellows agent. *Phytopathology,* **59,** 1555 (1969) (Abstr.)
96. Cohen, S. The occurrence in the body of *Bemisia tabaci* of a factor apparently related to the phenomenon of "periodic acquisition" of tomato yellow leaf curl virus. *Virology,* **31,** 180–83 (1967)
97. Cohen, S. *In vivo* effects in whiteflies of a possible antiviral factor. *Virology,* **37,** 448–54 (1969)
98. Cohen, S., Harpaz, I. Periodic, rather than continual acquisition of a new tomato virus by its vector, the tobacco whitefly *(Bemisia tabaci* Gennadius*). Entomol. Exptl. Appl.,* **7,** 155–66 (1964)
99. Cohen, S., Nitzany, F. E. Transmission and host range of the tomato yellow leaf curl virus. *Phytopathology,* **56,** 1127–31 (1966)
100. Converse, J. L., Nagle, S. C., Jr. Multiplication of yellow fever virus in insect tissue cell cultures. *J. Virol.,* **1,** 1096–97 (1967)
101. Cooper, P. D., Bellett, A. J. D. A transmissible interfering component of vesicular stomatitis virus preparations. *J. Gen. Microbiol.,* **21,** 485–97 (1959)
102. Costa, A. S. White flies as virus vectors. In *Viruses, Vectors, and Vegetation,* 95–119. (Maramorosch, K., Ed., Interscience, New York, 1969)
103. Crocker, T. T., Eastwood, J. M. Subcellular cultivation of a virus: growth of ornithosis virus in nonnucleate cytoplasm. *Virology,* **19,** 23–31 (1963)
104. Crowley, N. C. Factors affecting the local lesion response of *Nicotiana glutinosa* to lettuce necrotic yellows virus. *Virology,* **31,** 107–13 (1967)
105. Crowley, N. C., Harrison, B. D., Francki, R. I. B. Partial purification of lettuce necrotic yellows virus. *Virology,* **26,** 290–96 (1965)
106. Dadd, R. H., Mittler, T. E. Studies on the artificial feeding of the aphid *Myzus persicae* (Sulzer)—III. Some major nutritional requirements. *J. Insect Physiol.,* **11,** 717–43 (1965)
107. Dales, S., Silverberg, H. Controlled double infection with unrelated animal viruses. *Virology,* **34,** 531–43 (1968)
108. Dales, S., Gomatos, P. J., Hsu, K. C. The uptake and development of reovirus in strain L cells followed with labeled viral ribonucleic acid and ferritin-antibody conjugates. *Virology,* **25,** 193–211 (1965)
109. Davis, R. E., Whitcomb, R. F. Spectrum of antibiotic sensitivity of aster yellows disease in insects and plants. *Phytopathology,* **59,** 1556 (1969) (Abstr.)
110. Davis, R. E., Whitcomb, R. F. Evidence on possible *Mycoplasma*-like etiology of aster yellows disease: I. Suppression of symptom development in plants by antibiotics. (In preparation)
111. Davis, R. E., Whitcomb, R. F., Steere, R. L. Chemotherapy of aster yellows disease. *Phytopathology,* **58,** 884 (1968) (Abstr.)
112. Davis, R. E., Whitcomb, R. F., Steere, R. L. Remission of aster yellows disease by antibiotics. *Science,* **161,** 793–94 (1968)
113. Day, M. F. Mechanisms of transmission of viruses by arthropods. *Exptl. Parasitol.,* **4,** 387–418 (1955)
114. Day, M. F. The mechanism of the transmission of potato leaf roll virus by aphids. *Australian J. Biol. Sci.,* **8,** 498–513 (1955)
115. Day, M. F., Irzykiewicz, H., McKinnon, A. Observations on the feed-

ing of the virus vector *Orosius argentatus* (Evans), and comparisons with certain other jassids. *Australian J. Sci. Res., Ser. B,* **5,** 128–42 (1952)

116. Day, M. F., Venables, D. G. The transmission of cauliflower mosaic by aphids. *Australian J. Biol. Sci.,* **14,** 187–97 (1961)
117. Diener, T. O., Raymer, W. B. Potato spindle tuber virus: A plant virus with properties of a free nucleic acid. *Science,* **158,** 378–81 (1967)
118. Diener, T. O., Raymer, W. B. Potato spindle tuber virus: A plant virus with properties of a free nucleic acid. II. Characterization and partial purification. *Virology,* **37,** 351–66 (1969)
119. Dienes, L. Morphology and reproductive processes of L forms of bacteria. I. Streptococci and staphylococci. *J. Bacteriol.,* **92,** 693–702 (1967)
120. Dienes, L., Bullivant, S. Morphology and reproductive processes of the L-forms of bacteria. II. Comparative study of L-forms and *Mycoplasma* with the electron microscope. *J. Bacteriol.,* **95,** 672–87 (1968)
121. Dobroscky, I. D. Is the aster yellows virus detectable in its insect vector? *Phytopathology,* **19,** 1009–15 (1929)
122. Dobroscky, I. D. Morphological and cytological studies on the salivary glands and alimentary tract of *Cicadula sexnotata* (Fallén), the carrier of aster yellows virus. *Contrib. Boyce Thompson Inst.,* **3,** 39–58 (1931)
123. Dobroscky, I. D. Studies on cranberry false blossom disease and its insect vector. *Contrib. Boyce Thompson Inst.,* **3,** 59–83 (1931)
124. Doi, Y., Teranaka, M., Yora, K., Asuyama, H. Mycoplasma- or PLT group-like microorganisms found in the phloem elements of plants infected with mulberry dwarf, potato witches' broom, aster yellows, or paulownia witches' broom. *Ann. Phytopathol. Soc. Japan,* **33,** 259–66 (1967)
125. Domermuth, C. H., Nielsen, M. H., Freundt, E. A., Birch-Andersen, A. Ultrastructure of *Mycoplasma* species. *J. Bacteriol.,* **88,** 727–44 (1964)
126. Drake, D. C., Chapman, R. K. Evidence for long distance migration of the six-spotted leafhopper into Wisconsin. *Univ. Wisconsin Res. Bull. 261,* Part 1, 3–20 (1965)
127. Duffus, J. E. Possible multiplication in the aphid vector of sowthistle yellow vein virus, a virus with an extremely long insect latent period. *Virology,* **21,** 194–202 (1963)
128. Duffus, J. E. Radish yellows, a disease of radish, sugar beet and other crops. *Phytopathology,* **50,** 389–94 (1960)
129. Duffus, J. E., Gold, A. H. Transmission of beet western yellows virus by aphids feeding through a membrane. *Virology,* **27,** 388–90 (1965)
130. Duffus, J. E., Gold, A. H. Relationship of tracer-measured aphid feeding to acquisition of beet western yellows virus and to feeding inhibitors in plant extracts. *Phytopathology,* **57,** 1237–41 (1967)
131. Duffus, J. E., Gold, A. H. Membrane feeding and infectivity neutralization used in a serological comparison of potato leafroll and beet western yellows viruses. *Virology,* **37,** 150–53 (1969)
132. Dunnebacke, T. H., Kleinschmidt, A. K. Ribonucleic acid from reovirus as seen in protein monolayers by electron microscopy. *Z. Naturforsch.,* **22,** 159–64 (1967)
133. Eaton, M. D. Pleuropneumonia-like organisms and related forms. *Ann. Rev. Microbiol.,* **19,** 379–406 (1965)
134. Ehrman, L. A study of infectious hybrid sterility in *Drosophila paulistorum. Proc. Natl. Acad. Sci. U. S.,* **58,** 195–98 (1967)
135. Ehrman, L. Antibiotics and infectious hybrid sterility in *Drosophila paulistorum. Mol. Gen. Genet.,* **103,** 218–22 (1968)
136. Els, H. J., Verwoerd, D. W. Morphology of bluetongue virus. *Virology,* **38,** 213–19 (1969)
137. Esau, K. Phloem degeneration in celery infected with yellow leafroll virus of peach. *Virology,* **6,** 348–56 (1958)
138. Esau, K. *Plants, Viruses and Insects.* (Harvard Univ. Press, Cambridge, Mass., 110 pp., 1961)
139. Esau, K. Anatomy of plant virus

infections. *Ann. Rev. Phytopathol.*, **5,** 45–76 (1967)

140. Everett, T. R., Lamey, H. A. Hoja Blanca. In *Viruses, Vectors, and Vegetation,* 361–77. (Maramorosch, K., Ed.; Interscience, New York, 1969)
141. Field, E. J., Farmer, F., Caspary, E. A., Joyce, G. Susceptibility of scrapie agent to ionizing radiation. *Nature,* **222,** 90–91 (1969)
142. Fraser, K. B. Defective and delayed myxovirus infections. *Brit. Med. Bull.*, **23,** 178–84 (1967)
143. Frazier, N. W. A survey of the Mediterranean region for the beet leafhopper. *J. Econ. Entomol.*, **46,** 551–54 (1953)
144. Frazier, N. W., Posnette, A. F. Transmission and host range studies of strawberry green-petal virus. *Ann. Appl. Biol.*, **45,** 580–88 (1957)
145. Frazier, N. W., Sylvester, E. S. Half-lives of transmissibility of two aphid-borne viruses. *Virology,* **12,** 233–44 (1960)
146. Frederiksen, R. A. Simultaneous infection and transmission of two viruses in flax by *Macrosteles fascifrons. Phytopathology,* **54,** 1028–30 (1964)
147. Freitag, J. H. Negative evidence on multiplication of curly-top virus in the beet leafhopper, *Eutettix tenellus. Hilgardia,* **10,** 305–42 (1936)
148. Freitag, J. H. Host range of the Pierce's disease virus of grapes as determined by insect transmission. *Phytopathology,* **41,** 920–34 (1951)
149. Freitag, J. H. Mandibulate insects as vectors of plant viruses. *Proc. Intern. Congr. Entomol., 10th, Montreal,* **3,** 205–9 (1956)
150. Freitag, J. H. Interaction and mutual suppression among three strains of aster yellows virus. *Virology,* **24,** 401–13 (1964)
151. Freitag, J. H. Interaction between strains of aster yellows virus in the six-spotted leafhopper *Macrosteles fascifrons. Phytopathology,* **57,** 1016–24 (1967)
152. Freitag, J. H. Interactions of plant viruses and virus strains in their insect vectors. In *Viruses, Vectors, and Vegetation,* 303–25. Maramorosch, K., Ed., Interscience, New York, 1969)
153. Freitag, J. H., Severin, H. H. P. Ornamental flowering plants experimentally infected with curly top. *Hilgardia,* **10,** 263–302 (1936)
154. Freundt, E. A. Morphology and classification of the PPLO. *Ann. N. Y. Acad. Sci.*, **79,** 312–25 (1960)
155. Fukushi, T. Transmission of the virus through the eggs of an insect vector. *Proc. Imp. Acad. (Tokyo),* **9,** 457–60 (1933)
156. Fukushi, T. Plants susceptible to dwarf disease of rice plant. *Trans. Sapporo Natl. Hist. Soc.*, **13,** 162–66 (1934)
157. Fukushi, T. Studies on the dwarf disease of rice plant. *J. Fac. Agr. Hokkaido Imp. Univ.*, **37,** 41–164 (1934)
158. Fukushi, T. An insect vector of the dwarf disease of rice plant. *Proc. Imp. Acad. (Tokyo),* **13,** 328–31 (1937)
159. Fukushi, T. Retention of virus by its insect vectors through several generations. *Proc. Imp. Acad. (Tokyo),* **15,** 142–45 (1939)
160. Fukushi, T. Further studies on the dwarf disease of rice plant. *J. Fac. Agr., Hokkaido Imp. Univ.*, **45,** 83–154 (1940)
161. Fukushi, T. Relationships between propagative rice viruses and their vectors. In *Viruses, Vectors, and Vegetation,* 279–301. (Maramorosch, K., Ed.; Interscience, New York, 1969)
162. Fukushi, T., Shikata, E. Localization of rice dwarf virus in its insect vector. *Virology,* **21,** 503–5 (1963)
163. Fukushi, T., Shikata, E. Fine structure of rice dwarf virus. Virology, **21,** 500–3 (1963)
164. Fukushi, T., Shikata, E., Kimura, I. Some morphological characters of rice dwarf virus. *Virology,* **18,** 192–205 (1962)
165 Gálvez, G. E. Purification and characterization of rice tungro virus by analytical density-gradient centrifugation. *Virology,* **35,** 418–26 (1968)
166. Gamez, R., Black, L. M. Particle counts of wound tumor virus during its peak concentration in leafhoppers. *Virology,* **34,** 444–51 (1968)
167. Gamez, R., Black, L. M., MacLeod, R. Reexamination of the serological relationship between wound tumor virus and reovirus. *Virology,* **32,** 163–65 (1967)

168. Gamez, R., Chiu, R. J. The minimum concentration of a plant virus needed for infection of monolayers of vector cells. *Virology,* **34,** 356–57 (1968)
169. Gerola, F. M., Bassi, M. An electron microscopy study of leaf vein tumors from maize plants experimentally infected with maize rough dwarf virus. *Caryologia,* **19,** 13–40 (1966)
170. Gerola, F. M., Bassi, M., Lovisolo, O., Vidano, C. Virus-like particles in both maize plants infected with maize rough dwarf virus and the vector *Laodelphax striatellus* Fallén. *Phytopathol. Z.,* **56,** 97–99 (1966)
171. Giannotti, J., Devauchelle, G., Vago, C. Micro-organismes de type mycoplasme chez une cicadelle et une plante infectées par la phyllodie. *Compt. Rend. Ser. D,* **266,** 2168–70 (1968)
172. Giannotti, J., Marchoux, G., Vago, C., Duthoit, J. Micro-organismes de type mycoplasme dans les cellules libériennes de *Solanum lycopersicum L.* atteinte de stolbur. *Compt. Rend., Ser. D,* **267,** 454–56 (1968)
173. Giannotti, J., Morvan, G., Vago, C. Micro-organismes de type mycoplasme dans les cellules libériennes de *Malus sylvestris* L. atteint de la maladie des proliférations. *Comp. Rend., Ser. D,* **267,** 76–77 (1968)
174. Giannotti, J., Caudwell, A., Vago, C., Duthoit, J. Isolement et purification de micro-organismes de type mycoplasme à partir de vignes atteintes de Flavescence dorée. *Comp. Rend., Ser. D,* **268,** 845–47 (1969)
175. Giannotti, J., Devauchelle, G., Marchoux, G., Vago, C. Recherches sur le pléomorphisme des microorganismes de type mycoplasme chez les plantes atteintes de jaunisses. *Comp. Rend. Ser. D,* **268,** 1354–56 (1969)
176. Giannotti, J., Vago, C., Devauchelle, G., Marchoux, G. Recherches sur les microorganismes de type mycoplasme dans les cicadelles vectrices et dans les végétaux atteints de jaunisses. *Entomol. Exptl. Appl.,* **11,** 470–74 (1968)
177. Gibbs, A. J. Plant virus classification. *Advan. Virus Res.,* **14,** 263–328 (1969)
178. Gibbs, A. J., Harrison, B. D., Woods, R. D. Purification of pea enation mosaic virus. *Virology,* **29,** 348–51 (1966)
179. Giddings, N. J. Some interrelationships of virus strains in sugarbeet curly top. *Phytopathology,* **40,** 377–88 (1950)
180. Gill, C. C. Cyclical transmissibility of barley yellow dwarf virus from oats with increasing age of infection. *Phytopathology,* **59,** 23–28 (1969)
181. Gold, A. H., Duffus, J. E. Infectivity neutralization–a serological method as applied to persistent viruses of beets. *Virology,* **31,** 308–13 (1967)
182. Gomatos, P. J., Tamm, I. Animal and plant viruses with double-helical RNA. *Proc. Natl. Acad. Sci. U.S.,* **50,** 878–85 (1963)
183. Gomatos, P. J., Tamm, I., Dales, S., Franklin, R. M. Reovirus type 3: physical characteristics and interaction with L cells. *Virology,* **17,** 441–54 (1962)
184. Gourlay, R. N., Thrower, K. J. Morphology of *Mycoplasma mycoides* thread-phase growth. *J. Gen. Microbiol.,* **54,** 155–59 (1968)
185. Grace, T. D. C. Establishment of a line of mosquito (*Aedes aegypti* L.) cells grown *in vitro. Nature,* **211,** 366–67 (1966)
186. Grace, T. D. C. Tissue culture for arthropod viruses. *Trans. N. Y. Acad. Sci., Ser. II,* **21,** 237–41 (1959)
187. Granados, R. R. Chemotherapy of corn stunt disease. *Phytopathology,* **59,** 1556 (1969) (Abstr.)
188. Granados, R. R., Chapman, R. K. Identification of some aster-yellows virus strains and their transmission by the aster leafhopper, *Macrosteles fascifrons. Phytopathology,* **58,** 1685–92 (1968)
189. Granados, R. R., Chapman, R. K. Heat inactivation and interactions of four aster-yellows virus strains in their vector, *Macrosteles fascifrons* (Stål), *Virology,* **36,** 333–42 (1968)
190. Granados, R. R., Hirumi, H., Maramorosch, K. Electron microscopic evidence for wound-tumor virus accumulation in various organs of an inefficient leafhopper vector, *Agalliopsis novella. J. Invertebrate Pathol.,* **9,** 147–59 (1967)
191. Granados, R. R., Maramorosch, K.,

Shikata, E. *Mycoplasma:* Suspected etiologic agent of corn stunt. *Proc. Natl. Acad. Sci., U. S.,* **60,** 841–44 (1968)

192. Granados, R. R., Ward, L. S., Maramorosch, K. Insect viremia caused by a plant-pathogenic virus: Electron microscopy of vector hemocytes. *Virology,* **34,** 790–96 (1968)
193. Hackett, A. J. A possible morphologic basis for the autoinference phenomenon in vesicular stomatitis virus. *Virology,* **24,** 51–59 (1964)
194. Hackett, A. J., Sylvester, E. S., Richardson, J., Wood, P. Comparative electron micrographs of sowthistle yellow vein and vesicular stomatitis viruses. *Virology,* **36,** 693–96 (1968)
195. Hampton, R. O., Stevens, J. O., Allen, T. C. Mechanically transmissible *Mycoplasma* from naturally infected peas. *Plant Disease Reptr.,* **53,** 499–503 (1969)
196. Harpaz, I. Further studies on the vector relations of the maize rough dwarf virus (MRDV). *Maydica,* **11,** 18–23 (1966)
197. Harrap, K. A., Juniper, B. E. The internal structure of an insect virus. *Virology,* **29,** 175–78 (1966)
198. Harrison, B. D. Studies on the behavior of potato leaf roll and other viruses in the body of their aphid vector *Myzus persicae* (Sulz.). *Virology,* **6,** 265–77 (1958)
199. Harrison, B. D., Crowley, N. C. Properties and structure of lettuce necrotic yellows virus. *Virology,* **26,** 297–310 (1965)
200. Hartzell, A. Movement of intracellular bodies associated with peach yellows. *Contrib. Boyce Thompson Inst.,* **8,** 375–88 (1937)
201. Haselkorn, R. Physical and chemical properties of plant viruses. *Ann. Rev. Plant Physiol.,* **17,** 137–54 (1966)
202. Hayashi, Y., Bird, F. T. The use of sucrose gradients in the isolation of cytoplasmic-polyhedrosis virus particles. *J. Invertebrate Pathol.,* **11,** 40–44 (1968)
203. Hayashi, Y., Bird, F. T. Properties of a cytoplasmic-polyhedrosis virus from the white-marked tussock moth. *J. Invertebrate Pathol.,* **12,** 140 (1968)
204. Hayashi, Y., Kawase, S. Base pairing in ribonucleic acid extracted from the cytoplasmic polyhedra of the silkworm. *Virology,* **23,** 611–14 (1964)
205. Hayflick, L. Decontaminating tissue cultures infected with pleuropneumonia-like organisms. *Nature,* **185,** 783–84 (1960)
206. Hayflick, L., Ed. Biology of the mycoplasma. *Ann. N. Y. Acad. Sci.,* **143,** 1–824 (1967)
207. Hayflick, L., Chanock, R. M. *Mycoplasma* species of man. *Bacteriol. Rev.,* **29,** 185–221 (1965)
208. Hayflick, L., Stinebring, W. R. Intracellular growth of pleuropneumonialike organisms (PPLO) in tissue culture and *in ovo. Ann. N. Y. Acad. Sci.,* **170,** 433–49 (1960)
209. Herold, F., Bergold, G. H., Weibel, J. Isolation and electron microscopic demonstration of a virus infecting corn (*Zea mays* L.). *Virology,* **12,** 335–47 (1960)
210. Herold, F., Munz, K. Electron microscopic demonstration of virus-like particles in *Peregrinus maidis* following acquisition of maize mosaic virus. *Virology,* **25,** 412–17 (1965)
211. Herold, F., Munz, K. Virus particles in apparently healthy *Peregrinus maidis. J. Virol.,* **1,** 1028–36 (1967)
212. Herold, F., Trujillo, G., Munz, K. Virus-like particles related to Hoja Blanca disease of rice. *Phytopathology,* **58,** 546–47 (1968)
213. Hirumi, H., Granados, R. R., Maramorosch, K. Electron microscopy of a plant-pathogenic virus in the nervous system of its insect vector. *J. Virol.,* **1,** 430–44 (1967)
214. Hirumi, H., Maramorosch, K. Recovery of aster yellows virus from various organs of the insect vector, *Macrosteles fascifrons. Contrib. Boyce Thompson Inst.,* **22,** 141–52 (1963)
215. Hirumi, H., Maramorosch, K. Insect tissue culture: further studies on the cultivation of embryonic leafhopper tissues *in vitro. Contrib. Boyce Thompson Inst.,* **22,** 343–52 (1964)
216. Hirumi, H., Maramorosch, K. Insect tissue culture: use of blastokinetic stage of leafhopper embryo. *Science,* **144,** 1465–67 (1964)
217. Hirumi, H., Maramorosch, K. The *in*

vitro cultivation of embryonic leafhopper tissues. *Exptl. Cell Res.*, **36,** 625–31 (1964)
218. Hirumi, H., Maramorosch, K. Electron microscopy of wound tumor virus in cultured embryonic cells of the leafhopper *Macrosteles fascifrons*. Intern. colloquium on invertebrate tissue culture, 2nd, Istituto Lombardo: Fondazione Baselli, 203–47 (1967)
219. Hirumi, H., Maramorosch, K. Further evidence for a mycoplasma etiology of aster yellows. *Phytopathology*, **59,** 1030–31 (1969) (Abstr.)
220. Hirumi, H., Maramorosch, K. Mycoplasma-like bodies in the salivary glands of insect vectors carrying the aster yellows agent. *J. Virol.*, **3,** 82–84 (1969)
221. Hitchborn, J. H., Hills, G. J., Hull, R. Electron microscopy of virus-like particles found in diseased *Plantago lanceolata* in Britain. *Virology*, **28,** 768–72 (1966)
222. Hosaka, Y., Aizawa, K. The fine structure of the cytoplasmic-polyhedrosis virus of the silkworm, *Bombyx mori* (Linnaeus). *J. Insect Pathol.*, **6,** 53–77 (1964)
223. Howatson, A. F., Whitmore, G. F. The development and structure of vesicular stomatitis virus. *Virology*, **16,** 466–78 (1962)
224. Hsiung, G. D. Use of ultrafiltration for animal virus grouping. *Bacteriol. Rev.*, **29,** 477–86 (1965)
225. Huang, A. S., Greenawalt, J. W., Wagner, R. R. Defective T particles of vesicular stomatitis virus. I. Preparation morphology and some biologic properties. *Virology*, **30,** 161–72 (1966)
226. Huang, A. S., Wagner, R. R. Comparative sedimentation coefficients of RNA extracted from plaque-forming and defective particles of vesicular stomatitis virus. *J. Mol. Biol.*, **22,** 381–84 (1966)
227. Hunter, G. D., Kimberlin, R. H., Gibbons, R. A. Scrapie: a modified membrane hypothesis. *J. Theoret. Biol.*, **20,** 355–57 (1968)
228. Hutchins, L. M., Rue, J. L., Promising results of heat treatments for inactivation of phony disease virus in dormant peach nursery trees. *Phytopathology*, **29,** 12 (1939) (Abstr.)
229. Ie, T. S. An electron microscope study of tomato spotted wilt virus in the plant cell. *Neth. J. Plant Pathol.*, **70,** 114–15 (1964)
230. Ishihara, T. Families and genera of leafhopper vectors. In *Viruses, Vectors, and Vegetation*, 235–54. (Maramorosch, K., Ed., Interscience, New York, 1969)
231. Ishiie, T., Doi, Y., Yora, K., Asuyama, H. Suppressive effects of antibiotics of tetracycline group on symptom development of mulberry dwarf disease. *Ann. Phytopathol. Soc. Japan*, **33,** 267–75 (1967)
232. Izadpanah, K., Shepherd, R. J. Purification and properties of the pea enation mosaic virus. *Virology*, **28,** 463–67 (1966)
233. Jawetz, E. Agents of trachoma and inclusion conjunctivitis. *Ann. Rev. Microbiol.*, **18,** 301–34 (1964)
234. Jedlinski, H., Brown, C. M. Cross protection and mutual exclusion by three strains of barley yellow dwarf virus in *Avena sativa L. Virology*, **26,** 613–21 (1965)
235. Jensen, D. D. A plant virus lethal to its insect vector. *Virology*, **8,** 164–75 (1959)
236. Jensen, D. D. Influence of high temperatures on the pathogenicity and survival of western X-disease virus in leafhoppers. *Virology*, **36,** 662–67 (1968)
237. Jensen, D. D. Insect diseases induced by plant-pathogenic viruses. In *Viruses, Vectors, and Vegetation*, 505–25. (Maramorosch, K., ed., Interscience, New York, 1969)
238. Jensen, D. D., Whitcomb, R. F., Richardson, J. Lethality of injected peach western X-disease virus to its leafhopper vector. *Virology*, **31,** 532–38 (1967)
239. Jensen, S. G. Occurrence of virus particles in the phloem tissue of BYDV-infected barley. *Virology*, **38,** 83–91 (1969)
240. Joklik, W. K. The poxviruses. *Ann. Rev. Microbiol.*, **22,** 359–90 (1968)
241. Kanervo, V., Heinkinheimo, O., Raatikainen, M., Tinnilä, A. The leafhopper *Delphacodes pellucida* (F.) as the cause and distributor of the damage to oats in Finland. *Finnish State Agr. Res. Board 160*, 1–56 (1957)
242. Kaper, J. M. The small RNA viruses of plants, animals, and bacteria. A. Physical properties. In *Molecular Basis of Virology*, 1–133.

(Fraenkel-Conrat, H., Ed., Reinhold, New York, 1968)

243. Kassanis, B. Potato tubers freed from leaf-roll virus by heat. *Nature,* **164,** 881 (1949)

244. Kassanis, B. Interaction of viruses in plants. *Advan. Virus Res.,* **10,** 219–55 (1963)

245. Kawase, S., Miyajima, S. Infectious ribonucleic acid of the cytoplasmic-polyhedrosis virus of the silkworm, *Bombyx mori. J. Invertebrate Pathol.,* **11,** 63–69 (1968)

246. KenKnight, G. Chemotherapy of the peach rosette virus with antibiotics. *Phytopathology,* **45,** 348–49 (1955) (Abstr.)

247. Kennedy, J. S. Benefits to aphids from feeding on galled and virus infected leaves. *Nature,* **168,** 825–26 (1951)

248. Kimura, I., Shikata, E. Structural model of rice dwarf virus. *Proc. Japan Acad.,* **44,** 538–43 (1968)

249. Kirkpatrick, H. C., Ross, A. F. Aphid-transmission of potato leafroll virus to solanaceous species. *Phytopathology,* **42,** 540–46 (1952)

250. Kisimoto, R. Genetic variation in the ability of a planthopper vector, *Laodelphax striatellus* (Fallén) to acquire the rice stripe virus. *Virology,* **32,** 144–52 (1967)

251. Kisimoto, R., Watson, M. A. Abnormal development of embryos induced by inbreeding in *Delphacodes pellucida* Fabricius and *Delphacodes dubia* Kirschbaum, vectors of European wheat striate mosaic virus. *J. Invertebrate Pathol.,* **7,** 297–305 (1965)

252. Kitagawa, Y., Shikata, E. On some properties of rice black-streaked dwarf virus. *Mem. Fac. Agr. Hokkaido Univ.,* **6,** 439–45 (1969)

253. Kitagawa, Y., Shikata, E. Purification of rice black-streaked dwarf virus. *Mem. Fac. Agr. Hokkaido Univ.,* **6,** 446–51 (1969)

254. Kitajima, E. W. Electron microscopy of vira-cabeca virus (Brazilian tomato spotted wilt virus) within the host cell. *Virology,* **26,** 89–99 (1965)

255. Kitajima, E. W., Costa, A. S. Morphology and developmental stages of *Gomphrena* virus. *Virology,* **29,** 523–39 (1966)

256. Kleinschmidt, A. K., Dunnebacke, T. H., Spendlove, R. S., Schaffer, F. L., Whitcomb, R. F. Electron microscopy of RNA from reovirus and wound tumor virus. *J. Mol. Biol.,* **10,** 282–88 (1964)

257. Koch, R. Über die Milzbrandimpfung. Eine Entgegnung auf den von Pasteur in Genf gehaltenen Vortrag. Reprinted 1912 in Gesammelte Werke von Robert Koch, **1,** 207–31. (Leipzig). Cited in Riker, A. J., Riker, R. S. *Introduction to Research on Plant Diseases.* (John S. Swift Co., Inc., New York, 1936)

258. Kojima, M., Shikata, E., Sugawara, M., Murayama, D. Isolation and electron microscopy of potato leafroll virus from plants. *Virology,* **35,** 612–15 (1968)

259. Kunkel, L. O. Insect transmission of aster yellows. *Phytopathology,* **14,** 54 (1924) (Abstr.)

260. Kunkel, L. O. Studies on aster yellows. *Am. J. Botany,* **13,** 646–705 (1926)

261. Kunkel, L. O. Studies on aster yellows in some new host plants. *Contrib. Boyce Thompson Inst.,* **3,** 85–123 (1931)

262. Kunkel, L. O. Effect of heat on ability of *Cicadula sexnotata* (Fall.) to transmit aster yellows. *Am. J. Botany,* **24,** 316–27 (1937)

263. Kunkel, L. O. Heat cure of aster yellows in periwinkles. *Am. J. Botany,* **28,** 761–69 (1941)

264. Kunkel, L. O. Potato witches' broom transmission by dodder and cure by heat. *Proc. Am. Phil. Soc.,* **86,** 470–75 (1943)

265. Kunkel, L. O. Studies on cranberry false blossom. *Phytopathology,* **35,** 805–21 (1945)

266. Kunkel, L. O. Yellows diseases of plants. *Bull. Torrey Botan. Club,* **78,** 269–70 (1951)

267. Kunkel, L. O. Maintenance of yellows-type viruses in plant and insect reservoirs. In *The Dynamics of Virus and Rickettsial Infections* 150–63. (Hartman, F. W., Horsfall, F. L., Kidd, J. G., Eds., McGraw-Hill, New York, 1954)

268. Kunkel, L. O. Cross protection between strains of yellows-type viruses. *Advan Virus Res.,* **3,** 251–73 (1955)

269. Lanni, F. Viruses and molecular taxonomy. In *Plant Virology,* 386–426. (Corbett, M. K., Sisler, H. D.,

Eds., Univ. Florida Press, Gainesville, 1964)

270. Lee, P. E. Acquisition time and inoculation time in transmission of aster yellows virus by single leafhoppers. *Virology,* **17,** 394–96 (1962)
271. Lee, P. E. Electron microscopy of inclusions in plants infected by wheat striate mosaic virus. *Virology,* **23,** 145–51 (1964)
272. Lee, P. E. Viruslike particles in the salivary glands of apparently virus-free leafhoppers. *Virology,* **25,** 471–72 (1965)
273. Lee, P. E. Morphology of wheat striate mosaic virus and its localization in infected cells. *Virology,* **33,** 84–94 (1967)
274. Lee, P. E. Partial purification of wheat striate mosaic virus and fine structural studies of the virus. *Virology,* **34,** 583–89 (1968)
275. Lee, P. E., Chiykowski, L. N. Infectivity of aster-yellows virus preparations after differential centrifugations of extracts from viruliferous leafhoppers. *Virology,* **21,** 667–69 (1963)
276. Liebermeister, K. Morphology of the PPLO and L-forms of *Proteus. Ann. N. Y. Acad. Sci.,* **79,** 326–43 (1960)
277. Lin, S., Lee, C. Mycoplasma or Mycoplasma-like microorganism in white leaf disease of sugar cane. *Ann. Rept. Taiwan Sugar Exptl. Sta.,* 17–22 (1967–68)
278. Ling, K. C. Nonpersistence of the tungro virus of rice in its leafhopper vector, *Nephotettix impicticeps. Phytopathology,* **56,** 1252–56 (1966)
279. Ling, K. C. Nonpropagative leafhopper-borne viruses. In *Viruses, Vectors, and Vegetation,* 255–77. (Maramorosch, K., Ed., Interscience, New York, 1969)
280. Littau, V. C., Maramorosch, K. Cytological effects of aster yellows virus on its insect vector. *Virology,* **2,** 128–30 (1956)
281. Littau, V. C., Maramorosch, K. A study of the cytological effects of aster yellows virus on its insect vector. *Virology,* **10,** 483–500 (1960)
282. Lomakina, L. Y., Razvyazkina, G. M., Shubnikova, E. A. Cytological and histochemical changes in fat body of Cicada *Psammotettix striatus* Fall infected with winter wheat mosaic virus. *Vopr. Virusol., 1963:* 168–72 (1963)
283. Lu, Y. T., Shikata, E., Murayama, D. Isolation of northern cereal mosaic virus. *Mem. Fac. Agr. Hokkaido Univ.,* **6,** 335–39 (1968)
284. Lwoff, A., Tournier, P. The classification of viruses. *Ann Rev. Microbiol.,* **20,** 45–74 (1966)
285. MacCarthy, H. R. Aphid transmission of potato leafroll virus. *Phytopathology,* **44,** 167–74 (1954)
286. MacKinnon, J. P. Some factors that affect the aphid transmission of two viruses that persist in the vector. *Virology,* **20,** 281–87 (1963)
287. MacLeod, R. An interpretation of the observed polymorphism of potato yellow dwarf virus. *Virology,* **34,** 771–77 (1968)
288. MacLeod, R., Black, L. M., Moyer, F. H. The fine structure and intracellular localization of potato yellow dwarf virus. *Virology,* **29,** 540–52 (1966)
289. Maillet, P. L., Gourret, J., Hamon, C. Sur la présence de particules de type mycoplasme dans le liber de plantes atteintes de maladies du type "jaunisse" (aster yellow, phyllodie du trèfle, stolbur de la tomate) et sur la parenté ultrastructurale de ces particules avec celles trouvées chez divers insectes homoptères. *Comp. Rend. Ser. D.,* **266,** 2309–11 (1968)
290. Maniloff, J. H., Morowitz, H. J., Barrnett, R. J. Studies of the ultrastructure and ribosomal arrangements of the pleuropneumonia-like organism A5969. *J. Cell Biol.,* **25,** 139–50 (1965)
291. Maramorosch, K. The influence of temperature on the incubation of the wound-tumor virus in the insect *Agallia constricta. Phytopathology,* **39,** 14 (1949) (Abstr.)
292. Maramorosch, K. Effect of dosage on length of incubation period of aster yellows virus in its vector. *Proc. Soc. Exptl. Biol. Med.,* **75,** 744 (1950)
293. Maramorosch, K. Influence of temperature on incubation and transmission of the wound-tumor virus. *Phytopathology,* **40,** 1071–93 (1950)
294. Maramorosch, K. Direct evidence for the multiplication of aster-

yellows virus in its insect vector. *Phytopathology,* **42,** 59–64 (1952)

295. Maramorosch, K. Multiplication of aster yellows virus in its vector. *Nature,* **169,** 194–95 (1952)
296. Maramorosch, K. Incubation period of aster yellows virus. *Am. J. Botany,* **40,** 797–809 (1953)
297. Maramorosch, K. Mechanical transmission of curly top virus to its insect vector by needle inoculation. *Virology,* **1,** 286–300 (1955)
298. Maramorosch, K. Multiplication of aster yellows virus in *in vitro* preparations of insect tissues. *Virology,* **2,** 369–76 (1956)
299. Maramorosch, K. Reversal of virus-caused stunting in plants by gibberellic acid. *Science,* **126,** 651–52 (1957)
300. Maramorosch, K. Beneficial effect of virus-diseased plants on non-vector insects. *Tijdschr. Plantenziekten,* **64,** 383–91 (1958)
301. Maramorosch, K. Cross protection between two strains of corn stunt virus in an insect vector. *Virology,* **6,** 448–59 (1958)
302. Maramorosch, K. Viruses that infect and multiply in both plants and insects. *Trans. N. Y. Acad. Sci.,* **20,** 383–93 (1958)
303. Maramorosch, K. Friendly viruses. *Sci. Am.,* **203,** 138–44 (1960)
304. Maramorosch, K. Interrelationships between plant pathogenic viruses and insects. *Ann. N. Y. Acad. Sci.,* **118,** 363–70 (1964)
305. Maramorosch, K., Shikata, E., Granados, R. R. Structures resembling mycoplasma in diseased plants and in insect vectors. *Trans. N. Y. Acad. Sci.,* **30,** 841–55 (1968)
306. Maramorosch, K., Mitsuhashi, J., Streissle, G., Hirumi, H. Animal and plant viruses in insect tissues *in vitro. Bacteriol. Proc.,* 138 (1965)
307. Maramorosch, K., Shikata, E., Granados, R. R. Mycoplasma-like bodies in leafhoppers and diseased plants. *Phytopathology,* **58,** 886 (1968) (Abstr.)
308. Maramorosch, K., Shikata, E., Granados, R. R. The fate of plant-pathogenic viruses in insect vectors: electron microscopy observations. In *Viruses, Vectors, and Vegetation,* 417–31. (Maramorosch, K., Ed., Interscience, New York, (1969)
309. Marmion, B. P., Goodburn, G. M. Effect of organic gold salt on Eaton's primary atypical pneumonia agent and other observations. *Nature,* **189,** 247–48 (1961)
310. Martin, M. M. Purification and electron microscope studies of tomato spotted wilt virus (TSWV) from tomato roots. *Virology,* **22,** 645–49 (1964)
311. McElhaney, R. N., Tourtellotte, M. E. Mycoplasma membrane lipids: variations in fatty acid composition. *Science,* **164,** 433–34 (1969)
312. McEwen, F. L., Kawanishi, Y. Insect transmission of corn mosaic: laboratory studies in Hawaii. *J. Econ. Entomol.,* **60,** 1413–17 (1967)
313. McLean, G. D., Francki, R. I. B. Purification of lettuce necrotic yellows virus by column chromatography on calcium phosphate gel. *Virology,* **31,** 585–91 (1967)
314. McMartin, D. A. *Mycoplasma gallisepticum* in the respiratory tract of the fowl. *Vet. Record,* **81,** 317–20 (1967)
315. Melnick, J. L. Summary of classification of animal viruses. *Progr. Med. Virol.,* **11,** 451–53 (1969)
316. Melnick, J. L., McCombs, R. M. Classification and nomenclature of animal viruses. *Progr. Med. Virol.,* **8,** 400–9 (1966)
317. Mims, C. A., Day, M. F., Marshall, I. D. Cytopathic effect of semliki forest virus in the mosquito *Aedes aegypti. Am. J. Trop. Med. Hyg.,* **15,** 775–84 (1966)
318. Mitsuhashi, J. Infection of leafhopper and its tissues cultivated *in vitro* with *Chilo* iridescent virus. *J. Invertebrate Pathol.,* **9,** 432–34 (1967)
319. Mitsuhashi, J. Plant-pathogenic viruses in insect vector tissue culture. In *Viruses, Vectors, and Vegetation,* 475–503. (Maramorosch, K., Ed., Interscience, New York, 1969)
320. Mitsuhashi, J., Maramorosch, K. Inoculation of plant tissue cultures with aster yellows virus. *Virology,* **23,** 277-79 (1964)
321. Mitsuhashi, J., Nasu, S. An evidence for the multiplication of rice dwarf virus in the vector cell cultures inoculated *in vitro. Japanese J. Appl. Entomol. Zool.,* **2,** 113–14 (1967)
322. Miura, K., Fujii, I., Sakkaki, T.,

Fuke, M., Kawase, S. Double-stranded RNA from cytoplasmic-polyhedrosis virus of the silkworm. *J. Virol.*, **2**, 1211–22 (1968)
323. Miura, K., Kimura, I., Suzuki, N. Double-stranded ribonucleic acid from rice dwarf virus. *Virology*, **28**, 571–79 (1966)
324. Miyamoto, S., Miyamoto, Y. Notes on aphid-transmission of potato leafroll virus. *Sci. Repts. Hyogo Univ. Agr. Ser. Plant Protect.*, **7**, 51–66 (1966)
325. Morowitz, H. J., Tourtellotte, M. E. The smallest living cells. *Sci. Am.*, **206**, 117–26 (1962)
326. Moulder, J. W. The relation of the psittacosis group (Chlamydiae) to bacteria and viruses. *Ann. Rev. Microbiol.*, **20**, 107–30 (1966)
327. Murant, A. F., Goold, R. A., Roberts, I. M., Cathro, J. Carrot mottle—a persistent aphid-borne virus with unusual properties and particles. *J. Gen. Virol.*, **4**, 329–41 (1969)
328. Murphy, F. A., Coleman, P. H., Harrison, A. K., Gary, G. W., Jr. Colorado tick fever virus: an electron microscopic study. *Virology*, **35**, 28–40 (1968)
329. Murphy, F. A., Coleman, P. H., Whitfield, S. G. Electron microscopic observations of Flanders virus. *Virology*, **30**, 314–17 (1966)
330. Murphy, F. A., Fields, B. N., Kern Canyon virus: electron microscopic and immunological studies. *Virology*, **33**, 625–37 (1967)
331. Mussgay, M., Suarez, O. Multiplication of vesicular stomatitis virus in *Aedes aegypti* (L.) mosquitoes. *Virology*, **17**, 202–3 (1962)
332. Nagaraj, A. N., Black, L. M. Hereditary variation in the ability of a leafhopper to transmit two unrelated plant viruses. *Virology*, **16**, 152–62 (1962)
333. Nakai, T., Howatson, A. F. The fine structure of vesicular stomatitis virus. *Virology*, **35**, 268–81 (1968)
334. Nasu, S. Studies on some leafhoppers and planthoppers which transmit virus diseases of rice plant in Japan. *Kyushu Agr. Exp. Sta. Bull.* **8**, 153–349 (1963)
335. Nasu, S. Electron microscopic studies on transovarial passage of rice dwarf virus. *Japan J. Appl. Entomol. Zool.*, **9**, 225–37 (1965)
336. Nasu, S. Electron microscopy of the transovarial passage of rice dwarf virus. In *Viruses, Vectors, and Vegetation*, 433–48. (Maramorosch, K., Ed., Interscience, New York, 1969)
337. Newnham, A. G., Chu, H. P. An *in vitro* comparison of the effect of some antibacterial, antifungal and antiprotozoal agents on various strains of *Mycoplasma* (pleuropneumonia-like organisms: PPLO). *J. Hyg.*, **63**, 1–23 (1965)
338. Nielson, M. W. The leafhopper vectors of phytopathogenic viruses. Taxonomy, biology, and virus transmission. *U. S. Dept. Agr. ARS. Tech. Bull., 1382. (Washington, D.C.*, 1968)
339. Nocard, Roux, avec MM. Borrel, Salimbeni, Dujardin-Beaumetz. Le microbe de la péripneumonie. *Ann. Inst. Pasteur*, **12**, 240–62 (1898)
340. O'Loughlin, G. T., and Chambers, T. C. The systemic infection of an aphid by a plant virus. *Virology*, **33**, 262–71 (1967)
341. Oman, P. Criteria of specificity in virus-vector relationships. In *Viruses, Vectors, and Vegetation*, 1–22. (Maramorosch, K., Ed., Interscience, New York, 1969)
342. Orenski, S. W., Mitsuhashi, J., Ringel, S. M., Martin, J. F., Maramorosch, K. A presumptive bacterial symbiont from the eggs of the six-spotted leafhopper, *Macrosteles fascifrons* (Stå). *Contrib. Boyce Thompson Inst.*, **23**, 123–26 (1965)
343. Orenski, S. W., Murray, J. R., Maramorosch, K. Further studies on the feeding habits of aster-yellows virus-carrying corn leafhoppers. *Contrib. Boyce Thompson Inst.*, **23**, 47–50 (1965)
344. Osborn, H. T. Incubation period of pea mosaic in the aphid, *Macrosiphum pisi. Phytopathology*, **25**, 160–77 (1935)
345. Paliwal, Y. C. Changes in relative virus concentration in *Endria inimica* in relation to its ability to transmit wheat striate mosaic virus. *Phytopathology*, **58**, 386–87 (1968)
346. Paliwal, Y. C., Slykhuis, J. T. Localization of wheat streak mosaic virus in the alimentary canal of its vector *Aceria tulipae* Keifer. *Virology*, **32**, 344–53 (1967)
347. Peleg, J. Growth of arboviruses in monolayers from subcultured mos-

quito embryo cells. *Virology,* **35,** 617–19 (1968)

348. Peters, D. The purification of virus-like particles from the aphid *Myzus persicae. Virology,* **26,** 159–61 (1965)
349. Peters, D. Potato leafroll virus, its purification from its vector *Myzus persicae. Meded. L. E. B. Fonds* No. 45 (Veenman, H., Zonen, N. V., Wageningen, 100 pp., 1967)
350. Peters, D. The purification of potato leafroll virus from its vector *Myzus persicae. Virology,* **31,** 46–54 (1967)
351. Peters, D., Van Loon, L. C. Transmission of potato leafroll virus by aphids after feeding on virus preparations from aphids and plants. *Virology,* **35,** 597–600 (1968)
352. Piemeisel, R. L. Replacement control: changes in vegetation in relation to control of pests and diseases. *Botan. Rev.,* **20,** 1–32 (1954)
353. Pirone, T. P. Mechanism of transmission of stylet-borne viruses. In *Viruses, Vectors, and Vegetation* 199–210. (Maramorosch, K., Ed., Interscience, New York, 1969)
354. Ploaie, P. G. Virus-like particles in both plant and insect vector infected with clover phyllody virus. Proc. Conf. Czechoslovak Plant Virologists, 6th, 1967
355. Ploaie, P., Granados, R. G., Maramorosch, K. Mycoplasma-like structures in periwinkle plants with Crimean yellows, European clover dwarf, stolbur, and parastolbur. *Phytopathology,* **58,** 1063 (1968) (Abstr.)
356. Ploaie, P., Maramorosch, K. Electron microscopic demonstration of particles resembling mycoplasma or psittacosis-lymphogranuloma-trachoma group in plants infected with European yellows-type diseases. *Phytopathology,* **59,** 536–44 (1969)
357. Plus, N. Étude de la multiplication du virus de la sensibilité au gaz carbonique chez la Drosophile. *Bull. Biol. France Belg.,* **88,** 248–93 (1954)
358. Posnette, A. F., Ellenberger, C. E. Further studies of green petal and other leafhopper-transmitted viruses infecting strawberry and clover. *Ann. Appl. Biol.,* **51,** 69–83 (1963)
359. Prevec, L., Watanabe, Y., Gauntt, C. J., Graham, A. F. Transcription of the genomes of Type 1 and Type 3 reoviruses. *J. Virol.,* **2,** 289–97 (1968)
360. Protsenko, E. A. Electron microscopy of the aster yellows virus (*Leptomotropus callistephi* Ryzhkov). *Biol. Plantarum,* **1,** 187–91 (1959)
361. Průša, V., Jermoljev, E., Josef, V. Oat sterile-dwarf virus disease. *Biol. Plantarum,* **1,** 223–34 (1959)
362. Raymer, W. B., Diener, T. O. Potato spindle tuber virus: a plant virus with properties of a free nucleic acid. I. Assay, extraction, and concentration. *Virology,* **37,** 343–50 (1969)
363. Razin, S., Cosenza, B. J., Tourtellotte, M. E. Filamentous growth of mycoplasma. *Ann. N. Y. Acad. Sci.,* **143,** 66–76 (1967)
364. Reddy, D. V. R., Black, L. M. Production of wound-tumor virus and wound-tumor soluble antigen in the insect vector. *Virology,* **30,** 551–61 (1966)
365. Řeháček, J. The growth of arboviruses in mosquito cells *in vitro. Acta Virol.,* **12,** 241–46 (1968)
366. Richardson, J., Sylvester, E. S. Aphid honeydew as inoculum for the injection of pea aphids with pea-enation mosaic virus. *Virology,* **25,** 472–75 (1965)
367. Richardson, J. Sylvester, E. S. Further evidence of multiplication of sowthistle yellow vein virus in its aphid vector, *Hyperomyzus lactucae. Virology,* **35,** 347–55 (1968)
368. Rivera, C. T., Ou, S. H. Transmission studies of the two strains of rice tungro virus. *Plant Disease Reptr.,* **51,** 877–81 (1967)
369. Roberts, D. W., Granados, R. R. A poxlike virus from *Amsacta moorei. J. Invertebrate Pathol.,* **12,** 141–43 (1968)
370. Rochow, W. F. Transmission of barley yellow dwarf virus acquired from liquid extracts by aphids feeding through membranes. *Virology,* **12,** 223–32 (1960)
371. Rochow, W. F. The barley yellow dwarf virus disease of small grains. *Advan. Agron.,* **13,** 217–48 (1961)
372. Rochow, W. F. Variation within and among aphid vectors of plant

viruses. *Ann. N. Y. Acad. Sci.*, **105**, 713–29 (1963)

373. Rochow, W. F. Predominating strains of barley yellow dwarf virus in New York; changes during ten years. *Plant Disease Reptr.*, **51**, 195–99 (1967)
374. Rochow, W. F. Temperature alters vector specificity of one of four strains of barley yellow dwarf virus. *Phytopathology*, **57**, 344–45 (1967) (Abstr.)
375. Rochow, W. F. Specificity in aphid transmission of a circulative plant virus. In *Viruses, Vectors, and Vegetation*, 175–98. (Maramorosch, K., Ed., Interscience, New York, 1969)
376. Rochow, W. F., Ball, E. M. Serological blocking of aphid transmission of barley yellow dwarf virus. *Virology*, **33**, 359-62 (1967)
377. Rochow, W. F., Brakke, M. K. Purification of barley yellow dwarf virus. *Virology*, **24**, 310–22 (1964)
378. Rochow, W. F., Jedlinski, H., Coon, B. F., Murphy, H. C. Variation in barley yellow dwarf of oats in nature. *Plant Disease Reptr.*, **49**, 692–95 (1965)
379. Rosen, L. Reoviruses. *Virol. Monogr.*, **1**, 73–107 (1968)
380. Ruppel, E. G. Possible particles of beet western yellows virus in the intestines of viruliferous aphids. *Phytopathology*, **58**, 256–57 (1968)
381. Saikku, P., Von Bonsdorff, C. H. Electron microscopy of the Uukuniemi virus, an ungrouped arbovirus. *Virology*, **34**, 804–6 (1968)
382. Sakimura, K. Life history of *Thrips tabaci* L. on *Emilia sagittata* and its host plant range in Hawaii. *J. Econ. Entomol.*, **25**, 884–91 (1932)
383. Sakimura, K. The present status of thrips-borne viruses. In *Biological Transmission of Disease Agents*, 33–40. (Maramorosch, K., Ed., Academic Press, New York, 1962)
384. Sakimura, K. *Frankliniella fusca.*, an additional vector for the tomato spotted wilt virus, with notes on *Thrips tabaci*, another vector. *Phytopathology*, **53**, 412–15 (1963)
385. Samuel, G., Bald, J. G., Eardley, C. M. "Big Bud", a virus disease of the tomato. *Phytopathology*, **23**, 641–53 (1933)
386. Samuel, G., Bald, J. G., Pittman, H. A. Investigations on "spotted wilt" of tomatoes. *Australian Council Sci. Ind. Res. Bull. 44* (1930)
387. Sato, T., Kyogoku, Y., Higuchi, S., Mitsui, Y., Iitaka, Y., Tsuboi, M., Miura, K. A preliminary investigation on the molecular structure of rice dwarf virus ribonucleic acid. *J. Mol. Biol.*, **16**, 180–90 (1966)
388. Seecof, R. L. Deleterious effects on *Drosophila* development associated with the sigma virus infection. *Virology*, **22**, 142–48 (1964)
389. Seecof, R. L. The sigma virus infection of *Drosophila melanogaster*. *Current Topics Microbiol. Immunol.*, **42**, 59–93 (1968)
390. Seecof, R. L. Sigma virus multiplication in whole-animal culture of *Drosophila*. *Virology*, **38**, 134–39 (1969)
391. Severin, H. H. P. Evidence of nonspecific transmission of California aster-yellows virus by leafhoppers. *Hilgardia*, **17**, 21–59 (1945)
392. Severin, H. H. P. Longevity, or life histories, of leafhopper species on virus-infected and on healthy plants. *Hilgardia*, **17**, 121–37 (1946)
393. Severin, H. H. P., Freitag, J. H. Some properties of the curly-top virus. *Hilgardia*, **8**, 1–48 (1933)
394. Severin, H. H. P., Swezy, O. Filtration experiments on curly top of sugar beets. *Phytopathology*, **18**, 681–90 (1928)
395. Shatkin, A. J. Viruses containing double-stranded RNA. In *Molecular Basis of Virology*, 351–92. (Fraenkel Conrat, H., Ed., Reinhold, New York, 1968)
396. Shatkin, A. J., Sipe, J. D. RNA polymerase activity in purified reoviruses. *Proc. Natl. Acad. Sci. U.S.*, **61**, 1462–69 (1968)
397. Shepherd, R. J., Ghabrial, S. A. Isolation and some properties of the nucleic acid and protein components of pea enation mosaic virus. *Phytopathology*, **56**, 900 (1966) (Abstr.)
398. Shepherd, R. J., Wakeman, P. J. Romanko, R. R. DNA in cauliflower mosaic virus. *Virology*, **36**, 150–52 (1968)
399. Shikata, E. Electron microscopic studies on plant viruses. *J. Fac.*

Agr. Hokkaido Univ., **55**, Part 1, 1–110 (1966)

400. Shikata, E., Chen, M. Electron microscopy of rice transitory yellowing virus. *J. Virol.*, **3**, 261–64 (1969)
401. Shikata, E., Lu, Y. Electron microscopy of northern cereal mosaic virus in Japan. *Proc. Japan Acad.*, **43**, 918–23 (1967)
402. Shikata, E., Maramorosch, K. Electron microscopic evidence for the systemic invasion of an insect host by a plant pathogenic virus. *Virology*, **27**, 461–75 (1965)
403. Shikata, E., Maramorosch, K. Electron microscopy of pea enation mosaic virus in plant cell nuclei. *Virology*, **30**, 439–54 (1966)
404. Shikata, E., Maramorosch, K. Electron microscopy of wound tumor virus assembly sites in insect vectors and plants. *Virology*, **32**, 363–77 (1967)
405. Shikata, E., Maramorosch, K. Mycoplasmalike bodies in sieve pores of yellows diseased plants and in fatbody cells of two insect vectors. *Phytopathology*, **59**, 1559 (1969) (Abstr.)
406. Shikata, E., Maramorosch, K. Electron microscopy of insect-borne viruses *in situ*. In *Viruses, Vectors, and Vegetation*, 393–415. (Maramorosch, K., Ed., Interscience, New York, 1969)
407. Shikata, E., Maramorosch, K., Granados, R. R. Electron microscopy of pea enation mosaic virus in plants and aphid vectors. *Virology*, **29**, 426–36 (1966)
408. Shikata, E., Maramorosch, K., Ling, K. C., Matsumoto, T. On the mycoplasma-like structures encountered in the phloem cells of American aster yellows, corn stunt, Philippine rice yellow dwarf and Taiwan sugar cane white leaf diseased plants. *Ann. Phytopathol. Soc. Japan*, **34**, 208–9 (1968) (In Japanese. Abstr.)
409. Shinkai, A. Difference in rice dwarf virus transmitting ability of leafhoppers from various localities. *Ann. Phytopathol. Soc. Japan*, **21**, 127 (1956) (In Japanese. Abstr.)
410. Silverstein, S. C., Dales, S. The penetration of reovirus RNA and initiation of its genetic function in L-strain fibroblasts. *J. Cell Biol.*, **36**, 197–230 (1968)
411. Simons, J. N. Vector-virus relationships of pea-enation mosaic and the pea aphid *Macrosiphum pisi* (Kalt.). *Phytopathology*, **44**, 283–89 (1954)
412. Simons, J. N., Coe, D. M. Transmission of pseudo-curly top virus in Florida by a treehopper. *Virology*, **6**, 43–48 (1958)
413. Simpson, R. W., Hauser, R. E. Structural components of vesicular stomatitis virus. *Virology*, **29**, 654–67 (1966)
414. Simpson, R. W., Hauser, R. E. Structural differentiation of group A arboviruses based on nucleoid morphology in ultrathin sections. *Virology*, **34**, 568–70 (1968)
415. Singh, K. Grassy shoot disease of sugarcane. II. Hot-air therapy. *Current Sci.*, **37**, 592–94 (1968)
416. Sinha, R. C. Comparison of the ability of nymph and adult *Delphacodes pellucida* Fabricius to transmit European wheat striate mosaic virus. *Virology*, **10**, 344–52 (1960)
417. Sinha, R. C. Effect of age of vector and of abdomen punctures on virus transmission. *Phytopathology*, **53**, 1170–73 (1963)
418. Sinha, R. C. Recovery of potato yellow dwarf virus from hemolymph and internal organs of an insect vector. *Virology*, **27**, 118–19 (1965)
419. Sinha, R. C. Sequential infection and distribution of wound-tumor virus in the internal organs of a vector after ingestion of virus. *Virology*, **26**, 673–86 (1965)
420. Sinha, R. C. Response of wound-tumor virus infection in insects to vector age and temperature. *Virology*, **31**, 746–48 (1967)
421. Sinha, R. C. Recent work on leafhopper-transmitted viruses. *Advan. Virus Res.*, **13**, 181–223 (1968)
422. Sinha, R. C. Localization of viruses in vectors: serology and infectivity tests. In *Viruses, Vectors, and Vegetation*, 379–91. (Maramorosch, K., Ed., Interscience, New York, 1969)
423. Sinha, R. C., Chiykowski, L. N. Multiplication of aster-yellows virus in a nonvector leafhopper. *Virology*, **31**, 461–66 (1967)
424. Sinha, R. C., Chiykowski, L. N. Initial and subsequent sites of aster-yellows virus infection in a

leafhopper vector. *Virology,* **33,** 702–8 (1967)

425. Sinha, R. C., Chiykowski, L. N. Distribution of clover phyllody virus in the leafhopper *Macrosteles fasifrons* (Stål). *Acta Virol.,* **12,** 546–50 (1968)
426. Sinha, R. C., Reddy, D. V. R. Improved fluorescent smear technique and its application in detecting virus antigens in an insect vector. *Virology,* **24,** 626–34 (1964)
427. Slykhuis, J. T. Striate mosaic, a new disease of wheat in South Dakota. *Phytopathology,* **43,** 537–40 (1953)
428. Slykhuis, J. T. Vector and host relations of North American wheat striate mosaic virus. *Can. J. Botany,* **41,** 1171–85 (1963)
429. Slykhuis, J. T. Mites as vectors of plant viruses. In *Viruses, Vectors, and Vegetation,* 121–41. (Maramorosch, K., Ed., Interscience, New York, 1969)
430. Smith, K. M. A new virus affecting mangolds, sugarbeet and related plants. *Research,* **3,** 434 (1950)
431. Smith, K. M. A latent virus in sugarbeets and mangolds. *Nature,* **167,** 1061 (1951)
432. Smith, P. F., Sasaki, S. Stability of pleuropneumonialike organisms to some physical factors. *Appl. Microbiol.,* **6,** 184–89 (1958)
433. Spendlove, R. S., Lennette, E. H., Knight, C. O., Chin, J. N. Development of viral antigen and infectious virus in HeLa cells infected with reovirus. *J. Immunol.,* **90,** 548–53 (1963)
434. Spendlove, R. S., Lennette, E. H., Knight, C. O., Chin, J. N. Production in FL cells of infectious and potentially infectious reovirus. *J. Bacteriol.,* **92,** 1036–40 (1966)
435. Stairs, G. R. Inclusion-type insect viruses. *Current Topics Microbiol. Immunol.,* **42,** 1–23 (1968)
436. Stanley, N. F. Reoviruses. *Brit. Med. Bull.,* **23,** 150–55 (1967)
437. Staron, T., Cousin, M. T., Grison, C. Action de quelques antibiotiques sur des maladies végétales du type "jaunisse européene." *Compt. Rend., Ser. D.,* **267,** 2328–31 (1969)
438. Steere, R. L. Gel filtration of aster-yellows virus. *Phytopathology,* **57,** 832–33 (1967) (Abstr.)
439. Stegwee, D., Ponsen, M. B. Multiplication of potato leafroll virus in the aphid *Myzus persicae* (Sulz.). *Entomol. Exptl. Appl.,* **1,** 291–300 (1958)
440. Stoddard, E. M. Inactivating *in vivo* the virus of X-disease of peach by chemotherapy. *Phytopathology,* **32,** 17 (1942) (Abstr.)
441. Stoddard, E. M. Immunization of peach trees to X-disease by chemotherapy. *Phytopathology,* **34,** 1011–12 (1944) (Abstr.)
442. Stoddard, E. M. The X-disease of peach and its chemotherapy. *Conn. Agr. Exptl. Sta. Bull.* **506,** 1–19 (1947)
443. Storey, H. H. The inheritance by an insect vector of the ability to transmit a plant virus. *Proc. Roy. Soc. (London), Ser. B.* **112,** 46–60 (1932)
444. Storey, H. H. Investigations of the mechanism of the transmission of plant viruses by insect vectors. I. *Proc. Roy. Soc. (London), Ser. B,* **113,** 463–85 (1933)
445. Storey, H. H. Investigations of the mechanism of the transmission of plant viruses by insect vectors, III. The Insect's saliva. *Proc. Roy. Soc., B,* **127,** 526–43 (1939)
446. Story, G. E., Halliwell, R. S. Association of a mycoplasma-like organism with the bunchy top disease of papaya (*Carica papaya*) in the Dominican Republic. *Phytopathology,* **59,** 118 (1969) (Abstr.)
447. Streissle, G., Rosen, L., Tokumitsu, T. Host specificity of wound tumor virus and reoviruses and their serologic relationships. *Arch. Ges. Virusforsch.,* **22,** 409–16 (1968)
448. Streissle, G., Granados, R. R. The fine structure of wound tumor and reovirus. *Arch. Ges. Virusforsch.,* **25,** 369–72 (1968)
449. Stubbs, L. L., Grogan, R. G. Necrotic yellows: a newly recognized virus disease of lettuce. *Australian J. Agr. Res.,* **14,** 439–59 (1963)
450. Studdert, M. J., Pangborn, J., Addison, R. B. Blue-tongue virus structure. *Virology,* **29,** 509–11 (1966)
451. Sukhov, K. S., Sukhova, M. N. Interrelations between the virus of a new grain mosaic disease (Zakuklivanie) and its carrier *Delphax striatella* Fallén. *Compt. Rend. (Dokl.) Acad. Sci. URSS,* **26,** 479–82 (1940)
452. Sun, S. C., Sohol, R. S., Chu, K. C.,

Colcolough, H. L., Leiderman, E., Burch, G. E. *Mycoplasma gallisepticum* infection in mice and cynomolgus monkeys. *Am. J. Pathol.,* **53,** 1073–96 (1968)

453. Suzuki, N. Purification of single and double-stranded vector-borne RNA viruses. In *Viruses, Vectors, and Vegetation,* 557–78. (Maramorosch, K., Ed., Interscience, New York, 1969)

454. Swenson, K. Plant virus transmission by insects. In *Methods in Virology,* **I,** 267–307. (Maramorosch, K., Koprowski, H., Eds., Academic Press, New York and London, 1967)

455. Swenson, K. G. Plant susceptibility to virus infection by insect transmission. In *Viruses, Vectors, and Vegetation,* 143–57. (Maramorosch, K., Ed., Interscience, New York, 1969)

456. Sylvester, E. S., The latent period of pea-enation mosaic virus in the pea aphid, *Acyrthosiphon pisum* (Harris)—an approach to its estimation. *Virology,* **25,** 62–67 (1965)

457. Sylvester, E. S. Retention of inoculativity in the transmission of pea enation mosaic virus by pea aphids as associated with virus isolates, aphid reproduction and excretion. *Virology,* **32,** 524–31 (1967)

458. Sylvester, E. S. Virus transmission by aphids—a viewpoint. In *Viruses, Vectors, and Vegetation,* 159–73. (Maramorosch, K., Ed., Interscience, New York, 1969)

459. Sylvester, E. S., Richardson, J. "Recharging" pea aphids with pea enation mosaic virus. *Virology,* **30,** 592–97 (1966)

460. Sylvester, E. S., Richardson, J. Some effects of temperature on the transmission of pea enation mosaic virus and on the biology of the pea aphid vector. *J. Econ. Entomol.,* **59,** 255–61 (1966)

461. Sylvester, E. S., Richardson, J. Additional evidence of multiplication of the sowthistle yellow vein virus in an aphid vector—serial passage. *Virology,* **37,** 26–31 (1969)

462. Takahashi, Y., Sekiya, I. Adipose tissue of the green rice leafhopper, *Nephotettix cincticeps* Uhler, infected with the virus of the yellow dwarf disease of the rice plant. *Japanese J. Appl. Entomol. Zool.,* **6,** 90–94 (1962)

463. Tamura, A., Matsumoto, A., Higashi, N. Purification and chemical composition of reticulate bodies of the meningopneumonitis organisms. *J. Bacteriol.,* **93,** 2003–8 (1967)

464. Teakle, D. S., Steindl, D. R. L. Virus-like particles in galls on sugarcane plants affected by Fiji disease. *Virology,* **37,** 139–45 (1969)

465. Ting, W. E., Gold, A. H. Effects of aster-yellows virus infection on transport through plant stem sections. *Virology,* **32,** 570–79 (1967)

466. Tomita, K., Rich, A. X-ray diffraction investigations of complementary RNA. *Nature,* **201,** 1160–63 (1964)

467. Toyoda, S., Kimura, I., Suzuki, N. Purification of rice dwarf virus. *Ann. Phytopathol. Soc. Japan,* **30,** 225–30 (1965)

468. Tully, J. G. Biochemical morphological, and serological characterization of mycoplasma of murine origin. *J. Infect. Diseases,* **115,** 171–85 (1965)

469. Vago, C. Non-inclusion virus diseases of invertebrates. *Current Topics Microbiol. Immunol.,* **42,** 24–37 (1968)

470. Vago, C., Flandre, O., Culture prolongée de tissus d'insectes et de vecteurs de maladies en coagulum plasmatique. *Ann. Epiphyties,* **14,** *n° hors sér.* III, 127–39 (1963)

471. Valenta, V. Experiments on thermal inactivation of some European yellows viruses *in vivo. Biologia (Bratislava),* **14,** 146–49 (1959)

472. Valenta, V. Interference studies with yellows-type plant viruses. I. Cross protection tests with European viruses. *Acta Virol.,* **3,** 65–72 (1959)

473. Valenta, V. Interference studies with yellows-type plant viruses. II. Cross protection tests with European and American viruses. *Acta Virol.,* **3,** 145–52 (1959)

474. van Boven, C. P. A., Ensering, H. L., Hijmans, W. Size determination by the filtration method of the reproductive elements of group A streptococcal L-forms. *J. Gen. Microbiol.,* **52,** 403–12 (1968)

475. van Boven, C. P. A., Ensering, H. L., Hijmans, W. Size determination by phase-contrast microscopy of the reproductive elements of group A streptococcal L-forms.

J. Gen. Microbiol., **52**, 413–20 (1968)

476. van Kammen, A., Henstra, S., Ie, T. S. Morphology of tomato spotted wilt virus. *Virology*, **30**, 574–77 (1966)

477. Vasquez, C., Tournier, P. New interpretation of the reovirus structure. *Virology*, **24**, 128–30 (1964)

478. Vaughn, J. L. A review of the use of insect tissue culture for the study of insect-associated viruses. *Current Topics Microbiol. Immunol.*, **42**, 108–28 (1968)

479. Ver, B. A. Efficient filtration and sizing of viruses with membrane filters. *J. Virol.*, **2**, 21–25 (1968)

480. Verwoerd, D. W. Purification and characterization of bluetongue virus. *Virology*, **38**, 203–12 (1969)

481. Vidano, C. Il maize rough dwarf virus in ghiandole salivari e in micetoma di *Laodelphax striatellus* Fallén. *Atti Accad. Sci. Torino*, **100**, 731–48 (1966)

482. Vidano, C. Microecologia e moltiplicazione di MRDV nel vettore. *Atti Accad. Sci. Torino*, **101**, 717–33 (1967)

483. Vidano, C. Phases of maize rough dwarf virus multiplication in the vector *Laodelphax striatellus* Fallén. *Virology* (In press, 1970)

484. Vovk, A. M., Nikiforova, G. S., Issledovanie virusa stolbura v electronnom mikroscope. *Dokl. Akad Nauk USSR*, **102**, 839–40 (1955)

485. Wagner, R. R., Schnaitman, T. A., Snyder, R. M. Structural proteins of vesicular stomatitis viruses. *J. Virol.*, **3**, 395–403 (1969)

486. Watanabe, Y., Prevec, L., Graham, A. F. Specificity in transcription of the reovirus genome. *Proc. Natl. Acad. Sci. U.S.*, **58**, 1040–46 (1967)

487. Watanabe, Y., Millward, S., Graham, A. F. Regulation of transcription of the reovirus genome. *J. Mol. Biol.*, **36**, 107–23 (1968)

488. Watson, M. A., Sinha, R. C. Studies on the transmission of European wheat striate mosaic virus by *Delphacodes pellucida* Fabricius. *Virology*, **8**, 139–63 (1959)

489. Weiser, J. *Vagoiavirus*, gen. n., a virus causing disease in insects. *J. Invertebrate Pathol.*, **7**, 82–85 **(1965)**

490. Whitcomb, R. F. Bioassay of plant viruses transmitted persistently by their vectors. In *Viruses, Vectors, and Vegetation*, 449–62. (Maramorosch, K., Ed., Interscience, New York, 1969)

491. Whitcomb, R. F., Black, L. M. Synthesis and assay of wound-tumor soluble antigen in an insect vector. *Virology*, **15**, 136–45 (1961)

492. Whitcomb, R. F., Davis, R. E. Properties of the aster yellows agent. *Phytopathology*, **59**, 1561 (1969) (Abstr.)

493. Whitcomb, R. F., Davis, R. E. Evidence on possible mycoplasma-like etiology of aster yellows disease: II. Suppression of aster yellows in insect vectors. (In preparation)

494. Whitcomb, R. F., Davis, R. E., Purcell, R., Cohen, R., Steere, R. L. Aster yellows disease: mycoplasma-like properties of the infectious agent. (In preparation)

495. Whitcomb, R. F., Jensen, D. D. Proliferative symptoms in leafhoppers infected with western X-disease virus. *Virology*, **35**, 174–77 (1968)

496. Whitcomb, R. F., Jensen, D. D., Richardson, J. The infection of leafhoppers by western X-disease virus. I. Frequency of transmission after injection or acquisition feeding. *Virology*, **28**, 448–53 (1966)

497. Whitcomb, R. F., Jensen, D. D., Richardson, J. The infection of leafhoppers by western X-disease virus. II. Fluctuation of virus concentration in the hemolymph after injection. *Virology*, **28**, 454–58 (1966)

498. Whitcomb, R. F., Jensen, D. D., Richardson, J. The infection of leafhoppers by western X-disease virus. III. Salivary, neural, and adipose histopathology. *Virology*, **31**, 539–49 (1967)

499. Whitcomb, R. F., Jensen, D. D., Richardson, J. The infection of leafhoppers by western X-disease virus. IV. Pathology in the alimentary tract. *Virology*, **34**, 69–78 (1968)

500. Whitcomb, R. F., Jensen, D. D., Richardson, J. The infection of leafhoppers by western X-disease virus. V. Properties of the infectious agent. *J. Invertebrate Pathol.*, **12**, 192–201 (1968)

501. Whitcomb, R. F., Jensen, D. D.,

Richardson, J. The infection of leafhoppers by western X-disease virus. VI. Cytopathological interrelationships. *J. Invertebrate Pathol.*, **12,** 202–21 (1968)

502. Whitcomb, R. F., Richardson, J., Jensen, D. D. Anatomical and cytological notes on the leafhopper *Colladonus montanus* Van Duzee. (In preparation)

503. Wolanski, B. S., Francki, R. I. B. Structure of lettuce necrotic yellows virus. II. Electron microscopic studies on the effect of pH of phosphotungstic acid stain on the morphology of the virus. *Virology,* **37,** 437–47 (1969)

504. Wolanski, B. S. Structure of lettuce necrotic yellows virus. I. Electron microscopy of negatively stained preparations. *Virology,* **33,** 287–96 (1967)

505. Worley, J. F. Chains, filamentous forms and budding of mycoplasma-like organisms in aster yellows-infected plants. *Phytopathology,* **59,** 1561–62 (1969) (Abstr.)

506. Worley, J. F. Tissue distribution and replicative forms of a mycoplasma-like organism associated with aster yellows disease in *Nicotiana* and aster. *Phytopathology* (In press)

507. Wright, D. N. Nature of penicillin-induced growth inhibition of *Mycoplasma neurolyticum. J. Bacteriol.,* **93,** 185–90 (1967)

508. Yarwood, C. E., Resconich, E. C., Ark, P. A., Schlegel, D. E., Smith, K. M. So-called beet latent virus is a bacterium. *Plant Disease Reptr.,* **45,** 85–89 (1961)

509. Zazhurilo, V. K., Sitnikova, G. M. Interrelations between mosaic disease virus of winter wheat and its vector, *Deltocephalus striatus* L. *Dokl. Vses. Akad. Sel'skokhoz. Nauk.,* No. 11, 27–29 (1941) (*Proc. Lenin Acad. Agr. Sci., USSR)*

510. Zentmyer, G. A., Jr., Horsfall, J. G., Internal therapy with organic chemicals in treatment of vascular diseases. *Phytopathology,* **33,** 16–17 (1943) (Abstr.)

Note added in proof: A monograph by a tetracycline research group entitled "Effects of tetracycline antibiotics on various plants with mycoplasma disease" has recently come to our attention. Results of tetracycline treatments on several crop plants with suspected *Mycoplasma* disease are summarized in the 85-page text. Unfortunately, this report is not readily available since it is privately published. We hope that the group will soon make these results available to the wide scientific audience.

AUTHOR INDEX

SUBJECT INDEX

D

CUMULATIVE INDEXES

VOLUMES 6-15

INDEX OF CONTRIBUTING AUTHORS

INDEX OF CHAPTER TITLES

VOLUMES 6-15

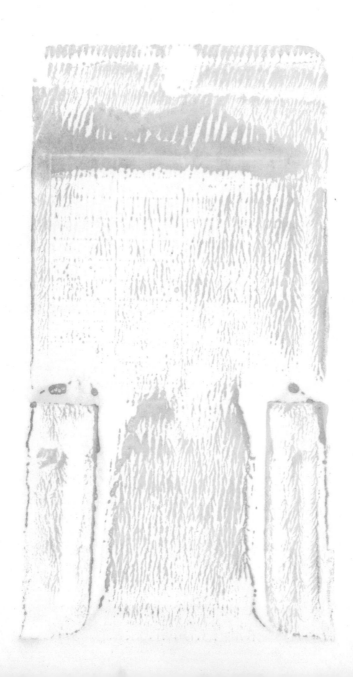

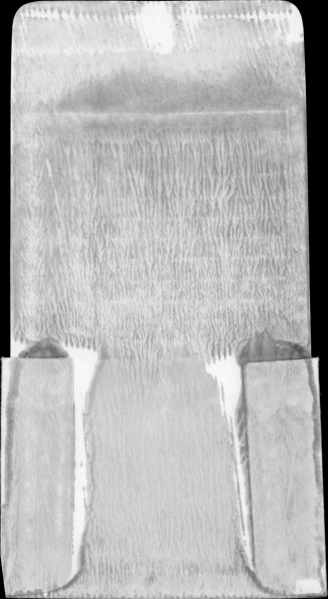

Date Due
